Plant Nutrition of Greenhouse Crops

Cees Sonneveld · Wim Voogt

Plant Nutrition of Greenhouse Crops

Springer

Dr. Cees Sonneveld
Tolhuislaan 35
3862 WK Nijkerk
Netherlands

Ing. Wim Voogt
Wageningen UR
Greenhouse Horticulture
Violierenweg 1
2665 MV Bleiswijk
Netherlands

ISBN 978-90-481-2531-9 e-ISBN 978-90-481-2532-6
DOI 10.1007/978-90-481-2532-6
Springer Dordrecht Heidelberg London New York

Library of Congress Control Number: 2009926976

Printed on acid-free paper

Springer is part of Springer Science+Business Media (www.springer.com)

Contents

Definitions

Adjusted mol weight mole weight adjusted on residual constituents, defined "molweight", expressed as: mol

C^+ Sum of valences of all cations in a solution, expressed as: mol l^{-1}

c Ion concentration in a solution, expressed as: mol l^{-1}

CEC Cation exchange capacity, electric charges on the surface of a material to adsorb cations, expressed as: mol kg^{-1}

EC Electrical conductivity at 25°, expressed as: dS m^{-1}

Hydroponics Cultivation in nutrient solution without any substrate other than the propagation material

Inert substrate Substrate that does not affect the status of the substrate solution as such that specific adjustments are required for compensation

m mille (10^{-3})

μ micro (10^{-6})

1:2 volume extract A specific extract prepared from 2 volumes of water to which so much field moist soil is added that the volume is increased with one volume

Residual salts Salts accumulated in the root zone of soils or substrates from fertilizers or irrigation water, because of an uptake lower than the addition

Root environment For soils *in situ* the soil depth in which the majority of the roots will be present. For greenhouses mostly 25 cm. For substrates usually the total substrate volume is taken into account

Salinity threshold	The maximum EC value in the root zone without any yield reduction, expressed as: dS m^{-1}
Soil solution	Solution extracted from soils at field capacity as defined in Section 3.3
Soilless cultivation	Cultivation other than in soils *in situ*
Substrate	Growing medium other than soils *in situ*
Substrate solution	Solution extracted from growing media at field moist condition as defined in Section 3.3
SYD	The slope of the salinity response function for values above the salinity threshold value in percents per unit EC, expressed as: %/dS m^{-1}
Uptake concentration	The ratio between the uptake of a mineral element and the water uptake by the crop, expressed as: mol l^{-1}
VPD	Vapour pressure deficit, expressed as: kPa

List of Permissions

Fig. 2.1	Modified by permission of the International Society Horticultural Science
Fig. 2.3	Reprinted by permission of Marcel Dekker
Fig. 3.6	Modified by permission of Marcel Dekker
Fig. 3.8	Modified by permission of Marcel Dekker
Table 4.1	Modified by permission of Springer
Table 4.3	Modified by permission of Springer
Table 4.8	Reprinted by permission of the Koninklijke Landbouwkundige Vereniging
Table 5.7	Reprinted by permission of Springer
Table 5.8	Reprinted by permission of the International Society Horticultural Science
Fig. 5.2	Modified by permission of the American Society Horticultural Science
Fig. 5.5	Reprinted by permission of Springer
Fig. 6.2	Reprinted by permission of the International Society Horticultural Science
Table 7.1	Reprinted by permission of Springer
Table 7.2	Reprinted by permission of the Koninklijke Landbouwkundige Vereniging
Table 7.5	Reprinted by permission of Springer
Table 7.6	Reprinted by permission of Marcel Dekker
Table 7.7	Reprinted by permission of Marcel Dekker
Table 7.10	Modified by permission of Springer
Table 7.11	Reprinted by permission of Marcel Dekker
Fig. 7.3	Reprinted by permission of Marcel Dekker
Table 8.2	Modified by permission of Springer
Table 8.3	Reprinted by permission of Springer
Table 8.6	Modified by permission of Springer
Fig. 8.1	Reprinted by permission of the International Society Horticultural Science
Fig. 8.4	Reprinted by permission of the Soil Science Society America
Table 9.4	Modified by permission of the Koninklijke Nederlandse Botanische Vereniging

Table 9.5	Modified by permission of the Journal Horticultural Science Biotechnology
Table 9.6	Modified by permission of the International Society Horticultural Science
Table 9.8	Modified by permission of the American Society Horticultural Science
Fig. 9.2	Modified by permission of Springer
Fig. 9.4	Modified by permission of the Journal Horticultural Science Biotechnology
Fig. 9.7	Modified by permission of Springer
Table 10.2	Modified by permission of the International Society Horticultural Science
Table 10.3	Modified by permission of Springer
Fig. 10.2	Reprinted by permission of Elsevier
Fig. 10.3	Modified by permission of Springer
Fig. 10.4	Modified by permission of Springer
Table 11.3	Reprinted by permission of Springer
Table 11.12	Modified by permission of Elsevier
Table 11.13	Reprinted by permission of the International Society Horticultural Science
Fig. 11.5	After Adams (2002). Modified after permission of Embryo Publications
Table 12.3	Reprinted by permission of the International Society Horticultural Science
Table 12.4	Reprinted after permission of Embryo Publications
Table 13.5	Modified by permission of the International Society Horticultural Science
Table 13.9	Reprinted after permission of Embryo Publications
Fig. 13.2	Reprinted by permission of the International Society Horticultural Science
Fig. 13.3	Reprinted by permission of Springer
Fig. 13.5	Reprinted by permission of Springer
Fig. 13.8	Modified by permission of Springer
Fig. 13.12	Modifies by permission of International Union of Soil Science
Fig. 14.1	Modified by permission of the International Society Horticultural Science
Fig. 14.5	After Boertje (1980). Reprinted by permission of the International Society Horticultural Science
Table 14.15	Modified by permission of the International Society Horticultural Science
Fig. 14.9	Modified by permission of Marcel Dekker
Table 15.3	Modified by permission of the International Society Horticultural Science

Fig. 15.3	Reprinted by permission of the Koninklijke Landbouwkundige Vereniging
Table 16.12	Reprinted by permission of the International Society Horticultural Science
Fig. 16.8	Reprinted by permission of the International Society Horticultural Science
Fig. 16.10	Modified by permission of Springer

Chapter 1
Greenhouse Horticulture

1.1 Introduction

Greenhouse cultivation has a long history and it is difficult to appoint where the first greenhouse was built. Such an appointment directly is hindered by a good definition of a greenhouse. However, independent of a precise definition, undoubtedly, one or more orangeries at castles or palaces will be mentioned, but no one can testify whether it really can be considered as the first greenhouse. The first developments of greenhouses focussed on commercial production of vegetables, fruits and flowers are dated at in the end of the 19th century and occurred mainly in the North-West area of Europe. In advance the production of the crops in greenhouses was mainly connected with the demand in the market for early fruits and vegetables. Another line in the development of greenhouses was the production of crops that could be grown hardly or not at all in the cool and wet climate conditions of North-West Europe.

At first the greenhouses were situated merely in areas of moderate climate throughout the year. This means climates where the temperatures do not fall too much below zero, to prevent crops from freezing during winter and where the temperatures do not rise too much during summer to avoid extremely high temperatures inside the greenhouse. Therefore, many greenhouse districts initially developed in coastal areas and on islands. The greenhouse area in the Westland district in The Netherlands and the greenhouses situated on the British Channel Islands were good examples of such developments.

After the Second World War, with the development of the European Community and the improvement of the transport, the greenhouse industry in North-West Europe was adversely affected by the competition of products from the Mediterranean countries. At that time many agricultural experts predicted that the greenhouse industry in North-West Europe would lose the competition with the South and finally should disappear and be taken over by field production in the Mediterranean areas. These expectations were merely based on the development of the North-American situation, where the vegetable production was situated in California and other Southern parts of the country in open fields and the big North-East market was supplied with fresh vegetables by road transportation. Up till now, in Europe the mentioned

C. Sonneveld, W. Voogt, *Plant Nutrition of Greenhouse Crops*,
DOI 10.1007/978-90-481-2532-6_1, © Springer Science+Business Media B.V. 2009

expectations did not come true. On the contrary, at the end of the 20th century greenhouse horticulture was a growing business all over the world and there are sufficient arguments to suggest that this development will be continued in the 21st century in many countries.

The strong development of greenhouses growing all over the world as came about in the second half of the 20 th century was affected by many factors. Among these following are mentioned as being the most important.

- Development of greenhouse construction. The simple glass construction like the lay flat systems and the wooden greenhouse constructions were replaced by metal constructions, possible furnished with heating and cooling systems suitable for a fully automatic climatic control.
- Breeding of new cultivars due to greenhouse cultivation of crops already grown in greenhouses and the increasing diversity of crops grown in this branch. The breeding of new cultivars contributed to increased yield, improved quality and diversity within a produce.
- The development of auxiliary systems, as used in modern greenhouses for climate control, irrigation, fertilization, biological pest control and growing technique.
- Substrate growing which ensured a better control of the root environment. The small root volume introduced with this growing method strongly improved the management of factors affecting the root development and root functions.
- Flexibility of the branch on the demand of the market and on the competition of products from elsewhere. This means quick adjustments on the demand and quick changes to a different product when the competition from elsewhere cannot be met.
- Successful operations of the greenhouse industry in increasing yields and decreasing costs. In this way the costs per unit produce was stabilized or increased only gradually.

Originally, the greenhouse industry operated in a supply market only, as was common for other agricultural branches. But with the improved transport abilities in the second half of the 20th century many horticulture products were transported from anywhere to all parts of the world. Thus, from this view point horticulture production in greenhouses had no longer arguments to operate as a supply market. The products of the greenhouse industry joined in free competition with field grown products from all over the world. In this way the greenhouse industry developed into a horticultural activity ready to be operative in the consumer market and to bring better and cheaper products on the market than those from the open field.

Many greenhouse products have a luxurious image, which especially is the case for greenhouse flowers. In this way the greenhouse industry has a strong relationship with living standards. Markets like this are characterized by diversity, quality and immediate answers to the demands of customers. The high demands on quality of such markets are not restricted to the product itself, but include also the production methods. Therefore, the greenhouse industry more and more will be confronted with factors like the environmental consequences of the production method and the conditions for the workers during the process. Increasing search for sustainable growing methods and a high standard for the conditions under which the work is carried out belong to the quality requirements as well.

1.2 Fertilization in Greenhouse Industry

In contrary of many other agricultural activities the costs of fertilization in the greenhouse industry are relatively low and amount to only a few percentages of the total costs. Thus, from economic view points were no arguments for a precise and careful application of plant nutrients. In the past an abundant use of fertilizers in the greenhouse industry was common practice and there was no interest by the growers to limitations in the use of fertilizers to prevent in this way the leaching of nutrients to the environment. However, in the last decades of the 20 th century environmental pollution became a subject of permanent attention by the governments of North-West European countries and was quickly followed by regulations from the European Community.

Measurements by the Dutch greenhouse cultivation learned that substantial quantities of nutrients can be transported to the deep ground water or surrounding surface water like ditches, canals and rivers. In Table 1.1 some data is summarized about nutrient leaching from greenhouse. The data is derived from studies carried out under conditions that not yet regulations were issued, roughly 1975–1980. Tomatoes were grown in soil as well in substrate systems; the soil was clayey loam and the substrate rock wool. The N mentioned with the residual factor reflects the undetected quantity, added but not traced within the study. One of the factors responsible for this undetected quantity, for example can be de-nitrification of N. The N efficiency was low, especially for the soil grown situation and the discharge of minerals to the environment high and estimated as unacceptable. Therefore, in the last decades of the 20th century extended studies were carried out to factors contributing to a minimum discharge of minerals, like the restrictions on the fertilizer use and an efficient water supply.

Besides the supply of minerals indeed restrictions on the quantity of drainage water played an important part in this field. However, restrictions in this field will

Table 1.1 Balance sheets of water and N as presented by Sonneveld (1993) for tomato growing in greenhouses under free drainage conditions. The quantities of water are expressed as $m^3\ ha^{-1}\ year^{-1}$ and the N as $kg\ ha^{-1}\ year^{-1}$

	Soil grown crop		Substrate grown crop	
Factors	Water	N	Water	N
Addition	12,950	2269	9691	1935
Uptake by crop	6700	609	7600	1110
Discharge by drainage	6250	1344	2091	825
Residual factor	0	316	0	0
Efficiency	0.52	0.27	0.78	0.57

cause problems with accumulation of salts and an unequal distribution of the moisture content in the soil. For substrate growing the problem was met by reuse of the drainage water and for soil grown crops by a switch to substrate cultivation or by an improvement of the supply of water and fertilizers. Reuse of drainage water and a more precise irrigation pattern strongly aggravate the salt accumulation in the root zone and set high demands on the water quality. These developments will be discussed in the proper chapters.

1.3 Nutrient Uptake

Despite a precise application and an efficient utilization of nutrients in the modern greenhouse industry, the required additions of nutrients will stay high in this horticultural branch. This is related to the high yields usually gained in greenhouses. Between the yield of crops and the uptake of nutrients often a close linear relationship was found, like shown for tomato and chrysanthemum in Fig. 1.1. The relationships shown for both crops differ strongly. This can be explained by several factors, of which are the most important ones: the characteristics of the crops, the mutual ratios of the ions in the external solution, the growing period, and the definition of the yield. Greenhouse crops are generally grown at high external nutrient concentrations and realise under these conditions an optimal nutrient status in the plant. A nutrient status in the external solution higher than required does not significantly affect the uptake (Sonneveld and Welles, 2005). However, the mutual ratios in the external solution of the ions will affect the uptake of a specific ion. The definition of the yield is also important. In the given examples for tomato the weight of the harvested fruits is defined, while for chrysanthemum the total weight of the shoots at harvest is taken into account. Furthermore, the growing period of a chrysanthemum

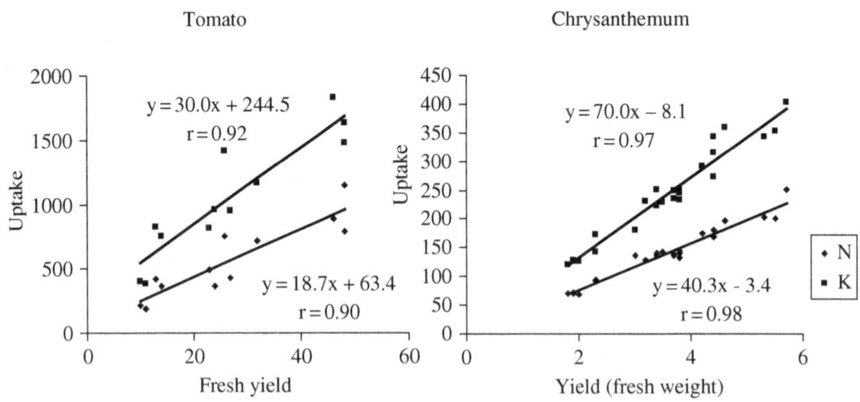

Fig. 1.1 Relationships between the yield of tomato and chrysanthemum (kg m^{-2}) and the uptake of N and K (kg ha^{-1}). The yield of tomato is expressed as fresh fruit production and those of chrysanthemum as the weight of shoots harvested

crop in greenhouses is much shorter than those of a tomato crop. Under the growing conditions in North-West Europe the duration of the growing periods are 11 and 3 months for tomato and chrysanthemum, respectively. Thus, annually one tomato crop will be grown, while for chrysanthemum over 4 crops are normally practiced. The annual uptake calculated with the data for a high yielding tomato crops will easily amount to 1100 kg N and 2000 kg K per ha and for a chrysanthemum cropping 700 kg N and 1100 kg K per ha, being quantities that strongly exceed the uptake of any field crop.

Besides the quantities of minerals absorbed by greenhouse crops, the ratios also will differ from those in the field. Marschner (1997) whose calculations are mainly based on data from Epstein (1965) presented average mole ratios between elements absorbed by plants sufficient for an adequate growth. This data was required from some field crops and he pointed out that the ratios between crops will considerably vary depending on the plant species. In comparison with the data presented by Marschner, ratios of mineral uptake by three greenhouse crops are presented in Table 1.2. The calculations were made for cucumber, lettuce and rose as being a vegetable fruit crop, a leafy vegetable crop and a flower crop, respectively. The data for greenhouse crops are mainly based on the results of De Kreij et al. (1992) and Sonneveld (1997). The differences between the greenhouse crops are substantial for some elements. Compared with the data of Marschner the most striking differences have been found for the generally high ratios for the elements K, Ca, P and S and the generally low ratios for Fe, Mn and Cu. The ratios for Mo greatly vary for the greenhouse crops presented and are high in comparison with the data of Marschner. The information about the uptake of this element for greenhouse crops is restricted and the relatively high ratios for these crops possibly can be explained by accidental additions of this element in the fertilization programmes. The required quantities of Mo are not yet exactly determined for greenhouse crops (Bloemhard and Van der

Table 1.2 Mol ratios of the average concentrations of mineral nutrients as absorbed by some greenhouse crops compared with the data of field crops as presented by Marschner (1997)

		Greenhouse crops		
Element	Field crops	Cucumber	Lettuce	Rose
N	1,000,000	1,000,000	1,000,000	1,000,000
K	250,000	568,000	604,000	359,000
Ca	125,000	247,000	168,000	174,000
Mg	80,000	82,000	52,000	57,000
P	60,000	92,000	94,000	76,000
S	30,000	65,000	43,000	73,000
B	2000	2000	1000	1600
Fe	2000	800	1000	800
Mn	1000	600	430	600
Zn	300	240	290	240
Cu	100	60	57	80
Mo	1	8	2	12

Lugt, 1995). Optimal concentrations mentioned in the literature differ strongly and young plants sometimes require higher concentrations than old ones (Roorda van Eysinga and Smilde, 1981).

1.4 Fertilization Programmes

In the past fertilization and irrigation in greenhouses was based on the experiences of growers. The addition of farm yard manure and other natural organic products was common practice, supplemented with fertilizers used for field crops. Often these fertilizers contained high NaCl contents. Formerly, this was mostly not a problem for field crops, because of the surplus of the precipitation in winter, by which the salt residues were leached from the root zone. However, since the natural precipitation was excluded by the greenhouse constructions salts could easily accumulate in the greenhouse soils to levels that reduced the growth of many crops. The salts accumulated during cultivation, especially in the top layers in the greenhouse soils, like shown in Table 1.3 (Van den Ende, 1952). The high salt content in the top layer at the end of the cropping period was a major hindrance for the start of a new crop. Therefore, such soils must be flooded before a new crop could be started. The salinity problem as presented resulted to special requirements for the irrigation and the fertilization in greenhouses, like the development of fertilizers with low residual salt contents better suited for the greenhouse situation and the development of irrigation systems with which it was possible to supply the crops with ample water during crop growth. With these new developed irrigation systems it was not only possible to wash out the residual salts from the soil after the cropping period, but it seemed possible to prevent high salt accumulations in the soil with an ample water supply during the cropping period. With the use of fertilizers easily soluble in water, the step to the addition of these fertilizers to the irrigation water was simple and the application of the so called fertilizer diluters was quickly established. By these systems the term fertigation was born and the addition of fertilizers in combination with the irrigation water became common practice.

The basis for a well controlled application of fertilisers to the irrigation water originated from the measurement of the electrical conductivity (EC) in flowing water streams by an apparatus placed on commercial greenhouse holdings in The

Table 1.3 NaCl contents of a clayey loam greenhouse soil over different depths at the end of a cropping period and after a flooding with 300 mm water. NaCl contents in mmol kg^{-1} dry soil

Depth	End of cropping	After flooding
0–5	87	3
5–10	30	6
10–15	15	6
15–20	12	6
20–25	12	5
25–30	12	6

After van den Ende (1952).

Netherlands. These so called "concentration meters" expressed the fertilizer concentration in the irrigation water as EC values (Sonneveld and Van den Ende, 1967). This manner of measurement of salt concentrations had already been practiced on laboratory level for many years, but had never been practiced in this way on greenhouse nurseries. Since that time, the measurement of the EC has become important as a determination for the total ionic concentrations in different solutions. Examples are the determinations of the EC in irrigation waters, in fertilizer solutions, in nutrient solutions, in drainage waters, and in soil and substrate extracts. Used in this way the determination of the EC became an important parameter for the control of fertilization management in the greenhouse industry. Another important factor in the management of the fertilization in the greenhouse industry was the development of soil testing methods, based on water extraction (Van den Ende, 1952). Originally, soil testing in greenhouses was used to check the salt and nutrient status on a yearly basis. The strong changes in the chemical composition of greenhouse soils and the increasing nutrient absorption of the crops resulted from the increasing yields introduced the need for a more frequent check on the nutrient status of the soil. Therefore, the so called "top dressing" samples were introduced as supplemental information about the development of the nutrient status of the soil during cultivation. The greenhouse soils, for example, were sampled and analysed every month and the application of fertilizer to the irrigation water was appointed on basis of this data. At the Glasshouse Crops Research Station at Naaldwijk, The Netherlands was an early eye for the development of a systematic fertilization on basis of such data, because in the beginning the greenhouse soil samples were analysed in a laboratory connected to this research station. The combination of plant nutrition research and the development of routine soil testing methods became a fruitful basis for the design of fertilization systems for greenhouse soils. Such systems have been developed by researchers of various research stations in The Netherlands, in cooperation with the workers of the advisory service (Breimer et al., 1988) and was made available for a wide range of crops and growing conditions (IKC, 1994).

The development of fertilization support systems especially was enlarged with the growing interest for substrate cultivation and it appeared to be a cultivation method for practical applications. The small root volumes used with this growing method are responsible for tremendous fluctuations in the salt and nutrient status of the root environment. Sonneveld (1981) calculated that in substrate systems in the root environment momentarily only few percentages are present of the total minerals required by the crop grown. Since then, the yield of the crops and along with this the mineral uptakes are strongly increased, while the substrate volumes made available to the crops and thus, the storage of minerals available in the root environment are only decreased. Utmost, the algorithm for the calculation of a nutrient solution for substrate growing is complicated, because all essential elements, at the least 14, must be taken into account. Such an algorithm is suitable for computerizing and the necessary outlines were presented (Sonneveld et al., 1999). Together with the results of an intensive control on the composition of the nutrient status in the root environment fully automatic systems for the nutrient supply of substrate grown crops have been designed. The necessity for an intensive control on the nutrient supply in substrate systems was accentuated in the last decades of the 20th century by the restrictions and regulations on emission of minerals to the environment. These developments strongly stimulated research to reduction of environmental pollution and the reuse of drainage water came into being on large-scale in substrate cultivation (Voogt and Sonneveld, 1997). In such systems, possible deviations from the optimum composition of nutrient solution in the root environment are not yet alleviated by the discharge of the drainage water, but are brought back into the system. Thus, depletions and accumulations easily occur in such systems and necessitate precisely tuned programmes, an intensive control on the salt and nutrient status in the root environment and a frequent feed back of the analytical data to the algorithm.

Parameters for the different systems used in the greenhouse industry are a broad view on the needs of minerals of greenhouse crops and the relations between the uptake of the different elements on the one hand and the nutrient status in the root environment on the other hand. Furthermore, the interactions of these relationships with the growing conditions, like irrigation, climate and type of soils and substrates are numerous and play an important role in the fertilization management.

In the greenhouse industry the nutrient status in the root environment is merely characterized by the composition of the soil solution and the substrate solution. The composition of these solutions sometimes can be determined by analysis of the solution directly from the field moist soil or from the substrate. When this is not possible or when it is too laborious the composition can be estimated by water extraction. Exchangeable quantities of nutrients are of secondary importance, because of the generally high nutrient status maintained in the root environment of greenhouse grown crops, while the interpretation of such data is more complicated than those derived from water extraction. The relationships between exchangeable nutrient concentrations in soils and substrates and the uptake by plants depend strongly on the adsorption capacity and the character of the adsorption complex of the soil and the substrate in which the plants are grown (De Vries and Dechering, 1960; Mengel and Kirkby, 1987; Roorda van Eysinga, 1965; Sonneveld and De Kreij, 1995).

1.5 Development of Greenhouse Horticulture

Last decades of the 20th century the greenhouse horticulture in high technological developed countries is characterized by a rapid changing technology with a degree innovative power. The technological applications were not solely directed on extension and higher productions, but primarily on the sustainability of the greenhouse industry. In this development main focuses were quality improvement of the produce and strong reductions on the environmental impact. With respect to last item, three subjects accompanied the research in the greenhouse industry: reduction of the use of energy, reduction of the use of plant protection chemicals and reduction of the emission of minerals, focussed on N and P. Solutions in these directions were found in high technological developments, like for example increasing yields, storage of energy over seasons, use of residual heat, co-generation of electricity, biological control of pests and diseases, and reuse of drainage water for substrate grown crops. The use of substrates in the greenhouse horticulture often has been

Picture 1.1 Modern greenhouses in The Netherlands

Picture 1.2 Greenhouses with plastic cover at the Mediterranean coast in Almeria Spain

criticized as "unnatural" and "industrial". However, it is just the industrial character that supplies excellent opportunities for further improvements to restrict environmental pollutions. Such developments true enough are developed and applied in a high technological greenhouse industry like exists in North-West Europe, North-America, Australia and Japan, but this does not mean that parts of this knowledge cannot be applied in less technological developed greenhouse cultivation, like

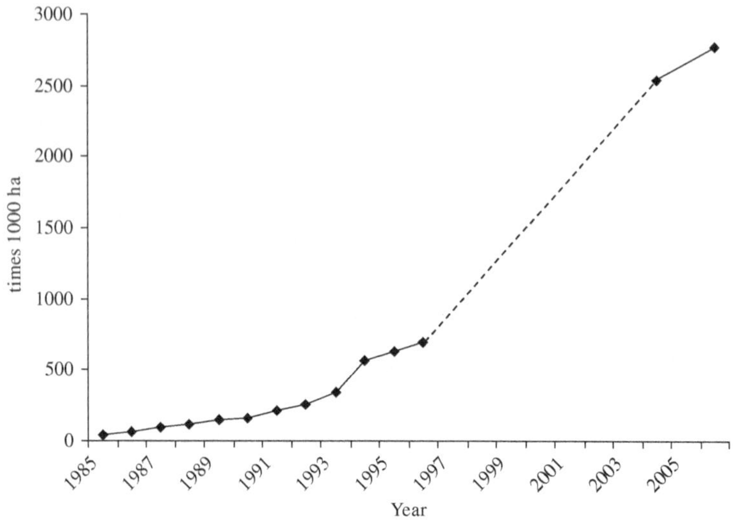

Fig. 1.2 The development of protected cultivation in China during the years 1985 till 1996. The areas concerned are tunnels and greenhouses, mainly covered by plastic. After Zhang (1999), and until 2006 following personal information of Zhang (2007, Personal information by e-mail)

Table 1.4 Areas (ha) of protected cultivation in the Mediterranean area in 2001

Type of greenhouses	Areas
Large plastic tunnels	168,265
Glasshouses	21,800
Low Tunnels	140,600
Total	330,665

After Pardossi et al. (2004).

protected cultivations under plastic. The development of protected cultivation under plastic is a strongly growing agricultural activity all over the world. The development of this type of protected horticulture in China is a good example of this. The development in this country, presented in Fig. 1.2 (Zhang, 1999), concerns mainly plastic tunnels and greenhouses. The figure shows the tremendous areas of protected cultivation that can be developed in short time when use is made of plastics. About 75% of the areas were tunnels and 25% greenhouses. The development in Europe is quite different. In the Mediterranean areas relatively more greenhouse constructions were found. The greenhouses amounted to 55% and the tunnels 45%, like shown in Table 1.4 (Pardossi et al., 2004). From the greenhouse constructions in the Mediterranean area about 10% is covered with glass, while in North-West Europe just only glass is used as cover on the greenhouses. Such differences have beside an economic reason also a technical basis. The low light intensity in North-West Europe requires maximum light transmission and this is much better with a glass than with a plastic cover.

References

Bloemhard C and Van der Lugt G 1995. Streeftraject molybdeen nog ruim. Vakblad voor de Bloemisterij 50(18), 37.

Breimer T Sonneveld C and Spaans L 1988. A computerized programme for fertigation of glasshouse crops. Acta Hort. 222, 43–50.

De Kreij C Sonneveld C Warmenhoven M G and Straver N A 1992. Guide values for nutrient element contents of vegetables and flowers under glass (third edition). Proefstation voor Tuinbouw onder Glas te Naaldwijk, The Netherlands, Series: Voedingsoplossingen voor Glastuinbouw no 15, 69 pp.

De Vries O and Dechering F J A 1960. Grondonderzoek. Uitgeverij Ceres Meppel, The Netherlands.

Epstein E 1965. Mineral metabolism. *In*: Bonner J and Varner J E (eds) Plant Biochemistry, Academic Press London, 438–446.

IKC 1994.Bemestingsadviesbasis Glastuinbouw 1994–1995. Informatie en Kennis Centrum Akker- en Tuinbouw, Naaldwijk, The Netherlands.

Marschner H 1997. Mineral Nutrition of Higher Plants. Academic Press, London.

Mengel K and Kirkby E A 1987. Principles of Plant Nutrition. International Potash Institute, Bern, Switzerland.

Pardossi A Tognoni F and Incrocci L 2004. Mediterranean Greenhouse Technology. Chronica Hort. 44(2), 29–34.

Roorda van Eysinga J P N L 1965. Bemestingsadviezen in de groenteteelt onder glas. Meded. Dir. Tuinbouw 28, 568–573.

Roorda van Eysinga J P N L and Smilde K W 1981. Nutritional Disorders in Glasshouse Tomatoes, Cucumbers and Lettuce. Centre for Agricultural Publishing and Documentation, Wageningen, The Netherlands, 130 pp.

Sonneveld C and Van den Ende J 1967. Bijmesten via de regenleiding met behulp van de concentratiemeter. Meded. Dir. Tuinb. 30, 54–60.

Sonneveld C 1981. Items for application of macro-elements in soilless cultures. Acta Hort. 126, 187–195.

Sonneveld C 1993. Mineralenbalansen bij kasteelten. Meststoffen, NMI Wageningen 1993, 44–49.

Sonneveld C and De Kreij C 1995. Standardisation of chemical analysis of growing media. Acta Hort. 401, 569–574.

Sonneveld C 1997. Mineralenopname bij teelten onder glas. Proefstation voor Bloemisterij en Glasgroente, Naaldwijk, The Netherlands. Internal Report 81, 35 pp.

Sonneveld C Voogt W and Spaans L 1999. A universal algorithm for calculation of nutrient solutions. Acta Hort. 481, 331–339.

Sonneveld C and Welles G W H 2005. Cation concentrations of plant tissues of fruit-vegetable crops as affected by the EC of the external nutrient solution and by humidity. Acta Hort. 697, 377–386.

Van den Ende J 1952. De betekenis van het chemisch grondonderzoek te Naaldwijk voor de bemesting bij teelten onder glas. Meded. Dir. Tuinbouw 15, 651–673.

Voogt W and Sonneveld C 1997. Nutrient management in closed growing systems for greenhouse production. *In*: Goto E et al., (eds) Plant Production in Closed Ecosystems. Kluwer Academic Publishers, Dordrecht, The Netherlands, 83–102.

Zhang Z 1999. Update development of protected cultivation in mainland China. Cronica Hort. 39(2), 11–15.

Chapter 2
Fertilizers and Soil Improvers

2.1 Introduction

In greenhouse industry fertilizers as well as soil improvers are widely used. Fertilizers are mainly applied to optimize the physical-chemical conditions of the root environment and are used for growing in soils in situ as well as for growing in substrates. Soil improvers are materials solely added to soils in situ primarily to maintain or improve its physical properties, but it also can improve its chemical and biological properties. Thus the difference between fertilizers and soil improvers is somewhat diffuse.

The optimization of the physical-chemical conditions by addition of fertilizers is focussed on the improvement of the availability of nutrients, and on control of the pH and the osmotic potential in the root environment. Since growth rate of crops and as a result yields are very high in greenhouses, the removal of nutrients is substantial. Thus, the application of most fertilizers is primarily essential to restore the nutrient status of the root environment. The pH level is controlled by the addition of specific fertilizers, but is also affected by the addition of the fertilizers added to improve the nutrient status. Upside down, the fertilizers used to control the pH contain nutrients, usually Ca and Mg. Control on the osmotic potential is realised by addition of extra fertilizers to decrease this value in the soil solution. A decreased osmotic potential (increased EC value) is sometimes required to reduce lush growth of crops under poor light conditions or to improve the quality of the harvested produce, being favourable effects of a decreased osmotic potential.

Soil improvers are usually applied as base dressing, whereby the nutrients added with them are taken into account with the application of the fertilizers required for the base dressing. Not al the nutrients in soil improvers are directly available. The N is gradually released during decomposition of organic matter and will be not taken into account directly with the addition. Fertilizers are used as base dressing as well for top dressings; they are applied for broadcasting as well for fertigation. In the greenhouse industry a vast quantity of the fertilizers is added as top dressings by fertigation, while with cultivation in substrates more or less all fertilizers are applied by fertigation. Therefore, the development of fertilizers due to fertigation has assumed substantial proportions, which resulted in a broad assortment of fertilizers suitable for this working method.

C. Sonneveld, W. Voogt, *Plant Nutrition of Greenhouse Crops*,

DOI 10.1007/978-90-481-2532-6_2,

In the following sections the composition and the application of different fertilizers used in the greenhouse industry will be discussed. Some materials used as soil improver also are used with the preparation of growing media. In the last case it is not defined as a soil improver, but as a growing medium constituent. In that capacity the use will be discussed in the chapter about the preparation of growing media, Chapter 11.

2.2 Fertilizers

In this section a review will be given of the fertilizers commonly used in greenhouse horticulture. The choice of the fertilizer types used in greenhouse industry sometimes differs from those for field crops, because of the fact that the choice for field crops is strongly determined by the price of the fertilizer. This scarcely is a factor in the greenhouse industry, because fertilizer costs represent only a minor fraction of the total costs of greenhouse industry. The characteristics on which the choice of fertilizers is based are high solubility and low residual salt contents. Furthermore, many fertilizers in the greenhouse industry are used for fertigation of soil grown crops and substrate cultures and therefore, must be free from insoluble material. Such residues are not harmful to crops, but enhance the clogging of drip irrigation systems used for fertigation. For fertigation with sprinkler systems, the blocking of the nozzles is less a problem, but insoluble residues easily precipitate on the crop which is visible as a residue on the produce.

The fertilizer industry often expresses the nutrient contents of fertilizers as oxides, like K_2O and P_2O_5. This way of expression has a historical basis, but generally fertilizers do not contain any oxides. However, in greenhouse horticulture more and more the nutrients are expressed as elements. For many purposes it is easier to work with mole weights, like with the calculation of nutrient solutions for substrate culture in Chapter 12. Therefore, in the present chapter besides the information supplied by the fertilizer producer about nutrient contents expressed as oxides also elemental contents and where informative mole weights will be given. The mole weights are calculated as defined by the international system of units Aylward and Findlay (1974). In some cases fertilizers are not made up of a single salt or compound, but contain residual constituents. This for example is the case with dissolved fertilizers, which self-evidently contain water as a constituent. In other cases the impurity of the compound plays a part. In such cases an adjusted mole weight will be calculated bases on the essential element or compound of the fertilizer in question. Such molweights will be reflected between quotation marks, like "molweight". A list of rounded atomic weights used in this book is added in Appendix A. The elemental contents are calculated on basis of following formula.

$$\%K = \frac{39.1}{47.1}\%K_2O \tag{2.1}$$

$$\%P = \frac{62}{142}\%P_2O_5 \tag{2.2}$$

$$\%Ca = \frac{40.1}{56.1}\%CaO \tag{2.3}$$

$$\%Mg = \frac{24.3}{40.3}\%MgO \tag{2.4}$$

$$\%S = \frac{32.1}{80.1}\%SO_3 \tag{2.5}$$

Beside the solid mineral salts, some fertilizers are available in soluble form. The percentage nutrients and the adjusted mole weight of such solutions depend on the concentration of the mineral salt in charge. They will be calculated with following formulae.

$$\%Nu_{sol} = \%Nu_{solid}Fraction_{salt} \tag{2.6}$$

In which

% Nu_{sol}= nutrient content in the solution in %
% Nu_{solid}= the nutrient content in the solid form in %
$Fraction_{salt}$= the fraction mineral salt in the solution

$$\text{“}molweight_{sol}\text{”} = \frac{molweight_{solid}}{Fraction_{salt}} \tag{2.7}$$

In which

$molweight_{sol}$= mole weight of the soluble form
$molweight_{solid}$= mole weight of the solid form
$Fraction_{salt}$ as defined under (2.6)

Among the fertilizers presented also some hydroxides, acids and carbonates are included. The less soluble forms of these chemicals are used in the greenhouse industry for pH adjustment of soils and substrates, while the soluble forms are used in fertilizer recipes and for the pH adjustment of nutrient solutions for substrate cultures. Acids are sometimes also used for adjustment of the pH of irrigation water with a high HCO_3 concentration, when used for drip irrigation of soil grown crops.

2.2.1 N fertilizers

N fertilizers used in greenhouse cultivation contain N as NO_3, NH_4 or urea. For growing in soil all forms are used, while for substrate growing mainly NO_3 is applied and NH4 also in small quantities. Urea is not used in substrate cultivation, because in substrate solutions urea will survive rather long and can be toxic to plants. Sometimes urea is used for pH stabilisation in the water used with drip

Table 2.1 N fertilizers used in greenhouse industry

Fertilizer	Chemical formula	% N	% others	Mole weight
Ammonium nitrate	NH_4NO_3	35		80
Nitrochalk	$NH_4NO_3 + CaCO_3$	21–27		
Ammonium nitrate solution	NH_4NO_3	35[1]		80[1]
Ammonium sulphate	$(NH_4)_2SO_4$	21	24 S	132.1
Calcium nitrate	$5[Ca(NO_3)_2.2H_2O].NH_4NO_3$	15.5	18 Ca	1080.5
Calcium nitrate solution	$Ca(NO_3)_2$	17[1]	24 Ca[1]	164.1[1]
Urea	$CO(NH_2)_2$	46		56
Nitric acid solution	HNO_3	22[1]		63[1]

[1]For the pure chemical. Real value depends on the strength of the solution, see text.

irrigation of potted plant cultures. In substrate growing NH_4 especially is added to nutrient solutions to control the pH, see Section 13.4. On calcareous soils the use of urea as well NH_4 is also effective for adjustment of the pH which is discussed in Section 15.7. In Table 2.1 a review is given of the N fertilizers commonly used in greenhouse industry.

2.2.2 *P fertilizers*

The P fertilizers used in greenhouses primarily consist of orthophosphates. The cheapest and most widely used forms are the Ca salts, from which only the mono form has a high solubility. The low concentrated fertilizers of this form contain a lot of gypsum and therefore, are seldom used in greenhouses. Calcium orthophosphates are never used for fertigation and substrate cultures, because these fertilizers mostly contain too much insoluble components. For greenhouse industry suitable P fertilizers are listed in Table 2.2.

Most P fertilizers contain F, which is toxic to some mono-cotyledon plants, especially many bulb and tuber crops, from which freesia is the most well known greenhouse crop (Roorda van Eysinga, 1974). For crops sensitive to F toxicity P fertilizers with a low content of F are on the market. Di-calcium phosphate ($CaHPO_4$) prepared for cattle feed is such a fertilizer, however, solely suitable for broadcasting. Some other P fertilizers are also produced with a low F content, which will be indicated on the packing.

In substrate cultivation beside H_3PO_4 completely soluble salts like $NH_4H_2PO_4$ and KH_2PO_4 are used. Recently the use of specific polyphosphates is introduced by some fertilizer producers. The main object of the use of polyphosphates is prevention of precipitation of Ca and Mg from the irrigation water and by this lessening of clogging of the irrigation system. The chemical formulations of these compounds are kept secret up till now. The concentrations recommended varies between 0.25 and 0.50 mmol P l^{-1} water.

Table 2.2 P fertilizers used in greenhouse industry

Fertilizer	Chemical formula	% P_2O_5	% P	% others	Mole weight
Super phosphate	$Ca(H_2PO_4)_2$	46	20	± 10 Ca	
Mono potassium phosphate	KH_2PO_4	51	22	28 K	136.1
Mono ammonium phosphate	$NH_4H_2PO_4$	60	26	12 N	115
Phosphoric acid	H_3PO_4	72[1]	32[1]		98[1]
Dicalcium phosphate	$CaHPO_4$	46	20	± 25 Ca	
Poly phosphate	Super FK[2]	16	6.7	22 K	
	Vitaphoska[2]	31	13	41 K	

[1]For the pure chemical. Real values depend on the strength of the solution, see text;
[2]trade marks, chemical composition not published.

2.2.3 *K fertilizers*

K fertilizers are listed in Table 2.3. For broadcasting with base dressings mostly the SO_4 salt is used. KCl is never used for soil grown crops in greenhouses. It is only used for special applications in substrate cultures, as a replacement for NO_3 when Cl is required in the nutrient solution and the concentration in the primary water is very low. To this purpose a very pure form is desired, because the Na content must be very low.

Table 2.3 K fertilizers used in greenhouse industry

Fertilizer	Chemical formula	% K_2O	% K	% others	Mole weight
Potassium sulphate	K_2SO_4	54	45	18 S	174.1
Potassium magnesium sulphate[1]	$K_2SO_4.MgSO_4$	29	24	20 S, 6 Mg	
Potassium nitrate	KNO_3	46	38	13 N	101.1
Potassium chloride	KCl	62	52	48 Cl	74.6
Potassium hydroxide	KOH	84[2]	70[2]		56.1[2]
Potassium bi-carbonate	$KHCO_3$	47	39		100.1
Potassium carbonate	K_2CO_3	68	56		138.2

[1]Qualities can differ, with minor deviations in nutrient contents;
[2]for the pure chemical. Real values depend on the strength of the solution, see text.

2.2.4 *Mg fertilizers*

Mg fertilizers used in greenhouse industry are listed in Table 2.4. Kieserite is the cheapest form of magnesium sulphate, but it is slowly soluble in cold water. Therefore, this fertilizer is solely used for broadcasting applications. Epsom salt is used when prompt dissolution is required, like for example with fertigation and the

Table 2.4 Mg fertilizers used in greenhouse industry

Fertilizer	Chemical formula	% MgO	% Mg	% others	Mole weight
Kieserite	$MgSO_4.H_2O$	27	16	21 S	
Epsom salt	$MgSO_4.7H_2O$	16	10	13 S	246.4
Magnesium nitrate	$Mg(NO_3)_2.6H_2O$	15	9	11 N	256.3
Magnesium nitrate solution	$Mg(NO_3)_2$	27[1]	16[1]	19 N[1]	148.3[1]

[1]For the pure chemical. Real value depends on the strength of the solution, see text.

preparation of nutrient solutions. When low SO_4 is wanted in nutrient solutions for substrate culture, the nitrate form is used. However, $Mg(NO_3)_2.6H_2O$ is very hygroscopic and thus mostly delivered as a concentrated solution.

2.2.5 *Ca fertilizers*

Many of the Ca fertilizers has a double function when used in soil growing, because these fertilizers are used as pH control in the soil. This is the case for $CaCO_3$ and $Ca(OH)_2$. They are also used for growing media to enhance the pH of acid substrate constituents, like peat and bark. $Ca(OH)_2$ is better soluble than $CaCO_3$, and thus sometimes used when a quick pH raise is wanted. $CaCO_3$ is the common form used for pH control. Many carbonates contain besides $CaCO_3$ also $MgCO_3$. The ratio between both components varies and depends on the requirements of the soil on Mg, a suitable form can be chosen. In Table 2.5 different Ca fertilizers are presented.

Liming of soil and growing media constituents induce following reactions depending on of the type of liming material.

$$Ca(OH)_2 + 2H_3O \rightarrow Ca + 4H_2O \tag{2.8}$$

$$CaCO_3 + 2H_3O \rightarrow Ca + 3H_2O + CO_2 \tag{2.9}$$

When Ca in a soluble form is desired, mostly $Ca(NO_3)_2$ is used, see Table 2.1. In some situations $CaCl_2$ is suitable instead of KCl for substrate grown crops.

Table 2.5 Ca fertilizers used in greenhouse industry

Fertilizer	Chemical formula	% CaO	% Ca	% others	Mole weight
Slaked lime	$Ca(OH)_2$	75	54		74.1
Limestone	$CaCO_3 + MgCO_3$	55[1]	39[1]	0–11 Mg[1]	
Calcium chloride	$CaCl_2.2H_2O$	38	27	47 Cl	147.1

[1]Maximum values, dependent on the type of lime stone that is used.

2.2.6 Micro Nutrients

In substrate cultivation the application of micronutrients is common practice, because most growing media has very low contents of minerals. Moreover, when they are available in the growing media the quantity of growing media is small and is possibly soon depleted by the crop. For soil grown crops the application is mostly not necessary, because most soils contain sufficient quantities of these elements to supply the crops. Some are available from natural origin, while others are brought in with the often intensive fertilization, with the irrigation water and with the soil improvers used during years in succession.

The availability of micro elements is strongly affected by the pH of the soil. High pH values reduce the uptake of all micro elements, except Mo from which the uptake is promoted by increasing pH (Lucas and Davies, 1961).

Micro nutrient fertilizers used in greenhouse cultures are listed in Table 2.6. In this table for many elements the SO_4 form is mentioned and best available in trade. However, the Cl as well the NO_3 forms of these are suitable too. The accompanying anion is mostly not important, because the quantities added are insignificant compared to the quantities of these ions available in the soil. However, for Fe the accompanying ion is a striking issue. Fe is supplied by so called artificial produced chelates, which are widely used in substrate cultivation as well for soil grown crops. The type of chelate used will be determined by the pH of the soil and the growing medium in which the plant is grown. See also Section 13.4 for more information about Fe chelates.

Table 2.6 Micro nutrients fertilizers used in greenhouse industry

Fertilizer	Chemical formula	% essential element	Mole weight	Remarks
Manganese sulphate	$MnSO_4.H_2O$	32 Mn	169	
Zinc sulphate	$ZnSO_4.7H_2O$	23 Zn	287.5	
Zinc sulphate	$ZnSO_4.6H_2O$	24 Zn	269.5	
Borax	$Na_2B_4O_7.10H_2O$	11 B	381.2	
Boric acid	H_3BO_3	17 B	61.8	
Copper sulphate	$CuSO_4.5H_2O$	25 Cu	249.7	
Sodium molybdate	$Na_2MoO_4.2H_2O$	40 Mo	241.9	
Iron chelate	Fe-EDTA	13 Fe	430[1]	pH < 6
Iron chelate	Fe-DTPA	3–11 Fe	1863–508[2]	pH < 7.5
Iron chelate	Fe-EDDHA	5–6 Fe	932–1118[2]	All pH values

[1]Calculated in basis of 1 mol Fe;
[2]dependent on Fe content.

2.2.7 Si fertilizers

Si is abundant present in the earth crust and mostly sufficient available to plants. It is not an essential element, but beneficial to some crops. For soil grown crops it is not necessary to add this element by fertilization, but for substrate grown crops the presence or the availability can be insufficient. Under these conditions Si is supplied as a fertilizer to substrates for crops with a beneficial reaction on this element. The application of Si during cultivation with the regular nutrient solution is complicated, since precipitation of Si compounds will occur and quickly block the nozzles of drip irrigation systems.

Si is not soluble in water, but some colloidal forms can survive as a stable gel, mainly in combination with Li and Na. These accompanying elements must be available in high concentrations to keep the compound stable, which made them toxic to plants. Thus, these compounds are not suitable as a fertilizer. Therefore, suitable Si compounds are stabilized with concentrated KOH. The mixture in which K and Si occur at a ratio 2:1 is best (Voogt and Sonneveld, 2001). This mixture is strongly basic and with the addition to nutrient solutions the basic reaction as well the K should be incorporated with the calculations of nutrient solutions. For further information about the application see Section 12.6. The compound in charge is a potassium meta silicate and it contains in mol ratios Si:K:OH = 1:2:2. It is delivered as different concentrated gels.

2.2.8 Other Elements

Other elements mentioned to be essential or beneficial for plant development are Ni, Na, Cl and Co (Marschner, 1995). With Ni and Co insufficient or not any research is carried out in greenhouse industry. It will be expected that at least for soil grown crops Ni and Co are naturally available from soil or from the impurities of fertilizers and soil improvers added. For substrate culture the addition of these elements will be considered, because of the continuous improvements of the purification of the fertilizers used for this growing method and the often total absence of minerals in some of the growing media used. Before application a further study to the necessity of these elements and the concentrations required should be carried out.

Na and Cl are abundant available in the biosphere and the essential quantities necessary are on micro nutrient level. In greenhouse horticulture, quantities of Cl higher than those essentially required are recommended (Voogt and Sonneveld, 2004). This is because of the interaction of Cl with the uptake of Ca. Fertilizers suitable to this purpose are mentioned already under the K and Ca fertilizers.

2.2.9 Compound Fertilizers

The fertilizer industry produces a lot of different compound fertilizers for soil grown crops, mostly containing guaranteed contents of N, P and K being the elements most widely applied in greenhouse industry. They can be separated in those meant for broadcasting and those suitable for fertigation. Last types often contain beside the elements mentioned, also Mg and micro nutrients. Besides the elements mentioned compound fertilizers mostly contain significant quantities SO_4 as a residue. The numbers with which compound fertilizers are characterized are

Table 2.7 Composition of specific compound fertilizers used for the fertilization with the preparation of substrates. The elements are expressed as mass %; in brackets the % P_2O_5, K_2O and MgO

	PG-mix fertilizers		
Elements	14+16+18	15+10+20	12+14+24
N	14	15	12
NO_3/NH_4	5.2/8.8	8.5/6.5	7.0/5.0
P	7.0(16)	4.4(10)	6.1(14)
K	14.9(18)	16.6(20)	19.9(24)
Mg	0.5(0.8)	1.8(3)	1.2(2)
S	7.6(19)	3.2(8)	5.6(14)
Fe	0.09	0.09	0.09
Mn	0.16	0.16	0.16
Zn	0.04	0.04	0.04
B	0.03	0.03	0.03
Cu	0.12	0.15	0.15
Mo	0.20	0.20	0.20

the percentage $N{:}P_2O_5{:}K_2O{:}MgO$. There are numerous different compound fertilizers, adjusted to crops and to application methods. The often high P contents are a drawback, because the addition of this element is usually not required by crops grown in soil. Most greenhouse soils contain more than enough P, because of the long term heavy supply of fertilizers and manures (Roorda van Eijsinga, 1971). With single fertilizers the application can better be focussed on the availability of the specific elements in the soil and the crop requirements. On the other hand the use of compound types is simple and easy to handle, because of the excellent granulation for the broadcasting types and a good solubility for the types due to fertigation.

For the base fertilization of substrates some special compound fertilizers are put into circulation and are widely used to that purpose for the preparation of naturally organic substrates. These fertilizers contain all required nutrient elements, including micro nutrients, except Ca. For this element presumably will be sufficiently supplied with the fertilizers added to adjust the pH. The ratios are focussed on an average requirement for different substrates and crops. A drawback of the use is the fixed ratios of the elements, which makes it difficult to adjust the fertilization of the substrates to specific requirements. In Table 2.7 the compound fertilizers, produced for the fertilization of naturally organic substrates, are listed.

2.2.10 Slow Release Fertilizers

Slow release fertilizers are characterized by a gradual release of nutrients. They usually consist of fertilizer granules coated by a synthetic film or in some cases by a sulphur covering. Among this type of fertilizers singular as well compound types are current. The nutrients of this type of fertilizers are gradually released during a

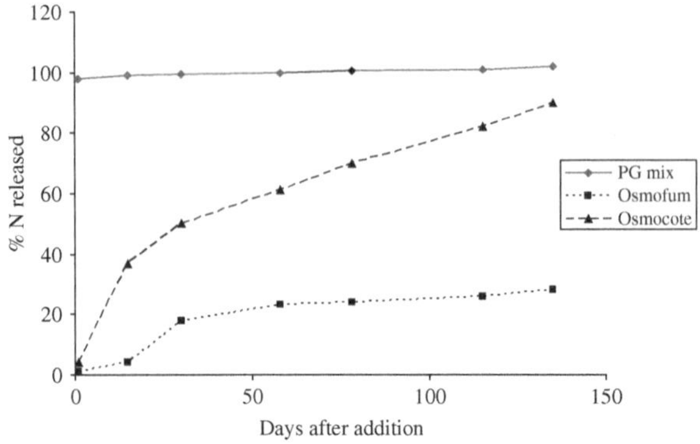

Fig. 2.1 Release of N in % of the total N from some modern slow release fertilizers in an incubation experiment with peat in comparison with a mineral compound fertilizer. After Prasad et al. (2004). *Modified by permission of the International Society Horticultural Science*

proposed period. The aim is that such happens in relation to the demand of the plant. However, this is not always achieved, because the release is not always in agreement with the proposal and depends much on the climatic conditions and the crops grown (Oertly, 1980). Moreover, the demand of the crop is not always predictable. The nutrient release kinetics of the fertilizers always follow a typical logarithmic related function, while the demands of crops more or less always follow an exponential function, at least in the beginning of the growing period.

In Fig. 2.1 some examples of the N release of some of such fertilizers are shown in comparison with the N release of a mineral (water soluble) fertilizer. The data are derived from Prasad et al. (2004). Osmocote (14+16+18) is a slow release inorganic fertilizer with a resin coating and Osmofum (4+3+8) is a mix of soy meal, bone meal, vinasse and cocoa shells. The release of the different elements is not equal in the mixtures.

Slow release fertilizers are not widely used in greenhouse industry, because the frequent irrigations offer possibilities for top dressings by fertigation at every time that the plant needs nutrient supply. Such top dressings can be carried out without noticeable extra input of labour. In special situations when fertigation is problematic, slow release fertilizers are used (Bos et al., 2002). This for example is the case for container grown plants with overhead irrigation, when too much water falls beside the containers. In such situations an important part of the nutrients is wasted with fertigation. Top dressings with fertigation are therefore not efficient and slow release fertilizers are used instead.

2.2.11 Organic Fertilizers

Organic fertilizers are produced from animal or plant material and are for that reason popular in the organic horticulture. They are sometimes used for traditional soil grown crops and to some extent as an amendment in organic substrates. There is a broad variety of source, some of these fertilizers are prepared from single

Fig. 2.2 Release of N from waste products of meat industry in an incubation experiment with greenhouse soil in comparison with mineral fertilizer (Proefstation, 1954)

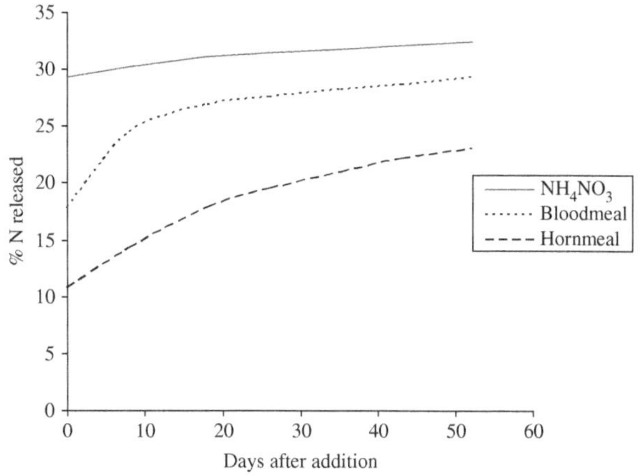

protein-like materials such as blood and slaughterhouse wastes, while others are prepared from animal manures or industrial biomass wastes. Some fertilizers are formulated with specific nutrient ratios from various sources and put into the market as organic fertilizer compounds. All these fertilizers have in common that the majority of the N and partly also P is present in organic form and is released gradually through decomposition by microbial activity of the soil. The rate of this so called mineralization process differs strongly among different products. As is clear from the data presented in Fig. 2.2, where the release of N is shown for blood meal and horn meal in comparison with NH_4NO_3 (Proefstation, 1954). This process also depends strongly on temperature, humidity, pH and available NO_3 in the soil. Some generally used organic fertilizers are listed in Table 2.8, together with the origin and the contents of major elements.

Table 2.8 Composition of some generally handled organic fertilizers produced from waste materials. The elements are expressed as mass %

Fertilizer	Constituents	N	P	P_2O_5	K	K_2O
Blood meal	Slaughterhouse blood	13	0	0	0	0
Bone meal	Slaughterhouse bones	5	7	16	0	0
Feather meal	Feathers and claws of chicken	13	0	0	0.4	0.5
Cow pellets	Cow manure	1.9	0.4	1	2.2	2.7
Chicken pellets	Chicken manure	2.2	0.3	0.8	1.0	1.2
Malt pellets	Brewery waste	5	1.4	3.1	4.0	4.8
Ricinus pellets	Castor oil industry waste	4	0.7	1.5	6.6	8
Vinasse potassium	Sugar beet industry	2	0	0	23.7	28.6

2.3 Soil Improvers

2.3.1 Organic Matter and Physical Characteristics of Greenhouse Soils

The physical characteristics of soils are mostly not a restriction for the employment of protected cultivation. In The Netherlands for example greenhouse industry is situated on very different soil types. In a series of 75 soil samples from greenhouses in The Netherlands the mass fraction organic matter of oven dried soils varied between 0.03 and 0.61, while the mass fraction clay (particles <0.002 mm) varied between 0.03 and 0.40 (Sonneveld et al., 1990). Furthermore, a relatively high salt content, mostly in combination with a high content of soluble Ca keeps the soil in a crumbly and flaky condition (Hilgard, 1919). The intensive tillage stimulates a loose structure further on. Therefore, the bulk densities of greenhouse soils mostly are lower than those of field soils. Between the fraction organic matter, determined by loss on ignition and the bulk density of greenhouse soils exists a closely relationship, like shown in Fig. 2.3 (Sonneveld, 1990). The characteristics of the relationships found for greenhouse soils and field soils are equal, but the parameters differ. This is shown in Fig. 2.4, where the relationship between the organic matter (loss on ignition) content of the 75 Dutch greenhouse soils of Fig. 2.3 and the bulk density is shown in comparison with the relationship found for these characteristics of field soils (Kortleven, 1970). It is understandable that the bulk density is strongly affected by the organic matter content, because the density of the mineral fraction is much higher than those of the organic fraction (Kipp et al., 2000; Klute, 1986). The functions for the relationships presented in Fig. 2.4 are following.

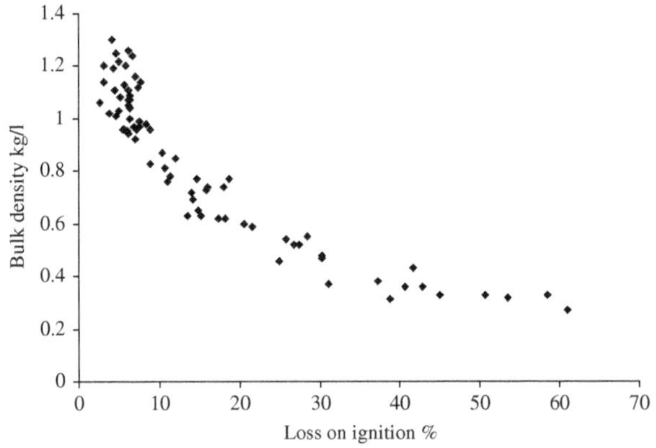

Fig. 2.3 Relationship between the % loss on ignition and the bulk density of greenhouse soils after Sonneveld (1990). *Reprinted by permission of Marcel Dekker*

For greenhouse soils:

$$\rho = \frac{1}{4.67 f_l + 0.69} \tag{2.10}$$

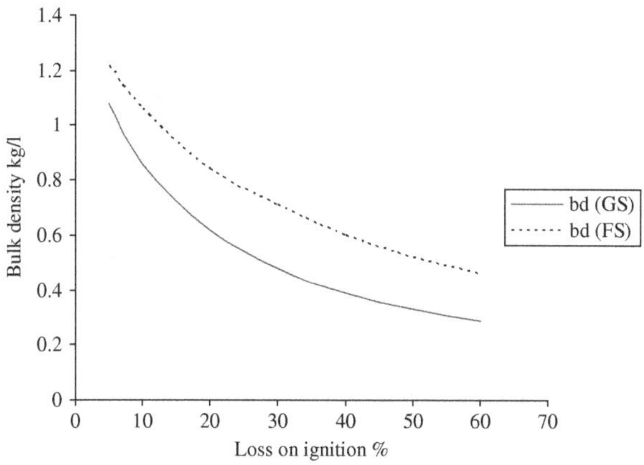

Fig. 2.4 Relationship between the % loss on ignition and the bulk density as found for the data in Table 2.1 for greenhouse soils (GS) in comparison with the relationship found for agriculture field soils (FS) by Kortleven (1970)

For field soils:

$$\rho = \frac{1}{2.52 f_I + 0.65} \tag{2.11}$$

In which:

ρ = bulk density in kg l^{-1}
f_I = mass fraction loss on ignition

The lower bulk density of the greenhouse soils includes a higher air volume.

In the foregoing formula organic matter contents and loss on ignition were not distinguished. Mostly, there are no significant differences between both characteristics. The fraction loss on ignition can be somewhat higher than the fraction organic matter, due to loss of adsorbed and structural water and the loss of CO_2 from carbonate. In a research with 75 samples of greenhouse soils Van den Ende (1988b) presented following relationship:

$$f_I = 1.013 f_H + 0.019 \tag{2.12}$$

In which:

f_I = mass fraction of loss on ignition
f_H = mass fraction organic matter

The differences between both characteristics are small and not important for practical use. Mostly the loss on ignition will be used, because of an easier determination method.

The role of organic matter in greenhouse soils is different. Next to the effect on the soil structure, it plays an important role in the water holding capacity of soils. In different researches a close relationship was found between the fraction loss on ignition and the water content of greenhouse soils. Van den Ende (1988a) and Sonneveld et al. (1990) presented comparable functions for this relationship, as will be shown in Section 3.3. The functions presented are operative for soils containing well decomposed organic matter. Soils with less decomposed organic matter will contain much more water in relation to the fraction organic matter, as found with the peat material used for substrate preparation (Kipp et al., 2000). In former time, when the irrigation was laborious and not frequently carried out, the water holding capacity of soil was an important characteristic. With the modern irrigation techniques in greenhouses it is of secondary importance. The same is true for the cation adsorption capacity of organic matter, which is less important due to the frequent top dressings with the modern fertigation techniques.

Besides the organic matter content the clay content is also important with respect to the water holding capacity of soils. However, the effect is less than those of the organic matter as is clear from formula (2.13), (Van den Ende, 1988b).

$$W_f = 2.374 f_H + 0.376 f_C + 0.134 \qquad (2.13)$$

In which:

W_f= the mass ration water/solid phase
f_H= mass fraction organic matter of oven dry soil
f_C= mass fraction clay of oven dry soil

From the formula it is clear that the effect of organic matter on the water holding capacity is about six times higher than those of clay.

Organic matter in soils, even the well decomposed form, gradually decomposes and some of the residual products of this process become soluble in the soil solution. These organic complexes can strongly affect the availability and the uptake of some micro nutrients. This maybe is one of the reasons that there is such a poor correlation between the results of water soluble micronutrients in soil solution and plant tissues. The soluble organic components in the soil solution can surely affect the availability (Marschner et al., 1987; Verloo, 1980). See also results presented in Section 10.8.

2.3.2 Soil Improvers

Soil improvers are widely used in greenhouse cultures to stabilize or increase the organic matter content of soils. Therefore, the organic matter is the main constituent of soil improvers. Besides an increased water holding capacity and cation exchange capacity, most soils show an improved structure by the addition of organic matter. The latter, especially is the case with clayey soil types, but also on loamy soils such effects can be expected. On sandy soils the addition of organic matter is merely important for an improved water holding capacity and an improved cation adsorption capacity. With heavy additions of organic matter, especially the more stable compounds like peaty materials, the organic matter content of sandy soils can become too high, with as consequence that the space between the aggregates are filled with organic matter, which hinders the vertical transport of water. Such effects can occur with sandy soils when organic matter contents increases over 10%. Therefore, it is recommended to increase the organic matter content on sandy soils not above 5%, being the optimum level for such soils. For some root crops, like radish and carrots, soils with low organic matter content are preferred. Such root crops grow best on pure sandy soils as they show an increased root branching with high organic matter contents, which is not appreciated and considered as a negative quality characteristic.

Soil improvers can be applied by mixing through the soils as well as by mulching. This depends on the crop, the soil type and the soil improver used. Generally, it is not recommendable to place an organic soil improver deep into the soil, where the penetration of air is difficult. This especially counts for soil improvers containing fresh organic material and for heavy soils with a poor penetration of air into deeper soil layers, like with soils with high clay contents. For an optimum decomposing process sufficient oxygen should be available, which always cannot be ensured in deeper soil layers.

Soil improvers frequently used in greenhouses are listed in Table 2.9. Beside the organic matter soil improvers often contain substantial quantities of mineral nutrients. Some soil improvers contain much residual salts, which can be a drawback. With the use of such soil improvers the osmotic potential of the soil solution can markedly be decreased which can require extra water supply to wash out the residual salts from the root zone. Only the most common types of soil improvers are mentioned. The contents of organic matter and mineral elements vary strongly, because of origin, preparation method and storage conditions. The data as listed in Table 2.9 are derived by comparison of the data of different authors,

Table 2.9 Composition of soil improvers used in greenhouse industry. The data roughly reflect the composition got by comparison of the data of authors mentioned at the bottom of the table. The data are expressed in kg dry matter, organic matter and total N, P and K supplied with 1000 kg fresh soil improver

Type	Dry matter	Organic matter	N	P	K
Farmyard manure	250	120	5.5	2	7
Cattle slurry	80	50	5.5	1.5	4.5
Chicken manure	600	400	26	11	18
Spent mushroom compost	350	180	7	2	8
Green compost	650	150	7	1.5	5.5
Peat	500	450	5	0	0
Bark	400	300	3	0.5	2

The data in this table are derived by comparison of data published by: Bokhorst (2005a and b); LNV (undated); Raviv et al. (2002); RHP (2008); Solbra (1979); Voogt (2008); Van der Wees (1993).

as is made clear at the bottom of this table and the differences found among the results of these authors learned that a variation of 50% will be taken into account. The water content with which the soil improver is delivered is a main characteristic, because it determines strongly the contents of valuable constituents by weight. High water contents suppress these contents on the fresh materials. The addition of soil improvers can be based on a quantity of dry matter. The constituents as expressed on the fresh material can be converted to contents based on dry material following formula (2.14).

$$C_D = \frac{C_F}{f_D} \tag{2.14}$$

In which

C_D= constituent expressed on dry matter
C_F= constituent expressed on fresh material
f_D= mass fraction dry matter

The water contents vary strongly among the types, but also within the type great variations occur. Cattle slurries easily has mass fraction of water higher than 0.95. Peat, chicken manure and compost easily have a mass fraction of water below 0.50, but in wet condition this quickly increases up to 0.75. The varying water content of solid soil improvers strongly affects the advisable quantity when based on fresh mass, since the water content usually is not always available at the moment of application. Addition on volume basis will affect the advisable quantity less than application on mass basis, because the water content of solid soil improvers scarcely affects the quantities of essential constituents in a volume. However, this is not in force for liquid manures. Application on volume basis requires a good definition of bulk density. Such a definition is developed by a European regulation (CEN, 2000, 2007). Last decennia the application of soil improvers is seriously embedded by governmental regulations, which restricts the application of soil improvers by limits for maximum additions. These limitations are on the one hand determined by the content of heavy metals, related to the maximum acceptable yearly additions of these metals as established for sewage sludge and composts (LNV, undated). On the other hand, the application is limited by regulations for maximum acceptable additions of N and P, as will be discussed in Section 16.6.

In history an application of 70–100 m^3 ha^{-1} farmyard manure or green compost every year or every second year was a normal practice to keep the soil structure of greenhouse soils in a good condition. The quantities applied nowadays are more focussed on the possibilities within the regulations. There is a great variation in the composition among the deliveries of soil improvers. Therefore a secure application within the limits set, often requires a chemical analysis of each delivery of a soil improver. This for example is nowadays obligatory within Dutch regulations.

The K and the P added with the soil improver should be directly taken into account on the total nutrient requirements. For P this usually solely concerns the base dressing, because this element is exclusively added as such. However, often the P required as base dressing is less than the quantity applied by some soil improvers, like animal manures and composts. Thus, applications of soil improvers can lead to accumulation of P in greenhouse soils, when the limits set are crossed. The addition of K with the soil improver directly affects the base dressing, but can work on the top dressing as well. Many soil improvers contain limited readily available N and thus, additional base dressing with N fertilizer is mostly required. Not available N is released during the decomposition process and possibly will be taken into account with the top dressings.

The organic N and P compounds in manures become available to plants with the decomposition of the organic matter in it by the micro biological activity in the soil. The rate of decomposition of the organic matter and henceforth the availability of N and P from the manures depends much on the characteristics of the soil improver (Sluijsmans and Kolenbrander, 1977), like the C:N ratio in particular. Besides, also the soil type and the growing conditions are important. In greenhouses the soil temperature, the moisture content and the porosity of the soil are mostly optimal for a quick decomposing process. Therefore, the decomposition in greenhouses will be faster in comparison with field conditions. For the calculation of the mineralization dynamics of organic matter, the simple and easy to handle one-dimensional model of Janssen (1984) proved to be useful. In this model the variation in the decomposition rate of all constituents of a soil improver is reduced to a single parameter, being the "initial age" of the material. The N mineralization calculated by this model is in good agreement with the results of incubation tests and field experiments (Marcelis et al., 2003). Thus, the N release of manures in time can be estimated with this model. With application of soil improvers with fresh organic material, specifically material with a high C:N ratio, temporal immobilization of mineral N is possible and addition of extra N by fertilizers with the base dressing can be necessary. The mineral N absorbed in advance will become available during the decomposition process.

The organic matter in different peat types has a relatively high stability, especially the black, well decomposed types. Such types of peat need to be well frozen in wet condition in the field before drying to ensure that it is suitable as a soil improver. Insufficient frozen black peat does not sufficiently absorb water after drying out and these properties are irreversible.

2.3.3 Contamination

In the commercial intercourse of soil improvers a great variation of waste products are available. The use of products from which the origin is not clear is not advisable. Ingredients of waste products from industries may be toxic to people or to plants. Even when these are not directly toxic to plants, the addition to soil can include problems in future. When for example waste products contain high contents of heavy metals, concentrations in soils can accumulate to undesirable levels by regular additions during years. This often does not directly affect plant growth, but the uptake of heavy metals by plants can increase to an unacceptable level in the produce for human consumption. Therefore, governmental regulations have set strict maximum limits for the contents of heavy metals in soil improvers. This especially is the case for different compost types, because it is well known that this material sometimes contains high concentrations heavy metals. The occurrence of these contaminants in composts differs strongly and depends much on the character of the waste left by local industries. Products like sewage sludge and municipal waste compost are well known as materials often contaminated and therefore, are not recommended as a soil improver in greenhouses. Bio waste composts and green composts commonly contain acceptable concentrations of heavy metals and are mostly suitable as a soil improver for greenhouse soils. However, the concentrations will be determined and are bound on limits. A review of the limits as has been set in the regulations within the different countries of the European Community is listed in Table 2.10 (Amlinger et al., 2004). The limits show great differences among countries. In the countries out of the European Union, like the USA, also limits are formulated, but these are often much higher than those within the European Community. The values for compost from bio waste as formulated within the European Community for organic production (EC Regulation, 1991) are added in the last column of Table 2.10. Manure and slurry from animal origin has sufficient low concentrations of most heavy metals. However, in some materials often high Cu and Zn concentrations can be determined, especially those derived from pig farms (CEN, 2004).

The application of soil improvers in relation to environmental consequences for soil grown crops and the use as a constituent for substrates will be discussed further on in the Chapters 16 and 11, respectively.

Table 2.10 Limits and mean values for total concentrations of heavy metals in green compost and bio waste compost within the European Community and values for compost as proposed for compost from bio waste (Amlinger et al., 2004). The concentrations are expressed as μmol kg^{-1} and as mg kg^{-1} dry matter

	Limit values within EuC countries						EuC organic growing[2]	
	Minimum		Maximum[1]		Mean[1]			
Elements	μmol	mg	μmol	mg	μmol	mg	μmol	mg
Cd	6.2	0.7	26.7	3.0	12.5	1.4	6.2	0.7
Cr	962	50	4808	250	1788	93	1346	70
Cu	393	25	9434	600	2248	143	1101	70
Hg	1.0	0.2	15.0	3.0	5.0	1.0	2.0	0.4
Ni	170	10	1704	100	801	47	426	25
Pb	217	45	1351	280	584	121	217	45
Zn	1147	75	22936	1500	6361	416	3058	200
As	67	5	668	50	307	23	–.–	–.–

[1]Some exceptional high values excluded;
[2]as proposed for compost from bio waste.

References

Amlinger F Pollak M and Favoino E 2004.Heavy metals and organic compounds from wastes used as organic fertilisers. Annex 2, Compost quality definition – Legislation and standards. Env.A.2./ETU.2001.0024, 21pp.

Aylward G H and Findlay T J V 1974. SI Chemical Data. Second edition. John Wiley and Sons, New York.

Bokhorst J 2005a. Compost. *In*:. Bodem & Bemesting Biokas. Louis Bolk Institute, Driebergen, 16–17.

Bokhorst J 2005b. Dierlijke mest. *In*:. Bodem & Bemesting Biokas. Louis Bolk Institute, Driebergen, 18–19.

Bos E J F Keijzer R A W Van Schie W L Verhagen J B G M and Zevenhoven M A 2002. Potgronden en substraten RHP. Stichting RHP, Naaldwijk, The Netherlands

CEN 2000. European committee for standardisation CEN/TC 223. Soil improvers and growing media – Determination of a quantity. EN 12580, 14pp.

CEN 2004. Draft discussion document for the ad hoc meeting on biowastes and sludges. 15–16 January 4004, Brussels, CEN/TC 223, N426, 22pp.

CEN 2007. European committee for standardisation CEN/TC 223. Soil improvers and growing media – Determination of a quantity for materials with a particle size greater than 60 mm. EN 15238, 14pp.

EC Regulation 1991. Regulation No. 2092/91/EEC 1991: Counsel regulation on organic production of agriculture products and indications referring thereto on agricultural products and foodstuffs. Last Amendment Counsel Regulation No. 1804/99/EEC, 19 07 1999, O.J. no. L222/1.

Hilgard E W 1919. Soils in the Humid and Arid Regions. Macmillan & Co, New York, 593pp.

Janssen B H 1984. A simple method for calculation decomposition and accumulation of 'young' soil organic matter. Plant Soil 76, 297–304.

Kipp J Wever G and De Kreij C 2000. International Substrate Manual – Analysis, Characteristics and Recommendations. Elsevier International Business Information, Doetinchem, The Netherlands, 94pp.

Klute A 1986. Methods for soil analysis. Part 1. Physical and mineralogical methods, second edition. Agronomy Monograph No 9 (Part 1). Am. Soc. Agron., Maidison, WI, 1188p.

Kortleven J 1970. Volumengewicht, porionvolumen en humusgehalte. Instituut voor Bodemvruchtbaarheid, Haren, The Netherlands. Rapport no 45.

LNV (undated). Zuiveringsslib, compost en zwarte grond; kwaliteit en gebruik in de landbouw. Ministerie van Landbouw, Natuurbeheer en Visserij, Den Haag, 27pp.

Lucas R E and Davis J F 1961. Relationship between pH values of organic soils and availabilities of 12 plant nutrients. Soil Sci. 92, 177–182.

Marcelis L F M Voogt W Visser P H B Postma J Heinen M De Werd R and Straatsma G 2003. Organische stof management in biologische kasteelt. Wageningen University Research, Rapport 70, 50pp.

Marschner H Römheld V and Cakmak I 1987. Root-induced changes of nutrient availability in the rhizospere. J. Plant Nutr. 10, 1175–1184.

Marschner H 1995. Mineral Nutrition of Higher Plants. Academic Press, London.

Oertly J J 1980. Controlled-release fertilizers. Fertilizer Res. 1, 103–123.

Prasad M Simmons P and Maher M J 2004. Release characteristics of organic fertilizers. Acta Hort. 644, 163–170.

Proefstation 1954. Proefstation voor de Groenten en Fruitteelt onder Glas te Naaldwijk, The Netherlands. Jaarverslag 1954, 28–29.

Raviv M Wallach R Silber A and Bar-Tal A 2002. Substrate and their analysis. *In*: Savvas D and Passam H (eds), Hydroponic Production of Vegetables and Ornamentals. Embryo Publications, Athens, 25–101.

RHP (2008). Personal Information of RHP Foundation, 's Gravenzande, The Netherlands.

Roorda van Eijsinga J P N L 1971. Determination of the phosphate status of soils in the Naaldwijk area for growing lettuce in glasshouses. Agriculture Research Reports 753. Centre Agric. Publish. Doc. Wageningen, 25 pp.

Roorda van Eysinga J P N L 1974. De opname van fluor door de wortel en de gevolgen daarvan voor diverse gewassen, in het bijzonder freesia. Theesis, Centrum voor Landbouwpublicaties en Landbouwdocumentatie, Wageningen, 83 pp.

Sluijsmans C M J and Kolenbrander G J 1977. The significance of animal manure as a source of nitrogen in soils. *In*: Proc. Intern. Seminar on Soil Environment and Fertility Management in Intensive Agriculture, Tokyo, 403–411.

Solbra K 1979. Composting of bark 1. Different bark qualities and their uses in plant production. Meded, Norsk Inst. Skogforskning 34, 285–323.

Sonneveld C 1990. Estimating quantities of water-soluble nutrients in soil using a specific 1:2 volume extract. Commun. Soil Sci. Plant Anal. 21, 1257–1265.

Sonneveld C Van den Ende J and De Bes S S 1990. Estimating the chemical composition of soil solutions by obtaining saturation extracts or specific 1:2 by volume extracts. Plant Soil 122, 169–175.

Van den Ende J 1988a. Water contents of glasshouse soils at field capacity and at saturation. 1. Relationships between water contents. Neth. J. Agric. Sci. 36, 265–274.

Van den Ende J 1988b. Water contents of glasshouse soils at field capacity and at saturation. 2. Estimating water contents from organic-matter and clay contents or loss-on-ignition. Neth. J. Agric. Sci. 36, 275–282.

Verloo M G 1980. Peat as a natural complexing agent for trace elements. Acta Hort. 99, 51–56.

Van der Wees A 1993. Meststoffen. Proefstation voor Tuinbouw onder Glas, Naaldwijk, The Netherlands. Plantenvoeding in de Glastuinbouw, Informatiereeks no 87, 13–25.

Voogt W and Sonneveld C 2001. Silicon in horticulture crops grown in soilless culture. *In*: Datnoff, L E, Snijder, G H and Korndörfer G H (eds), Silicon in Agriculture. Elsevier Science B.V., Amsterdam, 115–131.

Voogt W and Sonneveld C 2004. Interactions between nitrate (NO_3) and chloride (Cl) in nutrient solutions for substrate grown tomato. Acta Hort. 644, 359–368.

Voogt W 2008. Personal data bank of Wageningen University, The Netherlands.

Chapter 3
Soil Solution

3.1 Introduction

The characteristics of the soil solution in the root environment in the greenhouse industry differ much from those for field grown crops. This is caused firstly by the growing conditions in the greenhouse, which strongly differ from those in the field and secondly the function attributed to the soil solution with respect to plant development. One of the most striking differences between growing in the greenhouse and in the field is the exclusion of the natural precipitation in greenhouses, which offers opportunities for a full control of the water supply. Another difference is the heavy fertilizer application, related to the high nutrient uptake. In addition these application fertilizers are for the greater part added by fertigation. Furthermore, the irrigation and fertilizer addition not only has a function with respect to supply the plant with sufficient nutrients and water, but in greenhouses these actions are also a tool to control plant growth and produce quality. Sometimes, low osmotic potentials in the soil solution are maintained to prevent a lush growth or to improve fruit quality. Such effects on plant development, especially makes sense in substrate growing, where plants are grown in small rooting volumes and thus the composition of the soil (substrate) solution easily can be adjusted, for example on the demand of the crop under changing growing conditions. Thus, in principle it seems possible to supply plants under greenhouse conditions at the right time with the right quantity of water and nutrients, and losses of water and nutrients to the environment can be minimized. However, this is often frustrated by a heterogeneous water supply of irrigation systems, a heterogeneous water uptake by plants and accumulation of salts in the root environment from the irrigation water used. Thus, a precise matching on the demand by the water supply is hindered by an overdose of irrigation water to equalize the differences between wet and dry spots and to prevent too high accumulations of residual salts.

In the greenhouse industry an adequate management of water and nutrient supply is important. On the one hand to maintain optimal conditions for the plant in the soil solution with respect to plant nutrient uptake and to the requirements for the osmotic potential and on the other hand with respect to prevent leaching of nutrients and by this prevention of environmental pollution. Especially the high concentrations of nutrients in the soil solution contribute strongly to a high environmental pollution

C. Sonneveld, W. Voogt, *Plant Nutrition of Greenhouse Crops*,
DOI 10.1007/978-90-481-2532-6_3,

per area. This does not mean that the leaching of nutrients is high in relation to the total uptake. However, this item will be discussed further on in Chapter 6.

In the present chapter the osmotic potential of the soil solution will be discussed in relation to the prevailing moisture conditions during cultivation. Hereby, the connection will be discussed between the definition of the soil solution of soils in situ and those of substrates, because in the greenhouse industry substrate growing is important and will expand further on. Following the definition for soil solution the term "substrate" solution will be defined, being the solution extracted from substrates at moisture contents maintained during crop cultivation. Besides the osmotic potential, being a measure for the total of the different concentrations of mineral constituents, an impression will be given of the specific composition of the mineral constituents in the soil solution. Finally some guidelines about the role of the composition of the soil solution in relation to the mineral uptake of crops are presented.

3.2 Osmotic Potentials of Soil Solutions

In Table 3.1 the composition of soil solutions from field soils is given in comparison with those from greenhouse industry. In the comparison soil solutions as well as substrate solutions are taken into account. The most striking difference between the solutions derived from fields soils and those from greenhouses soils are the overall much higher nutrient concentrations in solutions from greenhouse. This especially holds for greenhouse soils where the EC in the solution is highest. Furthermore, it is obvious that in greenhouse cultivation nutrients contribute substantially to the total salt concentrations of soil and substrate solutions and thus to the osmotic potential.

Table 3.1 Ionic compositions of soil solutions. Ions expressed as mmol l^{-1} and EC as dS m^{-1}. The no's 1–5 are from field soils and the no's 6–9 from greenhouses

No[1]	K	Na	NH_4	Ca	Mg	NO_3	Cl	SO_4	HCO_3	P	EC
1	1.7	5.4	–	8.9	3.7	9.1	8.4	1.6	0.8	0.02	–
2	0.3	0.2	–	2.2	0.6	3.7	2.1	0.2	–	–	0.6
3	0.5	0.3	0.05	1.6	0.5	3.2	2.4	0.6	–	0.02	–
4	0.1	–	0.03	1.1	0.0	0.6	–	–	–	0.01	–
5	0.2	–	1.10	5.3	0.1	12.3	–	–	–	0.01	–
6	6.6	13.2	0.39	22.3	8.7	24.1	15.0	19.1	–	0.32	6.5
7	4.6	1.8	1.2	4.2	3.2	11.4	1.3	3.2	–	1.7	2.3
8	8.0	–	<0.5	10.0	4.5	23.0	–	6.8	–	1.0	4.0
9	5.0	–	<0.5	5.0	3.0	12.5	–	3.0	–	0.9	2.2

[1]Composition derived from: 1 – means of a historical series from Adams (1974); 2 – means of data of Qian and Wolt (1990); 3 – means of data of Peters (1990); 4 and 5 – data of Barraclough (1989) before and after top dressing with N, respectively; 6 – means of greenhouse soils by Van den Ende (1989) and Sonneveld et al. (1990); 7 – means peaty substrates of Sonneveld and Van Elderen (1994); 8 and 9 – recommended values for rock wool grown tomato and rose, respectively (Sonneveld, 1995).

This especially is the case in substrate systems where low saline primary water is used, and the osmotic potential is thus more or less solely brought about by nutrients. However, when water is used with a higher salinity level, and low osmotic potentials are desired in substrate cultivation, as indicated in Table 3.1 for tomato, the nutrient levels will be reduced to the required optimum for plant nutrition, while the osmotic potential will be lowerd further by accumulation of the residual salts from the saline water (Sonneveld, 1995). In the Chapters 7, 13 and 16 this item will be discussed in detail.

The most important characteristic of the soil solution for greenhouse cultivation is the determination of the EC, because the results of this determination in soil solutions is within the operational range for greenhouse cultivation closely linearly correlated with the osmotic potential of the soil solutions. Such a close relationship will be found, when the osmotic potential is solely build up by ions and ionic pairs of mineral salts. The relationship between the EC and the salt concentration is linear, over a relatively wide range. True enough, each ion has its own specific contribution to the EC (McNeal, et al., 1979; Sonneveld et al., 1966). In Fig. 3.1an impression is given of the relationships between the concentrations of different salts and the EC of a number of single salt solutions as found by Sonneveld et al. (1966). The contribution of a specific salt to the EC depends on factors like the valence of the ions, the dissociation constant, the activity of the ions and the ion pair formation and furthermore the temperature of the solution. In Fig. 3.1 the linear relationships between the salt concentrations and EC values of different single salt solutions are shown over a range up to 60 and 90 mmol l^{-1} for tertiary and binary salts, respectively. The

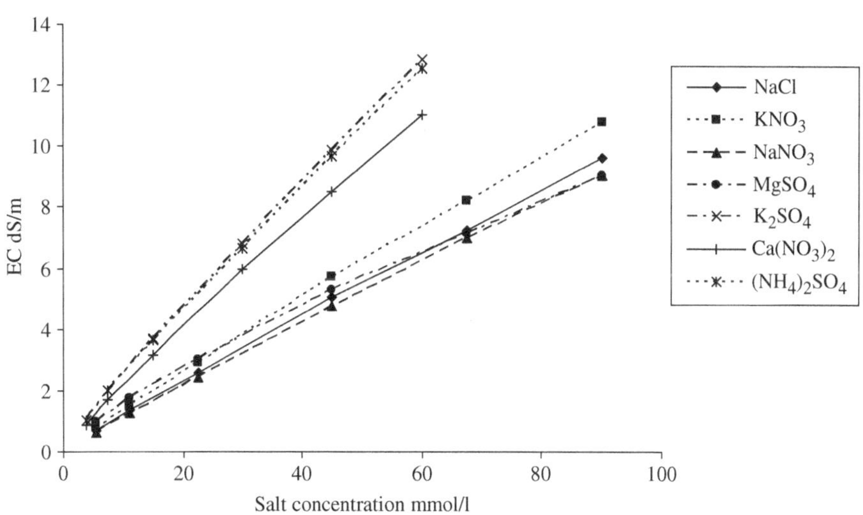

Fig. 3.1 The relationship between the concentrations of different salts and the EC, after Sonneveld et al. (1966) (Relationships calculated: NaCl, EC = 0.106c + 0.17; KNO_3, EC = 0119c + 0.24; $NaNO_3$, EC = 0.100c + 0.18; $MgSO_4$, EC= 0.094c + 0.73; $Ca(NO_3)_2$, EC = 0.180c + 0.39; K_2SO_4, EC = 0.208c + 0.45; $(NH_4)_2SO_4$, EC= 0.204c + 0.44)

relationships between salts differ strongly. With increasing solution temperature the EC of salt solutions increases also. Therefore, the EC is expressed at a standardised temperature, mostly 25°C. When the EC is measured at a different temperature, the value at the standard temperature can be approached by the temperature coefficient, which is the relative increase or decrease of the EC by an increase or decrease of 1°C, respectively. This coefficient is somewhat different for the temperature interval, the relation temperature and the salt composition of the solution. However, at a relation temperature of 25°C and no bigger deviations than 10°C a temperature coefficient of 2% is proved to be very suitable (Campbell et al., 1948; Sonneveld et al., 1966). Modern apparatus compensate the effect of the temperature deviation automatically.

For mixed salt solutions McNeal, et al. (1979) showed a linear segment method with which the contribution of various concentrations of different salts to the EC can be calculated. The method is suitable for concentrations up to 50 mmol l^{-1} for mono-valence and up to 25 mmol l^{-1} for bi-valence ions. The low intercepts given with these linear relationships point out that they are suitable for calculations until rather low concentrations.

For rough estimations the formula given by Sonneveld et al. (1999) can be used for mixed salt solutions.

$$EC \approx 0.1C^{+} \tag{3.1}$$

In which

EC = electrical conductivity of the solution in dS m^{-1}
C^{+} = the sum of valences of the cations in mmol l^{-1}

However, for a precise calculation of the EC from the ion composition the already mentioned method presented by McNeal et al. (1979) will be recommended.

For soil solutions, but also for various other mixed salt solutions like soil extracts, and natural waters a close relationship has been found between the osmotic potential and the EC, as shown with the data in Fig. 3.2. The relationships found for different solutions show strong similarity and a general formulation can be established as given in Eq. (3.2).

$$OP \approx -33.3EC \tag{3.2}$$

In which

OP = osmotic potential of the solution in kPa at 0°C
EC = electrical conductivity of the solution in dS m^{-1} at 25°C

For strongly diluted extracts like the 1:5 by weight soil extract, a somewhat different relationship has been found. The data agree very well with the results presented by Campbell et al. (1948).

The osmotic potential of the solution in the root environment in greenhouse cultivation appeared to be an important factor for growth regulation of crops. This was

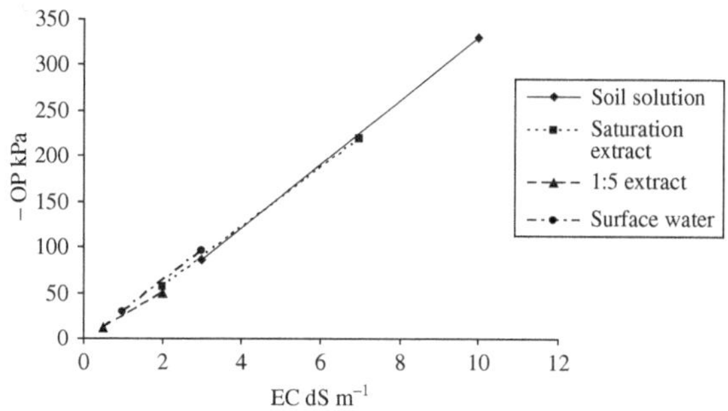

Fig. 3.2 The relationship between the EC and the osmotic potential (OP) of different solutions, after Van den Ende (1968) and Sonneveld and Van den Ende (1967). The *dots* show the interval of the data (Regression equations: Soil solution OP = –34.9 EC – 19; Saturation extract OP = –32.6 EC – 8; 1:5 extract OP = –24.5 EC – 0; surface water OP = –33.3 EC – 4)

not recognised from the beginning. In history, low osmotic potentials (high EC) in greenhouse soils were exclusively connected with the negative aspects of high salinity, like growth reduction and nutrient disorders (Riemens, 1951; Van den Ende, 1952). However, in greenhouses where crops easily show a lush growth often connected with a poor quality, also positive effects of a low osmotic potential in the soil solution were observed (Van den Ende, 1955). The lush growth of crops under greenhouse conditions especially appears at relatively high temperatures, reduced light intensity and ample water supply. Such conditions for example occur predominantly in winter in North-West Europe. Gradually, the osmotic potential of the soil solution became a tool for greenhouse growers to manipulate crop development. The cultivation in substrate as developed for various greenhouse crops especially enhanced the availability of water in the root environment by the usually low matrix potential in the substrates of such growing systems, which accentuate the need for the use of the osmotic potential as a tool for growth regulation. Substrate growing, as mentioned before, offers excellent perspectives for such a regulation, because of the controlability of the usually small rooting volumes.

In greenhouse crops disorders of a high osmotic potential (low EC) in the root environment are well known in vegetables as well as in flowers and covers a great variation of plant characteristics. Examples are: irregular colouring of tomato fruits (Sonneveld and Voogt, 1990), glassiness in lettuce (Maaswinkel and Welles, 1986) and aggravation of the occurrence of soft rot in Hippeastrum bulbs (Van den Bos, 1996). Guide values for required and acceptable concentrations of nutrients and residual ions in the root environment will be discussed in Chapter 7.

3.3 Moisture Contents

A drawback with the determinations in soil and substrate solutions is the lack of a good definition of the moisture status of soils and substrates for preparation of the solution. The moisture content of a soil fluctuates with the evaporation and the water uptake of the crops grown and the precipitation, irrigation and capillary rise.

This especially occurs for field crops grown without artificial irrigation or with low frequency irrigation schedules. In such cases, the moisture withdraw from the root zone between irrigations can be considerable, which for example directly will be reflected by a decrease of the osmotic potential of the soil solution. The fluctuations in greenhouse soils are restricted, because of the high frequency irrigation schedules maintained. This especially is the case in substrate systems, where the irrigation frequency under high transpiration conditions increases up to several times per hour.

3.3.1 Greenhouse Soils In Situ

For a wide range of soil types Van den Ende (1988a) found a close linear relationship between the water content of greenhouse soils cultivated with tomatoes and the water content at a pressure head of –6.3 kPa, as shown with formula (3.3).

$$w_f = 1.047 w_{-6.3} - 0.012 \quad r = 0.987 \tag{3.3}$$

In which:

w_f = mass ratio water/solid phase of field moist soil
$w_{-6.3}$ = mass ratio water/solid phase of soil at a pressure head of – 6.3 kPa

Thus, the water contents of the greenhouse soils grown with tomato were more or less equal to that at a pressure head of – 6.3 kPa.

Furthermore Van den Ende (1988b) found that the water content of the field moist soil was closely related to the loss on ignition, as given in following formula.

$$w_f = 2.617 f_l - 0.118 \quad r = 0.985 \tag{3.4}$$

In which:

w_f = mass ratio water/solid phase of field moist soil
f_l = mass fraction loss-on-ignition of oven dry soil

Sonneveld et al. (1990) also determined the relationship between the loss on ignition and the water content under growing conditions and found a comparable relationship for a series of 75 greenhouse soil samples. These samples were gathered from greenhouses with different crops, merely during the cultivation period. The mass fraction organic matter and clay of the soils varied from 0.03–0.61 and 0.03–0.40, respectively. The relationship is shown in Fig. 3.3, and the equation found is given in formula (3.5).

$$w_f = 2.821 f_l - 0.100 \quad r = 0.982 \tag{3.5}$$

Both formulae resulted in comparable values over a wide range of soil types. Thus, on basis of these formulae the field moist condition for greenhouse soils can be defined.

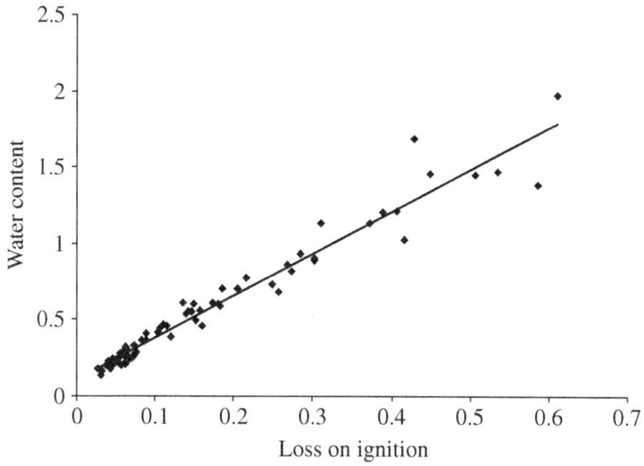

Fig. 3.3 The relationship between the loss on ignition (m/m) of greenhouse soils and the water content (g/g) at field moist conditions. After Sonneveld et. al. (1990). Regression equation: $y = 2.821x + 0.100$, $r = 0.982$

With the given formulae also the water volume can be calculated because the bulk density is also closely related with the loss-on-ignition fraction (Sonneveld, 1990), like already given in formula (2.10).

Combination of the formula (3.5) and (2.10) gives an equation for the water volume in greenhouse soils, as shown in Eq. (3.6).

$$wv_f = \frac{2.821\, f_l + 0.100}{4.67\, f_l + 0.69} \tag{3.6}$$

In which

$wv_f =$ volume fraction of water of field moist soil
$f_l =$ mass fraction loss-on-ignition of oven dry soil

This formula can be used to calculate roughly the current moisture condition of greenhouse soils under growing conditions and will be used as a standard when reference is made to the soil solution of greenhouse soils. This definition is true with a reasonable frequent irrigation and thus, the relation between loss on ignition and water content are in agreement with the formula presented as (3.5).

3.3.2 Substrates

For substrates no reasonable relationship between organic matter and water holding capacity will be expected, due to the great variation of materials used as a substrate or used as a substrate constituent and utmost the great variation of the quality within these materials. For example, a lot of substrates do not contain noteworthy organic matter, while they have a high water holding capacity. But even when substrates contain considerable quantities organic matter, like peaty substrates, the characteristics of the organic matter differ strongly and show a great variation in water holding capacity. In an investigation with peaty substrates (Sonneveld et al., 1974) a correlation coefficient of 0.809 was found between the mass fraction loss-on-ignition and the ratio moisture/solid phase at a pressure head of –3.2 kPa, which is considerably lower than the correlation coefficient found with greenhouse soils. Since then the variation in materials used to produce substrates is strongly increased. The pressure head of –3.2 kPa was chosen as being approximately the moisture content under growing conditions in that period. Later on, the moisture contents of substrates during cultivation became higher.

The growing conditions are another hindrance for a precise estimation of the water holding capacity. The moisture in most substrates is quite loosely bound and thus, the thickness of the substrate layer applied in the growing system will affect strongly the water holding capacity. Another factor is the irrigation method that plays an important role. When the water is supplied on the top, the water distribution in the substrate will differ strongly from the situation with water supply from the bottom. Thus, the definition of the water content at field capacity of a substrate not only depends on the characteristics of the substrate, but also on the growing conditions.

Wever (1995) compared the bulk densities and the water contents of a series of peaty growing media as found in practice for potted plants with the same characteristics measured at the laboratory following the CEN standard methods (CEN, 2006). The water content in the samples prepared at the laboratory following this method was measured at a pressure head of –1 kPa. The correlation coefficient between the bulk density as found in the field and measured at the laboratory was rather low ($r = 0.83$), but the average values had an acceptable agreement. The correlation coefficient for the water content found under field conditions and the content determined at –1 kPa at the laboratory was also rather low ($r = 0.83$), but on average the contents determined at the laboratory approached the field condition reasonably. Results of some calculations are listed in Table 3.2. The data in this table show that for a wide range of peaty substrates with a bulk density in the range from 50 to 300 kg m^3, that there is on average an acceptable agreement between the water contents of the growing media under field conditions and those found at the laboratory at –1 kPa. Thus, under growing conditions the water contents of the peaty growing

Table 3.2 Bulk density and water content of peaty substrates as determined at the laboratory at –1 kPa and comparable values of the bulk density and water content under field conditions, estimated by the regression equations found by Wever (1995)

Bulk density kg m^{-3}		Water content g g^{-1}	
Determined	Value estimated for practice	Determined	Value estimated for practice
50	54	10	9.6
300	261	3	3.9

media approaches on average reasonably the water contents at a pressure head of –1 kPa. The low correlation coefficient found for the relationships can be explained easily by the strong differences realised under growing conditions, as there are the different potting techniques, irrigation methods and frequencies, differences in time between latest watering and sampling errors and so on.

For some substrates other than peat, the water content at a pressure head of –1 kPa is not a good estimation of the water content under growing conditions. These substrates have lost already important parts of the water at such a relatively low suction. It seems that for these substrates the water content at free drainage after saturation is a better estimation for the water content under growing condition than at a pressure head of –1 kPa. For bulk material this free drainage situation can be compared with the determination of the water content at –0.3 kPa at the laboratory, being half of the height of the rings used for the standard method of CEN (2006). For pre-shaped material half of the height of the slabs or blocks should be considered as the pressure head of the free drainage condition. In Table 3.3 the water contents of a number of substrates is given at free drainage (leak out) condition and at –1 kPa, following Kipp et al. (2000). In mostly cases there is a considerable difference between both water contents. For pre-shaped materials like slabs and blocks of PU-foam and rock wool it should be concluded that the water content under growing conditions will be approached mostly better by the "leak out" condition than at –1 kPa, because at this pressure head an important part of the water is lost and the "leak out" condition approaches the situation in the field. The water content of expanded clay granules is already low at the "leak out" situation. Under growing conditions this substrate is usually placed in a water layer, which layer plays an important role in the uptake of water and nutrients. For the bulk materials the thickness of the substrate layer especially determines the water content at field capacity and thus at what pressure head the determination on the laboratory should be carried out.

Table 3.3 Relative water contents by volume of a series of substrates at a pressure head of –0.3 (leak out) and –1 kPa

Type of substrate	Pressure head	
	–0.3 kPa[1]	–1 kPa
Wood fibre	0.72	0.32
Expanded clay granules	0.19	0.13
Coco peaces	0.40	0.33
Coco dust	0.91	0.67
Perlite	0.44	0.31
PU foam slabs	0.60	0.06
PU foam pieces	0.58	0.07
Pumice	0.58	0.40
Rock wool slabs	0.94	0.42
Peat	Nd	0.79

[1] For bulk substrates, for pre-shaped substrates the pressure head will be half the height of the slabs.
After Kipp et al. (2000).

So with respect to a definition for "soil solution" following general conclusions are possible:

- For substrates retaining their water at a pressure head of –1 kPa or higher, the water content at –1 kPa should be considered as being the field capacity
- For substrates that have lost an important part of their water at –1 kPa, the water content at free drainage after saturation should be considered as being field capacity, because such substrates will be used in thin layers
- For very course substrates with a low water holding capacity placed in a water layer, this water layer at the bottom should be considered as being the "soil solution".
- For growing systems with a very restricted substrate volume and a high speed of the nutrient solution, like NFT and deep water culture, the circulating water can be considered as the "soil solution".
- For strongly different growing systems and growing conditions strongly different from the formulations described, specific definitions are required and should be formulated.

3.4 Changes in the Chemical Composition

The chemical composition of soil solutions will change strongly, mainly by factors like nutrient uptake by crops, leaching of nutrients by irrigation and supply of nutrients by fertilization. The grower often switches the concentrations of specific ions as well as the total ion concentration (EC) deliberating the requirements of the crop. For some crops the EC is increased strongly like at the start of fruit vegetable crops to promote an early fruit setting and to prevent a lush growth, as mentioned in Section 3.2. Such an increase is realised by use of accumulated residual salts in the soil left from the former crop cultivation, by the addition of extra nutrients, or by a combination of both factors. Later on in the growing cycle of such crops, when lower EC levels are required, the grower let them gradually decrease by means of over irrigation and by the nutrient uptake of the crop. When necessary, growers start the fertigation to prevent that the nutrient concentrations will be decreased until too low values, which negatively can affect fruit quality. In Fig. 3.4 the course of the cation composition of the soil solution is shown for the described situation, with a soil grown tomato cropping as an example. The cations were determined in the saturation extract, the concentrations of which are closely correlated to those in the soil solution. The NH_4 concentration is high at start, because of the steam sterilisation carried out just before the first sampling (see Chapter 10). The concentration gradually decreases with increasing microbiological activity in the soil. The concentrations of K, Ca and Mg were brought on the required high levels for tomato by base dressing. During the first months the concentrations of these cations gradually increase further on, by evaporation and action of capillary rise from the saturated zone, as long as there was no irrigation. When the irrigation started around

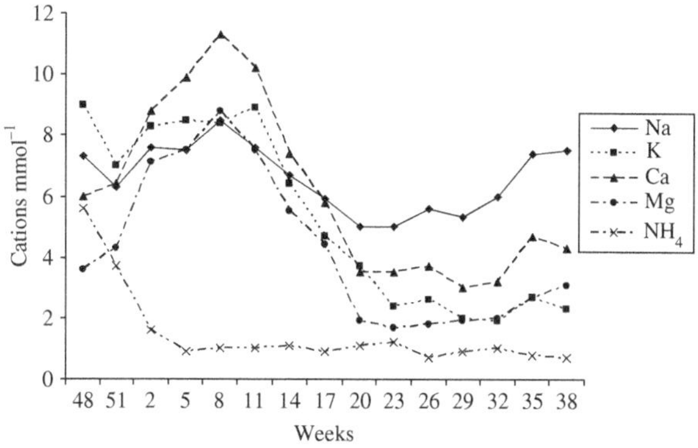

Fig. 3.4 The course of the cation contents as determined in the saturation extract in a greenhouse during a tomato cropping. The soil type was a sandy soil with 5% organic matter

week 8, the concentrations gradually decreased and the top dressing by fertigation started around week 14. Until the end of the cropping period at week 32, by fertigation 300 kg K and 35 kg Mg per ha was supplied. In the greenhouse concerned, water and fertilizers were supplied by sprinkler irrigation and the soil was sampled over a depth of 25 cm. The Na concentration is relatively high during the whole growing period, as a result of the high concentration of this ion in the irrigation water used.

A total different course of the analytical data in the soil solution can be expected for flower crops that does not require a low osmotic potential in the soil solution at start. Such crops often are started at a low fertilization level in the soil. An example of such a situation is given for a gerbera crop in Fig. 3.5, where the course of the anion concentrations together with the EC in the saturation extract is shown. After flooding of the soil in week 45, the salt contents were decreased. The base dressing

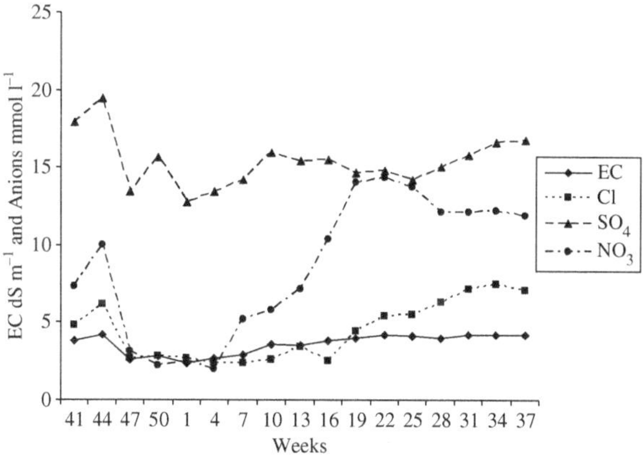

Fig. 3.5 The course of the EC and the anion concentrations of the saturation extract in a greenhouse during a gerbera cropping. The soil type was loam soil with 4% organic matter

with nitrogen in week 6 strongly increases the N concentration. After planting in the same week the irrigation during the cropping period scarcely covered the transpiration of the crop, so there was no leaching of minerals from the root zone. Top dressings were not given during the cropping period. From the irrigation water Cl and SO_4 accumulated in the second part of the growing period. The strong accumulation of NO_3 should be explained by mineralization or from capillary water ascended from soil layers below the root zone.

Both examples show that in greenhouses with soil grown crops by fertilization the osmotic potential (EC) of the root zone surely can be affected and by this the development of crops. However, the often high nutrient concentrations in the soil solution can involve heavily losses of minerals by leaching, which in turn strongly will pollute the deep ground water or surrounding surface water.

In substrate cultivation the control of salt and nutrient concentrations in the root environment has special effects in comparison with the control of these parameters in the soil solution. On the one hand the concentration can be adjusted much easier, because of the small rooting volume. On the other hand for the same reason, mistakes by errors and mismanagement are also more obvious. This directly follows from the quantity of water available in the rooting volume. For soil grown crops a quantity of 75–150 l m^{-2} can be calculated, while for substrate grown crop quantities between 4 and 12 l m^{-2} are calculated (Sonneveld, 1981a; Sonneveld, 2000). Thus, salt accumulations in the root environment in substrate growing are about 5–40 times more effective than in soil growing. Therefore, for substrate growing a precise control on the supply of fertilizers and the realized concentration in the root environment is very important. This especially is the case in closed growing systems, where the drainage water is re-used and mistakes are not washed out in the drainage water discharged.

3.5 Soil Solution and Uptake of Major Nutrients

Marschner (1997) showed an interesting general model for the uptake of macro nutrients in relation to the external concentration. K, P, NO_3 and SO_4 are supposed to be adsorbed at relative high quantities at low external concentrations and the uptake of Na, Mg and Ca are much more dependent of the external concentration. Comparable relationships has been found for substrate grown crops like for K and Mg with sweet pepper, eggplant and cucumber grown in rock wool (Sonneveld and Voogt, 1985), K and Ca with tomato grown in nutrient solution (Voogt, 1988), for P with cucumber grown in rock wool (Sonneveld, 1991) and for Ca with carnation grown in rock wool (Sonneveld and Voogt, 1986). The relationship differs for crops as is clear from Fig. 3.6, where the relationship is shown between the external and internal K concentration for different crops. In the low range of the external concentration the curve is very steep for sweet pepper and eggplant, while for cucumber the slope in this range is lower, but in the higher range a considerable concentration effect is noticed for the latter crop. Differences among the uptake of cations are clearly shown by the data in Fig. 3.7. The slopes are steep for Ca and Mg, while the

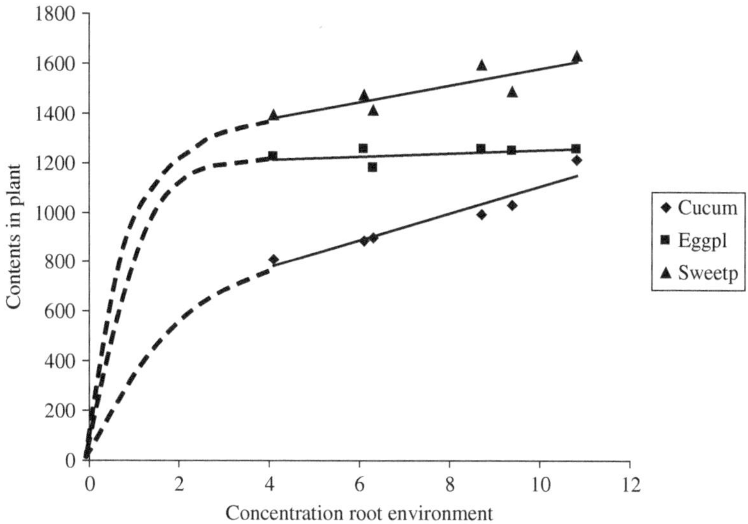

Fig. 3.6 The relationship between the K concentration in the root environment (mmol l^{-1}) and the K concentration (mmol kg^{-1} dry matter) in young leaves of rock wool grown cucumber, eggplant and sweet pepper. After Sonneveld and Voogt (1985). *Modified by permission of Marcel Dekker*

slope for the relationship for K is lower. In the lower range the relationship for K will become curvilinear, including a very steep slope to the point 0.0. With the supposition that in Fig. 3.7 all relationships end in the point 0.0, for K a convex model will occur, while for Ca and Mg a concave model will occur, like suggested in this figure. This is in full agreement with earlier supposed models (Sonneveld, 1991). This means that in a nutrient solution sufficient K will be adsorbed at relatively low concentrations in the root environment, while Ca and Mg are more dependent from a sufficient high concentration. Thus, in relation to the uptake the concentrations of

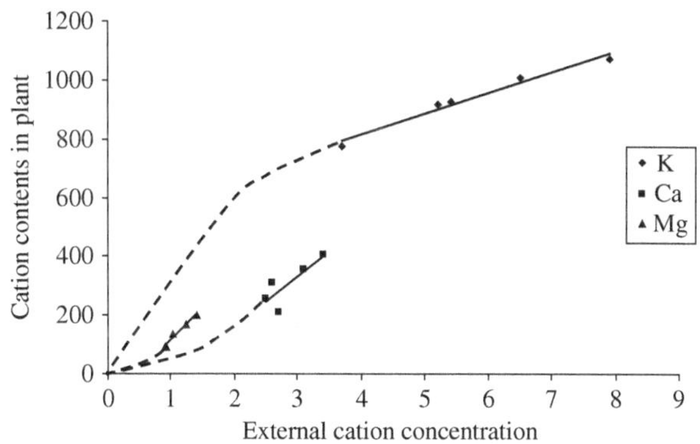

Fig. 3.7 Relationship between the external concentration of cations (mmol l^{-1}) and the cation concentration of young leaves (mmol kg^{-1} dry matter) of a carnation crop grown in rock wool. After Sonneveld and Voogt (1986)

Ca and Mg should be relatively much higher than the K concentration (Voogt and Sonneveld, 1997).

For soil grown crops comparable relationships will be expected for the relationship between external and internal concentrations as has been found for substrate growing. This has been demonstrated by Voogt (2002) for K with tomato. In the soil solutions of greenhouses the concentrations of nutrients are of the same magnitude as the solutions in the root environment of substrates, as already shown in Table 3.1. However, P is an exception, because of the relatively low concentrations found in soil solutions when compared with substrate solutions. Therefore, for most nutrients can be expected that the nutrient concentrations in substrate grown crops are reasonably in agreement with those found in soil grown crops. Such data have been found in a study with which the mineral composition of tomato and cucumber grown in soil or grown in rock wool were compared (Sonneveld, 1980). Results of this study are listed in Table 3.4. In soil grown crops often the Ca concentration is mostly somewhat higher than in substrate grown crops, because of the often higher Ca concentrations found in soil solutions. The K concentrations are mostly somewhat higher in substrate grown crops, which can be explained as a compensation for the lower Ca uptake. Because of the generally very high P availability in substrate solutions the P concentrations in the plant tissues are mostly highest for crops grown in substrate. The differences shown with Na and Cl will be explained by the quality of the irrigation water used. The Na and Cl concentrations in the irrigation water used for soil growing were mostly higher than those in the water used for substrate growing.

Table 3.4 Average mineral nutrient contents as has been found in young cucumber and young tomato leaves from greenhouse crops grown in rock wool or in soil. The crops on the holdings were sampled 3 till 5 times during the season. In the study for cucumber 9 and 3 and for tomato 3 and 5 holdings were incorporated with crops grown in rock wool and soil respectively. The contents are expressed as mmol kg^{-1} dry matter

	Cucumber		Tomato	
Elements	Rock wool	Soil	Rock wool	Soil
Na	48	52	57	87
K	683	552	1064	841
Ca	1177	1192	611	843
Mg	263	321	173	169
NO_3	286	221	257	250
N	4086	3750	3607	3343
Cl	90	293	130	231
P	210	161	174	145
SO_4	75	72	344	362
Mn	2.5	0.9	3.9	0.8
Fe	1.9	1.6	1.5	1.6
Zn	0.7	0.8	0.7	0.4
B	5.4	5.4	5.6	4.2

Data of Sonneveld (1980).

It is striking, that under conditions of sufficient and over-sufficient nutrient availability of macro nutrients in the root environment, great differences in external concentrations result in only small differences in the internal concentrations. This effect points to the strong control of plants on the uptake of major nutrients. Every plant type and even every plant part preferentially realises certain specific optimum concentrations. Such concentrations differ much between plants, which is clear from the differences between the mineral compositions of cucumber and tomato in Table 3.4. Both crops in this study were grown under more or less comparable conditions, but differ seriously for plant concentrations. Internal optimum concentrations are realised by the plant at a relative high absorption at low external concentrations and a relative reduced uptake at relative high external concentrations, like shown for K in Fig. 3.6. Thus, despite a specific high external concentration of a major element, many plants are able to survive at a relative strong restriction of the uptake of such an element under these conditions. This is clearly shown in a study where different crops were grown under addition of specific salts to the root environment (Sonneveld and Van den Ende, 1975). In the experiments of this study beside the standard application of major nutrients to the irrigation water, chlorides of Na, K, Ca and Mg were added in two concentrations. The binary salts (NaCl and KCl) in concentration of 12½ and 25 mmol l^{-1} and the tertiary salts $CaCl_2$ and $MgCl_2$ of 8⅓ and 16⅔ mmol l^{-1}, aiming at comparable osmotic potentials in the irrigation water with these additions. By these treatments the Na, Ca and Mg concentrations in the soil solution were increased with a factor 3 till 5 in comparison with the concentration in the standard treatment and for K with a factor 5–10. The effects on the cation uptake and on the yield of tomato and chrysanthemum are shown in Table 3.5 (Sonneveld and Van Beusekom, 1973; Sonneveld 1981b).

Table 3.5 Effects of the addition of specific salts to the root environment on the uptake of cations of soil grown tomato and chrysanthemum. The salts were added to the irrigation water at concentrations of 12½ and 25 mmol l^{-1} for the binary salts and 8⅓ and 16⅔ mmol l^{-1} for tertiary salts. For a further description of the experiment see text. The element contents are determined in young fully grown leaves and expressed as mmol kg^{-1} dry matter and yields in % of the standard

	Tomato					Chrysanthemum				
Salts added	*Na*	*K*	*Ca*	*Mg*	*Yield*[1] %	*Na*	*K*	*Ca*	*Mg*	*Yield*[1] %
Standard	261	939	1182	342	100	17	1606	456	276	100
NaCl 12½	670	749	1257	370	83	17	1529	454	280	66
NaCl 25	748	693	1322	358	66	39	1514	426	272	50
KCl 12½	187	1212	1137	313	80	13	2118	312	181	70
KCl 25	165	1714	1005	296	62	9	2338	244	136	58
$CaCl_2$ 8⅓	117	900	1496	263	82	17	1394	723	169	67
$CaCl_2$ 16⅔	122	949	1521	296	64	22	1240	823	148	54
$MgCl_2$ 8⅓	152	818	1157	580	82	17	1396	284	601	66
$MgCl_2$ 16⅔	135	719	1007	951	71	9	1176	212	819	45

[1]For tomato fruit weight and for chrysanthemum plant weight.
Data of Sonneveld and Van Beusekom (1973); Sonneveld (1981b).

The strong increases in the external concentrations are only partly reflected in the internal concentrations. The internal concentration increases in comparison with the standard with a factor between 1.3 and 3.0. The increased uptake of a specific cation reduces the uptake of other cations. Na was not absorbed by chrysanthemum and did not affect the level of other cations of this crop. Tomato absorbed significant quantities of Na, which reduces the K uptake. High K, Ca or Mg increases the uptake of these elements, and reduces the cations other than supplied in the overdose. It is remarkable that the sum cations (C^+) absorbed by the crops in this way remains more or less constant; showing an average of 4550 and 3123 for tomato and chrysanthemum, respectively, with no more deviation from the average than about 5%. Despite the great changes in cation uptake the yield of the crops was not strongly specifically affected, except the yield of chrysanthemum at the highest $MgCl_2$ concentration. The yield of tomato was only negatively affected by the decreased osmotic potential with 18 and 32% for the lowest and highest salt applications, respectively, and the yield of chrysanthemum with 33 and 46% respectively. However, at the highest $MgCl_2$ concentration the yield reduction was 55%. These results show that with the high nutrient concentrations in the soil solution many crops survive well with relatively great differences in the uptake of major elements. The composition of the soil solution reflects the uptake of nutrient elements quite well, but the relationships between external and internal concentrations are not linear.

3.6 Soil Solution and Uptake of Micro Nutrients

The quantities of micro nutrients in soil solutions are small in comparison with those found for major nutrients. In Table 3.6 average concentrations of micro nutrients are shown as has been found in greenhouse soil solutions (Sonneveld and De Bes, 1986) and in substrate solutions of rock wool (Sonneveld and Van Voorthuizen, 1988). The concentrations in the different growing media are of the same order of magnitude. Greatest difference has been found for Fe, which will be explained by the use of chelates in solutions used with rock wool substrates. Such complexes are used to

Table 3.6 Micronutrient concentrations as has been found in soil solutions of greenhouses (Sonneveld and De Bes, 1986) and in substrate solutions of rock wool (Sonneveld and Voorthuyzen, 1988). Average values of 75 and 90 Dutch sites respectively. Concentrations given as μmol l^{-1}

Elements	Soil	Rock wool substrate
Fe	4.0	22.6
Mn	15.1	8.1
Zn	5.9	11.5
B	62.0	55.3
Cu	2.2	1.1
Mo	0.5	–

keep Fe available to plants in substrate cultures, while such compounds in soils are only scarcely used.

The micronutrient concentrations in soil and substrate solutions are not always a good measure for the uptake of the crop. Apparently many factors affect the relationship between external and internal concentrations of micronutrients. The pH in the root environment and in the rhizosphere and organic compounds in the root environment are likely important factors. The great variation in these conditions in the root environment are probably responsible for the poor correlations often found between external and internal concentrations for micro nutrients under practical conditions (Marschner et al., 1987).

However, when micronutrients are applied to a more or less inert growing medium like rock wool, close correlations will be found between the concentrations in the nutrient solution supplied, the quantities in the substrate solution and the concentrations in the plant tissues (Sonneveld and De Bes, 1984). In Fig. 3.8 an example for such conditions is given for Mn with rock wool grown cucumber. However, even in more or less inert media like rock wool the availability in the root environment can be affected by the growing conditions, like shown for a gerbera crop in Fig. 3.9. The pH maintained in the root environment strongly affects the availability of the Mn supplied with the irrigation water (Sonneveld and Voogt, 1997).

The relationship between micronutrient concentrations in soil or substrate solution and in plants can be disturbed by soluble organic compounds. Such has been found for example for Cu and Zn. These elements can be bound at higher pH values on artificially produced organic (chelate) complexes that strongly, that plants are not able to absorb the Cu and Zn. Nevertheless, these elements are determined as soluble in soil or substrate solutions (Sonneveld and Voogt, 2001; Voogt and Sonneveld, 2009). Comparable complex formations also will be expected with natural organic compounds, which for example has been found for Cu with peat based organic complexes (Verloo, 1980). Then, the Cu bound on the soluble organic matter in soil or

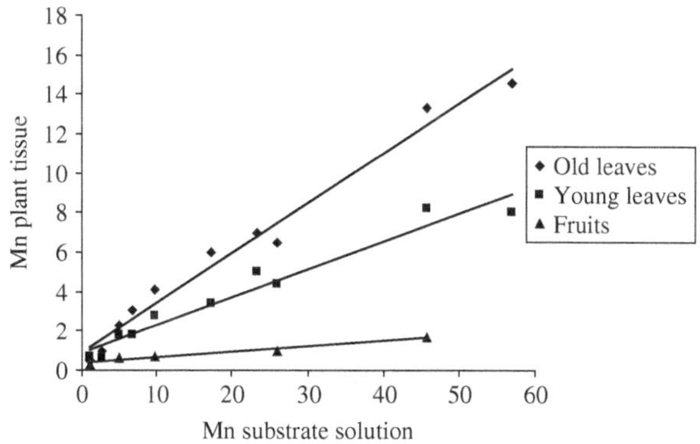

Fig. 3.8 The relationships between the Mn concentrations in the substrate solution (μmol l^{-1}) and in plant tissues (mmol kg^{-1}) of rock wool grown cucumber. After Sonneveld and De Bes (1984). *Modified by permission of Marcel Dekker*

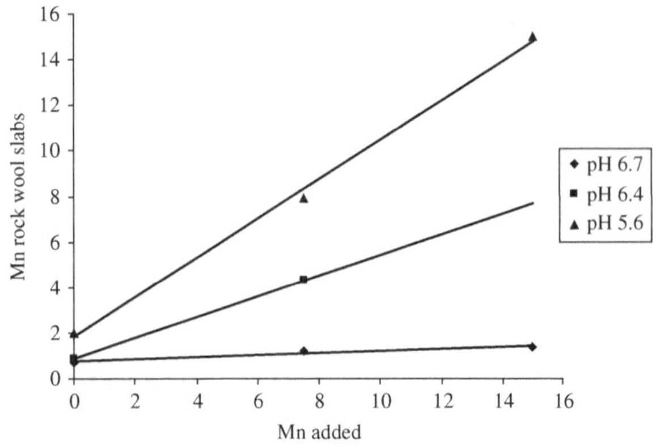

Fig. 3.9 The relationship between the Mn concentration added with the irrigation water (μmol l^{-1}) and the Mn concentration in the substrate solution of the rock wool slabs (μmol l^{-1}). Results of an experiment with rock wool grown gerbera, after Sonneveld and Voogt (1997)

substrate solutions is traced with the analysis, but not or only partly available to plants.

In view of the chemical and biological processes in the rhizosphere due to the activity of micro-organism or plant roots, simple relationships between external and internal micronutrient concentrations will not to be expected (Marschner, 1997). An important reason for this lack on correlation is also the great difference that exists between the availability of micro nutrients in the rhizosphere and the bulk soil represented in soil or substrate samples gathered for laboratory analysis. When in experiments close relationships are found between internal and external concentrations of micro nutrients mostly such results are only operative for the specific experimental conditions. More about the uptake of micro nutrients will be discussed in the Chapters 13 and 16 about nutrient management in substrate and soil grown crops, respectively.

References

Adams F 1974. Soil solution. *In*: Carson E W (ed) The Plant Root and Its Environment. University press of Virginia, Charlottesville, VA, 441–481.

Barraclough P B 1989. Root growth, macro nutrient uptake dynamics and soil fertility requirements of a high yielding winter oil seed crop. Plant Soil 119, 59–70.

Campbell R B Bower C A and Richards L A 1948. Change of electrical conductivity with temperature and the relation of osmotic pressure to electrical conductivity and ion concentration for soil extracts. Soil Sci. Soc. Am. Proc. 12, 66–69.

CEN 2006. EN 13041 A1. Soil improvers and growing media – Determination of physical properties – Dry bulk density, air volume, water volume, shrinkage value and total pore space, 21pp.

Kipp J A Wever G and De Kreij C 2000. International Substrate Manual – Analysis Characteristics Recommendations. Elsevier International Business information, Doetinchem, The Netherlands, 94pp.

Maaswinkel R M H and Welles G W H 1986. Factors influencing glassiness in lettuce. Neth. J. Agric Sci. 34, 57–65.

Marschner H Römheld V and Cakmak I 1987. Root-induced changes of nutrient availability in the rhizophere. J. Plant Nutr. 10, 1175–1184.

Marschner H 1997. Mineral Nutrition of Higher Plants, Second edition. Academic Press, London, 50–52.

McNeal B L Oster J D and Hatcher J T 1979. Calculation of electrical conductivity from soil solution composition data as an aid to in-situ estimation of soil salinity. Soil Sci. 110, 405–414.

Peters M 1990. Nutzungseinfluss auf die Stoffdynamik schleswig-holsteinischer Boden – Wasser-, Luft-, Nähr- und Schadstoffdynamik –. Dissertation, Schriftenreihe Institut für Pflanzenernährung und Bodenkunde Universität Kiel, nr 8, 293pp.

Riemens J M 1951. Verzouting en verdroging in het westen van Nederland, meer in het bijzonder in het Westland. Maandblad Landbouwvoorlichtingsdienst 1951, 56–59.

Qian P and Wolt D J 1990. Effects of drying and time of incubation on the composition of displaced solution. Soil Sci. 149, 367–374.

Sonneveld C Koornneef P and Van den Ende J 1966. De osmotische druk en het electrische geleidingsvermogen van enkele zoutoplossingen. Meded. Dir. Tuinb. 29, 471–474.

Sonneveld C and Van den Ende J 1967. De samenstelling van de zouten in het oppervlaktewater in het Zuidhollands glasdistrict. Meded. Dir. Tuinb. 30, 411–416.

Sonneveld C Van den Ende J and Van Dijk P A 1974. Analysis of growing media by means of 1:1½ volume extract. Comm. Soil Sci. Plant Anal. 5, 183–202.

Sonneveld C and Van Beusekom J 1973.Specifieke ioneffecten bij tuinbouwgewassen (Teeltjaar 1972). Proefstation voor Groenten en Fruitteelt onder Glas te Naaldwijk, Intern rapport no 617, 23pp.

Sonneveld C and Van den Ende J 1975. The effect of some salts on head weight and tipburn of lettuce and on fruit production and blossom-end rot of tomatoes. Neth. J. Agric. Sci. 23, 191–201.

Sonneveld C 1980. Gewasonderzoek op basis van plantensap. Praktijkonderzoek 1979. Proefstation voor Tuinbouw onder Glas te Naaldwijk, Intern verslag 1980, no 18, 37pp.

Sonneveld C 1981a. Items for application of macro-elements in soilless cultures. Acta Hort. 126, 187–195.

Sonneveld C 1981b. Specifieke zouteffecten bij chrysant (teelt 1979–1980), Proefstation voor Tuinbouw onder Glas te Naaldwijk, Internverslag nr 8, 18pp.

Sonneveld C and De Bes S S 1984. Micro nutrient uptake of glasshouse cucumbers grown on rockwool. Commun. Soil Sci. Plant Anal. 15, 519–535.

Sonneveld C and Voogt W 1985. Growth and cation absorption of some fruit-vegetable crops grown on rockwool as affected by different cation ratios in the nutrient solution. J. Plant Nutr. 8, 585–602.

Sonneveld C and De Bes S 1986. Grondonderzoek op basis van waterige extractie, Deel 2, Enkelvoudige lineaire correlaties. Proefstation voor Tuinbouw onder Glas. Intern rapport, no 31, 18pp.

Sonneveld C and Voogt W 1986. Supply and uptake of potassium, calcium and magnesium of spray carnations (*Dianthus caryophyllus*) grown in rockwool. Plant Soil 93, 259–268.

Sonneveld C and Voorthuyzen E 1988. Monsterfout en analysefout bij het chemisch onderzoek van voedingsoplossingen in steenwolmatten. Proefstation voor Tuinbouw onder Glas. Intern rapport, no 10, 13pp.

Sonneveld C 1990. Estimating quantities of water-soluble nutrients in soil using a specific 1:2 volume extract. Commun. Soil Sci. Plant Anal. 21, 1257–1265.

Sonneveld C Van den Ende J and De Bes S S 1990. Estimating the chemical composition of soil solutions by obtaining saturation extracts or specific 1:2 by volume extracts. Plant Soil 122, 169–175.

Sonneveld C and Voogt W 1990. Response of tomatoes (*Lycopersicon esculentum*) to an unequal distribution of nutrients in the root environment. Plant Soil 124, 251–256.

Sonneveld C 1991. Rockwool as a substrate for greenhouse crops. *In*: Bajaj Y P S (ed) Biotechnology in Agriculture and Forestry 17, High-Tech and Micropropagation I, Springer-Verlag, Berlin, 285–312.

Sonneveld C and Van Elderen C W 1994. Chemical analysis of peaty growing media by means of water extraction. Commun. Soil Sci. Plant Anal. 25, 3199–3208.

Sonneveld C 1995. Fertigation in the greenhouse industry. *In*: Proc. of the Dahlia Greidinger Intern. Symposium on Fertigation. Technion – Israel Institute of Technology, Haifa, 25 March–1 April 1995, 121–140.

Sonneveld C and Voogt W 1997. Effects of pH value and Mn application on yield and nutrient absorption with rockwool grown gerbera. Acta Hort. 450, 139–147.

Sonneveld C Voogt W and Spaans L 1999. A universal algorithm for calculation of nutrient solutions. Acta Hort. 481, 331–339.

Sonneveld C 2000. Effects of salinity on substrate grown vegetables and ornamentals in greenhouse horticulture. Thesis Wageningen University, Netherlands, 151pp.

Sonneveld C and Voogt W 2001. Chemical analysis in substrate systems and hydroponics – use and interpretation. Acta Hort. 548, 247–259.

Van den Bos A L 1996. EC in relatie tot het type substraat bij de bollenteelt van Amaryllus in een gesloten systeem.Proefstation voor de Bloemisterij en Glasgroente Rapport no 55, 30pp.

Van den Ende J 1952. De invloed van zout gietwater op de ontwikkeling van verschillende tuinbouwgewassen onder glas. Meded. Dir. Tuinbouw, 15, 884–903.

Van den Ende J 1955.De watervoorziening van tomaten, Meded. Dir. Tuinbouw 18, 904–917.

Van den Ende J 1968. Analysis of greenhouse soils by means of aqueous extracts. Proc. 6th Coll. Intern. Potash Inst. Florence, 246–255.

Van den Ende J 1988a. Water contents of glasshouse soils at field capacity and at saturation. 1. Relationships between water contents. Neth. J. Agric. Sci. 36, 265–274.

Van den Ende J 1988b. Water contents of glasshouse soils at field capacity and at saturation. 2. Estimating water contents from organic-matter and clay contents or loss-on-ignition. Neth. J. Agric. Sci. 36, 275–282.

Van den Ende J 1989. Estimating the composition of the soil solution of glasshouse soil. 1. Composition of soil solution and aqueous extracts. Neth. J. Agric. Sci. 37, 311–322.

Verloo M G 1980. Peat as a natural complexing agent for trace elements. Acta Hort. 99, 51–56.

Voogt W 1988. The growth of beefsteak tomato as affected by K/Ca ratios in the nutrient solution. Acta Hort. 222, 155.

Voogt W and Sonneveld C 1997. Nutrient management in closed growing systems for greenhouse production. *In*: Goto E et al. (eds), Plant Production in Closed Ecosystems, Kluwer Academic Publishers, Dordrecht, The Netherlands, 83–102.

Voogt W 2002. Potassium management of vegetables under intensive growth conditions. *In*: Pasricha N S and Bansal S K (eds) Potassium for sustainable crop production. Proc. Intern. Symp. On role of potassium in nutrient management for sustainable crop production in India. Intern. Potash Inst. Bern, 347–362.

Voogt W and Sonneveld C, 2009. The effects of Fe-chelate type and pH on substrate grown roses. Acta Hort. 819, 411–417.

Wever G 1995. Physical analysis of peat and peat-based growing media. Acta Hort. 401, 561–567.

Chapter 4
Soil and Substrate Testing to Estimate Nutrient Availability and Salinity Status

4.1 Introduction

In the greenhouse industry methods have been developed for the determination of the nutrient availability and salinity status of soils and substrates. As in other agriculture branches, soil testing has the aim to estimate the availability, including the solubility as well the quantity, of plant nutrients to enable the farmer to get maximum production with minimum fertilizer use. The success of the farmer thereby does not depend only on the precision of the method, but also on the knowledge of the requirements of the crop. Both the utility of the soil testing method and the fertilizer application in relation to the results to get maximum yield will be calibrated in fertilizer experiments. Until lately, farmers based their decision about the amount of fertilizer addition on the costs of the fertilizer and the profits of the expected yield increase. However, in recent years farmers also have to consider the environmental aspects in their decisions. Fertilizer applications should be focussed also on their effects to pollution of soil, water and air. Beside the availability of nutrients, the determinations of characteristics for the salinity status are important and interact with the fertilization programme considered. The definitions given so far are operative for greenhouse crops as well as for crops grown in the field. However, soil testing for greenhouse industry has some specific aspects which will be mentioned beforehand, because they are important in relation to the methods used. The aspects in view for greenhouses are following.

- Testing of soil and substrate is carried out frequently, often several times during the growing period of a crop. Thus, estimation of the release of nutrients over longer periods is not a requirement of the methods applied.
- Nutrients in the soil and the substrate solution form a substantial part of the total ion concentration of these solutions and should be taken into account in the judgement of soil salinity.
- The determination of the total salt status of greenhouse soils and substrates has a central position, because the osmotic potential in the root environment not only is a measure to prevent possible yield reduction, but also a tool for the farmer on crop development and produce quality.

C. Sonneveld, W. Voogt, *Plant Nutrition of Greenhouse Crops*,
DOI 10.1007/978-90-481-2532-6_4, © Springer Science+Business Media B.V. 2009

- Nutrient absorption by many crops in the greenhouse industry are that high that application of the total needs of nutrients of the crop as base dressing will lead to unacceptable high ion concentrations in the root environment. This especially is operative for crops grown in substrate, because of the small rooting volumes common with this growing method. Thus, application of the total fertilizers requirements as a base dressing is mostly impossible.
- For most crops frequent top dressings are possible and can be easily carried out when desirable with any irrigation by fertigation.

In view of these aspects it is understandable that the estimation of the composition of the soil and the substrate solution and by that extraction with water plays an important role in the routine soil testing for the greenhouse industry. Therefore, the Dutch developed and promoted such methods for their greenhouse industry since many years. It is true that water extraction only shows the activity of the elements determined, but considering the frequent determinations and top dressings for most nutrients there are not many arguments for the determination of the capacity of most nutrient elements. Different water extraction methods have been developed and the suitability was often judged in relation to the capability of the method under discussion to estimate the composition of the soil solution.

Soil solution needs some definition with respect to the water content of the soil or the substrate, because these will vary in relation to the growing conditions, especially to the water supply. The water supply in greenhouses during crop growth is characterized by a frequent irrigation, which means that the fluctuations in the water content are relatively small. Van den Ende (1988a) compared the water contents of a great number of greenhouse top soils, grown with tomato with the water content at a pressure head of –6.3 kPa. The study turned out that the water contents were approximately equal. In a later study with soils derived from greenhouses grown with various crops (Sonneveld, et al., 1990) approximately the comparable water contents in the greenhouse soils could be calculated on basis of the loss on ignition as were found in the former study (Van den Ende, 1988b). Therefore, the water content of greenhouse soils under growing conditions has been defined as being the water content at a pressure head at –6.3 kPa. Extended information is given in Section 3.3.

In the greenhouse industry al lot of crops are grown in substrate and the water contents realized are more or less artificial made and depend strongly on the type of substrate and the growing system. For course substrates, like mineral fibres, pumice, foam etc., losing their water at a very high pressure head, the water content at the leak out situation is considered as being the moisture condition at field capacity. For substrates holding their water at somewhat lower pressure head, like peat and peat related substrates, the water content at a pressure head of –1 kPa is considered as being the moisture content at field capacity (Kipp et al., 2000; Sonneveld and Van Elderen, 1994). Detailed information is presented in Section 3.3.

The different soil testing methods generally used in the greenhouse industry are discussed in following sections. Some methods are specifically directed at soil growing, while other methods are suitable for substrate growing. Information about specific suitability will be given in the description of the methods, while at the end of this chapter a review of the use of the methods will be presented in relation to growing medium and growing system.

4.2 Specific 1:2 by Volume Water Extract

The specific 1:2 by volume extract, henceforth called 1:2 extract, is prepared by filtration of a suspension obtained by adding sufficient field-moist soil to two volume parts of water so that the total volume is increased with one part (Sonneveld et al., 1990), see picture 4.1. When the soil is too dry, before the preparation of the extract some demineralised water will be added to the soil to restore field moist condition. The field moist condition of greenhouse soils is defined in Section 3.3 and agrees with the moisture content at a pressure head of –6.3 kPa. In advance this judgement should be compared with results of the sandbox method, but after some experience the judgement can be carried out visually. The suspension is shaken for 20 minutes. The method is exclusively recommended for greenhouse soils. For a detailed description of the preparation of 1:2 extracts reference is made to De Kreij et al. (2005).

Picture 4.1 Preparation of the specific 1:2 volume extract. Sufficient field-moist soil is added to two parts of water so that the volume is increased with one part

The EC and the concentrations of major nutrient, Na and Cl of the 1:2 extracts were closely correlated with those of the soil solutions. The relationship between the analytical data derived from the 1:2 extract and those from the soil solution are listed in Table 4.1 (Sonneveld et al., 1990). The close and linear relationship

Table 4.1 Regression equations for the relationships between EC and ionic concentrations in the soil solution (y) and those in the 1:2 extract (x) for a series of Dutch greenhouse soils. EC in dS m^{-1} and ions in mmol l^{-1}

Determination	Regression equation	r
EC	y = 3.12 x + 0.84	0.886
NH_4	y = 3.23 x + 0.05	0.782
K	y = 3.38 x – 0.80	0.922
Na	y = 4.04 x – 1.12	0.929
Ca	y = 2.53 x + 7.86	0.811
Mg	y = 3.48 x + 1.86	0.876
NO_3	y = 5.09 x + 0.14	0.899
Cl	y = 6.15 x – 2.04	0.952
SO_4	y = 1.47 x + 8.67	0.779
P	y = 1.78 x – 0.09	0.936

After Sonneveld et al. (1990). *Modified by permission of Springer*

between analytical data of the 1:2 extract allows a universal interpretation based on the composition of the soil solution.

The 1:2 extract, however, has some drawbacks, like the relatively high dilution of the soil solution and the fact that the dilution of the soil solution varies somewhat with the soil type in relation to the organic matter content. The dilution factor, the ratio of the water content of the 1:2 suspension to the water content of the field moist soil, decreased from 6 for mineral soils to 3.5 for soils with a high organic matter content (40%). The overall high water to soil ratio of the 1:2 suspensions bring about dissolution of sparingly soluble salts, mainly $CaSO_4$. This disturbs the estimation of the total ion concentration (EC) and the concentrations of Ca, Mg and SO_4 in the soil solution. Adjustments for soil type and sparingly soluble salts by dilution ratios brought the correlation coefficient above 0.95 for nearly all determinations listed in Table 4.1 (Sonneveld et al., 1990). The still low correlation coefficient found with NH_4 after these adjustment should be explained mainly by the low concentrations of this ion found in greenhouse soils. The average NH_4 concentration in the 1:2 extracts of the samples in the study was 0.10 mmol l^{-1}, and varied between 0.00 and 0.82.

A precise estimation of the EC of the soil solution especially is valuable with respect of the estimation of the osmotic potential, being one of the most important soil characteristics that affect crop development in greenhouses. Sonneveld et al. (1990) has found the equations denoted as formulae (4.1) and (4.2) with correlation coefficients 0.968 and 0.974, respectively.

$$EC_{ss} = 0.908dEC_{1:2} - 0.089dSO_{4(1:2)} + 0.68 \tag{4.1}$$

$$EC_{ss} = 2.744qEC_{1:2} - 0.284qSO_{4(1:2)} - 0.17 \tag{4.2}$$

In which

EC_{ss} = EC of the soil solution
$EC_{1:2}$ = EC of the specific 1:2 volume extract
$SO_{4(1:2)}$ = SO_4 concentration of the 1:2 extract in mmol l^{-1}
d = dilution factor, being the ratio between the water content of the 1:2 suspension and the water content of the field moist soil
q = the quantity of field moist soil with an under water volume of 1 litre in kg

Addition of the factor d as used in formula (4.1) has the drawback that this factor is difficult to determine, because the water content of the field moist soil should separately be determined. Therefore, formula (4.2) is better applicable, because the quantity of field moist soil used at preparation of 1:2 suspensions can be rather easily determined. The contribution of the d and q values to the increase of the correlation coefficient are more or less equal; which is understandable because d and q were highly correlated ($r = 0.894$).

The relationship between micro nutrient concentrations in the soil solution and in the 1:2 extract was also included in the study (Sonneveld and De Bes, 1986). However, for most of the micro nutrients the ratios between the concentrations in the 1:2 extract and the concentrations in the soil solution differed strongly from those of the macro nutrients and the correlation coefficients were generally much lower varying from 0.318 till 0.984. These results are no reason to suppose a simple and unequivocal interpretation of these elements in the 1:2 extract. The addition of micro nutrients for soil grown crops will be discussed in Chapter 16.

The 1:2 extract is also suitable as an estimator for quantities of water soluble nutrients in the root environment. This is due to the fact that the quantity of water present in the suspension with the preparation of the 1:2 extract is virtually independent on the soil type (Sonneveld, 1990), as is shown with the data of Table 4.2. From the fourth column of this table will be concluded that about 40 m^3 extract is prepared per 100 m^2 for soil types with an organic matter fraction varying between 0.05 and 0.30. Thus, with a concentration of 1 mmol l^{-1} of any element in the 1:2 extract a quantity of 40 mol of that element is water soluble available per 100 m^2 over a depth of 0.25 m.

Table 4.2 Quantities of water present in the suspension with the preparation of the 1:2 extract of different soil types, expressed as m^3 per 100 m^2 over a depth of 25 cm

Mass fraction organic matter	Dry weight per volume[1]	Water content 1:2 suspension[2]	Water volume 1:2 suspension[3]
0.05	1.08	1.40	37.8
0.10	0.86	1.62	39.1
0.20	0.62	2.64	40.9
0.30	0.48	3.46	41.5

[1] kg l^{-1}; [2] g per g dry matter; [3] m^3 per 100 m^2 greenhouse area.

For determinations related to salinity, like EC, Na and Cl, it is evident that calculations to concentrations in the soil solution are most suitable to the purpose. For plant nutrients calculations to available quantities per area also can be meaningful.

The 1:2 extraction has the advantage that it is a quick method and therefore very suitable for routine soil testing. It is used to that purpose for many years in the greenhouse industry. It was tested for greenhouse soils in situ and proved to be suitable for a wide range of soil types. The method was not tested for substrates, because the strongly different physical characteristics of these materials give no single reason for a successful application in this field. A haphazard application easily will leads to a misinterpretation of the analytical data obtained.

4.3 Saturation Extract

The saturation extract is prepared by filtration of a water saturated soil. Saturated soil is prepared by addition of demineralised water to field moist soil under continuous stirring with a spatula (Richards, 1954). The saturated condition is assessed by the glistening appearance of the soil paste if it reflects light and by the rapid disappearance of a diametrical groove drawn with the spatula. The use of field moist soil for the preparation of the saturated paste is preferred to air dry soil, because of the risk of denitrification when air dry soil is used (Van den Ende, 1989a).

The saturation extract has the advantage of a low water to soil ratio, closely related to the water content of soils at field capacity. Thus, the results are only slightly affected by sparingly soluble salts and allow an unequivocal interpretation for a wide range of soil types. The dilution factor, being the water content of the saturated paste to the water content of the field moist soil, varied between 1.8 for mineral soils and 1.5 for soils with a high (30%) organic matter content in studies with greenhouse soils (Van den Ende, 1988a). In another study a dilution factor of 2.0 was calculated for soils with 5% organic matter and 1.6 for soils high (40%) in organic matter (Sonneveld et al.,1990).

In Table 4.3 linear relationships are given for the relation between ionic concentrations in the saturation extract and those in the soil solution. The values of the correlation coefficients of these simple linear regressions are on the same level of about 0.95, as has been found with the 1:2 extract (Section 4.2) after correction for soil type and sparingly soluble salts. This involves a simple interpretation for the analytical data of the saturation extract, which is a big advantage. Therefore, the saturation extract is widely used all over the world, especially for the determination of soil salinity. However, the drawback for routine soil testing is the laborious preparation of the saturated soil suspension.

In soil salinity a rule of thumb is used that the salt concentration of the soil solution is twice that of the saturation extract (Maas and Hoffman, 1977). This factor of 2 between the both concentrations is too high for greenhouse soils, as can be derived from Table 4.3, where a factor of 1.6 is found for the EC, which is more in agreement with the moisture quotients given before. The most likely reason for this low quotient for the concentrations "field moist" and "saturated" is the frequent

Table 4.3 Regression equations for the relationship between EC and ionic concentrations in the soil solution (y) and those in the saturation extract (x) for a series of Dutch greenhouse soils. The EC is expressed as dS m^{-1} and ions in mmol l^{-1}

Determination	Regression equation	r
EC	y = 1.60 x – 0.18	0.958
NH_4	y = 1.33 x – 0.01	0.844
K	y = 1.54 x – 0.64	0.984
Na	y = 1.62 x – 0.42	0.978
Ca	y = 1.53 x + 1.03	0.946
Mg	y = 1.67 x + 0.02	0.943
NO_3	y = 1.82 x – 0.27	0.954
Cl	y = 1.99 x – 0.66	0.975
SO_4	y = 1.16 x + 2.76	0.926
P	y = 1.20 x – 0.03	0.954

From Sonneveld et al. (1990). *Modified by permission of Springer*

irrigation in greenhouses, through which the water content of the soils are continuous on a high level.

The saturation extract is used for soils in situ and accidentally applied for substrates. However, the saturated condition cannot always be easily discerned with substrates. Moreover, the pressure suction of the "field moist" condition between soil and substrate differ principally, as discussed in Section 3.3. Therefore, the right dilution of the substrate solution is not always achieved and misinterpretations are obvious.

4.4 Water Extracts Based on Weight Ratios

For routine testing of greenhouse soils formerly often use was made of different weight ratios of water to air dried soil. To this purpose ratios of 1:1, 1:2 and 1:5 are practised (Carpena et al., 1968; Van den Ende, 19688). Nowadays there is not much reason to practice them. The water to soil ratio in the suspension has no relationship with the water contents of soils under growing conditions, when used for different soil types. Therefore, the analytical data need adjustment to this water content to get an interpretation related to the chemical composition of the soil solution (Van den Ende, 1989b).

This especially counts when high ratios water to soil are used. For example, the ratio between the water content of a 1:5 suspension and the water content of field moist soil, varies for example between 25 for mineral soils and 5 for organic soils (mass fraction organic matter 0.4). Thus, the interpretation of the analytical data will be handicapped by this changing dilution in relation to the soil solution, but also an interpretation based on quantities is troubled, because of the differences of the bulk densities. Both problems especially occur in areas with strongly different soil types.

However, another drawback is the often high water to soil ratio in the suspension, responsible for a strong dissolution of sparingly soluble salts (Reitemeier, 1946). The ratios between the cations in the solution also will change with dilution, by exchange of cations in the solution and on the adsorption complex. Mono valence

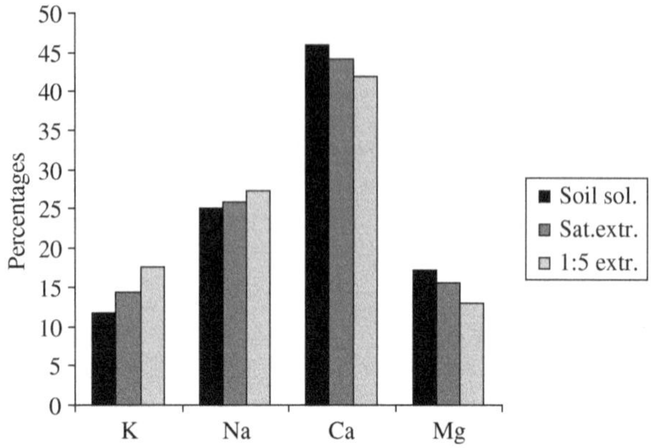

Fig. 4.1 Percentages of cations in soil solution, saturation extract and 1:5 w/w water extract calculated over average values of 75 greenhouse samples. After Van den Ende (1989b)

cations increase and bivalent cations decrease relatively with increasing dilution (Moss, 1963). The effects of this called "dilution and valence effect" of different dilutions with soil testing of greenhouse soils are shown in Fig. 4.1, after data of Van den Ende (1989b). The K and Na concentrations in the extracts increase and the Ca and Mg concentrations decrease relatively with increasing dilution from soil solution to 1:5 extract.

For interpretation in relation to the soil solution the water content of the soil under growing conditions should be known, and for interpretation in relation to quantities of water soluble nutrients the bulk density of the soil should be known. Both parameters will be estimated by the determination of the loss on ignition of the soil (Van den Ende, 1988b; Sections 2.3 and 3.3).

4.5 1:1½ Volume Water Extract

The 1:1½ volume extract is developed for peaty growing media. The extract is prepared by filtration of a suspension of 1 volume of fresh substrate and 1½ volume of water (Sonneveld and Van Elderen, 1994). The volume of the substrate is measured in a ring with a height 5 cm and a volume of at least 100 ml and pressed at 10 kPa. Before measurement of the volume the moisture content of the substrate will be judged and when the substrate is too dry, it will be adjusted with demineralised water to the moisture content at a pressure head of –3.2 kPa, see picture 4.2. In advance this judgement should be compared with results of the sandbox method, but after some experience the judgement can be carried out visually. This adjustment of the moisture content is especially important for substrates as delivered from the producer and not yet used for cultivation, because such material sometimes is very dry.

The analytical data of the 1:1½ volume extract of fifty peaty samples with widely varying characteristic were compared with those of the substrate solution at a pressure head of –1 kPa. The results of this comparison showed very close linear correlations between the data of both extracts for all the likely ions, as shown in Table 4.4. The correlation coefficients for the EC and the major elements varied between 0.912 and 0.992. The dilution effect for the different ions in the substratesolution varied

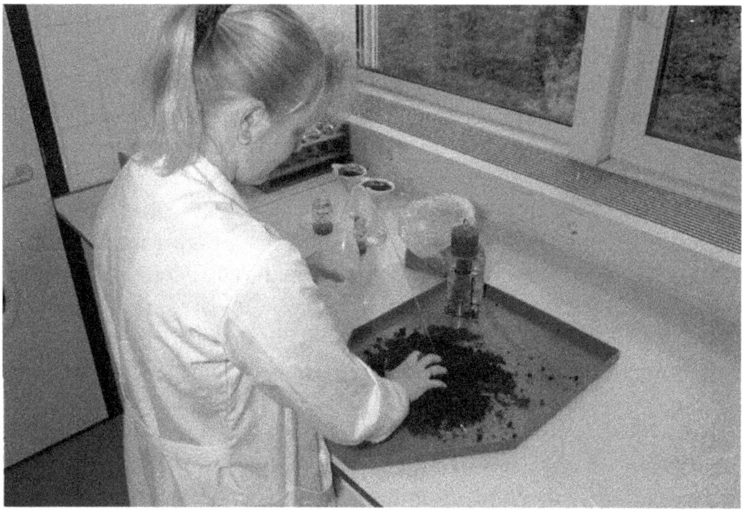

Picture 4.2 Preparation of the 1:1½ extract of peaty growing media. One volume of growing media is mixed with 1½ volume of water. The moisture contents of the growing media are adjusted to a pressure head of −3.2 kpa

Table 4.4 Regression equations for the relationships between the analytical data of the 1:1½ volume extract (x) and the substrate solution (y) of a series of peaty substrates. EC expressed as dS m^{-1}, major elements as mmol l^{-1} and micro elements as µmol l^{-1}

Determination	Regression equation	r
EC	y = 2.39 x + 0.17	0.982
Major elements		
NH_4	y = 2.63 x − 0.10	0.968
K	y = 2.52 x − 0.15	0.992
Na	y = 2.51 x − 0.01	0.938
Ca	y = 2.74 x + 0.60	0.982
Mg	y = 2.61 x + 0.53	0.961
NO_3	y = 2.80 x + 0.59	0.984
Cl	y = 2.76 x − 0.10	0.972
SO_4	y = 2.38 x + 0.52	0.912
P	y = 2.38 x + 0.19	0.954
Micro elements		
Fe	y = 2.58 x + 0.77	0.836
Mn	y = 3.51 x − 0.28	0.967
Zn	y = 3.14 x − 0.26	0.981
B	y = 1.44 x + 6.77	0.663
Cu	y = 1.38 x + 0.38	0.568
Mo	y = 0.79 x + 0.08	0.471

After Sonneveld C 1994. Unpublished data.

from 2.4 for the determinations of NH_4 and K and 3.2 for Ca (Sonneveld and Van Elderen, 1994). These low dilution factors are an advantage of the 1:1½ extract and implies that the method will not be strongly hindered by dilution effects, as is confirmed by the high correlation coefficients between the analytical data of the 1:1½ extract and the substrate solution. A drawback is the fact that it is not a universal method for all types of substrate. The 1:1½ volume extract method is just suitable for peaty substrates, which means mixtures in which peat is the main component. The presence of a volume fraction up to 25% of different other materials showed to be no hindrance for the application (Sonneveld and Van Elderen, 1994). However, the use of it for substrates with too much a different moisture characteristic, easily results in analytical data that induce misinterpretation.

With the same study the micro nutrients were determined, as shown at the bottom of Table 4.4 (Sonneveld and Voogt, 2009). For Fe, Mn and Zn the results are comparable with those of the major elements, only the regression coefficients of the equations for Mn and Zn were higher than for the major elements. This can be explained by preferential cation adsorption. For B, Cu and Mo the correlation coefficient are much lower than those found with the other elements, which could be explained by analytical errors, caused by the fact that the methods of determination of these elements in the 1:1½ extract was not adjusted to the low levels of these elements in this extract.

The 1:1½ method is based on the fact that substrates with natural organic material as main constituent contain a water content of about 50% by volume at a pressure head of –3.2 kPa and about 60% at –1 kPa. Thus, in the extraction suspension a ratio substrate to water exists of 1:2 v/v, while the dilution in relation to the substrate solution is about 2:0.6, which can be roughly expressed as 3:1. In this way the 1:1½ extract is suitable to express the analytical data as well on substrate volume as on substrate solution (Sonneveld and Voogt, 2009).

4.6 1:5 Volume Water Extract

A universal method for the determination of water soluble elements in substrates has been developed by CEN/TC 223, a European commission for standardisation of analytical methods for soil improvers and growing media. The extract is prepared by extraction of a suspension of 1 volume of substrate and 5 volumes of water (CEN, 2001a. The quantity of substrate used for the preparation of the suspension is based on the so called laboratory compacted bulk density. This bulk density is determined beforehand by filling a cylinder of about 1 litre, dimensions 100 mm diameter and 127 mm height with fresh substrate. The substrate in the cylinder is compacted in a special way by placement of a plunger of a certain weight on top of the filled cylinder, which is comparable with a pressure of about 0.9 kPa (CEN, 2007). A certain volume of the sample, CEN recommend 60 ml, is separated by weighing on basis of the compacted bulk density, determined beforehand. This quantity of substrate is mixed with 5 volumes of water, thus, 300 ml when the CEN recommendation is followed.

The striking characteristic of the CEN method is the high water to substrate ratio, which makes it possible to extract filtrate from the suspension by a simple filtration with all types of substrates. However, the method has the same drawbacks as mentioned before for high water to soil ratios. The dilution in the suspension in relation to the water holding capacity under growing conditions varies from 25 for substrate with a low water holding capacity to 6 for substrate with a high water holding capacity, like for example expanded clay and peat respectively. For an interpretation of the analytical data in relation to the substrate solution the water holding capacity of the substrate under growing conditions should be available (Sonneveld and Voogt, 2001, 2009).

4.7 Extraction of Pre-shaped Substrate by Water

The extraction methods discussed so far are just suitable for unformed material. The ultimate shape and density of such substrates under growing conditions is determined by the dimensions and the filling method of the containers applied in the growing system. However, this is not the case with pre-shaped materials like for example slabs of mineral wool and foams, because such materials got already the form for the growing conditions at the factory. Another situation arises with the so called slabs of pressed peat and coir. True enough, these slabs are pre-shaped on the factory, but the shape will surely significantly change when wetted under growing conditions. In this case the extract preparation offers different possibilities. The first is that the material on the laboratory is carefully pre-wetted with demineralised water up to the leek out condition and extracted as presented for the leek out situation at the end of this section for other pre-shaped substrates. The second option is that the material will be pre-wetted at the laboratory, while granulated by gently stroking of the substrate and by this handlings preventing as much as possible damages on the structure of the original material. After these treatments the substrate is suitable to be extracted by either the 1:1½ or the 1:5 volume methods.

For extraction of pre-shaped substrates a suitable peace of material is cut from the sample and the exact volume is calculated from the dimensions measured. The extraction can be carried out following the 1:5 volume method. With the 1:5 volume method an intensive mixing of substrate and water in the suspension should be ensured, which best can be carried out by unravelling the peace of substrate preventing as much as possible damage on the original components, like the fibres of mineral wool slabs.

Extracts of pre-shaped substrates also can be prepared after saturation to leak out condition. The leak out condition should be determined in a sub-sample. This sub-sample is immersed in water and leaked out until equilibrium is reached. The sample for the extract preparation at the leak out situation is mixed with water at the same water to substrate ratio as found in the sub-sample, and after an overnight storage of the material the extract will be gained by suction or by gently pressing of the substrate. The extraction at the leak out condition nicely links with the extraction during cultivation, when extract is gathered by suction from slabs more or less continuously in a leak out condition under growing conditions.

4.8 Soil and Substrate Solution

The soil and the substrate solution directly supply information about the ion concentrations of the plant root environment. Especially for determinations related to salinity the soil and the substrate solutions provide optimal information and therefore, are often used for research purposes in this field. However, it never has been employed for routine soil testing, because of its difficult and laborious preparation. The methods employed to extract the soil solution from field moist soils varies as listed by Fried and Broeshart (1967). The most suitable method is hydraulic pressing of the soil as described by Van den Ende (1989a). This method is less suitable for soils with low water contents at field capacity, like sandy soils with low organic matter fractions. In such cases the displacement method can be applied.

For substrates, however, the so called substrate solution is often used for routine testing of substrates. This especially is the case for substrates with high water contents at a high pressure head, like mineral fibres and foams (Sonneveld, 1995). During cultivation the crops are irrigated frequently in such substrates, sometimes dozens of times a day, which ensures a stable and high water content in the substrate. Under such conditions the substrate solution easily can be gathered, withdrawing it from the substrate by suction with the aid of a simple syringe. Also with peaty substrates the use of substrate solution is sometimes practiced, because the substrate solution at a pressure head of –1 kP, defined as the moisture condition at field capacity, can be easily pressed out. The method is applied with samples gathered from the greenhouse during cultivation. Before the sample is pressed a careful check on the right moisture condition is required, as well an adjustment with demineralised water when the moisture condition is too low.

Extraction of soil and substrate solution under practical conditions can occur by suction with the aid of cups produced with ceramic or artificial material. The cups are placed in the soil and the soil solution penetrates the wall of the cups as a consequence of the suction applied in the system. A drawback of this method is the accidental placement in the soil or the substrate. Therefore, in view of the great variation of concentrations of salts and nutrients in greenhouse soils and substrates, different placements are necessary scattered horizontally as well vertically. Another drawback is the possible adsorption of some elements that can occur by the material whereof the wall of the cup is produced. The right choice of cups is important to be ensured that the soil solution is not affected by the material from which they are produced (Shen and Hoffland, 2007).

Much research is carried out in hydroponics and the results of experiments are based on the solution in which the plants were grown. Such solutions can be considered as soil and substrate solutions. Many plant nutrition and salinity reactions in soil and substrate growing show a good agreement when compared with the reactions in hydroponics. However, when the matrixes of the soil or the substrate play a part in it, the reactions will differ from those in hydroponics. Such reactions especially occur in the rhizosphere of plant roots. This for example is sometimes the case for the uptake of micro nutrients.

The composition of the substrate solution also can be estimated from the composition of the water supplied to the plant and the drainage water. This method seemed to be very suitable, because it not only gives an acceptable estimation of the average concentrations in the root environment, but it also informs about lowest and highest concentrations in the root environment, generally supply and drainage in a substrate system, respectively. Especially for interpretation of the EC value knowledge about lowest and highest values are very useful, as will be discussed in Chapter 8.

4.9 Saturated Gypsum Solutions

The dilutions by water extraction of soils and substrates gives always raise to dissolution of sparingly soluble salts and by this an overestimation of the salt status. Gypsum is the most likely salt responsible for this effect. Therefore, some researchers recommended a saturated gypsum solution as extraction solution for the estimation of total soluble salt (Winsor et al., 1963), to mask the effect of gypsum on the determination of the salinity. This, sometimes lead to a remarkable improvement of the estimation of the salinity effect on crops (Fischer, 1992; Massey and Winsor, 1968), especially when high water to soil ratios are used with extraction. In Table 4.5 correlation coefficients are shown for the relationships between the yield of lettuce and the EC measurements in the growing media determined by extraction either with water or by a gypsum saturated solution (Massey and Winsor, 1968). The use of the gypsum solution did not increase the correlation coefficients up to the level found with the saturation extract. Therefore, the use of a saturated gypsum solution as extraction solution has not been found a wide application. Besides, the use of such a solution has drawbacks of which the most likely are following.

- It totally masks the contribution of gypsum to the osmotic potential of the soil solution, assuming that all soil and substrate solution are gypsum saturated under growing conditions. However, this is not the case (Sonneveld et al., 1990).
- The use of sufficiently narrow water to soil ratios became more customary last decennia, which prevent the trouble of overestimation for the greater part.
- The determination of the real concentration of Ca and SO_4 in the soil and substrate extracts is impossible, which blocks the opportunities for adjustment of the

Table 4.5 Correlation coefficients for the relationships between the yield of lettuce in two series of experiments and the EC determined by different extracts. The extracts are derived from a saturated soil suspension, a 2.5:1 v/v water to soil suspension and a 2.5:1 v/v gypsum saturated solution to soil suspension

Salinity test method	Series 5–8	Series 9–12
Saturated	0.954	0.982
1:2.5 v/v water	0.746	0.923
1:2.5 v/v water + gypsum	0.814	0.953

Withdrawn from Massey and Winsor (1968).

estimation of the osmotic potential of soil and substrate solutions on the dissolved gypsum.

- The lack of information about the real concentration of Ca and SO_4 in the extract blocks also the control on these elements by fertilization practices.

In studies with different extraction methods with water (Sonneveld et al., 1990; Sonneveld and Van Elderen, 1994) the suitability of narrow water to soil ratios and the measurement of SO_4 to improve the estimation of the EC of the soil solution is taken into account. The results of these studies are listed in Table 4.6. With a relative high water to soil ratio as used in the specific 1:2 volume extract a precise estimation of the EC of the soil solution will be gained when the dilution factor and the SO_4 concentration are added as explaining variables. Then equal and in this specific are added case even a little higher correlated estimation will be gained than with the saturation extract.

Table 4.6 Comparison of different equations and correlation coefficients to estimate the EC of the soil solution and substrate solution using either the EC of the saturation extract, different variables of the specific 1:2 volume extract or the EC of the 1:1½ volume extract

Equations used to estimate the EC of the soil solution	Correlation coefficients	Variables used
$1.60EC_{se} + 0.18$	0.958	EC_{se} = EC saturation extract
$3.12EC_{1:2} + 0.84$	0.886	$EC_{1:2}$ = EC of the 1:2 extract
$0.601\ dEC_{1:2} + 1.26$	0.944	d = dilution factor
$0.908\ dEC_{1:2} - 0.089\ dSO_{4_{1:2}} + 0.68$	0.968	$SO_{4(1:2)}$ = SO_4 concentration of the 1:2 extract
$2.39EC_{1:1\frac{1}{2}} + 0.17$	0.982	$EC_{1:1\frac{1}{2}}$ = the EC of the $1:1\frac{1}{2}$ volume extract

After Sonneveld et al. (1990); Sonneveld and Van Elderen (1994).

4.10 CAT Extraction

By CEN/TC 223, see the remarks about this commission in Section 4.6, a method for the determination of potentially available nutrients in substrates has been developed. The extract is prepared by extraction of a suspension of 1 volume of substrate and 5 volumes of CAT-extraction solution (CEN, 2001b). The quantity of substrate used for the preparation of the suspension is based on the so called laboratory compacted bulk density, the determination of which is described in Section 4.6. The CAT-solution is a solution of 0.01 mol l^{-1} $CaCl_2$ and 0.002 mol l^{-1} DTPA (diethylene tri-amine penta-acetic acid). The pH of this solution varies between 2.6 and 2.65, but the available pH buffer of the solution is weak.

The quantities of cations extracted by the CAT method will be much higher than the directly water available quantities. Thus, it rather reflects the "capacity" of these nutrients in a substrate than the "activity". This is clear from the data shown in

Table 4.7 Quantities of nutrients (mg l^{-1} substrate) extracted with water or CAT from four different substrates

Substrate	Extraction	NO_3	P	K	Cu	Mn
Composted	Water	25	236	1910	1.3	1.3
bark	CAT	14	290	2727	1.1	10.7
Fertilized	Water	83	31	114	0.0	0.3
clay/peat	CAT	73	35	148	0.7	9.8
Fertilized	Water	57	89	101	0.1	0.2
coarse peat	CAT	54	93	129	1.1	3.3
Composted	Water	67	48	126	0.0	0.4
wood fibre	CAT	65	55	152	1.0	8.5
Average	Water	58	101	563	0.4	0.6
	CAT	52	118	789	1.0	8.1

Withdrawn from CEN (2001a and b).

Table 4.7. Nutrients already dissolved in the substrate solution, like NO_3, are nearly not affected by the CAT solution. P is slightly affected maybe by the low pH buffer of the extraction solution and K by the cation exchange. Cation micro nutrients, like Cu and Mn, will become better soluble in the CAT extract by the low pH, cation exchange and complexion. Especially with Mn the quantities soluble in the CAT extract are strongly increased. This is understandable because most Mn under natural condition occur as manganese oxides, which solubility strongly depend on the pH (Fujimoto and Sherman, 1948; Leeper, 1947). The ratio Cu-CAT/Cu-water showed great variation. Cu can be bound strongly on organic matter, which varies greatly dependent on the type of material and pH (Verloo, 1980).

4.11 Exchangeable Cations

With CAT extraction for different nutrients mostly more cations are released than those present as exchangeable (Sonneveld and De Kreij, 1995). Therefore, for substrates a different method is developed for extraction of just the exchangeable quantities, based on the use of ammonium acetate (Knudsen et al., 1982). The extract is prepared from a suspension of 1:5 v/v fresh substrate and 0.5 mol l^{-1} ammonium acetate (NH_4Ac) solution, respectively. The volume is measured according to the method of CEN (1999a) and the moisture present in the fresh substrate is taken into account with the preparation of the suspension (Kipp et al., 2000). The NH_4Ac solution is buffered at a pH value of 4.65.

Obviously, the determination of exchangeable NH_4 is impossible in the NH_4Ac extract. However, the determination of exchangeable NH_4 in some substrates will be important, in view of the high concentration of this ion that can occur. In such cases instead of NH_4Ac an equivalent concentration of $BaCl_2$ is recommended (Kipp et al., 2000). For greenhouse soils NH_4Ac is also suitable for the determination of exchangeable cations. However, the determination of it is not obvious for greenhouse soils, because the actual mutual ratios of the cations in the soil solution are of more importance than the total available quantities. Such ratios best can be approximated by water extraction.

4.12 Phosphorus

With the water extractions like the saturation extract, the 1:2 volume extract and the 1:1½ volume extract close relationships were found between the P concentration of these extracts and those of the soil and substrate solutions. Thus, water extraction with a low water to soil and water to substrate ratio is a good method to get informed about the solubility of the P in soil and substrate solutions. However, it mostly does not give a good impression about the total available P and even less about the total storages in soils and substrates. This is clear from the data shown in Fig. 4.2, where the quantities of P extracted with different water to soil ratios of three different soils are shown. The P available in the soil solution, and those extracted with the saturation extract and the 1:2 volume extract is only a small fraction of the total water soluble quantities, extracted by a 1:100 w/w extract. Therefore, the determinations of P at low water to soil ratios only gives an impression of the solubility of P and are not precise estimators of the total water soluble quantities as is shown in Fig. 4.3 for the 1:2 volume extract. Many greenhouse soils are rich on water soluble P and most of the Dutch greenhouse soils shown in Fig. 4.3 contain between 1 and 5 mmol l^{-1} of soil. This means that in the top layer of 0.25 m of these soils a P storage is available between 75 and 400 kg P per ha.

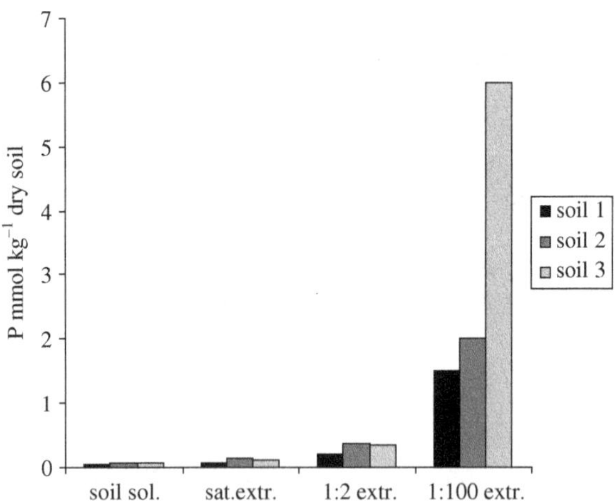

Fig. 4.2 Quantities of P (mmol kg^{-1} dry soil) extracted with different water to soil ratios, soil solution, saturation extract, 1:2 volume extract and 1:100 w/w extract, for three greenhouse soils (1 – peaty soil and 2 + 3 – loam soils)

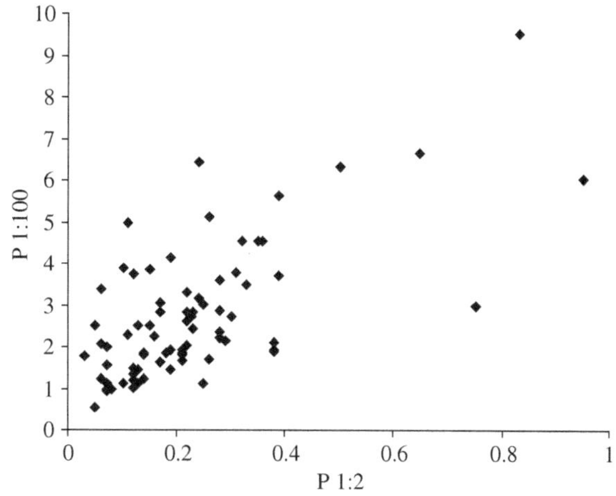

Fig. 4.3 Relationship between the P concentration in the 1:2 volume extract (mmol l^{-1} extract) and the water soluble P at an extraction ratio 1:100 w/w water to soil (mmol l^{-1} soil) for 75 greenhouse soils

In peaty substrates it has been found that often the bulk of the P is available in the substrate solution. In such cases with water extraction the P in the substrate behaves like other anions at dilution with water extraction (Sonneveld and Van Elderen, 1994). However, when clay was a constituent of the peaty substrates, the P behaves more like those in greenhouse soils (Sonneveld et al., 1974) and the concentration of it in the extract is more or less stable at low water to substrate ratios. Another factor that will play a role in P determination in substrate is the time delay between application and determination. Fertilized substrate mainly contains fresh orthophosphate as high soluble H_2PO_4 from fertilizers supplied. The precipitation and occlusion process to bind the P in the labile and non-labile pools is time consuming (Mengel and Kirkby, 1987), while the analysis usually is carried out shortly after addition.

In greenhouse industry often the quantities of labile and non-labile P mainly in soils and sometimes also in substrates are much higher than those that will be absorbed by crops. The determination of that pool is quite important and gives an impression of the "capacity" for long periods. The availability of the compounds should be checked more frequently by determination of the "intensity". Water extraction is the most obvious method for the determination of the "intensity" and for the "capacity" CAT, P-Al (NH_4-lactate-acetic acid) and other extraction methods are used (Alt and Peters, 1992). The suitability of such methods for soils will be discussed further on in Section 16.4.

4.13 pH

The pH is usually determined in a suspension of 1 volume part soil and 2 volume parts of demineralised water. The pH determined in this way showed substantial seasonal variations with field soils (De Vries and Dechering, 1960). Therefore, the determination of the pH in a solution of 1 mol l^{-1}KCl has been developed, the results of which are less sensitive for these variations. The seasonal variations for greenhouse soils are less than for field soils and so there is no urgency to use this determination for greenhouse soils, but even so used sometimes. The pH_{KCl} is generally lower than the pH determined in a water suspension. However, the differences for greenhouse soils are smaller than for field soils and will be discussed in more detail in Section 16.2.

For substrates CEN developed a method for the determination of the pH in a suspension of 1 volume part substrate and 5 volume parts of water (CEN, 1999b). The wide ratio between water and substrate is reason for deviations with the pH found under moisture conditions as realised in the field. This mainly occurs for substrates with a low buffer capacity. The pH determination for such substrate with water is always problematic and therefore the determination often is carried out with a standard nutrient solution (KIWA, 2003). The result will be discussed further on in Section 11.4.

Picture 4.3 Soil sampling in a greenhouse. Generally, the sampling depth is 0.25 m

4.14 Sampling

The results of the analytical data of soils and substrates are affected strongly by the method of sampling. The great variation in the chemical properties from spot to spot requires a special procedure to gather a sample that significantly reflects the composition of the soil and the substrate in the greenhouse. Variations in chemical properties of spots in a greenhouse area can be distinguished in systematic and in accidental components. Knowledge about the character of the variability is quite important with respect to the instructions to the sampler. Systematic components for example need specific attention and require as well specific actions which will be included in the instructions for the sampler. Systematic variation components can be either included as well avoided with the sampling, dependent on the expected reaction of the crop on such places. Places with a specific deviation, where the soil or the substrate does not contain plant roots will be avoided with the sampling. However, when plants have developed roots in such places mostly they will be systematically included in the sample. However, such will depend on the purpose of the sample and the plant reaction on the deviation, considering the deviations as discussed following and the reaction of the plant on an unequal distribution as presented in Chapter 8.

Examples of systematic variations in soils and substrates in the greenhouse industry are for example the distribution of salts and nutrients with the use of drip irrigation in soil grown crops. In Fig. 4.4 the distribution of NO_3 between the nozzles of a drip irrigation system is shown. The nozzles were placed near the plants and strong accumulations of salts and nutrients occur in the area between plants. Another example is the vertical distribution of salts shortly after fertilization as shown in Table 4.8.

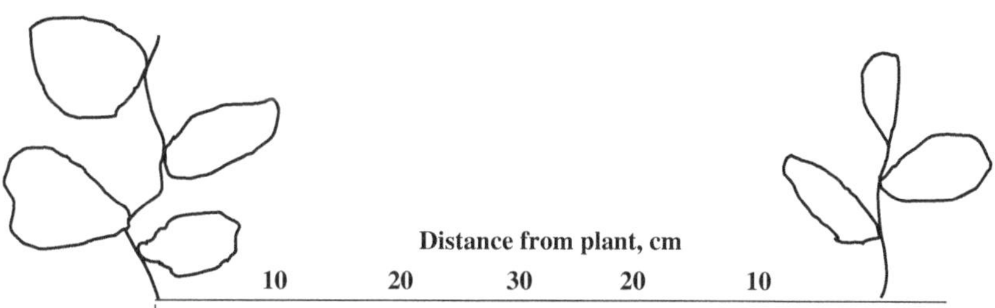

Depth, cm	Distance from plant, cm: 10	20	30	20	10
10	1.2	9.8	7.0		
20	0.8	2.5	3.3		
30	0.7	8.5	7.8		
40	0.7	2.2	2.9		

Fig. 4.4 NO_3 concentrations in the soil (mmol l^{-1} 1:2 volume extract) with drip irrigation at different distances of the irrigation spots. Tomato crop on clay soil, 6 months after planting. After Sonneveld et al. (1991)

Table 4.8 Vertical distribution of nutrients in greenhouse soils shortly after fertilization. Average values of three greenhouses. The concentrations are expressed as mmol l^{-1} 1:2 volume extract

Depth cm	N	P	K
0–8	5.2	0.44	4.2
8–16	4.4	0.15	2.9
16–24	2.9	0.14	1.5
24–40	2.1	0.05	1.0

After Sonneveld, 2009. *Reprinted by permission of the Koninklijke Landbouwkundige Vereniging*

Systematic differences in greenhouses can occur in many other situations, like the differences between growing beds and paths (Van den Ende and Knoppert, 1959; Van der Wees, 1983), irrigation furrows and the dry strips beside them, the vertical distribution of nutrients in the substrate of potted plants grown on flooded benches (Otten, 1994) and variations in pH in substrate systems caused by NH_4 application (Sonneveld and Voogt, 2001). When systematic differences are well known, the decision can be made to take different samples from the same site in such a way that the variation of the sampled object is reflected in the different samples. Other possibilities are an overall sampling with the purpose to get a rough estimation of the average chemical composition of the sampled object and a sampling of selected spots to estimate the composition of specific sites of the object from which crop response is expected. The choice of the sampling method will be made in relation to the purpose of the sampling and the expected reaction of the crop grown or the crop that will be grown in the soil or the substrate object sampled.

In experiments with salinity it was found that spots of high osmotic potential (low EC value) play a dominant role of crop reaction on salinity (Sonneveld and Voogt, 1990; Sonneveld and De Kreij, 1999). Thus, for this item spots of low EC values play a dominant part in the salinity effects of the crop. For nutrient uptake it was found that plants are able to absorb nutrients from high concentrated as well as from low concentrated spots. Thus, in such cases the total available quantity of nutrient in the rooted zone seems to be important (Sonneveld and Voogt, 2001). Thus, with the use of saline water in drip irrigation systems plants will react mainly on the lowest concentration in the spot under the dripper and only secondary on the accumulation of salts in the surrounded soil or substrate. However, nutrients accumulated in the surrounded soil or substrate volume are absorbed by plants and thus, are important for the crops grown.

Descriptions of sampling procedures are sparingly published for the greenhouse industry, perhaps because of the difficulties arising with the great variation in growing systems. In The Netherlands some guide-lines are published (De Kreij et al., 1999; Van den Bos et al., 1999; Van der Wees, 1993).

For an overall sampling of greenhouse soil it is recommended to gather 40 cores at random from the object. This number of cores is based on the theoretical fact that the error of an at random sample decreases with the square root of the number of sampling points. This means for soil sampling in formula:

$$s_n = \frac{s}{\sqrt{n}} \tag{4.3}$$

In which:

s = the standard deviation of the single cores (sub samples)
s_n = the standard deviation of a sample at n cores
n = the number of cores in the sample

The function presented in formula (4.3) is shown in Fig. 4.5. It is evident that the standard deviation at 40 sampling points is reduced to about 15% of the deviation at one sampling point and that a further increase of the sampling point is less effective. The strongest decrease, however, is reached up till 20 sampling points. So, in greenhouse industry 40 cores per sample is preferred, when the sampling is time consuming or difficult to carry out, 20 cores is considered to be sufficient. In such cases a careful handling of less sub samples is preferred above a higher number sub samples less carefully gathered.

For substrate sampling the European standardisation (CEN, 1999a) uses following formula to calculate the number of sub samples, being the number of sample points.

$$n_{sp} = 0.5\sqrt{V} \tag{4.4}$$

In which

n_{sp} = number of sampling points, with the restriction of $12 \geq n_{sp} \leq 30$
V = the nominal quantity of the sampled portion in m^3

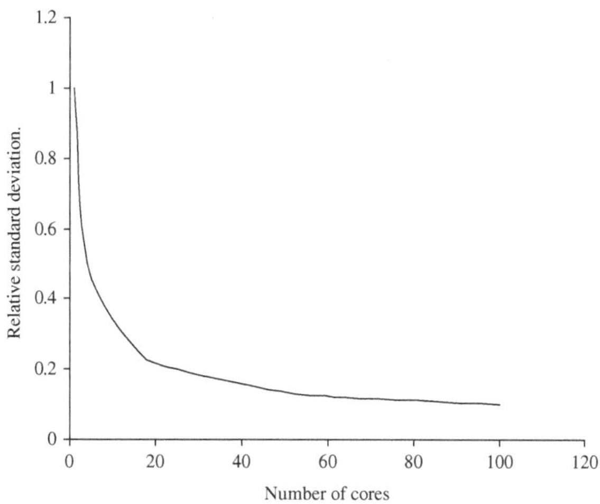

Fig. 4.5 Relationship between the number of cores from which a sample is composed with at random sampling and the standard deviation of the composed sample relative to the standard deviation of a sample composed of a single core, following formula (4.3)

The sampling depth recommended for the greenhouse industry for soil grown crops is mostly restricted to 0.25 m. In substrate growing the thickness of the substrate layer mostly is less than 0.25 m and thus, the sampling is carried out over the whole depth of the layer. With potted plants in flooded benches the upper 2 cm of the substrate is removed from the cores, because of the excessive high salt concentrations in it and the lack of roots in this layer.

In rock wool slabs, foam slabs and other pre-shaped substrates is it impossible to sample the substrate itself. Mostly, the nutrient solution in such substrates can be sucked very easily from the material by a simple syringe. The number of sampling points is the same as with ad random sampling of soils. It is impossible to gather nutrient solution with a syringe from coarse inert substrates, like pumice, perlite, vermiculite, and expanded clay granules. Sampling of the material at such is also problematic during cultivation. Therefore, in such substrates the free nutrient solution at the bottom of the containers or the drainage water will be sampled. In systems with circulating water, like NFT and deep water culture, the circulating solution is sampled. In substrate systems where the nutrient solution is reused, often sampling of the ingoing (supplied) and outgoing (drainage) solution is very useful and gives a good impression of the salt and nutrient status in the root environment (Sonneveld and Voogt, 2001), including highest and lowest salt and nutrient status.

Substrate material sometimes will be sampled on storage. Bulk material will be preferably sampled throughout the depth of the material, with which the top 50 mm is ignored. With the sampling it is important to preserve the characteristics of the material. Therefore, sampling by hand or shovel is more obvious than with an auger in such cases. With packed material is each sampling point a different randomly selected pack. With pre-shaped material a suitable part shall be cut from the slabs with a sharp knife or saw, without disturbing the characteristics of the material (CEN, 1999a).

4.15 Accuracy of Soil Testing

All handlings carried out to produce analytical data of soil testing contribute towards errors. Generally, the errors are distinguished as caused by factors outside and by factors inside the laboratory. The errors caused by factors outside the laboratory are strongly controlled by the handlings carried out with the sampling. The effect of the handlings with the sampling on the total error is mostly much greater than the effects caused by the laboratory handlings (Cline, 1944; Peck and Melsted, 1980; Vermeulen, 1960). Another source of deviations is the accidental laboratory on which the analysis is carried out. This factor is mostly ignored, because it is not evident when the measurement is carried out on only one laboratory. However, the casual handlings on the laboratory can substantially contribute to errors. This for example, was shown with a proficiency testing carried out with a new developed measurement of the laboratory compacted bulk density (CEN, 2007) with laboratories all over the world. This bulk density was due to the measurements of nutrients in substrates. The data were statistically analysed following the method of ISO (1994) and showed that the repeatability varied between 1.3 and 3.2%, while those for the reproducibility varied between 5.8 and 9.9% (De Kreij and Wever, 2005). The repeatability and reproducibility express the deviations within and between the laboratories, respectively. Herewith, is shown, that the deviations between laboratories can be much greater than those within laboratories.

With sampling the homogeneity of the soil, the working method of the sampler and the number of sampling points are factors that will strongly affect the accuracy of the results. Therefore, a careful instruction to the sampler as mentioned in Section 4.14 is very important to get a goal-directed and accurate sample. The accuracy within the laboratory is determined by the handling with the pre-treatment, the handlings of the analyst, the analytical methods applied and the accuracy of the apparatus involved. In modern routine laboratories with automatic apparatus, besides good instructions, also the choice of apparatus suitable to the purpose, is an important factor to get the required precision of the analytical data (Sonneveld and Voogt, 2009).

In The Netherlands research has been carried out to estimate the size of the errors with soil testing in greenhouses (Sonneveld, 1979). To this purpose some hundreds of greenhouse soils were sampled in duplicate and the samples were analysed also in duplicate at the laboratory. With these results besides the total error also the sampling error and the laboratory error could be estimated separately. The standard deviation was used as a measure for the errors.

The level of the analytical data and the standard deviation were linearly related. The functions calculated for such relationships mostly showed a positive intercept and thus the coefficients of variation were not a constant. Those for the total errors varied from 10 to more than 20% of the results and those resulting from the laboratory analyses varied roughly between 5 and 10%. The contribution of the sampling to the total standard deviation was 2–5 times higher than those of the handlings with the determination on the laboratory.

With substrate sampling and analysis more or less the same experience has been gained as with soil testing. Such has been found for sampling of nutrient solutions in rock wool slabs (Sonneveld and Voorthuizen, 1988). For this type of samples also a linear relationship was found between the total standard deviation and the level of the data, as shown for the determination of K in Fig. 4.6. The coefficients of variation of the total errors lay mostly between 10 and 20% of the results, just like for soil samples. Those for the laboratory analysis varied for most determinations between 2 and 10%. The contribution of the sampling to the standard deviation was sometimes more or less equal to those from the laboratory, but was mostly between 2 and 10 times higher.

The consequences of the effects of errors made by sampling and analyses are often underestimated. Therefore, adequate sampling procedures must be developed and results will be well tested and statistically analysed for the concerning sys-

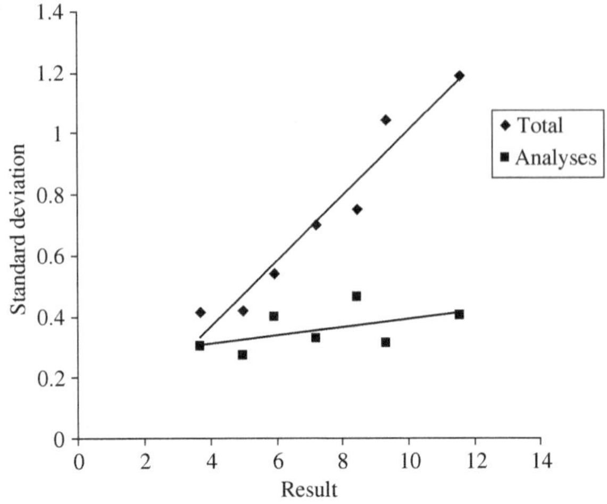

Fig. 4.6 Relationships between the result (mmol l^{-1}) of the K determination in nutrient solutions of rock wool slabs and the standard deviations following sampling and analysis on the laboratory. Total: s_t = 0.108 K – 0.065 and analysis: s_a = 0.013 K + 0.257 was calculated

tem. This is demonstrated by the data of Fig. 4.6. For the example presented first of all should be pointed to the fact that the system has been tested for results of the K determination between 3 and 12. Extrapolation of the regression equations in the low direction soon induces a contribution from the analyses to the total standard deviation higher than the total standard deviation itself, which is impossible. In Table 4.9 some results of the statistical analysis are summarized. The analyses of K were duplicated and the average values were used as the result. Thus, the total standard deviation was compounded following formula (4.5).

$$s_t = \sqrt{s_s^2 + \frac{1}{2}s_a^2} \tag{4.5}$$

In which

s_t = the total standard deviation
s_s = the standard deviation following from sampling
s_a = the standard deviation following the analysis on the laboratory

Table 4.9 Standard deviation as found for the K determination (mmol l^{-1}) in the nutrient solutions of rock wool slabs at two levels of results. s_t, s_a and s_m standard deviation for total, analyses and sampling, respectively. vc is the relative standard deviation (%)

Result	s_t	s_a	s_m	vc_t	vc_a	vc_m	P = 0.95	P = 0.997
3 mmol l^{-1}	0.258	0.296	0.151	8.6	9.9	5.0	2.48 - 3.52	2.23 - 3.77
12 mmol l^{-1}	1.230	0.413	1.195	10.2	3.4	10.0	9.55 – 14.45	8.23 – 15.67

Data derived from Fig. 4.6.

Furthermore, the data in Table 4.9 learns that in this specific case the relative total standard deviation is more or less stable over the range concerned, that the relative standard deviation for the analyses strongly increases with the decrease of the result and that this is upside down for the sampling for this determination.

One should be aware of the fact that the real value (μ) of an analytical result with a confidence level of P lies between limits following formula (4.6).

$$\mu = x \pm u_p s_t \tag{4.6}$$

In which:

μ = universal value of the result
x = analytical result as found in the sample
u_p = standard normal distributed unit corresponding with a confidence interval of P %
s_t = total standard deviation

A confidence interval of 0.95 and 0.997 corresponds with an u_p value of 2 and 3 respectively. Thus, keeping in mind that s_t often has a value between 10 and 20% of the result it easily can be calculated that the real value (μ) of an analytical result

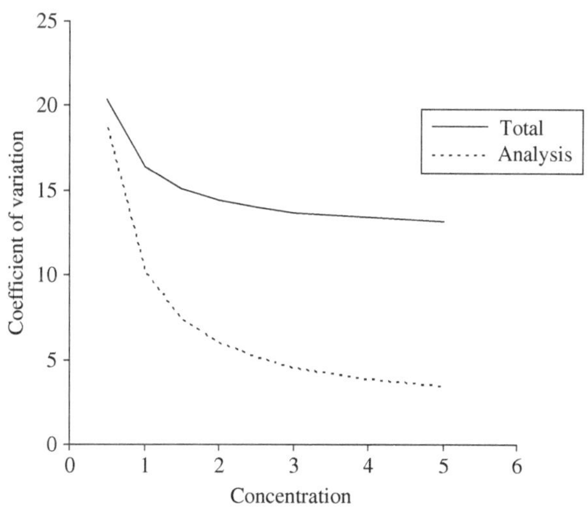

Fig. 4.7 Relationships between the concentration of Cl in nutrient solutions of rock wool slabs (mmol l^{-1}) and the coefficient of variation for the total error (s_t) and the error made by the determination on the laboratory (s_a). Calculated from the relationships $s_t = 0.124$ Cl + 0.040 and $s_a = 0.017$ Cl + 0.085

mostly lies between the result (x) ± 20 to 40% and nearly always lies between the result (x) ± 30 and 60%. In Table 4.9 are the confidence domains given in last two columns for the specific situation of the values in this table as an example. It shows a relatively favourable situation with a total standard deviation of about 10%. The precision of an analytical determination system can differ strongly and will be tested to the purpose. Such a system is suitable to the purpose, when the relative errors of the methods in the range employed are sufficiently low and this range includes the critical limits for plant development. An example is given in Fig. 4.7, where the relative standard deviation is shown in relation with the level of Cl determinations in nutrient solution of rock wool slabs. The coefficient of variation for the analysis on the laboratory increases strongly for values below 1 mmol l^{-1}, which resulted also in an increase of the total coefficient of variation. Thus, with the determination of Cl reasonable analytical data can be expected for values higher than 1 mmol l^{-1}, which is sufficiently distinct. The critical value for the system is higher as will be discussed in Chapter 7. The determination of Cl for the rock wool system was less precise than the K determination, because the total standard deviation is close to 15%.

4.16 Applications

In this section a review is presented of the applications of the soil and substrate testing methods in relation to systems and growing conditions. In this review only the common growing conditions are presented with general applied testing methods due to the situation. Specific situations will be discussed in the chapters belonging to such situations. Application before planting and during cultivation can differ in the subject under discussion and will be distinguished.

The following methods are suitable for the estimation of the nutrient and salinity status in the root environment of soil, substrate and hydroponics. When appropriate the section where the method is described is mentioned.

1. Specific 1:2 volume extract (4.2)
2. Saturation extract (4.3)
3. Substrate solution by suction with a syringe in the field (4.8)
4. Substrate solution by pressing at field capacity (4.8)
5. 1:1½ v/v extract (4.5)
6. 1:5 v/v extract (4.6)
7. Supplied nutrient solution
8. Drainage water
9. Supernatant solution on bottom of plant container
10. Circulation nutrient solution

Situation	Before cultivation	During cultivation*
Soil	1, 2	1, 2
Natural organic substrate, pre-shaped	5, 6	5, 6, 7+8
Natural organic substrate, loose	5, 6	5, 6, 7+8
Rock wool slabs and cubes	6	3, 7+8
Foam slabs and cubes	6	3, 7+8
Rock wool and foam loose	6	7+8, 9
Course grained mineral substrates	6	7+8, 9
Fine grained mineral substrates	6	6, 7+8
Hydroponics (without any substrate)	-.-	10

* Where 7+8 is mentioned, the compositions of two solutions are necessary; unless the composition of the nutrient solution supplied is known.

References

Alt D and Peters I 1992. Die $CaCl_2$/DTPA-Methode zur Untersuchung Gärtnerischer Erden auf Mengen- und Spurenelemente. Agribiol. Res. 45, 204–214.

Carpena O Guillén M G Fernandez F G and Caro M 1968. Saline soil classification using the 1:5 aqueous extract. 9th International Congress of Soil Science, Transactions vol. 1, 483–490.

CEN 1999a. European Committee for Standardisation CEN/TC 223. Soil improvers and growing media – Sampling. EN 12579, 11 pp.

CEN 1999b. European Committee for Standardisation CEN/TC 223. Soil improvers and growing media – Determination of pH. EN 13037, 9 pp.

CEN 2001a. European Committee for Standardisation CEN/TC 223. Soil improvers and growing media – Extraction of water soluble nutrients. EN 13652, 15 pp.

CEN 2001b. European Committee for Standardisation CEN/TC 223. Soil improvers and growing media – Extraction of calcium chloride/DTPA (CAT) soluble nutrients. EN 13651, 16 pp.

CEN 2007. European Committee for Standardisation CEN/TC 223. Soil improvers and growing media – Sample preparation for chemical and physical tests, determination of dry matter content, moisture content and laboratory compacted bulk density. EN 13040, 17 pp.

Cline M G 1944. Principles of soil testing. Soil Sci. 58, 275–288.

De Kreij C Van den Bos A L and Baas R 1999. Bemestingsadviesbasis substraten. Proefstation voor Bloemisterij en Glasgroente, Naaldwijk, The Netherlands, 145 pp.

De Kreij C Kavvadias V Assimakopoulou A Paschalidis Chr Paraskevopoulos A Genneadopoulou A and Lagopoulos D 2005. Fertigation: I. Methodology of the 1:2 volume water extract. Proc. Management, use and protection of soil resources, Sofia 15–19 May 2005, 144–147.

De Kreij C and Wever G 2005. Proficiency testing of growing media, soil improvers, soils and nutrient solutions. Comm. Soil Sci. Plant Anal. 36, 81–86.

De Vries O and Dechering F J A 1960. Grondonderzoek. Bedrijfslaboratorium voor Grond- en Gewasonderzoek, Oosterbeek, The Netherlands.

Fischer P 1992. Wie wirkt Gips auf Topfpflanzen? Deutscher Gartenbau 34, 2004–2005.

Fried M and Broeshart H 1967. The Soil-Plant System in Relation to Inorganic Nutrition. Academic Press, New York.

Fujimoto C K and Sherman G D 1948. Behavior of manganese in the soil and the manganese cycle. Soil Sci. 66, 131–145.

ISO 1994. ISO 5725-2 Accuracy (trueness and precision) of measurement methods and results-part 2. Basic method for the determination of repeatability and reproducibility of a standard measurement method. 42 pp.

Kipp J A Wever G and De Kreij C 2000. International Substrate Manual – Analysis Characteristics Recommendations. Elsevier International, Doetinchem, The Netherlands, 94 pp.

KIWA 2003. Appendices on analytical methods. BRL-K10001, Substrate materials for the product certificate for growing media, 62 pp.

Knudsen D Peterson G A and Pratt P F 1982. Lithium, sodium and potassium. *In*: Page A.L. (ed). Methods of Soil Analysis, Part 2, Chemical and Microbiological Properties, second edition. Agronomy Series Number 9 Part 2. American Society of Agronomy, Inc., Soil Sci. Soc. of America, Madison, WI.

Leeper G W 1947. The forms and reactions of manganese in the soil. Soil Sci. 63, 79–94.

Maas E V and Hoffman G J 1977. Crop salt tolerance – current assessment. J. Irr. Drainage Div. 103 IR2, 115–134.

Massey D M and Winsor G W 1968. Soil salinity studies. II: The relation of plant growth to salinity in soil and soil mixtures of differing physical properties. J. Sci. Fd Agric. 19, 332–338.

Mengel K and Kerkby E A 1987. Principles of Plant Nutrition, 4th edition. International Potash Institute, Bern. Special attention to Chapter 9, 403–426.

Moss P 1963. Some aspects of the cation status of soil moisture. Part I: The ratio law and soil moisture content. Plant Soil 18, 99–113.

Otten W 1994. Dynamics of water and nutrients for potted plants induced by flooded bench irrigation: Experiments and simulation. Thesis, Wageningen Agriculture University, The Netherlands, 115 pp.

Peck T R and Melsted S W 1980. Field sampling for soil testing. In: Walsh L N and Beaton J D (eds) Soil testing and plant analysis. Soil Sci. Soc. Amer., Madison Wisc., 67–75.

Reitemeier R F 1946. Effect of moisture content on the dissolved end exchangeable ions of soils of arid regions. Soil Sci. 61, 195–214.

Richards L A (ed) 1954. Diagnosis and improvement of saline and alkaline soils. USDA Agric. Handbook, no. 60 United States Department Agriculture, US Government Printing Office, Washington DC, 160pp.

Shen J and Hoffland E 2007. In situ sampling of small volumes of soil solution using modified micro-suction cups. Plant Soil 292, 161–169.

Sonneveld C Van den Ende J and Van Dijk P A 1974. Analysis of growing media by means of 1:1½ volume extract. Commun. Soil Sci. Plant Anal. 5, 183–202.

Sonneveld C 1979. The accuracy of analytical data of glasshouse soil testing. Neth J. Agric. Sci. 27, 297–304.

Sonneveld C and De Bes S 1986. Grondonderzoek op basis van waterige extractie. Deel 2. Enkelvoudige lineaire correlaties. Proefstation voor de Tuinbouw onder Glas te Naaldwijk, Intern Verslag nr 31, 18 pp.

Sonneveld C and Voorthuizen E 1988. Monsterfout en analysefout bij het chemisch onderzoek van voedingsoplossingen in steenwolmatten. Proefstation voor Tuinbouw onder glas, Naaldwijk, The Netherlands. Intern verslag, nr 10, 13 pp.

Sonneveld C 1990. Estimating quantities of water-soluble nutrients in soil using a specific 1:2 volume extract. Commun. Soil Sci. Plant Anal. 21, 1257–1265.

Sonneveld C Van den Ende J and De Bes S S 1990. Estimating the chemical composition of soil solutions by obtaining saturation extracts or specific 1:2 by volume extracts. Plant Soil 122, 169–175.

Sonneveld C and Voogt W 1990. Response of tomatoes (*Lycopersicon esculentum*) to an unequal distribution of nutrients in the root environment. Plant Soil 124, 251–256.

Sonneveld C Van den Bos A L Van der Burg A M M and Voogt W 1991. Fertigation in the greenhouse industry in The Netherlands. *In*: Fertigation /Chemigation, FAO, Rome, 186–193.
Sonneveld C and Van Elderen C W 1994. Chemical analysis of peaty growing media by means of water extraction. Comm. Soil Sci. Plant Anal. 25, 3199–3208.
Sonneveld C 1995. Fertigation in the greenhouse industry. *In*: Proc. of the Dahlia Greidinger Intern. Symposium on Fertigation. Technion – Israel Institute of Technology, Haifa, 25 March–1 April, 121–140.
Sonneveld C and De Kreij C 1995. Standardisation of chemical analysis of growing media. Acta Hort. 401, 569–574.
Sonneveld C and De Kreij C 1999. Response of cucumber (*Cucumis sativis L.*) to an unequal distribution of salts in the root environment. Plant Soil 209, 47–56.
Sonneveld C and Voogt W 2001. Chemical analysis in substrate systems and hydroponics – use and interpretation. Acta Hort. 548, 247–259.
Sonneveld C and Voogt W 2009. Determination of micro nutrients by water extraction and interpretation of the analytical data. Acta Hort. 819, 89–98.
Van den Bos A L De Kreij C and Voogt W 1999.Bemestingsadviesbasis Grond. Proefstation voor Bloemisterij en Glasgroente, Naaldwijk, The Netherlands, 54 pp.
Van den Ende J and Knoppert J 1959. Grondonderzoek ten behoeve van het bijmesten. Groenten en Fruit 14, 769.
Van den Ende J 1968. Analysis of greenhouse soils by means of aqueous extracts. Proc. 6th Coll. Intern. Potash Inst. Florence/Italy, 246–255.
Van den Ende J 1988a. Water contents of glasshouse soils at field capacity and at saturation. 1. Relationships between water contents. Neth. J. Agric. Sci. 36, 265–274.
Van den Ende J 1988b. Water contents of glasshouse soils at field capacity and at saturation. 2. Estimating water contents from organic matter and clay contents or from loss-on-ignition. Neth. J. Agric. Sci. 36, 275–282.
Van den Ende J 1989a. Estimating the chemical composition of the soil solution of glasshouse soils. 1. Composition of soil solution and aqueous extracts. Neth. J. Agric. Sci. 37, 311–322.
Van den Ende J 1989b. Estimating the chemical composition of the soil solution of glasshouse soils. 2. Relationships between the compositions of soil solution and aqueous extracts. Neth. J. Agric. Sci. 37, 323–334.
Van der Wees A 1983. Bemonstering en bemesting van grond. Groenten en Fruit, 38(26), 28–29.
Van der Wees A 1993. Bemonstering van grond en substraat. *In:* Plantevoeding in de Glastuinbouw. Proefstation voor Tuinbouw onder Glas te Naaldwijk. Infromatiereeks no 87, 40–43.
Verloo M G 1980. Peat as a natural complexing agent for trace elements. Acta Hort. 99, 51–56.
Vermeulen F H B 1960. Fehlerquellen bei der Bodenuntersuch. Stand und Leistung agrikulturchemischer und agrarbiologischen Forschung. VII, 80–85.
Winsor G W Davies J N and Massey D M 1963. Soil salinity studies. I Effect of calcium sulphate on the correlation between plant growth and electrical conductivity of soil extracts. J. Sci. Fd. Agric. Sci. 14, 42–48.

Chapter 5
Tissue Tests

5.1 Introduction

Tissue tests are widely used in horticulture practice and have in comparison with soil or substrate testing advantages as well disadvantages in comparison with soil testing. One of the main advantages of tissue tests is the certainty that analysed nutrients in plant tissues are really present in the tissue analysed, while analytical data of soil and substrate testing only estimate the availability of nutrients to plants. There is no guarantee that nutrients determined in the root environment as available quantities will be absorbed by plants. Numerous factors can hinder or stimulate the uptake by plants of nutrients determined with soil and substrate testing as being sufficiently available for that. The best known example of such a hindrance or stimulation is the root zone temperature. Low root zone temperature generally reduces shoot tissue concentrations, whereby Ca, P and Mg are mostly more affected than N and K. But shoot tissue concentrations also can be negatively affected by evidently too high root zone temperatures. The effects differ for crops and growing conditions (Daskalaki and Burrage, 1998; Ikeda and Osawa, 1984; Moustafa and Morgan, 1984). Huge effects on nutrient concentrations in plant tissues occur, when crops are grown at root temperature far below the optimum for crop development (Ali et al., 1994). Another example of a factor that can affect the uptake of some minerals is the use of root stocks. The effect of grafting apparently differs for crops and probably rootstocks. Baas (1998) found for rose a stimulation of the uptake of B by grafting, while Edelstein et al. (2005) reported for melons a reduced B uptake for grafted plants when compared with non grafted. The effects also differ for elements. Cabrera (2002) using the same rootstock "Natal Briar" with roses like Baas (1998) found besides a higher uptake of B also higher uptakes of Cl, Na and Mg, while P, K and Fe were significantly decreased by grafting.

Upside down, plant nutrient concentrations often do not supply information about the storage on hand in the root environment (Sonneveld and Welles, 2005). This especially is operative for the often high nutrient levels realised in the root environment for greenhouse crops. Thus, fertilization management based on tissue test have strong restrictions for greenhouse cultivation. This for example will be clear by the curve linear relationship often found between concentrations of nutrients in

C. Sonneveld, W. Voogt, *Plant Nutrition of Greenhouse Crops*,
DOI 10.1007/978-90-481-2532-6_5,

soil or substrates and in tissues of plants grown in it. An example has been shown in Fig. 3.6. In greenhouse cultivations the slope for the relationship between external and internal concentrations is low, especially in the range of high nutrient concentrations in the root environment. This indicates that small differences in plant concentrations can be the reflection of a broad range in concentrations in the root environment. In Fig. 5.1 some examples has been given for relationships between K and Ca concentrations in the root environment and in plant tissues. The most striking effect that determines the plant nutrient concentration is the characteristic of the crop itself. Most plants are able to realise a certain – more or less – optimal internal concentration of nutrients at relatively low external concentrations. A further increase of the external concentration affect the internal concentration only to a small extent, like shown for K for different crops in Fig. 5.1 and for Ca with the kohlrabi crop. It is also possible that there is no further increase of the internal concentration above a certain external concentration, like found for lily and hippeastrum with Ca; indicating that in these cases the internal concentration supplies no any information about the external concentration in a wide range. In the examples given in Fig. 5.1 all cation concentrations were proportionally increased. Under such conditions it is possible indeed, that with increasing external Ca concentration the internal concentration of the plant tissue decreases, which is caused by a reduced uptake or by a hindrance in the internal transport. This can be explained by ion competition, for example a stimulation of the K uptake by proportionally increasing concentrations will reduce the uptake or the transport of Ca (Adams and Ho, 1990; Adams, 1990; Bradfield and Guttridge, 1980 and 1982; Charbonneau et al., 1988; Sonneveld and Voogt, 1990). Comparable results were found with different fruit vegetable crops in a study in which the EC of the external solution was increased by a proportional increase of all nutrients (Sonneveld and Welles, 2005).

There is no distinct general preference for either the use of tissue tests or the use of soil and substrate testing. In some agriculture branches a preference exists,

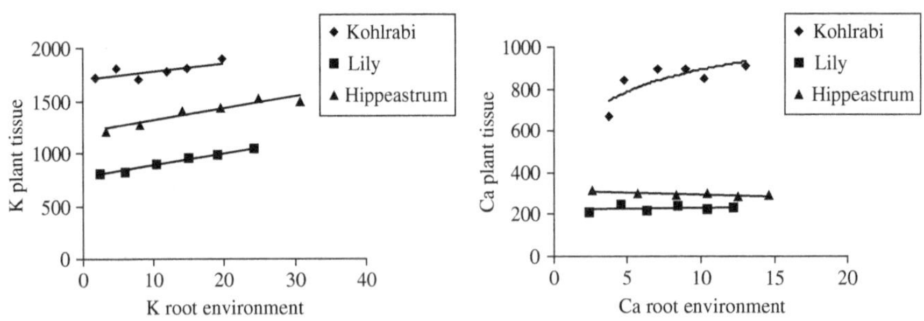

Fig. 5.1 The relationship between the K and Ca concentration in the root environment (mmol l^{-1} substrate solution) and in the above ground plant tissue (mmol kg^{-1} dry matter) for different substrate grown crops. The different K and Ca concentrations in the root environment were realised by a proportional increase of all cations. Data of Van den Bos (1994, 1996, 1997)

sometimes based on history and sometimes on an extended experience build up with one of the systems. However, in greenhouse industry analysis of soil and substrate is plainly prefered. This preference is understandable in view of the extended use and the various applications of the results of such analysis. In greenhouse industry analytical data of soils and substrates are not solely used as basis for fertilizer application, but for many different purposes, like management of the produce quality, leaching requirements for soils and substrates and regulation of the pH in the root environment during crop growth (Sonneveld, 2000; Sonneveld and Voogt, 2001). That does not alter the fact that in the greenhouse industry accidentally a good use is made of tissue tests, viz. to confirm the occurrence of deficiency and toxicity symptoms, to discover transport and distribution problems with nutrients within plants, and to discern effects of growing conditions that affect nutrient uptake, like climatic conditions and physical hindrances in the root environment. Such discernments will improve the fertilization management especially for those elements of which the uptake is strongly affected by growing conditions. This holds for example for Ca, of which the transport in plants is strongly affected by climatic conditions and for Mg and Fe of which the physical conditions in the root environment play an important part in the uptake. Tissue tests are a suitable tool to prevent failures of the interpretation of soil and substrate analysis under such conditions. Furthermore, tissue tests are required for food safety to check the contents of undesirable minerals, like high concentrations of NO_3 and heavy metals and last but not least for research purposes to assess the suitability of soil and substrate tests.

5.2 Distribution of Minerals Within Plants

It is widely determined that different plant organs show huge differences in mineral composition. This, for example, has been shown for more or less all mineral elements for roots, leaves, stems, flowers and fruits. Furthermore, leaf position and leaf age are very important with respect to the concentrations of minerals. In addition, great differences occur within plant organs, for example between distal and proximal parts of fruits, between bases and tops of leaves and between laminae and veins of leaves. In this section some examples of remarkable differences of distributions of minerals in plants will be given in relation to a specific behaviour of some elements in plant tissues.

Among the major elements, Ca shows the most striking effects of differences in the distribution within the plant. The distribution of this element in plants is strongly affected by transpiration and by the xylem and phloem transport. The phloem stream mostly does not contain any Ca or extremely low concentrations, while the xylem stream contains high Ca concentrations. Plant parts with a relatively large evaporation surface, like leaves, are mainly supplied by the xylem and thus, contain much Ca. Fruits, on the other hand, have a relatively low evaporation surface and are mainly supplied by the phloem and therefore contain less Ca in comparison with leaves. Old leaves with a long "evaporation time" and thus, a big "evaporation volume" have a much higher Ca concentrations than young leaves. In

Table 5.1 Concentrations of mineral elements (mmol kg^{-1} dry matter) in young and old tomato and cucumber leaves

Elements	Tomato leaves		Cucumber leaves	
	Young	Old	Young	Old
Na	56	52	48	56
K	1064	967	683	509
Ca	611	1501	1177	1820
Mg	173	255	263	449
N	3607	2393	4086	2593
P	174	194	210	129
Cl	130	45	90	73

After Sonneveld (1980).

this way, the differences in Ca concentrations found within a tomato plant by Adams (1990) varied between 1 and 600 mmol kg^{-1} in the dry matter of the distal ends of fruits and of the laminae of young leaves, respectively. The differences should even be more striking when the comparison was made with old leaves, because the Ca concentrations in old leaves can be double of those in young leaves. Differences in mineral concentrations between young and old leaves are shown by Sonneveld (1980) and summarized in Table 5.1. The most striking differences are the lower the low K and N and the high Ca and Mg concentrations in the old leaves in comparison with the young leaves. The unequal distribution of Ca among plant tissues, as brought about by the internal water transport processes, easily induces deficiency symptoms in therefore susceptible plant organs. Formerly, these phenomena often were denoted as "physiological disorders" and are well known with many crops in greenhouse cultivation. These disorders frequently are connected with climatic conditions. An example is Ca deficiency with cucumber, visible by necrosis on the edges of developing leaves, which especially occurs under humid climatic conditions (Bakker and Sonneveld, 1988; Bakker, 1990). Results are shown in Fig. 9.4.

The very uneven distribution of Mn and B in plant leaves is common knowledge. An example is given in Fig. 5.2 where the distribution of these elements is shown in relation to the distance from the leaf edges for lettuce (Bert and Honma, 1975). Both elements accumulate strongly in the leaf edges, which is in good agreement with the appearance of toxicity symptoms. Such symptoms are characterized by necroses at the leaf tops and edges. Comparable results have been found with plants with long small leaves (Kohl and Oertli, 1961; Benton Jones, 1970). Accumulation as shown for lettuce crops in Fig. 5.2 will depend strongly on the level of supply. A high supply induces strong accumulations in leaf tops, as shown for hippeastrum leaves in Table 5.2, while with lower supply the accumulation in leaf tops is restricted.

Substantial differences in the nutrient concentrations of leaves can occur by course of time in the growing season, like shown for B in sweet pepper leaves in Fig. 5.3 (Bloemhard, 1995). The pepper plants were grown with ample B supply in

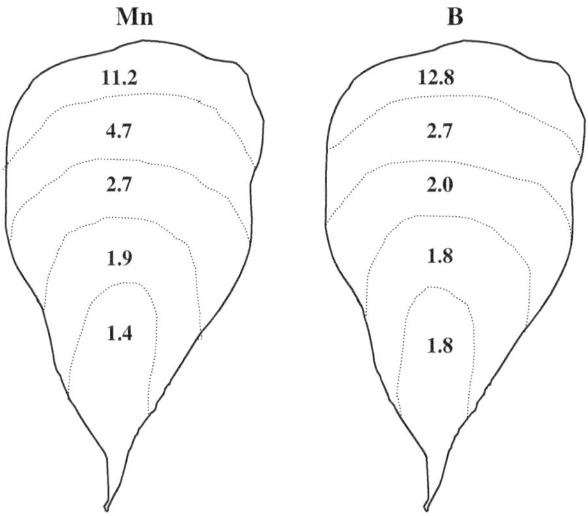

Fig. 5.2 Distribution of Mn and B (mmol kg^{-1} dry matter) in lettuce leaves, after Bert and Honma (1975). *Modified by permission of the American Society Horticultural Science*

the rock wool slabs. The crop was planted in week 5 and leaves of two weeks old were systematically sampled and analysed from week 7. The concentrations in the leaves decrease with the growing season and was lowest between the weeks 12 and 18. Despite a sufficient supply, often B deficiency occurs with sweet pepper in the period that the uptake is lacking. Growth rate and fruit load will play an important part in the course of the concentration.

The level of supply also will affect the organ in which the accumulation occurs. This is shown with the results of an experiment with Zn application in tomato, of which the results are listed in Table 5.3. The data in this table show, that at the lowest supply, being in the deficient range, the laminae received Zn preferentially to petioles. With sufficient and high supply Zn merely accumulates in the petioles. It was concluded that determination of Zn in petioles gave a better discernment for the Zn status of the plant than this determination in laminae (Sonneveld et al., 1986).

Table 5.2 Accumulation of B in hippeastrum leaves cv Appleblossom (mmol kg^{-1} dry matter) at a low and High B supply

	B supply	
Leaf parts	Low	High
Top	5.5	56.0
Middle	2.2	6.7
Basis	2.0	4.0

After Sonneveld (1991).

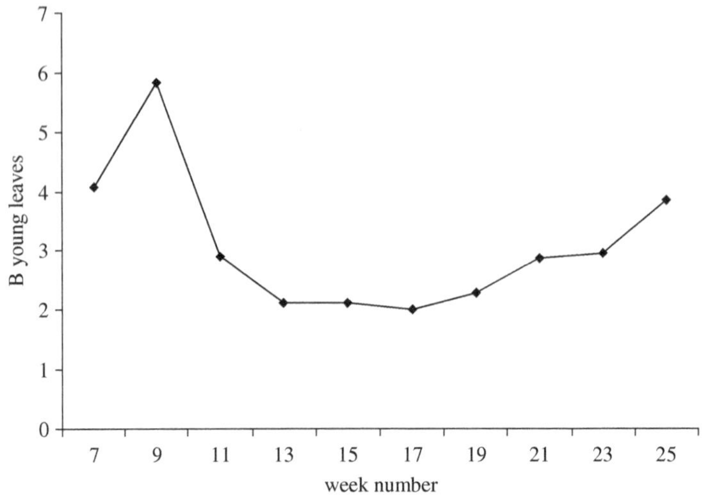

Fig. 5.3 The course of the B concentration of young sweet pepper leaves during the growing season (mmol kg^{-1} dry matter). Data of Bloemhard (1995)

Table 5.3 Accumulation of Zn in laminae and petioles of rock wool grown tomato (mmol kg^{-1} dry matter) in relation to the level of supply

Application (μmol l^{-1})	Laminae	Petioles
2.2	0.15	0.11
6.7	0.22	0.56
10.8	0.26	0.72
15.7	0.30	0.82
22.8	0.32	1.67

After Sonneveld et al. (1986).

5.3 Sampling

The results of the tissue analysis will be strongly affected by the plant part sampled, as will be clear from the foregoing section. Thus, the basis for suitable analytical results is clear-cut and unequivocal sampling instructions. The aim of the sampling should be well considered and established beforehand and the conclusions will be incorporated in the instructions. A routine sampling of a healthy crop due to get a general impression of the nutrient status of a crop differs greatly from a sampling of a crop due to confirm a supposed nutrient disorder.

In the first situation standard procedures can be followed in which a description is given what plant part will be sampled to give best information about the general nutrient status of the crop. This description can vary among crops and possible for the growth stage of the crop, but such instructions do not vary much for growing conditions or elements in charge. For crops showing no disorders following guidelines are given (De Kreij et al., 1992).

- For fruit bearing vegetables young fully grown leaves are sampled
- For leafy vegetables mostly whole plants are taken into the samples. When in this way too big samples are required, like with lettuce heads, parts of plants are taken in that way, that young and old leaves are proportionally represented in the sample
- With potted ornamentals the sample is composed of young fully grown leaves
- With cut flowers sometimes the sampling is adjusted to the crop. For roses for example the upper three five-leaves of mature peduncles are gathered. For carnation the fifth leaf pair of mature peduncles, or possible the fifth leaf pair of young shoots. For cymbidium the second leaf from the outside of young fully grown shoots. Anthurium is sampled by gathering leaves from which the bloom was recently harvested. With other cut flowers young fully grown leaves are gathered.

In the situation described of which the sampling is aimed at a general impression of the nutrient status of the crop, the samples will contain at least 20 plants or parts of 20 different plants. The sampling sites will be randomly scattered over the area in view.

Samplings carried out in relation to disorders of which is expected to be of nutritional origin are quite different from these described before. The sampling site in the field and the sampling place on the plant will be directed by the appearance of the symptoms. The sampling under these conditions will be discussed with the help of some examples.

- Magnesium deficiency mostly occurs in the oldest leaves of plants. Therefore, in this case not the young, but the elderly leaves will be sampled to confirm the disorder. Under certain conditions magnesium deficiency can occur in the middle of the plant (Sonneveld and Voogt, 1991) and then the sample should be composed out of leaves from this plant part. See the data in Table 5.4. In the young leaves no significant differences were found between the Mg concentrations in the disordered and healthy plants, while in the middle leaves the differences are most evident.
- The distribution of some elements will be dependent on the concentration of the element concerned. So, for B and Mn it is well known that with high and toxic levels accumulation occurs in the tops and edges of the leaves, like already shown

Table 5.4 Mg deficiency in tomato plants as affected by the Mg distribution in the leaves. Mg concentrations in mmol.kg^{-1} dry matter

	Mg deficiency	
Type of leaves	No	Heavy
Old	11.5	5.0
Middle	10.6	1.4
Young	7.4	7.0

After Sonneveld and Voogt (1991).

Table 5.5 B distribution in healthy and deficient sweet pepper leaves. Concentrations expressed in mmol kg^{-1} dry matter

	Plant condition	
Leaf part	Healthy	Deficient
Petiole	3.05	2.57
Basis	1.95	1.08
Top	1.86	0.77

After Sonneveld (1991).

in Fig. 5.2 (Kohl and Oertli, 1961; Benton Jones, 1970). In deficient leaves a different distribution is possible (Sonneveld, 1991) like shown in Table 5.5 for sweet pepper leaves. In the deficient leaves, the transport to the top is hindered. In this case the differences in B concentration are most evident when the tissue samples are gathered from the leaf tops, where the deficiency symptoms occurred.

- Well known are the problems with Ca deficiency in fruits (Adams and Ho, 1990; Millikan et al., 1971). In Fig. 5.4 the Ca distribution in sweet pepper fruits affected by blossom-end rot is shown in comparison with those in healthy fruits. The differences between the affected and the healthy fruits are most evident in the distal ends. The Ca deficiency, blossom-end rot, occurs also mainly in the distal end.

It will be clear from the foregoing examples that the sampling place on the plant is most important for the analytical results derived. Beside a sample of the disordered plants, plant organs or parts of plant organs, it is advisable to take also samples from comparable healthy plants, plant organs or parts of plant organs, respectively. Often with the aid of a comparison of data of such coupled samples a correct interpretation is possible. This especially makes sense, when the diagnosis of the symptoms is uncertain, the interpretation of the analytical data is weakly determined and for peculiar crops.

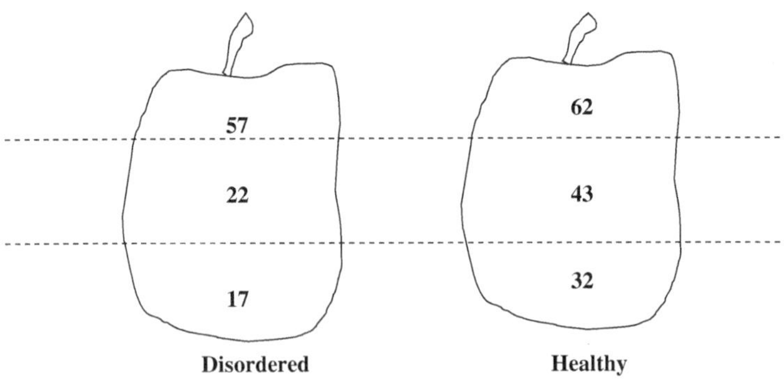

Fig. 5.4 Ca concentrations (mmol kg^{-1} dry matter) in different parts of sweet pepper fruits affected by blossom-end rot (disordered) and healthy fruits, after Sonneveld (1993)

Samples will be gathered in clean plastic bags and as soon as possible transported to the laboratory.

5.4 Pre-treatment and Extraction

The pre-treatment at the laboratory depend on the method of extraction that will be applied. Most plant tissues are severely contaminated with soil, dust, residues from plant protection sprays and residues from overhead irrigations. When the destruction is carried out in strong acid solutions for the determination of total concentrations of mineral elements, soil, dust and other precipitates also will be dissolved and determined by the analysis, which especially results in high concentrations of Fe and Al (Sonneveld and Van Dijk, 1982). For other micro elements the contaminations were less than for Fe and Al, but could significantly affect the results. With major elements no significant contamination could be traced, but will be surely imaginable with overhead fertigation. Thus, cleansing of tissue samples is recommended in the first place for the determination of micro nutrients. Most suitable for this treatment is a solution of 0.1% Teepol. A solution of 0.1 mmol l^{-1} hydrochloric acid is also suitable, but has the disadvantage that afterwards no reliable determination of Cl is possible. Determinations by plant sap analysis will not be affected by contaminations of soil and dust, because no strong destructions are carried out. Soluble mineral salts can affect the results, but cleansing for it is just recommended under specific contaminations, when for example soluble salts are expected. When washed, sticky water from the cleansing process can dilute the pressed sap.

For the determination of total minerals, the samples are dried at 80°C after the cleansing process and milled in a grinder. With these handlings also contamination can occur and therefore, the handlings will be carried out with apparatus free of the elements whereupon the samples will be analysed. Likely contaminations in these processes are Cu and Zn, for example from galvanized trays used for the drying and Cu and Zn fittings of grinders.

For the determination of total concentrations of all nutrients, except N, an extraction with aqua regia is most likely (CEN, 2001a), while for N the Kjeldahl (CEN, 2001b) and Dumas (CEN, 2001c) methods are recommended.

Besides the determination of total nutrient concentrations in plant tissues other weaker extraction methods are employed. Among such methods the preparation of plant saps is the only method focussed on a universal use. Other plant extraction methods were mostly developed in relation to the determination of Fe, because of improvement of the poor interpretation possibilities of the determination of total Fe. Suggestions that most of the Fe in plant tissues is inactive, has given occasion to proposals to determine the so called "active" Fe. To this purpose different extraction methods for the determination of Fe in plant tissues were tested on their suitability, from which diluted solutions of HCl and o-phenanthroline are best known (DeKock et al., 1979; Katyal and Sharma, 1980; Manzanares et al., 1990). The results of such methods did not always give an improvement of the interpretation for Fe in plant tissues and if it did, the interpretation was not unambiguous (Lang and Reed, 1987).

Plant sap analysis has an ancient history (Emmert, 1930), but was never applied on a large scale. Later on it is mostly developed as a quick test (Scaife and Bray, 1977) for specific requirements, because the analytical data did not show improvements for general interpretations in comparison with the analytical results of total analysis. The relationship between quantities extracted by total analysis and those extracted by plant sap varies strongly dependent on crop, type of plant tissue and element (compound) determined (Sonneveld and De Bes, 1983). For easily soluble elements and compounds, like K, Na, Cl and NO_3 quantities extracted with both methods are more or less equal and mostly linearly related. However, the quantities of other elements extracted by both methods will differ strongly. Great differences between quantities extracted with a total destruction and those extracted with plant sap analysis has been found for Ca, Mg, P, N, and some micro nutrients. It is understandable that quantities of elements being a part of the plant tissue matrix are not solved in the plant sap and thus not extracted by the preparation of it. However, this does not mean that in such cases no relationship exists between quantities extracted by a total destruction and those extracted by plant sap analysis. Such is clear from the data presented in Fig. 5.5, where the relationship between analytical data of total Ca contents of carnation leaves are shown in relation to those analysed by plant sap analysis. The relationships show an exponential character, which model can be explained by the fact that with low Ca uptake all Ca becomes part of the matrix of the plant tissue. Only with a higher Ca uptake part of this element remains free in the plant sap. Such relationships are quite different from those shown in Fig. 5.6, where an example is given for the K concentrations in tomato tissues. The relationship is linear and shows that the quantities of K extracted from the different plant tissues by plant sap analysis is about equal to that extracted by total destruction of the dried material.

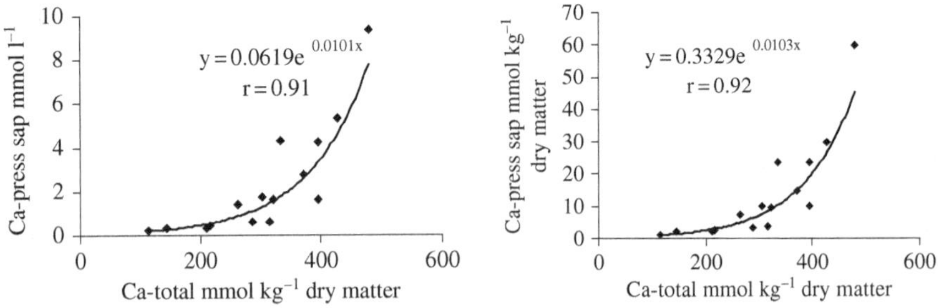

Fig. 5.5 The relationship between Ca concentrations of carnation leaves, determined through digestion of dried material and that determined through plant sap analysis, expressed on plant sap (*left*) or on dry matter (*right*). After Sonneveld and Voogt (1986). *Reprinted by permission of Springer*

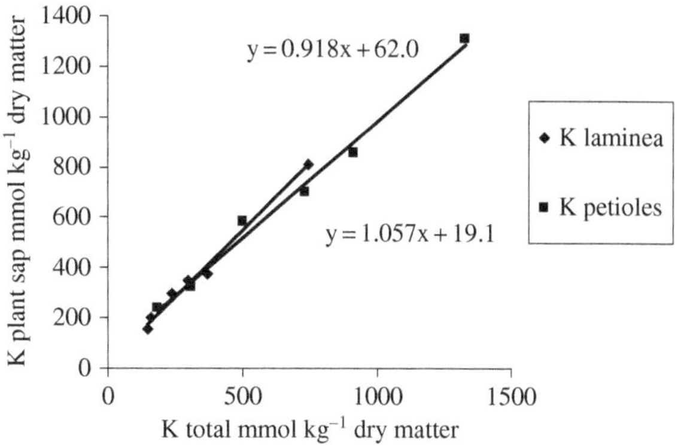

Fig. 5.6 Relationships between the K concentration of tomato tissues determined by destruction of the dry matter and by plant sap analysis (Sonneveld, 1984)

5.5 Interpretation

As already discussed, results of tissue test in greenhouse horticulture are mainly used as a confirmation of nutrient disorders in plants. Incidentally it also is used as a quick test to determine the plant nutrient status to control fertilizer application. However, interpretations of tissue tests are fairly complicated, because they are not unequivocal. Great differences occur between the requirements for optimum production, and for deficient and toxic levels of plant nutrients of different crops, but even for different cultivars of the same crop. Furthermore, the growing conditions of the crop, the growth stage and the plant part sampled will be of great influence on the results of the analytical results and thus, on the interpretation.

In Table 5.6 some examples of data of optimum levels are listed for some greenhouse crops as published by De Kreij et al. (1992). The data in this table clearly demonstrate the differences between crops, but the range given for the different elements per crop show that plants can grow well within a relatively wide range of nutrient concentrations in their tissue. The data in Table 5.6 of lettuce are derived from whole heads and those of the other crops of young fully grown leaves, including the petioles unless otherwise stated. The use of young leaves is often practised with sampling of healthy crops due to a general view on the nutrient status. In Appendix B guide values for optimum concentrations of nutrients of some other greenhouse grown crops are listed.

However, sometimes, plants are sensitive for nutrient disorders in a specific plant part, as discussed in Section 5.3. Another example is shown in Table 5.3, where the Zn concentrations in the petioles show a better discernment of the deficient and toxic condition than in the laminae. Such occurs in different situations, especially for elements with an unequal distribution in plant parts, as shown in examples before. Ca deficiency for example will occur in the younger parts of leaves and especially distal ends of fruits. B and Mn toxicity generally become visible in the tops of older leaves, where these elements are accumulated, while the deficiency symptoms of these elements appear in the younger leaves. It will be clear that the sampling and consequently the interpretation in such cases are focussed on the tissue where the

Table 5.6 Optimum levels of nutrient concentrations in greenhouse crops as given by De Kreij et al. (1992). The determinations are carried out by total destruction of dried material and expressed as mmol kg^{-1} dried material

	Guide values for optimal production			
Elements	Tomato[1]	Lettuce[2]	Rose[1]	Chrysanthemum[1]
K	900–1300	2000–3500	800–900	650–1550
Ca	400–800	200–300	250–450	250–750
Mg	150–200	100–300	90–160	120–400
N	2000–3000	3000–4000	1700–2800	2800–3600
P	100–150	200–250	100–200	100–200
S	400	80–100	–	–
Fe	1.5–2.0	3.0–4.0	1.0–2.7	5.0
Mn	1.0–3.0	1.0–2.0	1.1–2.7	0.4–4.5
Zn	0.6[3]	0.5–3.0	0.3–0.8	0.2–1.5
B	5.0–7.0	3.0–4.0	2.8–5.6	2.3–7.4
Cu	0.10	0.10–0.25	0.08–0.25	–
Mo	0.03–0.06	0.03–0.06	–	–

[1]Determined in young fully grown leaves;
[2]determined in whole heads;
[3]determined in petioles.

disorder will appear. However, no interpretations are available for all plant types and plant parts. With the interpretation of symptoms of disorders, it will be very helpful when comparable samples are gathered from healthy and disordered plants separately as a reference, as discussed in Section 5.3. In such cases, a careful considered sampling method is very important.

An example of the necessity of specific interpretations is the difference in sensitivity among cultivars of lettuce to Mn toxicity, as listed in Table 5.7 (Sonneveld and Voogt, 1975). The different cultivars were grown on a steam sterilised soil and the Mn concentration of whole heads was determined as well the severity of the symptoms. The degree of the symptoms showed great differences, while the Mn concentrations in the tissue of the cultivars only slightly differed and did not reflect the degree of toxicity. Also for other crops great genetic differences of sensitivity for Mn toxicity among cultivars were found (Horst, 1983; Foy et al., 1988; Sonneveld and Koningen, 1973).

Table 5.7 Index figures for Mn toxicity and Mn content of lettuce heads (mmol kg dry matter) for different cultivars. Index for Mn toxicity: 1–3 light, 4–6 moderate and 7–10 severe symptoms

Cultivars	Mn toxicity	Mn in heads
Blackpool	9.0	14.2
Rapide	8.0	14.1
Noran	7.2	12.2
Deciso	6.0	12.0
Deci-minor	4.0	13.2
Plenos	0.2	14.6

Results after Sonneveld and Voogt (1975). *Reprinted by permission of Springer*

Table 5.8 Average values, standard deviation and coefficient of variation of optimal K and Ca concentrations of tomato plant tissues with various dry matter contents. The concentrations on the dry matter are expressed as mmol kg^{-1} and those of the plant sap as mmol l^{-1}

Determination	Plant tissue	Average value	Standard deviation	Coefficient of variation
K dry matter	laminae	1026	239	23
K plant sap	laminae	116	8.4	7.2
K dry matter	petioles	1937	522	27
K plant sap	petioles	131	5.3	4.0
Ca dry matter	laminae	842	180	21
Ca plant sap	laminae	47.3	9.5	20
Ca dry matter	petioles	712	84	12
Ca plant sap	petioles	20.3	7.1	35

Results after Sonneveld and De Bes (1988). *Reprinted by permission of the International Society Horticultural Science*

Another factor that can affect the interpretation of tissue tests is the dry matter content of the tissue. For K it was found that the K concentration of the plant sap was a better index for the K status of the plant than the K concentration of the dry matter. In experiments with tomato grown under different climatic conditions tomato plants were obtained with varying dry matter contents (Sonneveld and De Bes, 1988). The optimum K status over the climatic conditions, which means over different dry matter contents, for laminae and petioles was much more stable for the plant sap concentrations than for the dry matter concentrations, which was shown by the low coefficient of variation for the optimum values. Results of these experiments are listed in Table 5.8. Comparable results have been found for other crops too (Cassidy, 1970; Sugiyama et al., 1985). The results for the Ca determination do not justify preference for plant sap analysis. The coefficient of variation for the optimal values of Ca was surely not lower than those for dried material analysis, as follows from Table 5.8. An explanation for this different behaviour is maybe the fact that K in plants is more or less completely dissolved in the plant sap, while Ca often for the greater part is fixed to the dry matter. In many plants K is the dominant cation in the plant sap and mainly determines the osmotic potential and thus, plants strive after a constant concentration in their sap (Plaut et al., 2005; Sonneveld and Voogt, 2008). Therefore, it is understandable that for interpretation K is preferably expressed on the water content of the plant (De Kreij et al., 1992).

Mineral concentrations of dry matter and plant sap concentrations can be converted to each other following formula (5.1).

$$K_d = \frac{100 - DM}{DM} K_p \tag{5.1}$$

In which

DM = dry matter content of the fresh material
K_p = K concentration of the plant sap
K_d = K concentration of the dry material

Results of such calculations procure information about the differences between quantities of elements soluble in plant sap in relation to total quantities absorbed by plants. For nutrients except K is no prove that plant sap analysis forms a basis for closer interpretations of analytical data of tissue tests.

5.6 Symptoms of Nutrient Disorders

Symptoms of nutrient disorders in greenhouse crops do not differ in principle from those in field crops. Most of the characteristics of nutrient disorders found with field crops also occur in greenhouse crops. So there is no reason for an extended description of these symptoms. However, the specific climatic conditions in greenhouses sometimes induce symptoms accompanied by specific characteristics. Therefore, a short description will be given of the most characteristic symptoms of the nutrient disorders with possible relation to greenhouse conditions. The symptoms differ for crops, growing conditions and growth stage and often cannot unambiguously be described. Sometimes a diagnosis requires comparison of different descriptions and pictures of several crops to make good supposition and even then a tissue test can be necessary to confirm the diagnosis. Especially the diagnosis for the innumerable exotic flower crops grown in greenhouses offers problems in the recognition of nutrient disorders and requires often research to discover the real cause of the problem and the conditions under which they appear.

For the following description of symptoms authors brought in their experiences while reference is made to the presentation of De Kreij (1993). Furthermore, the books of Roorda Van Eysinga and Smilde (1980, 1981) and Winsor and Adams (1987) contain many pictures of nutritional disorders of greenhouse crops. Symptoms of many crops are described also by Chapman (1966).

K – In advance the deficiency appears with coarse chlorotic spots on young leaves. With serious deficiency yellow colouring on the edges of older leaves appear, which later on become necrotic. The symptoms can occur over the full length of the plant.

Ca – In leaves Ca deficiency will appear in the young leaves or tops of the plant. Coarse yellow spots appear on the edges of the young leaves and the growing point can die. The deficiency in leaves mostly is connected with humid growing conditions and a high growth rate. Ca deficiency in fruits sometimes becomes in advance visible by glassy spots, later on changing into the well known necrotic or soft spots on the blossom end of the fruits. It is strongly related to the transport of this element in the plant and a low humidity in the greenhouse aggravates the disorder in the fruit. Ca deficiency is a phenomenon frequently occurring in greenhouse cultivation with a great diversity of symptoms and will extensively discussed in Chapter 9.

High Ca concentrations in fruits promote the appearance of gold specks and green spots on fruits, which appear mostly in the upper part of the fruit. Pictures of different Ca disorders of vegetable fruit crops are presented in chapter 9 and by Savvas et al. (2008).

Mg – Deficiency is characterized by course yellowing in the older leaves. The leaves are cracking and thick.

N – The deficiency of N is characterized by growth reduction, often thin tops and a pale colour of the leaves, with heavy deficiency older leaves prematurely fall from plant.

P – Deficiency is shown by a strong growth reduction. The colour of the leaves is dark green with sometimes a reddish glow. This glow also can appear on the stems. Older leaves will prematurely fall from the plant.

Toxic symptoms occur in specific crops and cultivars with an insufficient control on the P uptake. Leaves and especially cotyledons can show discoloured spots and even chlorosis.

S – S deficiency looks often like N deficiency, showing a pale colour of the leaves over the whole plant.

Fe – Many greenhouse crops can suffer from Fe deficiency. The most characteristic symptom is chlorosis in the tops of the plant. In serious deficiencies the symptoms survive in older leaves too. However, in advance the problem occurs in the top and is grown out with the age of the leaves.

Fe toxicity can occur in crops, but has not been found often in greenhouse crops. The "bronzing" going together with the toxicity occur mainly in water logged soils.

Mn – The symptoms of Mn deficiency occur just like those of Fe deficiency in the tops of plants, where also chlorosis occurs. Therefore, it is often difficult to distinguish the symptoms of Mn deficiency from those of Fe. Utmost, both deficiencies are strongly aggravated by comparable conditions, being a high pH in the root environment. With rose leaf drop occur (Voogt and Sonneveld, 2009).

Mn toxicity is also well known in greenhouse culture. It frequently occurs in soils after steam sterilisation, which will be discussed in Chapter 10. The symptoms will be found mainly in the old leaves merely by necrosis on the tops of long narrow leaves or on the edges of different shaped leaves. In some crops necrosis alongside the veins and yellowing of the small veins is visible. In rose dropping of older leaves occur, while purple spots on the stems are shown (Sonneveld and Koningen, 1973).

Zn – Zn deficiency is characterized by misshapen form of the plant, small leaves and shortening of the internodes, mostly described as "little leave" and "rosetting", respectively. The symptoms often go together with chlorosis.

Zinc toxicity in greenhouse cultivation is well know and often occurred in combination with galvanized greenhouse constructions; dripping of condensation water and the use of rain water for irrigation caught in galvanised gutters. Nowadays, the constructions often are made of aluminium and the symptoms disappeared. Zn toxicity is characterized by necrosis alongside the veins, in advance it often induces Fe or Mn chlorosis.

B – Deficiency of B causes misshapen growing points and fruits. The symptoms in fruits are quite specific, fruits sometimes burst, dry out and seeds become visible. The tissues become cracking, tops will die off and leaves are misshapen when developed. Brown blockings are sometimes visible in the veins of leaves.

The typical symptoms of B toxicity are the appearance of chlorosis and necrosis on the edges or tips of mature leaves.

Cu – Cu deficiency is well known from the use of high moor peat as a substrate. The complex formation in this growing medium between the soluble organic matter and the Cu ion requires high Cu applications to ensure a sufficient uptake of this element in peat substrates. The symptoms of deficiency are characterized by a pale colour of the whole plant, misshapen fruits, an insufficient flower formation and when developed misshapen flowers in ornamental crops. With serious deficiencies old leaves die off.

Cu toxicity can induce Fe deficiency and thus chlorosis occurs in young leaves. Cu is accumulated in the roots and so with tissue analysis due to determine Cu toxicity, often the roots should be analysed for a sound conclusion.

Mo – Mo deficiency merely develops in young plants and the symptoms are often very characteristic. The so called whiptail formations in young leaves are well known by cauliflower. In other crops it induces wilting in older leaves of young plants or chlorosis in mature leaves. Mo deficiency especially occurs at low pH values in the root environment.

Mo toxicity so far is not known in mature greenhouse crops. It is peculiar that in fodder crops high Mo concentrations are rather dangerous for animals than toxic to crops (Marschner, 1997).

For some major elements no toxicity symptoms are mentioned, like for K, Mg, N and S. Generally, in such cases, no specific symptoms are known at high applications of these elements. However, sometimes too high concentrations of these elements induce deficiency of other elements. This is caused by ionic competition, like can happen with high applications of K, Ca or Mg. Best known are those for high additions of K, which aggravate the occurrence of Ca or Mg deficiency. Furthermore, high applications of all macro elements induce a low osmotic potential (high EC) in the root environment and symptoms of high salinity can occur. These symptoms are less specific, but are generally characterized by discolouring and necrosis in oldest leaves (Sonneveld, 1978).

References

Adams P 1990. Effect of salinity on the distribution of calcium in tomato (*Lycopersicon esculentum*) fruit and leaves. *In*: Van Beusichem M L (ed) Plant Nutrition – Physiology and Application. Kluwer Academic Publishers, Dordrecht, 473–476.

Adams P and Ho 1990. Effect of salinity on calcium transport in tomato (*Lycopersicon esculentum*). *In*: Van Beusichem M L (ed) Plant Nutrition – Physiology and Application. Kluwer Academic Publishers, Dordrecht, 469–472.

Ali I A Kafkafi U Sugimoto Y and Inanaga S 1994. Response of sand-grown tomato supplied with varying ratios of nitrate/ammonium to constant and variable root temperatures. J. Plant Nutr. 17, 2001–2004.

Baas R 1998. Wat doet calcium met de kwaliteit van rozen? Vakblad voor de Bloemisterij 53(5), 44–45.

Bakker J C and Sonneveld C 1988. Calcium deficiency of glasshouse cucumber as affected by environmental humidity and mineral nutrition. J. Hort. Sci. 63, 241–246.

Bakker J C 1990. Effect of day and night humidity on yield and fruit quality of glasshouse tomatoes (*Lycopersicon esculentum* Mill.). J. Hort. Sci. 65, 323–331.

Benton Jones J 1970. Distribution of fifteen elements in corn leaves. Commun. Soil Sci. Plant Anal. 1, 27–33.

Bert J S and Honma S 1975. Effect of soil moisture and irrigation method on tipburn and edgburn severity in greenhouse soils. J. Amer. Soc. Hort. Sci. 100, 278–282.

Bloemhard C 1995. Boriumgehalten in het gewas bij paprika. Proefstation voor Bloemisterij en Glasgroente, Naaldwijk, The Netherlands. Proefverslag nr 14, 12pp.

Bradfield E G and Guttridge C G 1980. Night-time salt concentration and calcium transport in tomato fruits. *In*: Long Ashton Res. Sta. Report 1980, 61.

Bradfield E G and Guttridge C G 1982. Uptake and movement of calcium in plants. *In*: Long Ashton Res. Sta. Report 1982, 136–137.

Cabrera R 2002. Rose yield, dry matter partitioning and nutrient status response to rootstock selection. Sci. Hort. 95, 75–83.

Cassidy N G 1970. The distribution of potassium in plants. Plant Soil 32, 263–267.

CEN 2001a. EN 13650. Soil improvers and growing media – extraction of aqua regia soluble elements, 17pp.

CEN 2001b. EN 13654-1 Soil improvers and growing media – determination of nitrogen-Part 1: Modified Kjeldahl method, 10pp.

CEN 2001c. EN 13654-2 Soil improvers and growing media – determination of nitrogen-Part 2: Dumas method, 8pp.

Chapman H D 1966. Diagnostic Criteria for Plants and Soils. University California, Division Agriculture Sciences, Berkeley, CA, 793pp.

Charbonneau J Gosselin A and Trudel M J 1988. Influence de la conductivité électrique de la solution nutritive sur la croissance et le développement de la tomate de serre cultivée avec ou sans éclairage d'appoint. Can. J. Plant Sci. 68, 267–276.

Daskalaki A and Burrage S W 1998. Solution temperature and the uptake of water and nutrients by cucumber (*Cucumus sativus L.*) in hydroponics. Acta Hort. 458, 317–322.

DeKock P C Hall A and Inkson R H E 1979. Active iron in plant leaves. Ann. Bot. 43 737–740.

De Kreij C Sonneveld C Warmenhoven M G and Straver N A 1992. Guide values for nutrient element contents of vegetables and flowers under glass (third edition). Proefstation voor Tuinbouw onder Glas te Naaldwijk, The Netherlands, Series: Voedingsoplossingen voor Glastuinbouw no 15, 69pp.

De Kreij, 1993. Plantevoeding, gebrek- en overmaatverschijnselen. *In*: Proefstation voor Tuinbouw onder Glas te Naaldwijk, The Netherlands, Informatiereeks No 87, Plantenvoeding in de Glastuinbouw (third edition), 75–79.

Edelstein M Ben-Hur M Cohen R Burger Y and Ravina I 2005. Boron and salinity effects on grafted and non grafted melon plants. Plant Soil 269, 273–284.

Emmert E M 1930. A method for the rapid determination of phosphate in fresh plant tissue. Plant Physiol. 5, 413–417.

Foy C D Scott B J and Fisher J A 1988. Genetic differences in plant tolerance to manganese toxicity. *In*: Graham R D et al. (eds), Manganese in Soils and Plants, Kluwer Academic Press, Dordrecht, 293–307.

Horst W J et al., 1983. Factors responsible for genotipic manganese tolerance in cowpea. Plant Soil 72, 213–218.

Ikeda H and Osawa T 1984. lettuce growth as influenced by N source and temperature of the nutrient solution. Sixth Intern. Congr. Soilless Culture Lunteren 1984 Proc., 273–284.

Katyal J C and Sharma B D 1980. A new technique of plant analysis to resolve iron chlorosis. Plant Soil 55, 105–119.

Kohl H C and Oertli J J 1961. Distribution of boron in leaves. Plant Physiol. 36, 420–424.

Lang H J and Reed D W 1987. Comparison of HCl extraction versus total iron analysis for iron tissue analysis. J. Plant Nutr. 10, 795–804.

Manzanares M Lucena J J and Garate A 1990. Iron II determination in leaves of strawberry (Fragaria vesca). *In*: Van Beusichem M L (ed) Plant Nutrition – Physiology and Applications. Kluwer Academic Publishers, Dordrecht, 805–808.

Marschner H 1997. Mineral Nutrition of Higher Plants. Academic press, London.

Millikan C R Bjarnason E N Osborn R K and Hanger B C 1971. Calcium concentration in tomato fruits in relation to the incidence of blossom-end rot. Austr. J. Exp. Agric. Anim. Husb. 11, 570–575.

Moustafa A T and Morgan J V 1984. The effect of root zone warming and air temperature on dry matter accumulation, nutrient uptake and nutrient composition of spray chrysanthemum in NFT. Sixth Intern. Congr. Soilless Culture Lunteren 1984 Proc., 401–419.

Plaut Z Grave A Yehezkel Ch and Matan E 2005. Osmotic adjustment of tomato fruits and leaves grown in sand and irrigated with saline water. Acta Hort. 697, 285–291.

Roorda van Eysinga J P N L and Smilde K W 1980. Voedingsziekten bij chrysant. Proefstation voor Tuinbouw onder Glas, Naaldwijk, The Netherlands, 42pp.

Roorda van Eysinga J P N L and Smilde K W 1981. Nutritional Disorders in Glasshouse Tomatoes, Cucumbers and Lettuce. Centre for Agricultural Publishing and Documentation, Wageningen, The Netherlands, 130pp.

Savvas D Ntasi G and Passam H C 2008. Plant nutrition and physiological disorders in greenhouse grown tomato, sweet pepper and eggplant. Europ. J. Sci. Biotechn. Vol 2, 45–61.

Scaife M A and Bray B G 1977. Quick sap tests for improved control of crop nutrient status. ADAS Q. Rev. 27, 137–145.

Sonneveld C and Koningen A 1973. Mangaanvergiftiging bij roos. Vakblad Bloemisterij 28(30), 11.

Sonneveld C and Voogt S J 1975. Studies on the manganese uptake of lettuce on steam-sterilised glasshouse soil. Plant Soil 42, 49–64.

Sonneveld C 1978. Normen voor het zoutgehalte van gietwater. *In*: Waterbehandeling en Waterkwaliteit voor de Glastuinbouw. Proefstation voor de Groenten en Fruitteelt onder Glas, Naaldwijk, The Netherlands. Informatiereeks no 50, 2–5.

Sonneveld C 1980. Gewasonderzoek op basis van plantesap. Practijkonderzoek 1979. Proefstation voor Tuinbouw onder Glas te Naaldwijk. Intern verslag no 18, 37pp.

Sonneveld C and Van Dijk P A 1982. The effectiveness of some washing procedures on the removal of contaminants from plant tissue samples of glasshouse crops. Commun. Soil Sci. Plant Anal. 13, 487–496.

Sonneveld C and De Bes S S 1983. Relationship between analytical data of plant sap and dried material of glasshouse crops. Commun. Soil Sci. Plant Anal. 14, 75–87.

Sonneveld C 1984. Effecten van verschillende kationen verhoudingen bij vleestomaten geteeld in voedingsfilm. Proefstation voor Tuinbouw onder Glas te Naaldwijk, Intern verslag no 58, 9pp.

Sonneveld C De Bes S S and Voogt W 1986. Zinc uptake and distribution in tomatoes grown in rockwool. Soilless Cult. 2(2), 49–60.

Sonneveld C and Voogt W 1986. Supply and uptake of potassium, calcium and magnesium of spray carnations (*Dianthus caryophyllus*) grown in rockwool. Plant Soil 93, 259–268.

Sonneveld C and De Bes S S 1988. Interpretation of analytical data of tissue tests. Acta Hort. 222, 147–153.

Sonneveld C and Voogt W 1990. Response of tomatoes (*Lycopersicon esculentum*) to an unequal distribution of nutrients in the root environment. Plant Soil 124, 251–256.

Sonneveld C 1991. Borium in wortelmilieu en plant. Proefstation voor Tuinbouw onder Glas, Intern verslag nr 19, 18pp.

Sonneveld C and Voogt W 1991. Effects of Ca-stress on blossom-end rot and Mg-deficiency in rockwool grown tomato. Acta Hort. 294, 81–88.

Sonneveld C 1993. Gewasonderzoek. *In*: Proefstation voor Tuinbouw onder Glas te Naaldwijk, The Netherlands, Informatiereeks No 87, Plantenvoeding in de Glastuinbouw (third edition), 62–68.

Sonneveld C 2000. Effects of salinity on substrate grown vegetables and ornamentals in greenhouse horticulture. Thesis Wageningen University, Netherlands, 151pp.

Sonneveld C and Voogt W 2001. Chemical analysis in substrate systems and hydroponics – use and interpretation. Acta Hort. 548, 247–259.

Sonneveld C and Welles G W H 2005. Cation concentrations of plant tissues of fruit-vegetable crops as affected by the EC of the external nutrient solution and by humidity. Acta Hort. 697, 377–386.

Sonneveld C and Voogt W 2008. Nutrient concentrations of plant tissues of greenhouse crops as affected by the EC of the external solution. Acta Hort. 779, 313–320.

Sugiyama N Shidachi K and Okada K 1985. Evaluation of critical K concentration expressed on tissue water basis in spinach. J. Jpn. Soc. Hort. Sci. 54, 336–343.

Van den Bos A L 1994. EC in relatie tot het type substraat bij koolrabi in een gesloten teeltsysteem. Proefstation voor Tuinbouw onder Glas, Intern Verslag, 27, 13pp.

Van den Bos A L 1996. EC in relatie tot het type substraat bij amaryllis in een gesloten teeltsysteem. Proefstation voor Bloemisterij en Glasgroente, Rapport 55, 30pp.

Van den Bos A L 1997. EC in relatie tot het type substraat bij lelie in een gesloten teeltsysteem. Proefstation voor Bloemisterij en Glasgroente, Rapport 81, 34pp.

Voogt W and Sonneveld C 2009. The effects of Fe-chelate type and pH on substrate grown roses. Acta Hort. 819, 411–417.

Winsor W and Adams P 1987. Diagnosis of Mineral Disorders in Plants. Vol 3 Glasshouse Crops. Her Majesty's Stationary Office, London.

Chapter 6
Water Uptake and Water Supply

6.1 Introduction

The water uptake and the water supply do not directly affect the mineral absorption of plants. However, many connections exist between the management of minerals and water. The most evident of those connections are the following.

- Water plays an important part with the transport of minerals in the root environment, thus, from surroundings to the root surface, and it is also important for the nutrient transport in the plant.
- The water management is also important with the vertical transport of salts in the root zone and with this for the leaching of residual salts and nutrients from the root zone to the deep ground water or surrounding surface water. Thus, it plays an important part in the operational efficiency of the management of minerals in greenhouse soils and substrates.
- The additions of nutrients in the greenhouse industry occur mainly with top dressings by fertigation, which means a combined addition of water and nutrients.
- Water often contain minerals, which partly can be estimated as residual salts, like Na and Cl, but for another part is appreciated as nutrients, like Ca, Mg, and SO_4, which directly affect the addition of nutrients in fertilization programmes. However, last group will be denoted as residuals when the concentrations in the irrigation water exceed those of the uptake concentrations.
- The ratio between the uptake of nutrients and water, denoted as uptake concentration, is used as a basis for nutrient supply. Despite, that the uptake concentration has no real physiological basis in the nutrition of plants, it offers valuable information. This because of the experience that the variations in the uptake concentrations are less than those of the absolute values of nutrient uptakes (Savvas and Lenz, 1995).

The factors mentioned are more than enough reasons to add a chapter about water supply to this book about plant nutrition. The water supply in the greenhouse industry is solely carried out by artificial irrigation and thus the control of it can easily be included in the management of the greenhouse. The water use and the water supply will not entirely be discussed in this chapter, but some guide lines in relation to the above mentioned factors will be presented.

C. Sonneveld, W. Voogt, *Plant Nutrition of Greenhouse Crops*,
DOI 10.1007/978-90-481-2532-6_6,

6.2 Water Uptake by Plants

Water uptake by plants in greenhouses is entirely studied by Stanghellini (1987). In this study it was concluded that the transpiration rate is almost proportional to the leaf area and that in the greenhouse climate the radiation, the ambient temperature and the humidity play a prominent part. To a minor extent the temperature of the greenhouse cover and of the soil surface also play a part.

On basis of the factors affecting the water absorption of plants, models have been developed with simple and quickly measurable parameters. The stipulation that the parameters should be simple and quickly measurable is suggested by a practical suitability to irrigate on basis of the everyday water absorption of the crop. Nowadays, when many crops are grown in substrate systems, these requirements on the parameters are accentuated, because of the small water storage in the root environment of these systems and the need to keep this storage on a reasonable level. In such systems often water is supplied several times per hour and thus, the development of a method furnishing quick and preferably secure estimations of the water uptake are evident. Such estimations are developed by De Graaf (1988) on basis of the global radiation measured outside the greenhouse, the use of the heating system in the greenhouse, the air temperature in the greenhouse and the plant size. In formula (6.1) the relationship between the water uptake of the crop and the parameters mentioned is given.

$$E = \frac{h}{m}\left\{aT_gR + b\sum_{i=1440}^{i=1} \min_i(T_t - T_a)\right\} \tag{6.1}$$

In which

E = estimated water uptake of the crop l m^{-2} day^{-1}
h = actual height or size of the crop
m = minimum height or size of the crop with which the maximum transpiration can be realized. When h>m a value of 1.0 should be used for the quotient h/m
a = emperical crop specific factor
R = the global radiation measured outside the greenhouse in kJ cm^{-2} day^{-1}
T_g = the relative light transmission of the greenhouse
b = specific crop factor attributable to heating
$\min_i$ = the successive minutes during the day that there is a difference between the ambient temperature in the greenhouse and the heating tubes
T_t = the temperature of the heating tubes
T_a = the ambient temperature in the greenhouse

The water uptake as calculated with the formula presented can be divided into the part that is used by the plant for transpiration and the part that is used for plant growth. The last mentioned part is small in comparison with the first mentioned. In Fig. 6.1 the differences between both parts of the water uptake is shown as has been found for a tomato crop (Voogt et al., 2006b) for a period in early spring and a period in summer under Dutch conditions. Another part of water use is the

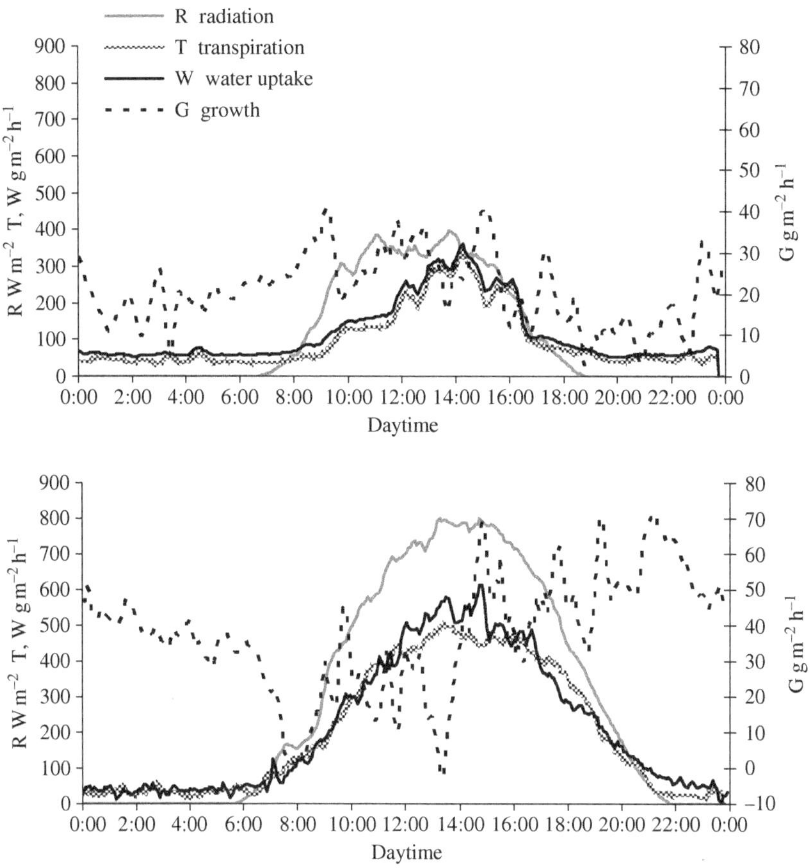

Fig. 6.1 Typical daily pattern of the measured total water uptake, transpiration and plant growth of a tomato crop in g m^{-2} h^{-1}, as affected by the radiation measured in W m^{-2} over 5 minutes intervals, for a day in March (*top*) and a day in July (*down*). Both figures represent the averages of a period of 7 days. Results Voogt et al. (2006b)

evaporation from the soil and substrate surfaces. This only is of interest when soil and substrate surfaces are uncovered and especially when the whole soil surface regularly is irrigated, like with sprinkler irrigation and ebb and flow systems. This evaporation is often included in the measurements of the transpiration of the crop, because the water evaporated is added with the irrigation. It depends on the manner of the measurements whether the evaporation is included or excluded in the crop factor. Mostly the difference between the transpiration and the water use including the evaporation is small, like demonstrated by De Graaf and van den Ende (1981) and shown in Fig. 6.2.

The advantage of the model presented affords the possibility of *ad hoc* applications by introduction of crop specific factors. Examples of crop specific factors are listed in Table 6.1. Most of these values are related to the global radiation measured

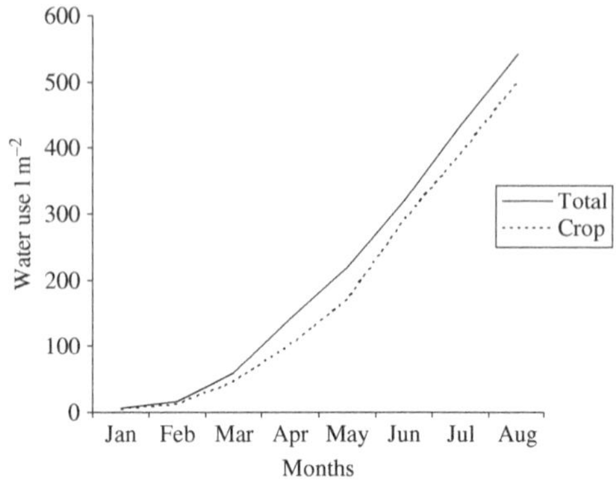

Fig. 6.2 Cumulative water use during a soil grown tomato cropping, including the evaporation of the soil surface and the water uptake of the crop in l m^{-2} (De Graaf and Van den Ende, 1981). *Reprinted by permission of the International Society Horticultural Science*

outside the greenhouse. Therefore, the values are adjusted with light transmission of the greenhouse construction. These adjusted values will be used and the light transmission of the concerned greenhouse will be fit in the calculations, like done in formula (6.1). For the crop factor "b" attributed to the heating a value of 0.22 10^{-4} was found for tomato under Dutch growing conditions (De Graaf, 1988), while these factor also was used for cucumber (De Graaf and Esmeijer, 1998).

The minimum plant height with which the maximum transpiration is realised, the factor m also will be known. This is a factor represented for the leaf area index (LAI), being an important factor in plant transpiration (Stanghellini, 1987). For practical application a plant is estimated as being mature with respect to the transpiration when LAI >3 (Voogt et al., 2006a). However, the LAI cannot be measured under growing conditions and therefore, is often estimated by the plant size. The actual plant size is related to the minimum size with maximum transpiration capacity. This value m in formula (6.1) is related to a certain plant height. For a row crop like tomato m is estimated on a height of 1.5 m. For crops with a high planting

Table 6.1 Crop type factor "a" {M(a)} measured in a greenhouse with a light transmission (Tr_R) and related to global radiation measurements outside the greenhouse. The data are recalculated to a full light transmission in the greenhouse {C(a)}

Crop	M(a)	Tr_R	C(a)	Source
Tomato	1.78	0.65	2.74	De Graaf (1988); De Graaf (1993)
Sweet pepper	1.70	0.65	2.62	De Graaf (1988)
Cucumber	2.00	0.65	3.08	De Graaf and Esmeijer (1998)
Radish	1.54	0.65	2.37	Sonneveld (1995)
Radish	1.83	0.65	2.82	Van der Burg (1994)
Chrysanthmum	2.28	0.65	3.51	De Graaf (1988)
Rose[1]	–	–	3.40	Baas and Van Rijssel (2006)
Cymbidium	0.6	0.72	0.83	Voogt and Van Winkel (2008)
Alstroemeria	1.1	0.75	1.46	Voogt and Van Winkel (2005)

[1]Related to the radiation inside the greenhouse.

density, like chrysanthemum, a height of 0.25 m was suggested (De Graaf, 1993). However, Voogt et al. (2000) mentioned a height of 0.4 m for this crop in a more recent publication.

De Graaf (1988) supposed that addition of the results of measurements of the stoma resistance and the vapour pressure deficit should improve the estimation of the crop transpiration, like suggested by Stanghellini (1987) and Marcelis (1987). Therefore, Baas and Van Rijssel (2006) studied in an experiment the effect of different factors on the transpiration of a full grown rose crop. They measured the global radiation inside the greenhouse, the energy from heating under the canopy and the vapour pressure deficit (VPD_{air} and $VPD_{laef\text{-}air}$). They concluded that the transpiration can be estimated from the global radiation and the contribution of the heating and that the addition of the VPD_{air} or $VPD_{laef\text{-}air}$ did not improve the estimation of the transpiration.

In modern greenhouses often artificial lighting and screening is applied. The screen can be a thermal screen used for energy saving which for the greater part transmit the light, or a full screen used for day length adjustment which exclude all radiation. In both cases the global radiation (R) must be corrected according to the screening time intervals and a specific reduction factor for the radiation of the actual screens. This factor (s_t in formula 6.2) is zero for the full screen. In case of artificial lighting the effective radiation also is taken into account. Therefore, the estimation of the transpiration by the model presented in formula (6.1) is modified. Voogt et al. (2006a) presented a formula for these adjustments, like shown in formula (6.2).

$$E = \frac{h}{m}\left[a\left\{T_g\left(\sum_{k=1440}^{k=1}\frac{\mathrm{min_k}}{1440}R + \sum_{m=1440}^{m=1}\frac{\mathrm{min_m}}{1440}s_t R\right) + \sum_{n=1440}^{n=1}\frac{\mathrm{min_n}}{1440}R_a\right\} + b\sum_{i=1440}^{i=1}\mathrm{min_i}(\mathrm{T_t}-\mathrm{T_a})\right] \tag{6.2}$$

In which:

$\mathrm{min_k}$ = time that no screen in the greenhouse is used in min day^{-1}
$\mathrm{min_m}$ = time that only the thermal screen in the greenhouse is used in min day^{-1}
$\mathrm{min_n}$ = time that the artificial lighting is in operation in min day^{-1}
s_t = the relative light transmission of the screen in the greenhouse
R_a = effective radiation from artificial lighting during operation in kJ cm^{-2} day^{-1} as calculated from formula (6.2a)
The other parameters as indicated with formula (6.1).

R_a can be calculated following installed (used) capacity (De Graaf and Spaans, 1998; De Graaf et al, 2004; Houter, 1996), following formula (6.2a).

$$R_a = 0.75\ P\ 3.6\ h_a \tag{6.2a}$$

In which:

R_a = as indicated with formula (6.2)
P= installed capacity in W m^{-2}
h_a= hours of operation

The formulae presented are based on optimal growing conditions concerning water supply, plant nutrition and climatic conditions in the greenhouse. However, some factors not mentioned in the formulae are well known as affecting the water use of greenhouse crops. The best known are salinity in the root environment and the CO_2 concentration of the air.

With respect of salinity the suppositions are often based on the misunderstanding that a low osmotic potential (high EC) in the root environment reduces the water uptake by plants. An abrupt increase of the salinity decreases the water absorption indeed (Van Ieperen, 1996), but over long periods big differences in water use of plants grown at different salinities have not necessarily been found, as long as the transpiration capacity, like for example the leaf area, is not strongly affected by the salinity. This has been found for tomato as shown in Table 6.2 (Sonneveld, 2000) and for radish (Sonneveld and Van den Bos, 1995). Despite that the yield of the tomato was significantly reduced by an increased EC in the root environment as shown in Table 6.2, the water absorption was not affected. Thus, the well known wilting after a heavy osmotic shock only occur for a short period and plants quickly adjust their water adsorption and will not show a further water stress. However, this does not alter the fact that in many cases a reduced water adsorption has been observed with increasing salinity (Baas et al., 1995; De Kreij and Van den Berg, 1990; Yaron et al., 1969). This is caused by adaptation of plants to salinity stress, like a reduction of the leaf area, and not directly by a hindrance of water absorption (Lagerwerff and Eagle, 1962).

With respect to increased CO_2 concentrations it has been found that the transpiration is affected only to a small extent, which under practical conditions was restricted to some percentages (Nederhof, 1994). Eggplant was most sensitive and showed periodical a reduction of 15% over certain periods, but over the whole growing season a reduction of 4% was calculated.

Table 6.2 Water absorption of tomato plants grown at different EC values in a recirculation nutrient solution. The growing period was from June until September

EC dS m^{-1}	Relative yield	Water used l m^{-2}
1.5	99	258
2.5	100	264
3.5	96	276
4.5	89	252

Data from Sonneveld (2000).

6.3 Variations of Uptake and Supply

Many different irrigation systems are available in the greenhouse industry. They can be globally distinguished in following groups.

- Systems due to spot or strip irrigation, like drip irrigation systems and mini sprinklers. This group is characterized by local wet spots or strips in the greenhouse where the water is supplied, while the remaining part of the surface stays dry. The crop is not wetted during irrigation (Van den Ende and De Graaf, 1974).
- Systems with which the whole area of the greenhouse is irrigated, like with high level sprinkler irrigation, see Picture 6.1. The spray lines with nozzles are placed in top of the greenhouse. The whole greenhouse area and also the crop canopy are wetted with any irrigation.
- Circulating systems, in which the water is continuously circulated in a thin layer, like in (NFT) nutrient film technique systems (Graves, 1983) or in thick layer, like in deep flow cultures (Maloupa, 2002). The plant grows directly in the water stream or water layer.
- Ebb and flow systems for plants grown in containers or in pre-shaped blocks. With these systems the water is supplied in a thin layer for a relatively short period. During this period, sufficient water is absorbed by the substrate and afterwards the system is drained. This system is often used for potted plants on tables and in gullies and for plant propagation on concrete floors.

Picture 6.1 Irrigation by overhead sprinkling in a greenhouse with soil grown chrysanthemum

The water uptake among individual plants varies strongly, as well as the local water supply of irrigation systems. This has been shown by studies of Van der Burg and Hamaker (1987) and of Van Schie et al. (1982) for drip irrigation systems. In the study of Van der Burg and Hamaker (1987) the water supply and the drainage water was measured on different spots in a greenhouse grown with tomato on rock wool slabs during a period from March until September. The spots measured held two tomato plants and both plants were supplied with a dripper. The average coefficient of variation of the water supply increased during the season from 4.8 till 13.2%, while the coefficient of variation of the water uptake was not increased during the season and on average amounted to 10.2%. Because the measured spots held two plants and thus, also two drippers, follow that for a single dripper or plant the coefficient of variation increases with a factor $\sqrt{2}$. This resulted in a value for the coefficient of variation from 6.8 till 18.7% for the water supply and 14.4% for the water uptake. The drainage water is affected by the variation of the uptake by the plant as well as by the variation of the supply. Those variations are independently and thus, the variation of the drainage water will be calculated as following formula (6.3).

$$s_d = \sqrt{s_s^2 + s_u^2} \tag{6.3}$$

In which

s_d = the standard deviation of the drainage water
s_s = the standard deviation of the water supply
s_u = the standard deviation of the water uptake of the plant

Results of such calculations derived from the data in Fig. 6.3 are summarized in Table 6.3. Under practical conditions it can be supposed for example that a plant can use water from the plants left and right besides to compensate a possible shortage. In this case the coefficient of variation will be divided by a factor $\sqrt{3}$ resulting from an average reaction of 3 plants together. Furthermore, plants will not suffer from drought, thus the drainage always will be ≥ 0. This will be approximated with a confidence limit of for example 1% ($P \leq 0.01$), which means that on average not more than one on a hundred plants will suffer from drought. This agrees with a standard normal distributed unit T of 2.33. In this way the drainage will be calculated in relation to the water uptake by formula (6.4).

$$d \geq T\frac{s_d}{\sqrt{p_n}} \tag{6.4}$$

In which

T = standard normal distributed unit agreeing with a determined confidence limit
d = the quantity of drainage water, required as oversupply
p_n = the number of plants that can equalize the water mutually and the value mostly will vary between 1 and 3
s_d = standard deviation of the drainage water as given in formula (6.3)

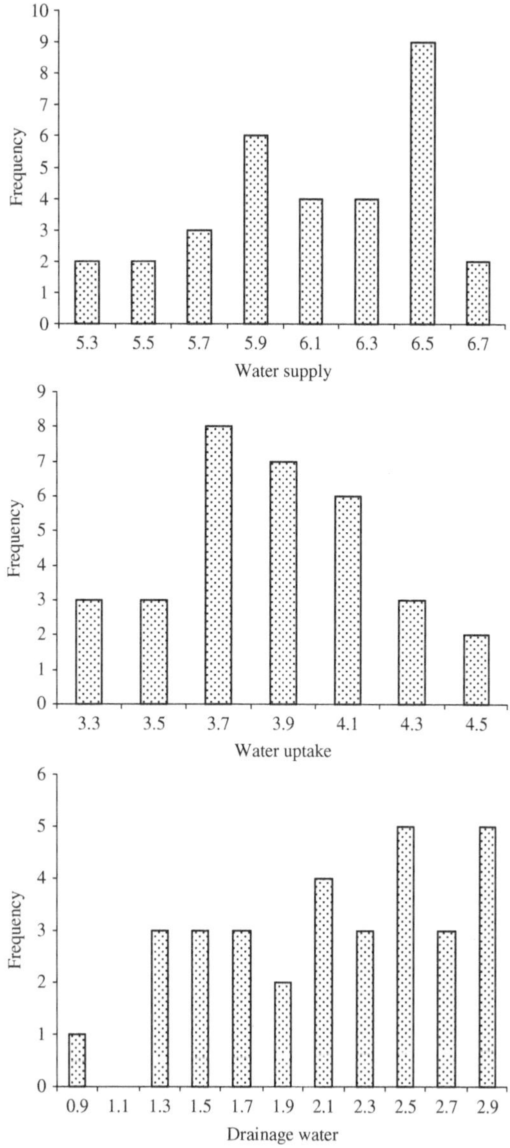

Fig. 6.3 Variation in water supply of drippers, water uptake by tomato plants grown in rock wool and drainage water as has been found on 32 sites in a greenhouse. The quantities are expressed as l m^{-2}. Data derived from Van der Burg and Hamaker (1987)

In this formula s_d will be calculated from formula (6.3) in which s_s will be found by iteration, when the water uptake is determined. Calculations of the required drainage with formula (6.4) were carried out for a water uptake of 2 l, a vc_u = 10% and p_n values 1–3. The required oversupply can be expressed on the supply, like shown in Fig. 6.4, and this fraction is independent of the level of the supply. In Fig. 6.4 it is presented as the "inequality fraction". The value strongly increases with the coefficient of variation of the irrigation system, especially in the single plant

Table 6.3 Average values and variation, standard deviation (s) and coefficient of variation (cv), calculated from the data in Fig. 6.3 for the data measured at the spots of two plants and as estimated for a single plant. The water quantities are given as $l\ m^{-2}$ and $l\ plant^{-1}$ for the two plant spots and single plant situation, respectively

	Spots of two plants			Single plant		
Parameter	Supply	Uptake	Drainage	Supply	Uptake	Drainage
Average	6.1	3.9	2.2	3.1	1.9	1.1
s	0.40	0.32	0.57	0.28	0.23	0.40
cv	6.6	8.4	26.7	9.3	11.9	37.4

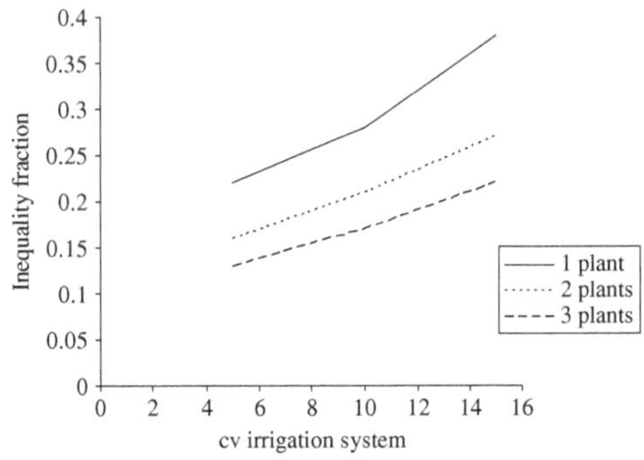

Fig. 6.4 Inequality fraction as affected by the coefficient of variation of the drip system and the number of plants that mutually can profit from the water supplied. The confidence limit for the drainage water ≥ 0 is put on $P = 0.01$ and the coefficient of variation of the water uptake of the plant is put on 10%. See also the text

situation and soon reaches a value of 40%. This agrees well with experiences under practical conditions, because the coefficient of variation of drippers easily increases to values over 20% (Van Schie et al., 1982). The plant factor (p_n) depends on factors like the growing conditions and the plant age. Plants need time to adjust their root system on the water supply. For example, in the situation of a row crop with one nozzle per plant, in the beginning each plant is dependent of the water supply of the one nozzle placed near the plant. Later on when the root system extends, mutual utilization of the water supplied by three or possible more nozzles can be supposed.

Variation in the water supply does not occur just with drip irrigation systems, but also were found with sprinkler irrigation systems. Sonneveld (1995) reported work of Heemskerk, in which the distribution of the water supply by sprinkler irrigation systems were measured on small sites with areas of 0.2 × 0.25 m. The coefficient of variation of the precipitation in the sites was 22%. In later measurements coefficients of variation for the water release of sprinkler installation up to 20% were common practice (Heemskerk et al., 1997). Thus, also with sprinkler irrigation systems an ample water supply is necessary to equalize effects of dry spots. Calculations in a model with an unequal distribution of the water supply of a sprinkler irrigation system for a chrysanthemum crop learned that with a coefficient of variation of the sprinkler irrigation system of 27% an overdose of water 22% is necessary to supply all plants sufficiently with water (Assinck and Heinen, 2001). This resulted to a leaching fraction of 20% of the water supplied, which is in good agreement with experiences in practice (Sonneveld, 1993b). A different option in the calculations

Table 6.4 Chemical composition of precipitated materials from drippers of different greenhouse holdings. Loss on ignition in % of the dry matter and elements as mol kg^{-1} dried material. Where – is given the element was no substantial part of the precipitated materials

Sample nr	Loss on ignition	Ca	P	Fe	Al	Si	S
1	42	–	2.7	3.6	–	0.4	–
2	21	–	0.2	0.4	0.4	7.6	–
3	26	5.6	5.4	–	–	–	–
4	17	5.6	–	–	–	–	5.8

of Assinck and Heinen was the addition of so much water, that nearly no leaching occur, with as a consequence that part of the plants suffer from drought. Last option mostly is not accepted for economic reasons in the greenhouse industry.

Besides the variation inherent in the design and the technical lay-out of the irrigation system, the unequal water distribution is strongly aggravated by clogging of drippers and nozzles. This clogging often is caused by precipitation of constituents from the primary water used and from the fertilizers added or from the growth of micro organisms. In an investigation the composition of precipitated materials was analysed from drippers of different greenhouse holdings in The Netherlands (De Bes, 1986). In the precipitated material substantial quantities of ortho-P, Ca, Fe, Al, Si, S and organic material were found, as follows from the data listed in Table 6.4. The origin of some constituent can be explained by the addition of fertilizers, like P and Ca. A combination of these elements easily precipitates at higher pH (> 6.5) values. Fe, Al and Si are highly represented in dust and clay particles, but Ca, Fe and Si are often available in primary water and can be a precipitate also from this origin. Organic matter can occur in the primary water, especially in surface water. This soluble organic matter will precipitate by changes of the pH or by addition of cations with the fertigation practices. In well water organic matter easily can occur by the development of bacteria. The growth of some specific species is strongly promoted by high methane concentrations sometimes found in this type of water (De Kreij et al., 2003). The bacterial slime build up in such cases also is traced as the cause for clogging and is determined in the analysis as organic material.

Logic components that occur in the precipitates in are $Ca_3(PO_4)_2$ or $CaHPO_4$ with possible equivalent compounds of Fe or Al, furthermore H_4SiO_4, $Fe(OH)_3$, $Al(OH)_3$ and $CaSO_4$, are likely. All compounds possible will precipitate together with crystalline water, which will be part of the loss on ignition determined. Sample nr 3 in Table 6.4 for example, clearly shows a precipitate of $CaHPO_4$ and sample nr 4 a precipitate of $CaSO_4$. The other samples existed of more complex components.

6.4 Water Quality

The quality of the irrigation water with respect to the mineral composition affects the water supply. When the concentration of any mineral is higher than the uptake concentration, the residual salt accumulates in the root environment and will be

Picture 6.2 A basin for storage of rain water

leached by extra water supply. Na and Cl are the ions often abundantly present in water, but sparingly absorbed by most greenhouse crops. Therefore, these ions often determine the leaching requirements. However, in specific cases other ions, like Ca, Mg or SO_4 also can control the need of leaching. In Table 7.13 some examples of the composition of irrigation water are listed in comparison with the uptake concentrations of minerals of two greenhouse crops.

Approximate concentrations for irrigation water with which it is possible to grow greenhouse crops without salt accumulation in the root environment are listed by Sonneveld (1993a). A review of such data is presented in Table 6.5. The precise concentrations acceptable for specific crops vary, because of the great differences in the uptake concentrations. For most crops Na is more critical than Cl, because of the lower uptake of this element by most crops. From data presented by Sonneveld (2000) for Na and Cl at low concentrations in the root environment (< 5 mmol l^{-1})

Table 6.5 Limits for ion concentrations in primary irrigation water whereby no accumulation of the ions mentioned occurs in the root environment

Minerals in the water	Limits
Na mmol l_{-1}	0.2–1.0
Cl	0.3–1.5
Ca	0.7–2.0
Mg	0.3–0.7
SO_4	0.5–1.5
Mn μmol l^{-1}	5–15
B	10–20
Zn	3–5

Data after Sonneveld (1993a).

for Na a range can be given from virtually 0.0 mmol l^{-1} for a rose crop until 1.8 mmol l^{-1} for a winter grown radish crop. The value of 1.8 mmol l^{-1} for winter grown radish is quite exceptional and connected with the low water uptake under Dutch climate conditions in winter. For most crops the uptake of Na and Cl is significantly increased at higher concentrations in the root environment, with exception of crops with a very low uptake, like rose. The uptake does not only differ among crops, but can differ also for cultivars, like shown in Fig. 6.5 for the uptake of Cl with alstroemeria (Sonneveld, 1988). The uptake also will be affected by the concentration of other ions. This is shown by the results of an iris crop grown on pumice and irrigated with water of a NaCl concentration of 8 mmol l^{-1} (De Kreij and Van der Burg, 1998). Some data of this experiment are summarized in Table 6.6. The uptake of Na and Cl is much higher at a low fertilization level than at a standard fertilization, especially the transport of Na to the shoot is increased at low fertilization. The uptake concentration for Na was doubled, but as a result of the low fertilization the growth of the crop was strongly reduced. Comparable data are shown in Table 7.7 with bouvardia. The higher fertilization level reduced the Na uptake and thereby the toxic effect of the high sodium uptake. Often such effects merely are attributable to K uptake, which strongly aggravates or reduces the Na uptake with low or high K concentrations in the root environment, respectively; see also Satti et al. (1996). Comparable data has been found by changing of the Cl/NO_3 ratio with rock wool grown tomatoes (Voogt and Sonneveld, 2004), shown in Fig. 6.6. High Cl/NO_3 ratio went in hands with a low NO_3 concentration in the root environment, which strongly aggravated the uptake of Cl. Thus, extra uptake of Na and Cl can be realised by specific interventions in the fertilization. However, such interventions need very close control on the nutrient status in the root environment, because strong interventions in the nutrient status in the root environment easily induce yield reduction. Therefore, under practical growing conditions seldom uptake concentrations can be realised much higher than those presented in Table 6.5.

The salinity status of a soil during cultivation often is strongly related to the water supply, like shown for a cucumber crop in Fig. 6.7. The EC of the saturation extract is followed during a whole growing season, together with the water supply.

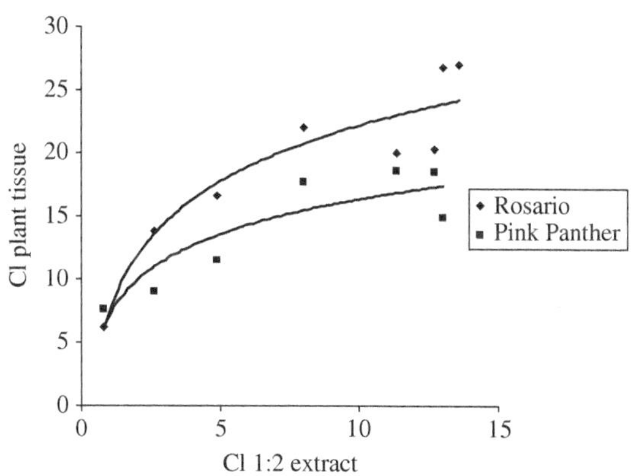

Fig. 6.5 Relationships between the Cl concentration in the root environment (mmol Cl l^{-1} of the 1:2 volume extract) and the Cl concentration in the leaves (mmol kg $^{-1}$ dry matter) for different alstroemeria cultivars grown in soil. Data after Sonneveld (1988)

Table 6.6 Effects of the fertilization level on the uptake of Na and Cl by an iris crop, grown in pumice with irrigation water of a NaCl concentration of 8 mmol l^{-1}

	Shoot		Bulb		Uptake concentration	
Fertilization level	Na	Cl	Na	Cl	Na	Cl
Low	292	314	322	181	1.99	1.49
Standard	87	270	260	149	0.89	1.26

After De Kreij and Van der Burg (1998).

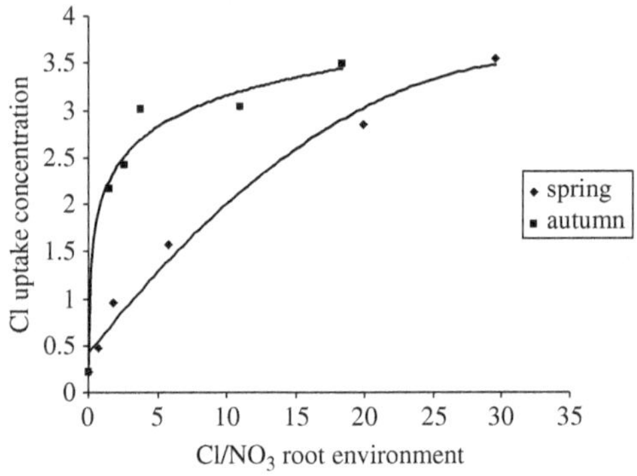

Fig. 6.6 Cl uptake concentration as affected by the Cl/NO_3 ratio for rock wool grown tomato. Data after Voogt and Sonneveld (2004)

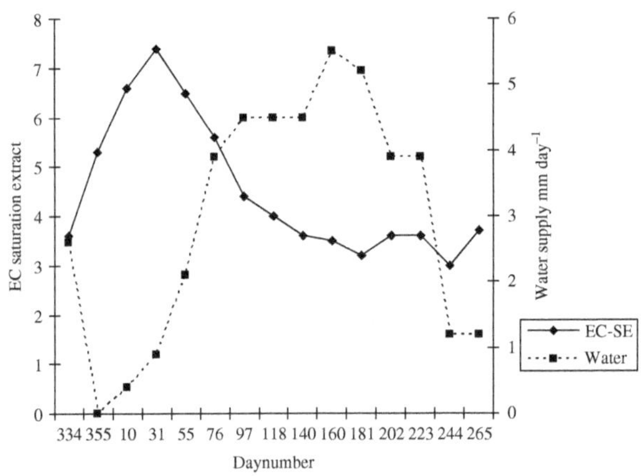

Fig. 6.7 Course of the EC of the saturation extract in the top soil (0.25 m) and the water supply in a greenhouse with soil grown cucumber. After Sonneveld (1969)

At start in winter a base fertilization is given and the water supply is low in that period. The EC in the root environment increases further on during the first period, because of sparingly water supply and salt accumulation from the deeper soil layers. In spring when the ample water supply is started the EC decreases by leaching of salts, because the water supply was 5–6 l m^{-2} day^{-1}, while the transpiration on average will be estimated on 2–3 l m^{-2} day^{-1}. The EC decreases only until a certain required level and is maintained on that level later on by regular top dressings with nutrients. In substrate systems salt accumulation will occur quickly, because of the restricted root volume. Nice examples of salt accumulation are shown by Savvas et al. (2005a) in a closed hydroponics system grown with cucumber for 120 days. The NaCl concentration in the primary water used were 0.8, 5, 10 and 15 mmol l^{-1} and after 120 days the concentrations in the drainage water were accumulated to about 8, 30, 45 and 55 mmol l^{-1}, respectively. The uptake concentration of Na and Cl increased significantly, but the high concentrations in the root environment ensure a substantial yield reduction of 12% per unit EC value measured in the drainage water (Savvas et al., 2005b). Such data underlines the need of leaching of salts by extra water supply during cultivation in substrate systems.

The leaching requirements will be calculated with formula (7.4). This formula is mainly based on the ion concentration in the primary water and this in the drainage water, supposing that the concentration added with the fertilizer application is under control and relatively low. Formula (7.4) can be applied for a specific salt as well for the EC as a measure for the total salt concentration. The concentration of the drainage water used in this formula reflects the concentration accepted as the maximum concentration in the drainage water or in the soil solution that leaves the root zone. The accumulation of salts depends much on the growing system, whereby the irrigation system plays a substantial part. With an overhead sprinkler system the whole area is irrigated and the salts will be washed down to deeper soil layers, or to a possible drainage system by which it is transported to the deep ground water and the surrounding surface water, respectively. With spot or strip irrigation part of the salts accumulate on the dry spots or paths during cultivation and at the end of the cropping period an unequal distribution of salts will occur. Local accumulations in the top layers between drippers also can occur in substrates when the surface of the substrate is not covered and furthermore, the accumulation follows its path to the drainage (Van Noordwijk and Raats, 1980). But even when substrates are wrapped in plastic bags, local accumulations can be substantial and will be highest in the end of the growing season (Van der Burg and Sonneveld, 1987). The measurements of the EC of the nutrient solution extracted from rock wool slabs in greenhouses under such conditions varied between values of about 2.5 and 8.0 dS m^{-1}. Salt accumulations in the root zone belong to growing systems under practical conditions in greenhouses, it is more or less impossible to prevent it and it is strongly promoted by a local supply of the irrigation water (Heinen, 1997), like also shown in Fig. 6.8 (Voogt and Van Winkel, 2009). The salt accumulation shows a contradictory pattern with the moisture content and the EC in the 1:2 volume extract varies from about 0.5 till 4.0 dS m^{-1}, which can be calculated to 2.4–13.3 in the soil solution. It can be expected that the differences in soils generally will be of the same size as found with substrates.

The leaching requirements as mentioned before and calculated with formula (7.4) are operative for soil as well substrate growing. The management during crop cultivation is already discussed in Section 7.9, but can differ for substrate and soil growing. The main difference between soil and substrate is the root volume in which the plant is grown. The small volumes used in substrate growing on the one hand

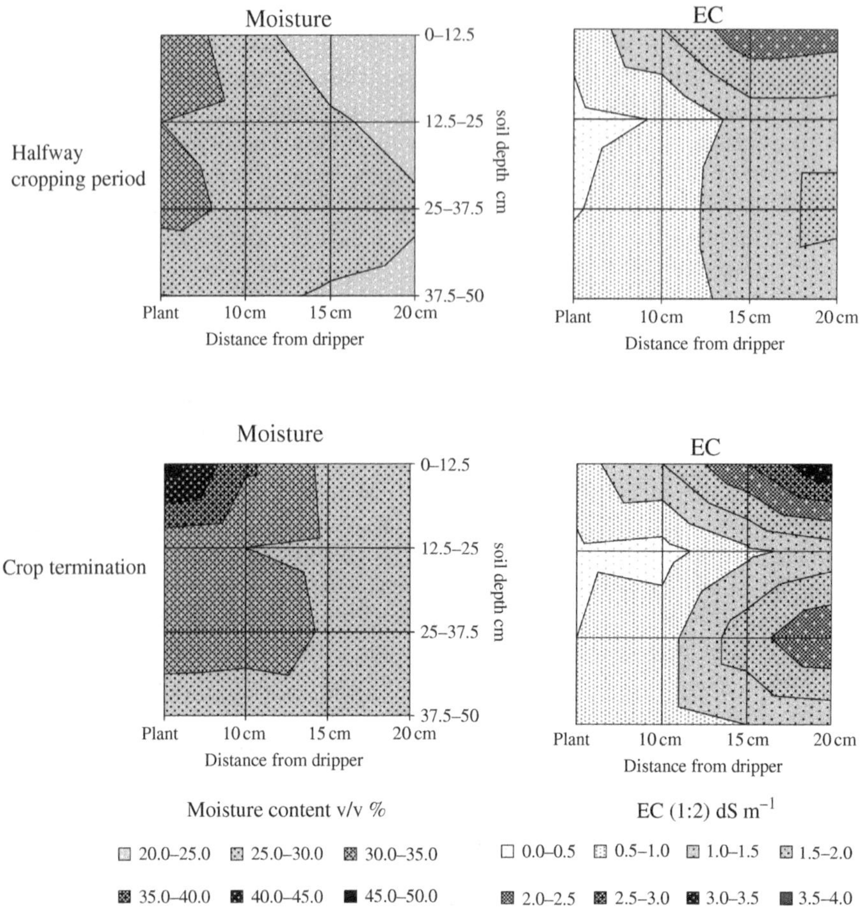

Fig. 6.8 The two dimensional pattern of EC (1:2 volume extract) and volumetric moisture content distribution in the soil profile with tomato irrigated with drip irrigation, determined longitudinal and in soil layers of 12.5 cm depth, determined halfway and at the end of a long term tomato cropping (after Voogt and Van Winkel, 2009)

accentuate the need of leaching during cultivation, but on the other hand it opens excellent possibilities to control the salinity status.

Beside the leaching requirements during cultivation to prevent too high salt accumulations in the rooting zone, for soil grown crops leaching of accumulated salts can be necessary before starting a new crop. The need for this leaching not only depends on the accumulated salt concentration, but also on the consecutive crops. When a crop ends with a very unequal distribution of salts in the soil profile and the next crop covers the whole area of the greenhouse with a high planting density, an intensive flooding is necessary to prevent an unequal start of the next crop. This for instance is the case in crop rotations of fruit vegetables followed by leafy vegetables, radish and a lot flower crops. When in crop rotation comparable crops are succeeding, like fruit vegetable in rows, at the start a somewhat higher salinity status of the soil is often desirable and a heavy flooding is not necessary.

The flooding of soils is based on the replacement of the soil solution by fresh water and movement of the actual soil solution out of the root zone. The quantity of water required to that purpose depends on the depth of the root zone, the water holding capacity of the soil and the dispersion factor. This dispersion factor depends on the soil type and is high for soils with a course granule structure, like clayey soils. The quantity of water necessary for leaching can be calculated by the formula (6.5).

$$wv_l = wv_f f_d d \tag{6.5}$$

In which

wv_l = water necessary for flooding in mm
wv_f = volume fraction of water of the soil at field capacity as defined in formula (3.6)
f_d = dispersion factor
d = depth of the soil layer that will be washed out in mm

The Dutch advisory service maintained for the soils used in the greenhouse industry dispersion coefficients varying from 1.25 to 2.0 for sandy and clayey soils, respectively. Some results of their calculations used as recommendations for growers are summarized in Table 6.7 (Van den Bos et al., 1999).

In some well water under anaerobic conditions soluble Fe is present as bivalent Fe ions, which can be harmful to plants. This bivalent Fe easily oxidizes to trivalent Fe, when the water comes into contact with O_2. This process can be simply quickened by aeration or chemical oxidation systems (Ten Cate, 1978). The trivalent Fe precipitate as $Fe(OH)_3$. The chemical reaction is presented in formula (6.6).

Table 6.7 Water quantities required for leaching (wv_l) of a greenhouse soil over a depth of 250 mm, dependent on soil type, volume fraction of water at field capacity (wv_f) and dispersion factor (f_d)

Soil type	% clay	% organic matter	wv_f	f_d	wv_l
Sand	<7	<3	0.18	1.25	60
Sand	<7	3–5	0.23	1.25	75
Sand	<7	5–7	0.28	1.25	90
Sand	<7	7–10	0.32	1.25	100
Sand	<7	10–20	0.35	1.25	120
Loamy sand	<7	<5	0.36	1.25	120
Loamy sand	<7	5–10	0.44	1.25	140
Loam	<17	<5	0.26	1.25	90
Loam	<17	5–10	0.32	1.25	110
Loam	<17	>10	0.35	1.25	120
Clayey loam	>17	<20	0.44	1.75	200
Clay	>17	>20	0.52	2.00	230
Clayey peat	–	–	0.60	1.25	200

Van den Bos et al. (1999).

$$4Fe^{2+} + O_2 + 18H_2O \rightarrow 4Fe(OH)_3 \downarrow +8H_30^+ \quad (6.6)$$

With the oxidation H_3O ions are released, as follows from formula (6.6). This can be responsible for a strong decrease of the pH in the water under treatment, followed by a retardation of the oxidation process. Therefore it is important that the pH of the water is sufficiently buffered, to react with the released H_3O ions. The pH buffer when available in well water mainly exists of HCO_3. When the Fe is oxidized, a good filtration is necessary to separate the precipitated Fe from the water. In Fig. 6.9 effects of irrigation water with overhead spraying on plants are shown in relation to the HCO_3 concentration and the soluble Fe concentration in the irrigation water (Van den Ende, 1970). Despites, soluble Fe is not highly toxic to plants, it has different drawbacks when the concentrations pass certain thresholds. Summarizing, following situations will occur, when the irrigation water contain substantial concentrations of soluble Fe.

- Leaf scorching when sprayed over the crop, which especially occur when the Fe can be insufficiently oxidized. For limits see Fig. 6.9
- Contamination of greenhouse construction, crops and harvested produce when used for overhead spraying. The limits are denoted in Fig. 6.9
- Clogging of irrigation systems as mentioned in Section 6.3. This occurs with all Fe soluble in primarily irrigation water and the clogging get worse with increasing concentrations.

See also Section 7.8 about effects of Fe in irrigation water.

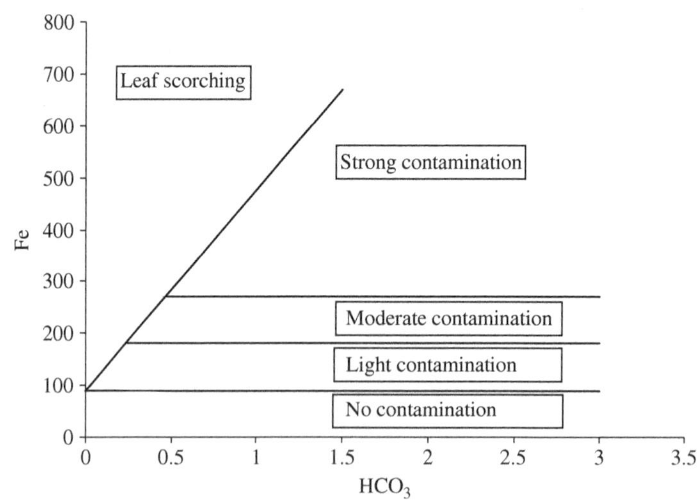

Fig. 6.9 Effects of soluble Fe (µmol l^{-1}) in the irrigation water on crops when used for overhead spraying in relation to the HCO_3 concentration (mmol l^{-1}). After van den Ende (1970)

6.5 Water Supply

The water supply during crop cultivation ultimately results from three main factors being:

- The uptake of the crop and the evaporation of the surface as discussed in Section 6.2
- The inequality resulted from differences by the plant uptake as well as by the distribution of the irrigation system as discussed in Section 6.3
- The leaching requirements determined by the water quality and the accepted salt accumulation in the root zone, as discussed in Section 6.4 and will be calculated by formula (7.4)

Thus, the water supply during cultivation can be formulated as follows in formula (6.7).

$$S_w = \frac{E}{(1-I)(1-LF)} \tag{6.7}$$

In which

S_W = water supply
E = estimated transpiration of the crop as denoted in the formulae (6.1) and (6.2)
I = inequality factor as discussed in Section 6.3 and calculated with the aid of formula (6.4)
LF = leaching fraction as discussed in Section 6.4

The inequality factor and the leaching fraction are multiplied in formula (6.7), because both factors are mutually independent.

Despite the reliable relationships found between the water uptake of plants and the different explaining climatic factors, under practical conditions serious errors occur, like shown in Fig. 6.10. The reason of such errors will be explained by different minor factors as presented by De Graaf (1995) and inaccuracies and deviations of the models developed. Such errors especially in substrate systems easily can deregulate the water management, because of the small water storage available in the root environment. Therefore, for substrate grown crops a system for water supply has been developed with a feed back on the amount of drainage water (De Graaf, 1988). To this purpose the amount of drainage water will be measured and dependent on the results the estimated water requirement is adjusted. With frequent irrigations, like often occur with substrate growing, the length of the interval between the irrigation periods is adjusted. With less frequent irrigations better the length of the irrigation period can be adjusted.

For soil grown crops the measurement of the amount of drainage water is for the greater part impossible. Therefore, for soil grown crops in situ measurements of the moisture condition of the soil are valuable with the possibilities of adjustment of the results calculated by the models developed. The traditional tensiometer offers such possibilities. Drawbacks of these apparatus are the difficulties with the

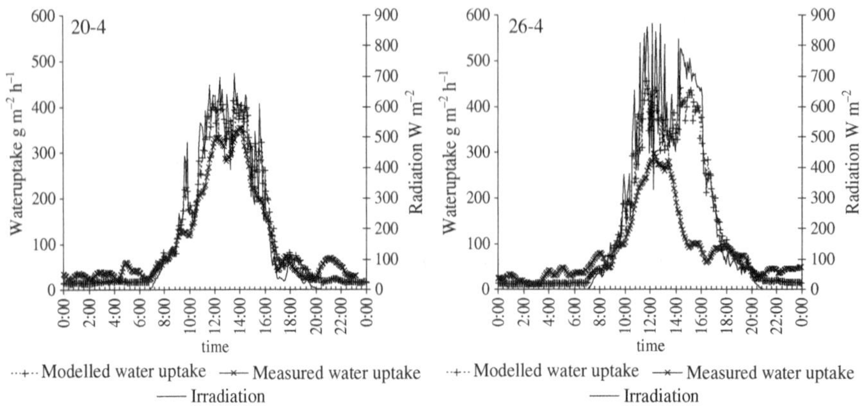

Fig. 6.10 Water uptakes in relation to the global radiation of rock wool grown tomato during two days in April. The water uptake was calculated following the model after Voogt et al. (2006b) and measured with an automatic balance. On the *left* a good agreement between measured and calculated uptakes, while on the *right* this agreement is disturbed

interpretation of the results, the susceptibility for errors and the accidental placement. Last factor is due to placement on a possible dry or wet spot, originated by irregular distribution of the irrigation water and this of the water uptake by the plant. These facts emphasize the need for alternative methods to get an acceptable view on the average situation. Nowadays, electronic measurements become available, like the WET sensor (Balendonck et al. 2005), based on measurement of a dielectric constant of the substrate (Hilhorst, 2000). Results of such measurements are promising, like shown in Fig. 6.11. The WET sensor showed reliable results during a period

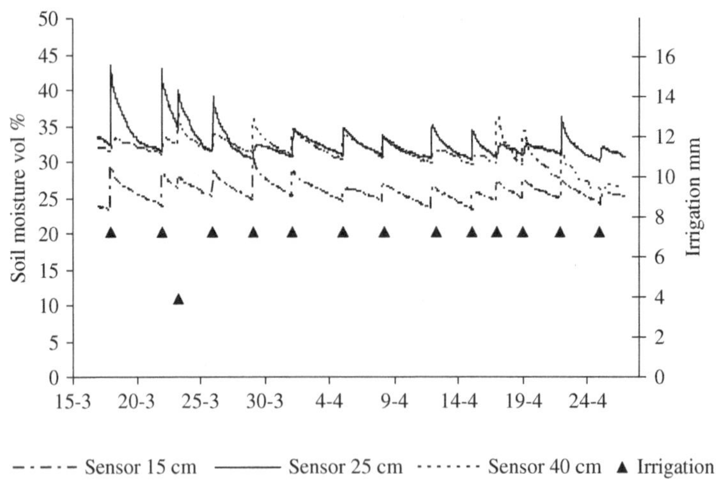

Fig. 6.11 Water contents measured in a loamy sand soil in a greenhouse at three depths with the WET sensor and the water supplied by irrigation. Data derived from Voogt and Balendonck (2006)

of 40 days. Although the experience with such systems is restricted, the advantage of this method is the direct translation of the measured value to the moisture content of the soil and the slight susceptibility for errors. The drawback of the errors of the measurement by the placement on an accidental spot, as mentioned for the tensiometer will be operative too.

The strong development in the electronic measurements and computerized management in the water use and water supply is a helpful tool in the validation and calibration of the theoretical models developed for water supply. The frequent measurements in systems in action offer excellent opportunities for the development of self-learning models. With such models the estimations of factors attributed to the parameters in existing models can be improved and effects of new parameters can be estimated (Elings and Voogt, 2008).

References

Assinck F B T and Heinen M 2001. Modelverkenning naar het effect van niet-uniform verdeelde watergiften op de opname van chrysanten onder glas. Alterra Instituut Wageningen, Report 393, 36pp.

Baas R Nijssen H M C Van den Berg T J M and Warmenhoven M G 1995. Yield and quality of carnation (*Dianthus caryophyllus* L) and gerbera (*Gerbera jamesonii* L) in a closed nutrient system as affected by sodium chloride. Sci. Hort. 61, 273–284.

Baas R and Van Rijssel E 2006. Transpiration of glasshouse rose crops: Evaluation of regression models. Acta Hort. 718, 547–554.

Balendonck J Bruins M A Wattima M R Voogt W and Huys A 2005. WET-sensor pore water EC calibration for three horticultural soils. Acta Hort. 691, 789–796.

De Bes S S 1986. Tussentijds overzicht "Verstoppende materialen druppelsystemen". Internal report, 1986, 4pp.

De Graaf R and Van den Ende J 1981. Transpiration and evapotranspiration of the glasshouse crops. Acta Hort. 119, 147–158.

De Graaf R 1988. Automation of the water supplied of glasshouse crops by means of calculating the transpiration and measuring the amount of drainage water. Acta Hort. 229, 219–231.

De Graaf R 1993. Verdamping en watervoorziening. Proefstation voor Tuinbouw onder Glas, Plantevoeding in de Glastuinbouw, Informatiereeks 87, 210–223.

De Graaf R 1995. Influence of moisture deficit leaf-air and cultural practices on transpiration of glasshouse roses. Acta Hort. 401, 545–552.

De Graaf R and Esmeijer M H 1998. Comparing calculated and measured water consumption in a study of the (minimal) transpiration of cucumbers grown on rockwool. Acta Hort. 458, 103–111.

De Graaf R and Spaans L 1998. Watergeven bij roos onder glas. Proefstation voor Bloemisterij en Glasgroente Naaldwijk, Intern Verslag 171, 14pp.

De Graaf R Roos M and Blok C 2004. De waterhuishouding bij belichte komkommers. Praktijkonderzoek Plant en Omgeving Naaldwijk, Rapport (June 2004), 36pp.

De Kreij C and Van den Berg Th J M 1990. Nutrient uptake, production and quality of *Rosa hybrida* in rockwool as affected by electrical conductivity of the nutrient solution. *In:* M L van Beusichem (ed). Plant Nutrition – Physiology and Applications. Kluwer Academic Publishers, Dordrecht, 519–523.

De Kreij C and Van der Burg A M M 1998. Effect van Natrium chloride en EC op productie en kwaliteit van iris. Proefstation voor Bloemisterij en Glasgroente Naaldwijk, Rapport 158, 24pp.

De Kreij C Van der Burg A M M and Runia W T 2003. Drip irrigation clogging in Dutch greenhouses are affected by Methane and organic acids. Agric. Water Manag. 60, 73–85

Elings A and Voogt W 2008. Performance of the Intkan model for greenhouse crop transpiration. Acta Hort. 801, 1221–1228.

Graves C J 1983. The nutrient film technique. Hort. Rev. 5, 1–44.

Heemskerk M J Van Os E A Ruijs M N A and Schotman R W 1997. Verbeteren van watergeefsystemen voor grondgebonden teelten. Proefstation voor Bloemisterij en Glasgroente Naaldwijk, Rapport 84, Deelrapport 1: Inventarisatie en verbetering regenleidingsystemen, 61pp.

Heinen M 1997. Dynamics of water and nutrients in closed, recirculating cropping systems in greenhouse horticulture. Proefschrift Landbouwuniversiteit Wageningen, 270pp.

Hilhorst M A 2000. A pore water conductivity sensor. Soil Sci. Soc. Am. Proc. J. 64, 1922–1925.

Houter B 1996. Constanten voor verdampingsprogramma. Priva De Lier The Netherlands. Internal communicatie data sheet, lpp

Lagerwerff J V and Eagle H E 1962. Transpiration related to ion uptake by beans from saline substrates. Soil Sci. 93, 420–430.

Maloupa E M 2002. Hydroponic systems. *In*: Savvas D and Passam H (ed) Hydroponic Production of Vegetables and Ornamentals. Embryo Publications, Athens, 143–178.

Marcelis L F M 1987. Simulatie van de waterhuishouding en de fotosynthese van kasgewassen. Doctoraalonderzoek Theoretische Teeltkunde. Landbouwuniversiteit Wageningen, 124pp.

Nederhof E M 1994. Effects of CO_2concentration on photosynthesis, transpiration and production of greenhouse vegetable crops. Landbouwuniversiteit Wageningen, Thesis213pp.

Satti S M E Al-Yhyai R A and Al-Said F 1996. Fruit quality and partioning of mineral elements in processing tomato in response to saline nutrients. J. Plant Nutr. 19, 705–715.

Savvas D and Lenz F 1995. Närhstoffaufnahme von Aubergine (*Solanum melongena L.*) in Hydrocultur. Gartenbauwissenschaft 60(1), 29–33.

Savvas D Pappa V A Kotsiras A and Gizas G 2005a. NaCl accumulation in a cucumber crop grown in a completely closed hydroponic system as influenced by NaCl concentration in irrigation water. Eur. J. Hort. Sci. 70, 217–223.

Savvas D Pappa V A Gizas G and Maglaras L 2005b. Influence of NaCl concentration in the irrigation water on salt accumulation in the root zone and yield in a cucumber crop grown in a closed hydroponic system Acta Hort. 697, 93–99.

Sonneveld C 1969. Het verloop van grondanalysecijfers (1965–1966): Onderzoek op basis van het verzadigingsextract. Proefstation voor de Groenten- en Fruitteelt onder Glas Naaldwijk. Intern Rapport, 6pp.

Sonneveld C 1988. The salt tolerance of Alstoemeria (*Alstroemeria, X*). Acta Hort. 228, 317–326.

Sonneveld C 1993a. Watervoorziening en waterkwaliteit bij teelten onder glas. *In*: Van Keulen H and Penning de Vries F T W (eds), Agrobiologische Thema's 8, 41–48. CABO-DLO Wageningen.

Sonneveld C 1993b. Mineralenbalansen bij kasteelten. Meststoffen, NMI Wageningen, 44–49.

Sonneveld C 1995. Fertigation in the greenhouse industry. *In*: Proc. of the Dahlia Greidinger Intern. Symposium on Fertigation. Technion – Israel Institute of Technology, Haifa, 25 March–1 April 1995, 121–140.

Sonneveld C and Van den Bos A L 1995. Effects of nutrient levels on growth and quality of radish (*Raphanus sativus* L.) grown on different substrates. J. Plant Nutr. 18, 501–513.

Sonneveld C 2000. Effects of salinity on substrate grown vegetables and ornamentals in greenhouse horticulture. Thesis Wageningen University, Netherlands, 151pp.

Stanghellini C 1987. Transpiration of greenhouse crops – an aid to climate management. Wageningen University, Thesis, 150pp.

Ten Cate H R 1978. Ontijzeren van gietwater. *In*: Proefstation voor de Groente- en Fruitteelt onder Glas te Naaldwijk, Waterkwaliteit en Waterbehandeling voor de Glastuinbouw, Series: Informatiereeks no 50, 12–14.

Van den Bos A De Kreij C and Voogt W 1999. Bemestingsadviesbasis grond. Proefstation Bloemisterij en Glasgroente Naaldwijk. Brochure PPO. 171, 54 pp.

Van den Ende J 1970. Kwaliteitsnormen voor het gietwater. Bedrijfsontwikkeling1 nr 7, 45–51.
Van den Ende J and De Graaf R 1974. Comparison of methods of water supply to hothouse tomatoes. Acta Hort. 35, 61–70.
Van der Burg A M M and Hamaker Ph, 1987. Variatie in waterafgifte druppelaars en wateropname. Groenten en Fruit 42 no 19, 30–33.
Van der Burg A M M and Sonneveld C 1987. Variation of the EC values in rockwool slabs. Glasshouse Crops Research Station Naaldwijk, Annual Report, 20.
Van der Burg A 1994. Watergift beter afstemmen op waterverbruik. Groenten en Fruit/Glasgroenten 4 no 47, 25.
Van Ieperen W 1996. Consequences of diurnal variation in salinity on water relations and yield of tomato. Thesis: Wageningen Agric. Univ., 176 pp.
Van Noordwijk M and Raats P A C 1980. Drip and drainage systems for rockwoll culture in relation to accumulation and leaching of salts. Proc. Fifth Intern. Congr. Soilless Culture, Wageningen, 279–287.
Van Schie J J Voogt W Sonneveld C and Beekmans G 1982. Voorlopige resultaten druppelbevloeiingsonderzoek. Vakblad Bloemisterij 37 no 30, 32–33.
Voogt W Kipp J A De Graaf R and Spaans L 2000. A fertigation model for glasshouse crops grown in soil. Acta Hort. 537, 495–502.
Voogt W and Sonneveld C 2004. Interactions between nitrate (NO_3) and chloride (Cl) in nutrient solutions for substrate grown tomato. Acta Hort. 644, 359–368.
Voogt W and Van Winkel A 2005. Onderzoek naar verdamping van alstroemeria. Unpublished data.
Voogt W and Balendonck J 2006. Ideale meter voor bodemvocht bestaat nog niet. Vakblad Bloemisterij 61 no 5, 42–43.
Voogt W Van Winkel A and Steinbuch F 2006a. Evaluation of the "fertigation model", a decision support system for water and nutrient supply for soil grown greenhouse crops. Acta Hort. 718, 531–538.
Voogt W and Van Winkel A 2008. De voedingsopname van cymbidium tijdens het jaar. Wageningen University Research, Report July 2008, 23pp.
Voogt W and Van Winkel A 2009. Vocht en verdeling van nutrienten in de bodem onder invloed van de watergift en mulching in biologische teelt. Wageningen UR Greenhouse Horticulture, Report, 40pp.
Voogt W Van Winkel A and Warmenhoven M 2006b. Resultaten van weeggoten hydrion III. Wageningen UR Greenhouse Horticulture, Internal Report, 65pp.
Yaron B Ziesling N and Halevy A H 1969. Response of Baccara roses to saline irrigation. J. Amer. Soc. Hort. Sci. 94, 481–484.

Chapter 7
Salinity and Water Quality

7.1 Introduction

The impact of salinity on greenhouse grown crops, especially when grown in substrate systems, differs from the impact of salinity on crops grown under field conditions. The most striking difference between greenhouse and field conditions is the overall much higher concentrations of nutrients in greenhouse soils and substrates. This especially holds where high ion levels are knowingly maintained in the soil or substrate solution to control plant growth under poor light conditions or to improve quality of the produce (Sonneveld, 2000). Thus, in greenhouse cultivation nutrients contribute substantially to the osmotic potential of the solution in the root environment. This especially is the case in substrate systems when water of a low salinity status is used and thus, the osmotic potential is more or less solely brought about by nutrients. Furthermore, factors strongly affecting salinity effects on crops, like the climatic conditions in the greenhouse and the addition of water, are artificially controlled and therefore, differ much from those under field conditions. These factors induce special requirements for the management of salinity under greenhouse conditions.

In studies on salinity effects on crops, osmotic and specific effects were distinguished from the beginning (Bernstein, 1976; Hayward and Long, 1940). This differentiation is still actual and is operative for greenhouse conditions too. Osmotic effects on plants are determined by the osmotic potential of the external solution independent of the composition of the osmotic solutes and generally restricted to effects without any essential disturbance of ion uptake by plants or distribution within plants. Specific effects can be distinguished into two groups namely effects through nutrition and effects through toxicity. Specific salinity effects through nutrition imply that crop growth is affected by disturbance of uptake or distribution of ions essential for plant growth. Toxic salinity effects occur by excess uptake of an osmotically significant ion. This means that toxicity of essential micro elements like B, Mn, Cu, etc. and non essential micro elements like F, Li, Se, Cr, etc., not being of osmotic significance, are beyond the scope of these considerations.

The in the classification presented sodicity problem, remains out of scope. Bernstein (1976) considered this problem as "salinity related", but there is no reason to consider it as not belonging to the field of salinity. Sodicity is related to an

C. Sonneveld, W. Voogt, *Plant Nutrition of Greenhouse Crops*,
DOI 10.1007/978-90-481-2532-6_7,

unbalanced ratio between (bi)carbonate and bivalent cations, mainly Ca and Mg (Bernstein, 1975; Richards, 1954). In principle, when in water the (bi)carbonate concentration in equivalence exceed the (Ca + Mg) concentrations, the water causes sodicity. When used for irrigation, the carbonate precipitates with the present bivalent cations in the soil or the substrate. The bivalent cations added with the irrigation water are insufficient to cover this need and thus, bivalent cations mainly Ca and Mg are withdrawn from the soil/substrate solution and the adsorption complex. The remaining mono valence cations from the irrigation water, mostly Na, survive in the soil and substrate solution and on the adsorption complex. Soils and substrates affected by this phenomenon are chemically characterized by high Na and low Ca and Mg concentrations. They are physically characterized by dispersion (Hilgard, 1919) and a rapid decomposition of the organic matter.

It is not always possible to distinguish clearly between osmotic and specific salinity effects. Often combinations of osmotic and specific effects occur. Combined effects can be expected between nutritional and toxic effects. In such cases a decrease of the uptake of a nutrient is often accompanied by an increase in the uptake of ions involved in salinity (Bernstein, 1964). For most crops and growing conditions the osmotic effects predominate (Bernstein, 1976). The best-known phenomenon of osmotic effects is the wilting of plants with suddenly decreased osmotic potential, related to a lost or a reduced osmotic gradient for water absorption. This, however, is not the most common symptom. In practice, with increasing salinity the osmotic potential of the solution in the root environment decreases only slowly and plants can adjust in due time for it (Bernstein, 1961, 1963; Nukaya, 1983; Van den Ende et al., 1975). It is likely that the adaptations by plants are responsible for the growth reduction, as has been suggested already by Bernstein (1976). These adjustments are diverse and there is no simple unequivocal mechanism underlying the growth reduction of crops caused by osmotic effects (Greenway and Munns, 1980; Lüttge and Smith, 1984). In many studies a great diversity in the osmotic adjustments are traced, for example increased dry matter contents which includes increased concentrations of soluble solids, increased absorption of Na, Cl or nutrients and metabolic adjustments of the organic solutes (Gulzar et al., 2003; Hsiao and Läuchli, 1986; Meloni et al., 2001; Plaut et al., 2005). The character of the adjustments differ among crops (Sonneveld and Voogt, 2008), but also the type of salinity and the growing conditions affect the character of the adjustments (Veen and Kleinendorst, 1986).

On the whole, effects of salinity strongly depend on crop, cultivar, rootstock and growing conditions like temperature, humidity, CO_2 enrichment, irrigation method and fertilization level. In Fig. 7.1 a relational diagram shows different aspects of salinity. In this diagram the osmotic potential (EC) of the soil solution is affected by fertilization and salinity. In principle, the osmotic potential is made up of nutrients and residual ions, like Na and Cl. The latter originate mainly from the irrigation water. In irrigation water also excessive concentrations of Ca, SO_4 and Mg can act as residual salts and decrease the osmotic potential. Thus, the osmotic potential in the soil and the substrate solution is determined by nutrients as well by residual salts. Nutrients are considered as such as long as their concentrations can be considered as with the range for normal plant nutrition. Nutrients above this level will be

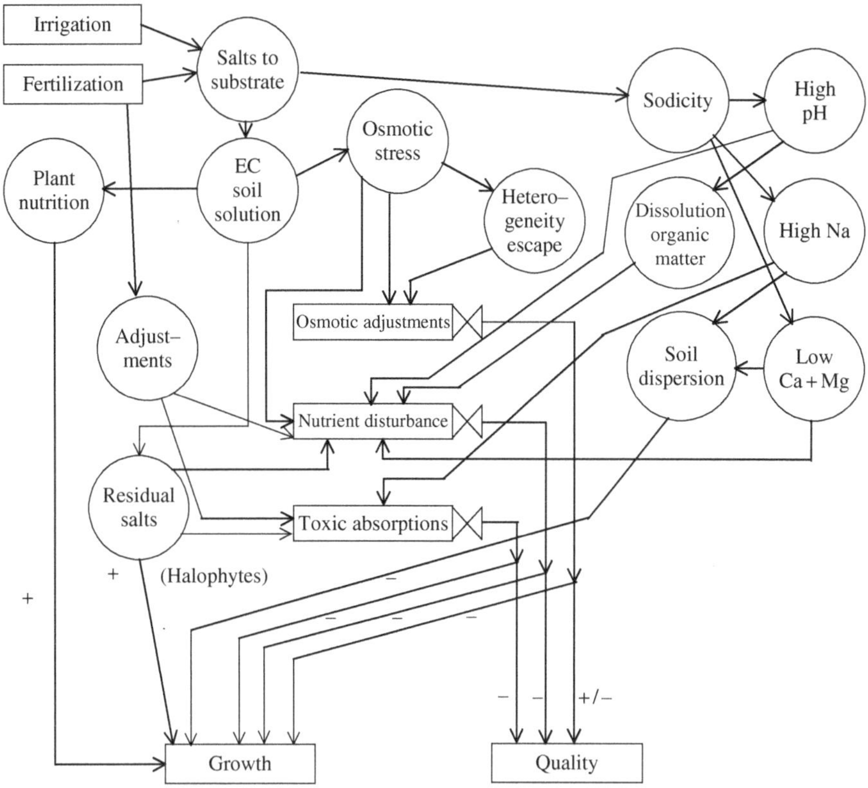

Fig. 7.1 Relational diagram for salinity effects on greenhouse crops. An effect is mainly negative (−), mainly positive (+) or positive as well negative (−/+)

considered as osmotics and affect together with the residual salts the physiological processes in plants by osmotic adjustments, nutrient disturbance or toxicity. Growth and quality of crops can be affected negatively as well as positively. Sodicity follows a different pathway and the effects on the development of plants are exclusively negative.

7.2 Salinity Models

In an extensive review Maas and Hoffman (1977) have presented an analysis of salt tolerance phenomenon on basis of a simple model. The model is characterized by two parameters, the salinity threshold value (c_t) and the salinity yield decrease value (SYD). The model is shown in Fig. 7.2A and is described in the following equation.

$$\begin{aligned} Y_r &= 1 & 0 &\leq c_{ss} \leq c_t \\ Y_r &= 1 - SYD(c_{ss} - c_t) & c_t &< c_{ss} \leq c_z \\ Y_r &= 0 & c_{ss} &\geq c_z \end{aligned} \tag{7.1}$$

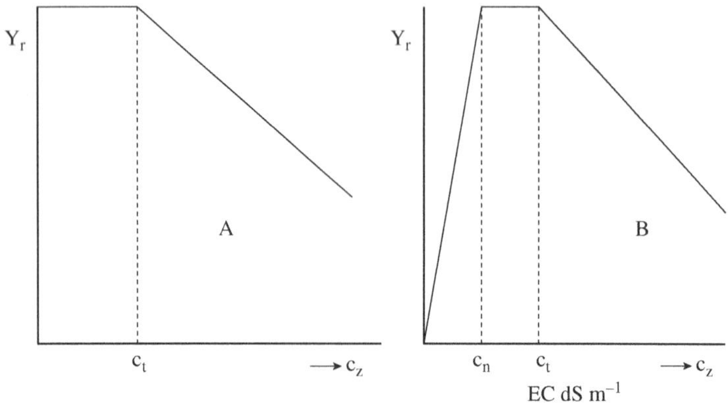

Fig. 7.2 Relationships between the EC value in the root environment and the yield of crops according to the model of Maas and Hoffman (1977) (**A**) and the adjusted model according to Sonneveld (1991) (**B**)

In which

Y_r = relative yield in relation to the yield under non saline conditions
c_{ss} = ion concentration in the soil and the substrate solution
c_t = salinity threshold value, being the maximum root zone concentration without yield reduction
c_z = root zone concentration beyond which the yield is zero
SYD = salinity yield decrease value, being the slope of the salinity response function between c_t and c_z in % per unit EC in dS m^{-1}

Soil salinity is generally expressed as the EC of the saturation extract (EC_e). For substrate often the EC of the substrate solution is used. For soils a close relationship exists between the EC of the soil solution and that of the saturation extract. For greenhouse soils the ratio calculated between the average EC of soil solution and that of the saturation extract of a big series of samples was about 1.6 (Van den Ende, 1989; Sonneveld et al., 1990).

In the model of Maas/Hoffman the EC caused by plant nutrients is not taken into account, because the model starts with a maximum yield at EC = 0. This is understandable because the model has been developed for field crops. Generally, in open fields fertilization with N, P and K has a marginal effect on the EC of the soil solution and when there is a significant effect of these nutrients, it never is a long term effect. In greenhouse soils plant nutrients substantially contributes to the EC of the soil solution. Sonneveld et al. (1990) found that one third of the total ion concentration consisted of N and K. In substrate systems and hydroponics plant nutrients play an even more important role, because the EC sometimes is exclusively determined by nutrients. Thus, the Maas/Hoffman model needs refinement with respect to the EC related to plant nutrients. Such a model is given by Sonneveld (1991) and shown in Fig. 7.2B.

$$
\begin{array}{ll}
Y_r = < 1 & 0 \leq c_{ss} \leq c_n \\
Y_r = 1 & c_n < c_{ss} \leq c_t \\
Y_r = 1 - SYD(c_{ss} - c_t) & c_t < c_{ss} \leq c_z \\
Y_r = 0 & c_{ss} \geq c_z
\end{array}
\tag{7.2}
$$

In which

c_n = minimum total ion concentration of plant nutrients necessary for optimum growth expressed as EC in dS m^{-1}
other parameters as given with equation (7.1)

Yield response curves do not always show the linear relationship as predicted by the models given. In particular with dramatic yield reductions the response is often non linear, as shown by Van Genugten and Hoffman (1984). In addition, a piecewise discontinuous path as suggested by the presented models is not to be expected for biological processes like salinity stress in plants. Therefore, Sonneveld et al. (2004) tried to find a continuous response curve for the relationship between the EC in the root environment and the yield characteristics of vegetables and ornamentals grown in greenhouse. Best results were obtained with an exponential curve following Eq. (7.3).

$$
Y_r = A + BQ^{EC_{ss}} + CEC_{ss} \tag{7.3}
$$

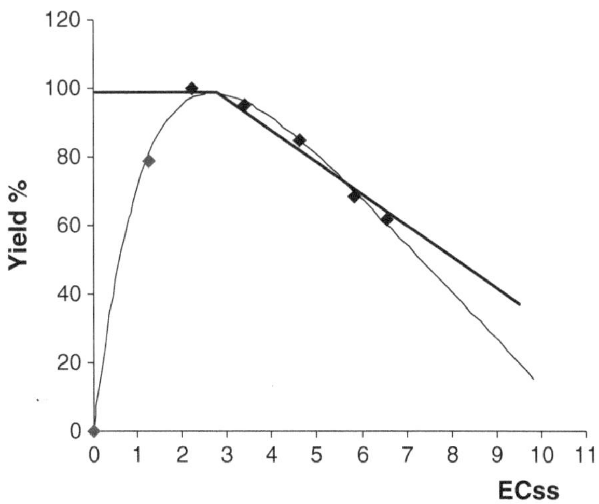

Fig. 7.3 The relationship between the EC_{ss} and the yield of lettuce with a linear and an exponential approach
The exponential function fitted is: $Y = 152.9 - 153.1 \times 0.439^{EC_{ss}} - 13.99EC_{ss}$. A high value of Q in the exponential fitting is responsible for a curvilinear SYD value, as shown for the lettuce crop in this figure. The threshold value for the exponential and linear fitting are equal at about $EC_{ss} = 2.6$ for the presented data. Derived from Sonneveld et al. (2004). *Reprinted by permission of Marcel Dekker*

In which

A, B, Q and C = empirical determined parameters
Y_r = relative yield in relation to the yield under non saline conditions
EC_{ss} = EC of the solution in the root environment

Calculation of an exponential response curve as suggested by Eq. (7.3) requires that sufficient data over a relatively broad range of salinities are available. The data set should contain also observations in the domain of the low EC values, which means below optimum yield. When insufficient observations are available, calculation of an exponential function will not be possible. Then estimations with the piecewise linear model may give acceptable results, if one can be sure that observations affected by yield reductions by too low nutrient supply are not present in the data set or are removed. In Fig. 7.3 an example of both response models is shown.

7.3 Osmotic Effects

The most common effect on plant growth by osmotic stress in the root zone is the reduction of vegetative growth resulting in decreased shoot weight, which often coincides with a reduction of leaf area, plant height and stem thickness. The assessment of osmotic stress is not unequivocal and will depend on the objective of crop production. From physiological point of view total dry matter production of plants is an obvious estimator. This, however, is often not the best estimator for the effect of salinity on the economic production volume of crops especially for greenhouse crops. Fresh weight and quality characteristics are more important for the greenhouse industry than the dry matter production. Therefore, the impact of salinity is strongly depended on the demand of the market and this complicates the estimation of the effect on the economic production volume. Especial for flower crops, the economic value can differ from the fresh weight production, because the price paid by the consumer often depends more on the appearance of the plant or the flower than on the plant size or flower size. Such is clear from the data listed in Table 7.1, where the SYD value of the fresh flower weight for two crops were compared with

Table 7.1 Comparison between the SYD values for the reduction of economic value and for fresh flower weight

	SYD values (% per dS m^{-1})	
Crop and cultivar	Economic value	Fresh flower weight
Anthurium cultv. A	23.1	21.9
Anthurium cultv. B	29.8	34.3
Hippeastrum	11.2	16.3

Results derived from Sonneveld and Voogt (1983). *Reprinted by permission of Springer*

Table 7.2 SYD values for soil grown greenhouse crops as has been found by Sonneveld (1988) in comparison with values reviewed by Maas and Hoffman (1977) and Maas (1986). The data are expressed on basis of the EC of the saturation extract (EC_e)

Crop	Sonneveld Exp. A[1]	Exp. D[1]	Maas and Hoffman, and Maas
Bean	13.9	14.7	19
Celery	–	7.7	6.2
Cucumber	11.7	8.8	13
Lettuce	3.1	4.6	13
Pepper	13.1	12.6	14
Radish	4.1	–	13
Spinach	+0.9	1.2	7.6
Tomato	5.7	6.5	9.9

[1]Different experiments as described by Sonneveld (1988). *Reprinted by permission of the Koninklijke Landbouwkundige Vereniging.*

the reduction of the economic value (Sonneveld and Voogt, 1983). The relationships between salinity and the different plant characteristics show for anthurium a better agreement than for hippeastrum. Comparable effects were shown with vegetables, where with increasing salinity negative aspects by yield reduction can contradict with positive aspects on quality (Adams, 1991; Mizrahi et al., 1988; Ohta et al., 1991; Verkerke et al., 1993).

Salinity related yield reductions found in experiments with soil grown greenhouse crops (Sonneveld, 1988) are listed in Table 7.2. The data are given in comparison with the data gathered in reviews by Maas and Hoffman (1977) and Maas (1986). The SYD values from literature are generally higher than those found under Dutch greenhouse conditions, especially for lettuce, radish and spinach. These crops were grown in winter and early spring in the Dutch experiments, which points to the fact that the growing conditions will affect the severity of the salinity stress. Such has been found by other researchers too (An et al., 2005; Maas, 1986; Magistad et al., 1943). Comparable results were found with greenhouse crops (Sonneveld, 2000) from which an example with *Cyclamen* is given in Table 7.3. In this experiment the plants grown under standard growing conditions, a whitewash screen combined with a movable screen at a radiation intensity of 600 W m^{-2}, were compared

Table 7.3 SYD values for shoot fresh weights between EC 1.5 and 4.0 dS m^{-1} in the root environment of summer grown *Cyclamen* at different screening regimes

Screening regime Fixed screen	Movable screen	Cultivars 'Julia'	'Louisa'
Whitewash	At 600 W m^{-2}	0	+4.6
No	At 300 W m^{-2}	−5.4	−4.0
No	At 600 W m^{-2}	−10.6	−4.2
No	At 800 W m^{-2}	−12.0	−9.9

After Verberkt (1997).

Table 7.4 Comparison of SYD values for soil grown and substrate grown crops. The data are recalculated to soil and substrate solution, respectively

Crops	Soil grown	Substrate grown
Cucumber	7.0	5.7
Sweet pepper	8.2	7.6
Tomato	4.6	4.8
Carnation	2.9	3.9
Gerbera	7.4	9.8

with levels of less extreme screening. It is clear that under standard screening the growth of the *Cyclamen* was not or even positively affected by salinity. With less screening, thus higher radiation, the SYD values increased. It is not yet clear which factor is responsible for a higher salt tolerance, because factors that determine the climatic conditions, as there are light intensity, humidity and temperature, are mostly strongly related (Sonneveld, 2000). Therefore Maas and Hoffman (1977) suggested to distinguish between "hot and dry" and "cool and humid" climates, from which crops are under latter conditions more tolerant than under first.

Crops grown in substrate do not seem to be more salt sensitive than crops grown in soil. This is shown by the data listed in Table 7.4, where the Salinity Yield Decrease values for soil and substrate grown crops do not show a systematic difference. SYD values for soil grown crops (Sonneveld, 1988) were recalculated to values for soil solution and compared with the SYD values for substrate grown crops, expressed on substrate solution. (Sonneveld and Van der Burg, 1991; Sonneveld et al., 1999). The SYD values for soil grown crops were expressed on soil solution by dividing the SYD values based on saturation extract by 1.6, in agreement with the findings of Sonneveld et al. (1990).

Salinity can affect the quality of the produce of greenhouse crops positively as well negatively. So it has been found that the storage and consumption quality of fruits like cucumber, pepino, strawberry, sweet pepper and tomato is improved by an increased EC in the root environment. A better colour and a longer shelf live are reported (Janse, 1985; Sonneveld and Van Beusekom, 1974a; Sonneveld and Van der Burg, 1991). Only at a very high EC the shelf life was shortened (Mizrahi, 1982). Furthermore, increased sugar and acid contents and improved taste have been found with increased EC (Adams, 1991; Awang, et al., 1993; Cornish 1992; Mizrahi and Pasternak, 1985; Mizrahi et al., 1988; Ohta et al., 1991; Petersen et al., 1998; Pluda et al., 1993; Verkerke et al., 1992). For sweet pepper a reduced russeting of the fruits was noticed by an increased EC (Van Uffelen and Bakker, 1987). A good example of effects of increased EC values on the fruit quality of tomato is presented with the data in Table 7.5. The relatively small increase of the EC value induced already a slightly lower yield, an improved shape index, a quicker colouring, a longer shelf life, and a higher EC, acid content and sugar content in the fruit sap. A high EC reduced also the occurrence of glassiness in lettuce (Maaswinkel and Welles, 1986), improved the tuber development of winter grown radish and reduced the sponginess of a spring/summer grown crop (Sonneveld and van den Bos, 1995) and in dill and thyme the essential oil concentration was increased (Udagawa, 1995). Negative effects of increasing salinity have been found by a reduced Ca uptake (Adams and

Table 7.5 Effects of an increased EC value in the root environment on yield and quality of tomato fruits

Characteristics	EC 2.6	EC 3.5
Number of fruits, m^{-2}	224	222
Fruit yield, kg m^{-2}	12.7	11.9
Average fruit weight, g	56	54
Fruit shape index	6.4	6.6
Colouring in days	4.4	4.1
Shelf life in days	17.5	19.2
EC fruit sap dS m^{-1}	5.8	6.2
Acids in fruit sap, mmol l^{-1}	75	84
Refraction in fruit sap, % Brix	4.8	5.0

After Sonneveld and Welles (1988). *Reprinted by permission of Springer*

Ho, 1990; Bernstein 1975, 1976; Geraldson, 1957; Semer-Olsen and Gislerød, 1980; Shear, 1975) or an inadequate xylem transport and redistribution of Ca (Adams and Ho, 1992; Ehret and Ho, 1986; Geraldson, 1957). In cut flowers a reduced flower size, length and thickness of stems, leaf loss and leaf discolouring was found (Baas et al., 1994; De Kreij and Van Os, 1989; Ploegman, 1976; Urban et al., 1995).

From the results mentioned it should be clear that for greenhouse crops the maximum production sometimes may conflict with the optimum levels for quality. However, for greenhouse production high quality standards must be aimed at to be able to compete with field products in the market. Therefore, sometimes higher EC values are maintained than wanted for maximum production. Such EC values can be realised by accumulation of residual salts, or by addition of nutrients above the levels necessary for an optimal nutrient uptake of the crop (Sonneveld, 1995).

7.4 Specific Salinity Effects

The discrimination between osmotic and specific effects is debatable, because of the choice of a proper reference. Generally, the osmotic potential of the reference treatment is made with soluble organic compounds with high molecular weights, because such compounds are hardly absorbed by crops. The most common compound used to that purpose is polyethylene glycol (PEG). However, the use of such compounds is questionable when looking at long term salinity. With long term salinity it is generally not the availability of water at the low (negative) osmotic potential that is responsible for the growth reduction, but rather the adjustments made by the crop to escape from it. Part of the adjustments made by crops will consist of extra ion uptake, promoted by the higher concentration of ions responsible for the salinity, which are not available when compounds like PEG are used to decrease the osmotic potential. This and other objections against osmotic comparisons are discussed by Sonneveld (2000). It was concluded that it is sometimes difficult to distinguish clearly between osmotic and specific effects. Much more it is difficult to distinguish between toxicity and nutrient disturbance (Bernstein,

Table 7.6 Yield of bouvardia (kg m^{-2}) as effected by different EC values in the root environment (dS m^{-1}) realised by addition of nutrients and NaCl, Na or Cl. (mmol l^{-1})

Experiment 1		Experiment 2	
Treatment EC/NaCL	Yield	Treatment EC/Na or Cl	Yield
2.2/0	1.31	2.0/0	6.49
2.8/5	1.20	2.0/10Cl	6.08
3.4/10	0.99	3.0/10Cl	5.90
4.0/15	0.84	2.0/10Na	3.40
4.6/20	0.78	3.0/10Na	5.27
4.6/0	1.18	3.0/0	6.32

After Sonneveld et al. (1999). *Reprinted by permission of Marcel Dekker*

1964), being the effects mainly determining the specific salinity effects as mentioned before.

A good example of a specific salinity in which toxicity and nutrient disturbance cannot be well distinguished is given by Sonneveld et al. (1999), in two experiments with bouvardia. In the first experiment a series of EC values in the root environment was realised by addition of nutrients, with or without NaCl, as shown in Table 7.6. The flower weights are decreased by increasing conductivity. This, however, was specifically the case where the EC partly was increased with NaCl, which follow from comparison between the treatments 4.6/20NaCl and 4.6/0NaCl in experiment 1. In the second experiment it was investigated which ion was responsible for the specific sensitivity. In this experiment the EC was increased solely by nutrients or either by Na or by Cl. From the yields it is clear, that there was a specific reduction by addition of Na, which follows from the low yield at treatment 2.0/10Na. However, when the same concentration of Na was given with a higher EC, which include a higher concentration of nutrients, the yield is improved despite the higher EC. The data of the tissue analysis listed in Table 7.7 shows the Na effects. The high Na at the low nutrient status 2.0/10Na, strongly reduces the uptake of all nutrients in comparison with the reference treatment 2.0/0 with 20–40% and strongly increases the Na concentration in the tissue. An increased nutrient supply, treatment 3.0/10Na, strongly increases the nutrient concentrations in the tissue more or less up till the

Table 7.7 Mineral concentrations (mmol l^{-1}) in the total top mass of bouvardia as affected by the addition of Na in the root environment. The data are derived from experiment 2, presented in Table 7.6

	Minerals in plant tissue							
Treatments	N	NO_3	P	K	Ca	Mg	Na	Cl
2.0/0	2164	433	192	907	363	126	7	64
2.0/10Na	1685	93	128	559	213	85	151	82
3.0/10Na	2176	389	202	838	339	157	96	64

After Sonneveld et al. (1999). *Reprinted by permission of Marcel Dekker*

level of the reference treatment, but it also reduces the Na concentration. So, it is not possible to explain what factor is responsible for the strong growth reduction, the high sodium concentration in the tissue, the low nutrient concentrations in it or a combination of both factors.

Specific sensitivity to NaCl is a well known phenomenon (Greenway and Mums, 1980), but is not frequently found in greenhouse crops. In experiments with cucumber specific NaCl sensitivity was found sometimes (Sonneveld and van Beusekom, 1974a; Sonneveld and Van der Burg, 1991), but was not always manifested (Sonneveld and Voogt, 1978). Other examples of specific sensitivity of NaCl are bouvardia, as mentioned before, aster (Sonneveld et al., 1999) and anthurium (Voogt, 1981). In the experiment with last crop the addition of NaCl was compared with a mixed salt addition to the irrigation water. The results of the experiment are shown in Fig. 7.4. Yield correlated better with Cl than with the EC, indicating that the crop is specific sensitive for NaCl. Most experiments are carried out with the addition of NaCl and then it is not clear whether Na or Cl is responsible for the specific yield reduction. In the example of bouvardia described before, it was made clear in a further study that Na was the disturbing ion. However, this is not always the case.

A reduced Ca uptake and transport at high EC values in the root environment as mentioned in Section 7.3 again is an example by which no sound discrimination between osmotic and specific salinity effects is possible. The cause merely is an osmotic effect, but the plant suffer from a nutritional disorder. A disturbed Ca regulation in plants is one of the most common specific symptoms under saline conditions in greenhouse cultivation. It causes blossom-end rot in tomato (Adams and El-Gizawy, 1986; Adams and Ho, 1993), eggplant (Savvas and Lenz, 1994) and pepper (Sonneveld and Van Beusekom 1974b; Sonneveld and Van der Burg 1991). Other Ca disorders promoted by high salinity are tipburn in lettuce (Sonneveld and Van Beusekom, 1974a; Wiebe, 1967), blackheart in celery (Geraldson, 1957) and tipburn in Chinese cabbage (Van Berkel, 1987). Ca application can be too high as well, which demonstrated specific growth reduction and leaf necrosis with cucumber (Abed, 1973; Shimida, 1973; Sonneveld and Voogt,

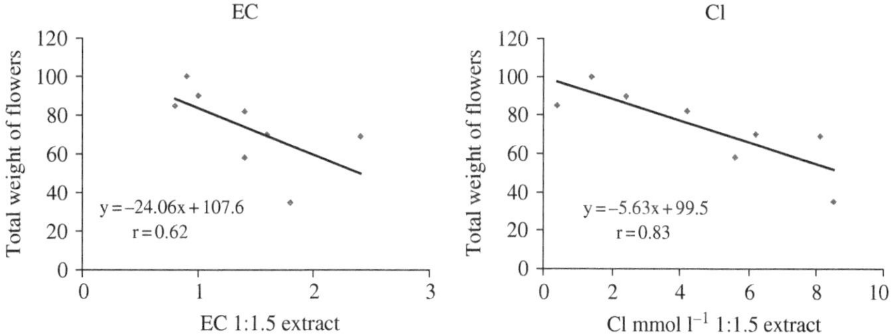

Fig. 7.4 The relationship between the EC and the Cl concentration in the 1:1½ extract on the one hand and the yield of flowers on the other hand with an anthurium crop. Data of Voogt (1981)

1978). In tomato, pepper and eggplant gold speck (De Kreij et al., 1992) green spot (Janse and De Kreij, 1989; Voogt and Sonneveld, 1985) and calix browning (Maaswinkel, 1988) are demonstrated, respectively as described in detail in Chapter 9. These symptoms are reduced with high EC values in the root environment.

High K and Mg in the root environment mostly do not result in specific symptoms in plants. However, extremely high Mg applications cause specific chlorotic and necrotic symptoms in cucumber (Abed, 1973; Sonneveld and Voogt 1978). Another, more common effect, is the competition of K and Mg with the uptake of Ca. An example of such an effect is given in Table 7.8 for a chrysanthemum crop (Sonneveld, 1981). Different chloride salts were added to the irrigation water in osmotically equal ionic concentrations. Sodium is scarcely absorbed by the crop and affects the uptake of other cations only little. Such was seriously the case with the addition of other cations. In the results of the chrysanthemum presented the flower weight was decreased strongly by the decreased osmotic potential, but the differences between the salts were small. Thus, the differences in cation uptake did not seriously affect the growth. However, with too strong reductions such ion competitions easily cause cation deficiencies in greenhouse crops, from which Ca disorders as mentioned are known best. Specific reductions in the Ca uptake by K and Mg were demonstrated also in other crops (Sonneveld and Voogt, 1978; Sonneveld, 1979), but such surely does not occur for all crops (Sonneveld and Van den Ende, 1975).

The P level is low in most soil solutions and from osmotic view point unimportant. Salinity, however, may interact with P. This interaction depends on plant species and concentration (Cerda et al., 1977; Cerda and Bingham, 1978; Maas and Nieman, 1978). For a restricted number of field grown plant species P can become toxic when this nutrient is excessively available, often only for some cultivars (Howell and Bernhard, 1961). Such effects surely will be expected in substrate growing, where generally much higher P concentrations are available in the substrate solution in the root environment than in soil solutions. P affects the uptake and distribution of Ca by plants. With a reduced uptake at low P levels, blossom-end rot in tomato was aggravated (De Kreij, 1996) and with high levels of P the Ca

Table 7.8 Effects of the addition of different cations to the irrigation water on growth and cation uptake of chrysanthemum "Dark Westland"

Treatments		Yield	Leaf analysis mmol kg^{-1}			
Salts added	Concentration mmol l^{-1}	Top weight g per stem	Na	K	Ca	Mg
No	0	130	9	1688	404	288
NaCl	25	67	30	1555	401	280
KCl	25	70	9	2417	244	128
$CaCl_2$	16⅔	76	13	1258	830	144
$MgCl_2$	16⅔	63	9	1322	172	798

Data of Sonneveld (1981).

uptake in tomato was increased and by this the occurrence of gold speck (Voogt and Sonneveld-Van Buchem, 1989).

Specific salinity effects at high concentrations of NO_3 and SO_4 are unknown. High SO_4 is mentioned as being responsible for a reduced Ca uptake (Bernstein, 1976). This, however, was not confirmed with addition of Na_2SO_4 when compared with other Na-salts (Sonneveld and Van den Ende, 1975). Specific effects of these ions also were not found in hydroponics with varying NO_3:SO_4 ratios (Nukaya et al, 1991) under conditions that the NO_3 supply is sufficiently ensured (Nukaya and Hashimoto, 2000). Replacement of NO_3 by Cl in rock wool grown tomatoes strongly aggravates the Ca uptake (Voogt and Sonneveld, 2004). It is not yet clear what phenomenon is responsible for the increased uptake of Ca, the high Cl, the low NO_3 or a combination of both changed concentrations.

7.5 Mineral Absorption

From the foregoing section it will be clear that the uptake of mineral elements can be strongly affected by salinity. First of all, with salinity the concentrations of Na and Cl are often relatively strongly increased in the soil and substrate solutions and by this the absorption of these elements by plants. NaCl often is abundantly present in water all over the world and therefore, in irrigated areas Na and Cl commonly contribute substantially to salinity. Commonly, the concentrations of Na and Cl in the plants and in the root environment are closely related. The relationships can be approached by linear regression, but in some cases an exponential function shows a better approach (Sonneveld et al., 1999). For most crops the uptake of Cl is relatively higher than Na. In Fig. 7.5, the relationship between Na and Cl concentrations in the root environment are shown in relation to the concentrations in the plant shoots of lily (Baas and Van der Berg, 1996).

Despite the uptake of extra Na and Cl with increasing salinity the uptake of nutrients is mostly not seriously affected. This is shown with the data of Table 7.9 where

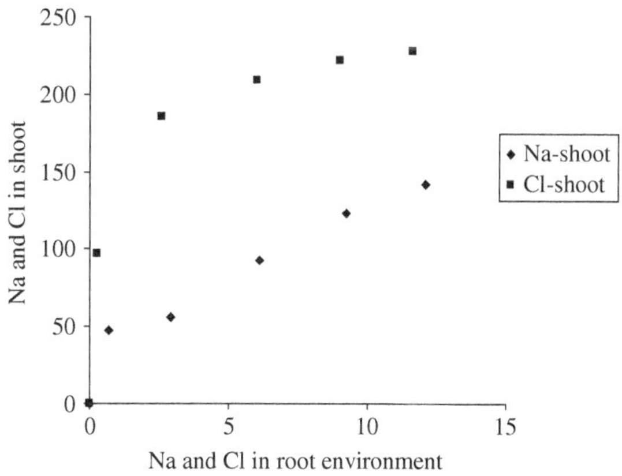

Fig. 7.5 The relationship between Na and Cl concentrations in the substrate solution (mmol l^{-1}) and Na and Cl concentration in the shoot of lily cv "Connecticut King" (mmol kg^{-1} dry matter). After Baas and Van der Berg, (1996)

Table 7.9 Mineral contents of tomato, sweet pepper and cucumber leaves (mmol kg^{-1} dry matter) grown in soil as affected by NaCl applications in the irrigation water. The plants were irrigated with water containing 5 and 30 mmol l^{-1} NaCl for the standard and the saline treatments, respectively

Crop	Water	Na	K	Ca	Mg	N	Cl	P	S
Tomato	Standard	259	941	1132	360	2850	732	87	389
	Saline	704	765	1242	354	2821	1024	92	395
Sweet pepper	Standard	4	1458	848	609	3257	96	68	175
	Saline	26	1394	857	564	3289	370	77	202
Cucumber	Standard	61	824	1180	547	3186	614	148	150
	Saline	257	656	1115	669	3300	1248	177	94

Data after Sonneveld and Van den Ende (1975); Sonneveld and Voogt (1978) and Sonneveld (1979).

the mineral contents of different crops at high and low NaCl are listed. The greatest differences in mineral uptake under standard and saline conditions occur with the uptake of Na and Cl. Some crops like sweet pepper have a strong resistance against the uptake of Na. In this process particularly the pith cells play a decisive role in the retarding of the Na from the shoot (Blom-Zandstra et al., 1998). Rose also showed an effective exclusion of Na (Baas and Van der Berg, 2004; Sonneveld et al., 1999). For tomato and cucumber, crops with a high uptake of Na, was found that the K uptake was substantially reduced. Such high uptakes are mostly not harmful to crops. In some cases it seems possible that Na relieves K in plants (Besford, 1978).

An increase of all nutrients in equal ratios in the root environment can results in a stimulation of the K uptake, while the uptake of Ca and of Mg is reduced. Such an effect is shown by the results listed in Table 7.10. Relatively big differences occurred in the suboptimal range of the EC value from 0.75 to 2.5. Despite that the EC values in this experiment were realised by addition of nutrients in equal ratios, the uptake of K is increased and those of Ca and Mg reduced with increasing EC. This easily promotes Ca deficiency, especially in combination with high EC values whereby the Ca transport in plants is hindered to parts sensitive to Ca deficiency. Therefore, high EC values easily induce Ca deficiency in the distal end of tomato fruits (Adams, 1990). However, effects as presented in Table 7.10 depend on crop and growing conditions, but are often manifest for vegetable fruit crops (Sonneveld and Welles, 2005). For other crops different effects are demonstrated. A relative

Table 7.10 Analytical data (mmol kg^{-1} dry matter) of laminae of young tomato leaves as affected by different EC values of equal ratios of nutrients

	EC values		
Elements	0.75	2.5	5.0
K	658	953	1080
Ca	858	794	587
Mg	274	161	160

After Sonneveld and Voogt (1990). *Modified by permission of Springer*

increase of the concentrations of all nutrients in the external solution above the optimum concentration can lead to a gradual increase of the uptake of all cations, like shown by Sonneveld (2002). However, related to the increase of the external concentrations, the increase of the concentrations in the plant tissues often are marginal in this domain.

7.6 Interpretation

The interpretation of salinity in the root environment under greenhouse conditions differs strongly from those under field conditions. One of the most striking differences is that under greenhouse conditions the interpretation is focussed on too high as well on desired values of the EC. From the previous paragraphs can be concluded that the EC value in the root environment to obtain an optimal effects on plant production has a broad range. This, not only depend on the plant type, cultivar grown and the objective of production, but also on the growing conditions. Important factors among the growing conditions that affect the interpretation of the EC are the method of water supply, the fertilization, the salt distribution in the root zone, and the climatic conditions, as radiation level, temperature, humidity and CO_2 level.

Gale and Zeroni (1984) stated that many greenhouse plants can be grown at an osmotic potential in the root environment of –100 till –200 kPa, which is equivalent to an EC of 3–6 dS m^{-1}. These values are mentioned by authors in relation to so called "controlled environment agriculture" (CEA) conditions. These conditions were characterized by a constant ion concentration and high water potential around the roots, a low evaporation potential, moderate temperatures and carbon dioxide enrichment, which all are factors aggravating salinity resistance. Such conditions can be realised in greenhouses in North-West Europe and comparable climatic zones all over the world from autumn till early spring and then the limits from 3 to 6 dS m^{-1} as acceptable for greenhouse productions, are in agreement with findings in these areas (Sonneveld and Welles, 1988). For summer conditions, however, such values seemed to be too high for most crops, to get maximum productions.

For some vegetable crops maximum yield is not the only goal for greenhouse productions, as the produce quality is at least important as the maximum yield. Therefore, despite the fact that maximum productions for substrate grown fruit vegetable crops can be expected with EC values not higher than 2.5–3.0 dS m^{-1} (Sonneveld and Van der Burg, 1991) values of 2.5–5.5; 2.0–4.5 and 1.8–3.6 for tomato, cucumber and sweet pepper are recommended, respectively, (De Kreij et al., 1997a, b, c,) to ensure a satisfying quality level of the produce. For soil grown vegetable crops information about salinity threshold values has been sparingly gained. Sonneveld (1988) mentioned values for saturation extracts between 3.8 and 4.7, comparable with values in the soil solution of 5.9 and 7.3 (Sonneveld et al., 1990). For soil grown tomato, cucumber and sweet pepper crops EC values of 1.4–1.7; 1.0–1.6 and 1.1–1.4 dS m^{-1} in the 1:2 volume extract (Van den Bos et al., 1999) corresponding with 5.2–6.1; 4.0–5.8 and 4.3–5.2 dS m^{-1} in the soil solution

(Sonneveld et al., 1990) are maintained, respectively to ensure a satisfied quality level of the produce.

In contradiction to fruit vegetables, for flower crops the EC values recommended are mostly close to the salinity threshold values, since high EC values are mostly unfavourable for the produce quality. Salinity threshold values found for cut flowers grown in substrate were often between 1.0 and 2.5 dS m^{-1} (Sonneveld et al., 1999, 2004). EC values recommended for rose, *Anthurium*, carnation and *Hippeastrum* varied between 1.2–2.7, 0.7–1.3, 1.5–3.3 and 1.5–3.3, respectively (De Kreij et al., 1997d, e, f, g). For soil grown cut flowers threshold values varying between 3.4 and 4.0 dS m^{-1} in the saturation extract have been found (Sonneveld, 1988). These values correspond with values of 5.3 and 6.2 in the soil solution, respectively (Sonneveld et al., 1990). Values recommended for these crops grown in greenhouse soils are 1.0–1.4, 0.9–1.4 and 0.8–1.4 dS m^{-1} in the 1:2 volume extract (Van den Bos et al., 1999) for rose, carnation and *Hippeastrum* respectively. These values correspond with values of 4.0–5.2, 3.6–5.2 and 3.3–5.2 in the soil solution respectively.

For a series of potted plants grown in peat substrates threshold values between 0.7 to beyond 3.8 dS m^{-1} were found (Sonneveld, 2000), while recommended values are between 0.25 and 1.2 dS m^{-1} in the 1:1½ volume extract (De Kreij et al., 1999), corresponding with 0.8 and 3.0 in the substrate solution, respectively (Sonneveld and van Elderen, 1994).

EC values in the soil solution recommended to crops grown in soil generally are much higher than those recommended in the substrate solution. This is especially the case with cut flowers, which can be explained by the fact that for these crops grown in substrate no specific high EC values are recommended like this is done with vegetable crops to improve plant condition or fruit quality. The general high EC values recommended for soil grown crops are in agreement with the high threshold values found with these crops in comparison with substrate grown crops. The reason for these much higher threshold values may be attributed to the great spatial diversity and the big root volume available for soil grown crops. Through that soil grown plants are ensured of an escape from local too high EC values in the root environment (See Chapter 8). The difference found between the salinity threshold values of soil and of substrate cultivation seems to be in contradiction with the comparison of the SYD values discussed in Section 7.4. However, the reaction of plants to salinity escape for threshold values can differ from the escape of SYD values.

The great variation in crops, especially as found with flower crops, and growing conditions in the greenhouse industry is responsible for the wide range of interpretation values for the osmotic potential in the root environment. In view of this great variation, it will be not surprising when among them are crops specific sensitive to a certain ion, mostly Na, eventually Cl. This especially is operative for potted plants, presenting the greatest variation of plants. In a study with six cut flower crops (Sonneveld et al., 1999), two of them, bouvardia and aster, showed to be specific sensitive to sodium chloride. The bouvardia crop showed the extreme sensitivity directly with the harvest of the first flush of flowers (see Section 7.4). The aster crop showed no salinity effect at all with the harvest of the first flush of flowers, while the re-growth after the first harvest was strongly specifically disturbed by addition

Table 7.11 Flower weight of the first flush and re-growth afterwards of aster grown in substrate as affected by EC ($dS\ m^{-1}$) and NaCl concentration ($mmol\ l^{-1}$) in the root environment

EC/NaCl	Flower weight kg m^{-2} (%)	Re-growth %
1.8/0	0.35(100)	86.5
2.4/5	0.38(109)	75.8
3.0/10	0.35(100)	63.5
3.5/15	0.36(103)	52.2
4.2/20	0.37(106)	39.7
4.2/0	0.39(111)	75.8

After Sonneveld et al. (1999). *Reprinted by permission of Marcel Dekker*

of NaCl. This disturbance started directly from the lowest NaCl level in the root environment, being 5 mmol l^{-1} NaCl, like is shown in Table 7.11. Thus, also the growth stage affects the occurrence of salinity disorders. Interpretation of specific salinity effects of crops should be based on the concentration of that specific ion in the solution in the root environment, and will be diagnosed over the whole growing period of crops involved.

7.7 Fertilization and Salinity

When desired EC values in the root environment are higher than necessary for an optimal uptake of plant nutrients, there will be an interaction between the concentration of nutrients supplied and the salinity level. When no specific salinity effects occur, the type of salts supplied to increase the EC above the EC necessary for optimal nutrient uptake (c_n in equation 7.2) the interval between c_n and the desired EC level can be filled up with any salt available. In such cases mostly residual salts present in the irrigation water are utilised to let those accumulate in the root environment up to the desired EC value.

The nutrient level desired in the root environment depend on crop, and growing conditions, but it is well known that plants are able to take up sufficient nutrients, especially N, P and K, for optimal growth at relatively low concentrations in the root environment (Clement et al., 1978; Ingestad, 1970; Massey and Winsor, 1980; Siddiqi et al., 1998; Voogt and Sonneveld, 2004; Wild et al., 1987). The very low concentrations sometimes presented in these studies are often found under specific growing conditions and not universal applicable. Often, the experiments were carried out in substrate or hydroponics systems mostly with a high circulation rate, which is not suitable for commercial grown crops. Sonneveld (2000), concluded from a number of experiments in substrate systems that for an optimum nutrient uptake in commercial substrate systems the total nutrient status in the root environment for most crops should be on a level of about 1.5 dS m^{-1}. It can be expected that for soil grown crops comparable concentrations are required, because of the fact that in protected cultivation top dressings are frequently carried out. However, the nutrient concentrations recommended to growers for soil grown crops give occasion to higher EC values in the soil solution for nutrient requirements than those recommended for substrate systems (Van den Bos et al., 1999).

Table 7.12 Concentrations of macro elements in the root environment recommended for a tomato crop grown in a substrate system when no NaCl accumulates from the irrigation water and when the maximum acceptable accumulation of NaCl occurs. The EC maintained is 4.0 dS m^{-1}

	Accumulation of NaCl	
Elements	Without	Maximum
K	8.0	4.0
Na	0	22.0
Ca	10.0	4.0
Mg	4.5	1.5
NO_3	23.0	10.5
Cl	0	22.0
H_2PO_4	1.0	0.5
SO_4	6.5	2.0

With substrate systems for tomato with a recommended average EC in the substrate solution of 4.0 dS m^{-1} and a recommended total nutrient concentration agreeing with 1.5 dS m^{-1} give space for an accumulation of residual salts of up to 2.5 dS m^{-1}. When Na and Cl are accumulated in equivalently equal concentrations the maximum tolerable concentration in the root environment for both ions is about 22 mmol l^{-1}. Thus, the nutrient levels maintained in the root environment strongly affects the salt accumulation allowed. In Table 7.12 the consequences for the nutrient concentrations for a tomato crop grown in substrate are shown in case of maximum tolerable accumulations of Na and Cl from the irrigation water in comparison with the use of rain water for irrigation and the EC is exclusively realised with nutrients. For crops with which no high EC in the root environment is required, an interaction of decreased nutrient levels with an increased salinity as described does not exist. Interactions the other way round exist for crops with which a higher nutrient supply is desired with an increasing salinity, to control undesirable uptake of toxic ions and improve the nutrient uptake. This for example is shown with the bouvardia crop (Sonneveld et al., 1999), data of which are listed in Tables 7.6 and 7.7.

7.8 Water Quality

Salinity in greenhouses is merely caused by accumulation of residual salts from the irrigation water. Other factors that will contribute to it are the application of fertilizers and the use of soil improvers. Ions present in the irrigation water will accumulate in the root environment when the concentration is higher than the apparent uptake concentration. This means, that ions like Na and Cl absorbed in relatively low concentration but often abundant present in irrigation water strongly contribute to salinity. Ions like NO_3, NH_4, K and HPO_4 are mostly present in low concentrations in the irrigation water, but are absorbed in relatively high concentrations. Thus, such ions do not contribute in the salinity from the irrigation water and when they do, it originates from over-fertilization. Ca, Mg and SO_4 true enough are absorbed in notably quantities, but the concentrations in irrigation water easily exceed the apparent uptake concentrations. This is demonstrated by the data listed in Table 7.13, where ion concentrations of some types of irrigation water are shown in comparison

Table 7.13 Composition of irrigation waters in comparison with apparent uptake concentrations of greenhouse crops. All concentrations are given as mmol l^{-1}

	Composition of irrigation water			Apparent uptake concentrations	
Elements	Rain water[1]	Surface water[2]	Well water[3]	Tomato	Rose
K	0.04	0.4	1.2	6.1	1.9
Na	0.59	4.6	13.0	0.4 – 1.4	0.0 –0.1
Ca	0.14	3.0	10.0	2.2	0.9
Mg	0.07	0.8	1.0	0.9	0.3
N	0.28	0.4	–	9.6	5.2
Cl	0.66	4.1	10.0	0.6 – 1.6	0.1 –0.2
S	0.22	1.8	14.1	1.2	0.4
P	–	0.02	–	1.1	0.4

[1] From coastal areas in the Netherlands;
[2] from ditches in greenhouse districts in the western part of The Netherlands;
[3] deep well water from South Tunesia.

with uptake concentrations of two greenhouse crops; tomato and rose representing high and low apparent uptake concentrations among these crops, respectively (Sonneveld, 2000).

Salinity in the root environment in greenhouses will be controlled by leaching of salts, which can be carried out after crop cultivation, but often is necessary during crop growth. Especially in the small root volumes available to plants in substrate grown crops salt accumulation occurs quite quickly and thus, a regular drain off can be required during crop growth to prevent high accumulations. The leaching fraction, defined as the ratio between the quantity of drainage water and the quantity of water supplied, can be calculated by Eq. (7.4).

$$LF = \frac{c_w - c_f - c_u}{c_d - c_u} \tag{7.4}$$

In which

LF = leaching fraction
c_w = concentration of a certain ion in the irrigation water, mmol l^{-1}
c_f = the concentration increase of that ion from the fertilizer, addition mmol l^{-1}
c_u = the apparent uptake concentration of the crop grown, mmol l^{-1}
c_d = the accepted concentration of that ion in the drainage water, mmol l^{-1}

Equation (7.4) is not only suitable for calculations on the basis of specific ions, but also for calculations based on total salt concentrations (EC). The apparent uptake concentration is represented by the sum of all ions; an "apparent EC uptake".

The concentrations in the irrigation water, the concentrations added by the fertilizer applications and the apparent uptake concentrations are more or less given values in the circumstances. The concentration accepted in the drainage water should be appointed by the grower.

For tomatoes in the foregoing section it was calculated that an average Na and Cl concentrations in the root environment of 22 mmol l^{-1} should be acceptable. The average ion concentrations mostly can be estimated from the average of the input and output of the growing system (Sonneveld and Voogt, 2001), like given in Eq. (7.4).

$$c_{ss} \approx \frac{1}{2}(c_s + c_d) \tag{7.5}$$

in which

c_{ss} = average concentration accepted in the root environment, mmol l^{-1}
c_s = concentration in the solution supplied (c_w + c_f), mmol l^{-1}, see formula (7.4)
c_d = accepted concentration in the drainage water

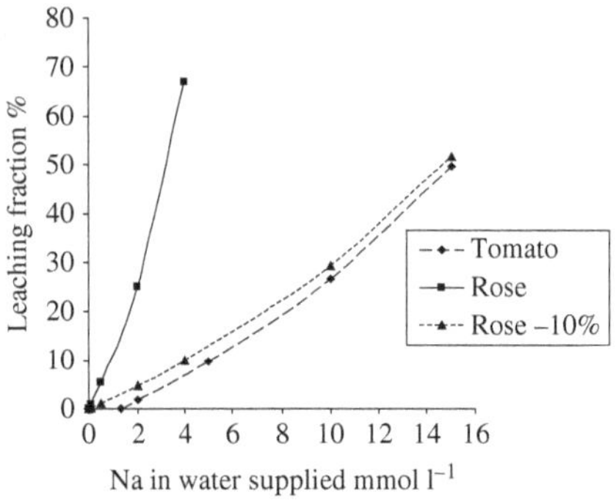

Fig. 7.6 Relationship between the Na concentrations of the water supplied and the leaching fractions necessary to keep the salinity level in a substrate system within preset limits. For tomato, rose and rose −10% an average Na accumulation in the root environment is accepted of 22, 5 and 22 mmol l^{-1} respectively. At an average accumulation of NaCl up to 22 mmol l^{-1} for rose a growth reduction of 10% will be calculated

By means of Eqs. (7.4) and (7.5) the relations between the concentration in the solution supplied and the required leaching fraction can be calculated. In Fig. 7.6 results of such calculations are shown in case that Na and Cl are accumulated in the root environment of a substrate system. Na is used as indicator, because this ion generally will accumulate quicker than Cl in the root environment, because of the general lower uptake of Na than those of Cl. For tomato a required average EC value of 4 dS m^{-1} and for rose an acceptable EC value of 2.1 dS m^{-1} is used, being the salinity threshold value (c_t) for this crop (Sonneveld et al., 1999). For rose also calculations for an EC value of 4.0 dS m^{-1} are made as an option, being the value at which a growth reduction of 10% will occur (Sonneveld et al., 1999).

The apparent uptake concentration of Na for tomato and rose were estimated on 1.3 and 0.0,[1] respectively (Sonneveld and Van der Burg; 1991; Sonneveld et al., 1999). In this way the consequences of mineral contents in irrigation water can be calculated and the results will be taken into consideration. In such considerations factors have to be taken into account like, salt sensitivity of crops, climatic conditions, salt accumulations in the root environment, leaching fractions, irrigation methods, temporal and spatial variations, yield reductions, quality aspects of the produce, and effects of environmental pollution. In the examples presented the salt accumulation in the water storage of the root environment at the start is not taken into account, because for crops with a long growing season this accumulation in the water storage is negligible in relation to the total water use. When relatively big water storages are maintained with small root volumes in substrate systems for short growing periods adjusted calculations are possible. Under such conditions more complicated calculations can be carried out (Carmassi et al., 2005; Savvas, 2002).

In Table 7.14 global guide values are given for the irrigation water quality. The guidelines are divided in three groups. The standards mentioned under group 1 are representative for water suitable for all crops and growing conditions in protected cultivation, with scarcely need for leaching requirements. The water corresponding with the standards mentioned under group 2 as well is suitable for all crops and growing conditions in protected cultivation, but with some extra water supply, up to 20%, is necessary to meet salt accumulation. Water quality of group 3 can be used in greenhouse cultivation under condition that a high leaching fraction can be realised (20–30%). Leaching of salts can be realised by extra water supply during

Table 7.14 Guide values for water quality. The water of group 1 is required if no or nearly no leaching (< 5%) is accepted; the values of group 2 are required for moderate leaching conditions (5–20%) and for group 3 high leaching requirements are necessary (> 20%)

		Group		
Characteristic	Unit	1	2	3
EC	dS m^{-1}	< 0.5	< 1.0	< 1.5
pH		See text		
Na	mmol l^{-1}	< 0.5	< 3.0	< 5.0
Cl	mmol l^{-1}	< 0.5	< 3.0	< 5.0
Ca	mmol l^{-1}	< 1.5	< 2.5	< 3.5
Mg	mmol l^{-1}	< 0.7	< 1.25	< 2.0
SO_4	mmol l^{-1}	< 0.7	< 1.25	< 2.0
HCO_3	mmol l^{-1}	< 5.0	< 7.5	< 10.0
Fe	μmol l^{-1}	< 10	See text	
Mn	μmol l^{-1}	< 10	See text	
B	μmol l^{-1}	< 15	< 25	< 50
Zn	μmol l^{-1}	< 3	< 5	< 10
Cu	μmol l^{-1}	< 1	< 1.5	< 3

[1] Being the rounding off from the average < 0.05

crop cultivation or possible after cropping periods, as discussed in Section 6.4. Last option is operative for soil grown crops, with a substantial rooting volume. The small root volume of substrate grown crops hardly offers possibilities for salt accumulation without a strong decrease of the osmotic potential of the soil solution. Thus, for substrate grown crops the extra water supply for leaching of residual salts merely will be realised during cropping. Therefore, the water of group 3 mostly is not recommended for substrate grown crops, because of the high leaching requirements during crop cultivation and the thereby heavy nutrient losses to the environment. Recently more detailed guide values for the quality of irrigation water focussed on EC, Na and Cl are published by Voogt (2008).

The EC values to accept are connected with the thereby mentioned concentrations of Na, Cl, Ca, Mg and SO_4. The pH of irrigation water is of low importance, because the pH can be adjusted by application of acids. The quantity of acid necessary to control the pH mainly depends on the HCO_3 concentration. The HCO_3 limits are not related in the first place by a lack of possibilities of acid dosing, but earlier with too high concentrations of accompanying cations. Mostly by Ca and Mg but sometimes by Na, being the worst case. Among the acids used for pH correction HNO_3 can be used in highest concentrations and in minor concentrations H_3PO_4 and H_2SO_4 in agreement of the anion uptake of the crop and the concentration already available in the primary water. Acid dosing at high carbonate waters is quite common for substrate cultivation. For soil grown crops, however, seldom applied. High carbonate concentrations surely increase the pH value of soils. Irrigation of 1000 mm water with 1 mmol HCO_3 is equivalent with 50 g $CaCO_3$ per m^2.

The concentration of micro elements has nothing to do with salinity, but mostly with toxicity. Zn in irrigation water often occurs when it has been in contact with galvanised pipe lines during transport, storage basins or greenhouse gutters. When used for soil grown crops the toxicity is strongly restricted because Zn is adsorbed on the clay and humus particles. On long terms, however, it causes soil pollution. The same will be the case with Cu, which is specifically adsorbed on humus particles (Verloo, 1980). Adsorption processes as described for soil also occur by substrates with adsorption complexes, like peaty substrates and coir material. Especially with cultivation in inert substrates or with hydroponics Zn and Cu toxicity can be expected with high concentrations of these elements. The toxicity of Mn depends strongly on the pH maintained in the root environment. With increasing pH this element easily precipitates in the form of manganese oxides in soil, substrates and even hydroponics (Sonneveld and Voogt, 1980). pH values below 6.0 are beneficial for the availability of Mn and by this the risk of toxicity at high concentrations. High concentrations of Fe are mostly not toxic to crops, but Fe precipitates easily and blocks irrigation systems like the nozzles of trickle irrigation systems. For this purposes the water should be more or less free of any Fe, but at least it should be $< 10\ \mu mol\ l^{-1}$ as given in Table 7.14. With sprinkler irrigation systems blocking of the nozzles is mostly not a problem, but with high concentrations ($> 100\ \mu mol\ l^{-1}$) pollution by precipitated Fe on the crop can be a problem. For ornamental crops even lower concentrations can be a handicap, because of a lower "pollution acceptance" for ornamentals. At a low pH of the water Fe concentrations > 100 μmol l^{-1} can cause leaf burning (Sonneveld et al., 1991; Ten Cate, 1978). See also Section 6.4 for detailed information about Fe in irrigation waters.

Other elements that can be traced in irrigation water as being a problem for greenhouse industry are B, F and Br. B can accumulate in the root zone when the concentration exceed the uptake concentration and will become toxic to crops (Bingham and Garber, 1970). F is especially toxic to bulb and tuber crops (Roorda van Eysinga, 1974). Br mostly originated from soil desinfestation with CH_3Br (Van den Bos, 1993). The uptake of Br by consumption crops will be mainly controlled because of limits for human health and is only toxic for specific sensitive crops. Acceptable concentrations for B are listed in Table 7.14. Rough limits for F and Br are presented by Sonneveld et al. (1991). For F the recommendation for bulb and tuber crops is < 25 μmol l^{-1} and for other crops < 50 μmol l^{-1}. For Br concentrations < 40 μmol l^{-1} are desirable when used for soil grown crops. For substrate grown crops there is no experience with this element, but it is likely that the limit will be lower for this type of growing.

7.9 Dealing with Salinity

Greenhouse production in many countries is focussed on a consumer market and the drive of the greenhouse industry to supply such a market should be the delivery of better and possible cheaper products than those from field grown crops. Moreover, consumer markets are characterized by diversity, quality and immediate answers to demands of often luxurious products. Added to this, the increasing concern of the customers to environmental issues presses the growers to search for sustainable production methods too.

Greenhouse industry offers excellent possibilities for a great diversity of products with an optimal control on quality and environmental consequences. Therefore, this industry more and more should formulate the effects of different factors that determine growth and quality of crops grown, in relation to environmental consequences. One of the main factors in this respect is under discussion in the present chapter. Salinity in greenhouse production not only affect yield, but offers possibilities for control of produce quality. Furthermore, with a close control on the salinity the environmental consequences can be restricted. Such controls easier can be optimised in systems of substrate growing than of soil growing. The smaller root volume of substrate systems offers excellent possibilities for short term adjustments of salinity in relation to other growing factors. Therefore, salinity in greenhouse industry should not be considered as just a handicap connected with yield reduction and a need for leaching of salts with environmental pollution by nutrients in his track. Mild salinity may be a tool to control crop development and produce quality. Important factors for the management of salinity are the salinity threshold values, the required levels of nutrients and the ion distribution within the root environment. The effects of these factors are strongly different for crops, but depend on other growing factors too, as discussed before. The exact effects of these factors

Fig. 7.7 Relation diagram for factors affecting salinity effects on plants

in relation to salinity is often insufficient known and at present mainly the direction of various factors interacting with salinity is traced, like shown by Sonneveld (2000) and presented in Fig. 7.7. When optimising salinity in greenhouse industry the strategic and tactical decisions should be in line with the expected management. Strategic factors in salinity management are e.g. the soil or substrate type, the growing system, and the irrigation system, because these factors can be nearly not or not at all changed and adjusted during the growing season. Other factors have a more tactical character, like the primary water quality, the acceptable EC levels, the nutrient level in relation to the residual salts and the decision about reuse of drainage water, because such factors have opportunities for adjustment during crop growth. Some factors are suitable for full management during crop cultivation,

Table 7.15 N efficiency simulated in relation to crop, growing system and interpretation of the EC in the root environment (Sonneveld, 2000). The EC of the uptake solution is derived from Sonneveld and Voogt (2001) and the EC of the supplied solution and of the drainage water are calculated using Eqs. (7.4) and (7.5). The conditions taken into account are a leaching fraction of 25%, a NaCl concentration in the solution supplied of 2 mmol l^{-1} and an average nutrient concentration in the root environment agreeing with an EC of 1.5 dS m^{-1}

Crop	Drainage system	N efficiency	Interpretation value	EC uptake	EC supplied	EC drainage
Tomato	Free	0.51	$EC_{SS} = 4.0$[1]	1.3	2.4	5.6
Tomato	Reuse	0.98	$EC_{SS} = 4.0$[1]	1.3	2.4	5.6
Rose	Free	0.52	$EC_{SS} = 2.1$[2]	0.6	1.2	3.0
Rose	Reuse	0.50	$EC_{SS} = 2.1$[2]	0.6	1.2	3.0
Rose	Free	0.52	$EC_{S} = 2.1$[3]	0.6	1.2	3.0
Rose	Reuse	0.86	$EC_{S} = 2.1$[3]	0.6	2.1	6.6

[1]EC of the substrate solution as a required value;
[2]EC of the substrate solution as an acceptable value;
[3]EC of the supplied solution as an acceptable value.

like the rate and the frequency of the water supply and the fertilizer application. The effects of the management of the water and fertilizer use in relation to salinity aspects can be tremendously, especially in substrate systems. This is shown in Table 7.15, where results are summarized of simulations of different water supply strategies in relation to water quality on salt accumulation and N use efficiency (Sonneveld, 2000). In all cases a minimum nutrient status corresponding with an EC_{SS} of 1.5 in the root environment is taken into account. For tomato EC_{SS} is fixed to 4.0 dS m^{-1} with respect to quality demands (De Kreij et al., 1997a) and simultaneously this level offers the opportunity for alleviation of the effects of accumulation of residual salts in the root environment. Henceforth, it also increases the possibilities of reuse of drainage water and increase the efficiency of water and nutrients. The nutrient efficiency is reflected by the N efficiency, expressed as the ratio of N uptake and N supply. For rose no improvement of nutrient use efficiency is possible when an $EC_{SS} = 2.1$ dS m^{-1} is set as a maximum average value in the root environment, which is in agreement with the salinity threshold value found for this crop (Sonneveld et al., 1999). However, it has been found that crops, with respect to salinity, especially react on the lowest EC value in the root environment (Sonneveld and Voogt, 1990; Sonneveld and De Kreij, 1999). This can be interpreted as the EC of the supplied solution (EC_S). Starting from this interpretation high salt accumulations in the root environment by reuse of drainage water are acceptable, as calculated with $EC_S = 2.1$ dS m^{-1} for rose, followed by a strong improvement of nutrient efficiency. The results of the simulation as presented in Table 7.15 appoint to great differences of nutrient efficiency in relation to the water and nutrient management, and the interpretation of determinations of the EC value in the root environment. In experiments it has been found that especially for tomato, but also for cucumber with big spatial differences between EC values in the root environment the growth is mainly determined by the lowest EC value. See also Section 8.2. For other crops results of such experiments are not yet available. Thus, for rose cannot be estimated, what the effect will be on the growth of such a high EC of 6.6 dS m^{-1} of the drainage water, being the result of the last simulation in this survey.

For soil grown crops the same questions arise with respect to interpretation. Especially with the use of trickle irrigation big spatial differences occur. With this irrigation system the differences are most manifest for spots under and between the nozzles, see Section 8.1. The phenomenon of spatial variation is quite common in protected cultivation and with all types of irrigation systems, horizontal as well vertical differences will be found (Sonneveld and Heinen, 1997). Under such conditions a good tuning between the place and method of sampling and interpretation of analytical results is crucial.

References

Abed A H 1973. Einfluss von Salzkonzentration und Konzentrationsänderung in der Nährlösung auf Assimilation und Transpiration von Gurkenpflanzen. Dissertation Technischen Universität Hannover, 117pp.

Adams P and El-Gizawy A M 1986. Effect of salinity and watering level on the calcium content of tomato. Acta Hort. 190, 253–259.

Adams P 1990. Effect of salinity on the distribution of calcium in tomato (*Lycopersicon esculentum*) fruit and leaves. *In:* Van Beusichem LM (ed) Plant nutrition – Physiology and application. Kluwer Publishers, Dordrecht, 473–476.

Adams P and Ho L C 1990. Effect of salinity on calcium transport in tomato (*Lycopersicon esculentum*). *In*: Van Beusichem L M (ed) Plant Nutrition – Physiology and Application. Kluwer Publishers, Dordrecht, 469–472.

Adams P 1991. Effects of increasing salinity of the nutrient solution with major elements or sodium chloride on yield, quality and composition of tomatoes grown on rockwool. J. Hort. Sci. 66, 201–207.

Adams P and Ho L C 1992. The susceptibility of modern tomato cultivars to blossom-end rot in relation to salinity. Hort. Sci. 67, 827–839.

Adams P and Ho L C 1993. Effects of environment on the uptake and distribution of calcium in tomato on the incidence of blossom-end rot. Plant Soil 154, 127–132.

An P Inanaga S Li X J Eneji A E and Zhu N W 2005. Interactive effects of salinity and air humidity on two tomato cultivars differing in salt tolerance. J Plant Nutr. 28, 459–473.

Awang Y R Atherton J G and Taylor A J 1993. Salinity effects on strawberry plants grown in rockwool. II Fruit quality. J. Hort. Sci. 68, 791–795.

Baas R Nijssen H M C Van der Berg T J M and Warmenhoven M G 1994. Yield and quality of carnation (*Dianthus caryophyllus*) and gerbera (*Gerbera jamesonii* L.) in a closed nutrient system as affected by sodium chloride. Sci. Hort. 61, 273–284.

Baas R and Van der Berg T J M 1996. Invloed NaCl and EC op lelie tijdens de trek. Proefstation voor Bloemisterij and Glasgroente, Aalsmeer. Rapport 25, 18pp.

Baas R and Van der Berg D 2004. Limiting nutrient emission from a cut rose closed system by high-flux irrigation and low nutrient concentrations? Acta Hort. 644, 39–46.

Bernstein L 1961. Osmotic adjustment of plants to saline media. I Steady state. Am. J. Bot. 48, 909–918.

Bernstein L 1963. Osmotic adjustment of plants to saline media. II Dynamic phase. Am. J. Bot. 50, 360–370.

Bernstein L 1964. Effects of salinity on mineral composition and growth of plants. Plant Anal. Fert. Probl. IV, 25–45

Bernstein L 1975. Effects of salinity and sodicity on plant growth. Ann. Rev. Phytopath. 13, 295–312.

Bernstein L 1976. Physiological basis of salt tolerance in plants. Proc. Intern. Symp. Genetic Control Diversity in Plants. Lahore, Pakistan, March 1976. Plenum Press, New York, 283–290.

Besford R T 1978. Effect of sodium in the nutrient medium on the incidence of potassium-deficiency symptoms in tomato plants. Plant Soil 50, 427–432.

Bingham F T and Garbar M J 1970. Zonal salinization of the root system with NaCl and boron in relation to growth and water uptake of cornplants. Soil Sci. Soc. Am. Proc. 34, 122–126.

Blom-Zandstra M Vogelzang S A and Veen B W 1998. Sodium fluxes in sweet pepper exposed to varying sodium concentrations. J. Exp. Bot. 49, 1863–1868.

Carmassi G Incrocci L Maggini R Malorgio F Tognoni F and Pardossi A, 2005. Modeling salinity build-up in recirculating nutrient solution culture. J. Plant Nutr. 28, 431–445

Cerda A Bingham F T and Hoffman G J 1977. Interactive effect of salinity and phosphorus on sesame. Soil Sci. Soc. Am. J. 41, 915–918.

Cerda A and Bingham F T 1978. Yield, mineral composition and salt tolerance of tomato and wheat as affected by NaCL and P nutrition. Agrochimica 12, 140–149.

Clement C R Hopper M J and Jones L H P 1978. The uptake of nitrate by *Lollium perenne* from flowing nutrient solution. J. Exp. Bot. 29, 453–464.

Cornish P S 1992. Use of high electrical conductivity of nutrient solution to improve the quality of salad tomatoes (*Lycopersicum esculentum*) grown in hydroponic culture. Austr. J. Exp. Agric. 32, 513–520.

De Kreij C and Van Os P C 1989. Production and quality of gerbera in rockwool as affected by electrical conductivity of the nutrient solution. 7th International Congress on Soilless Culture. Flevohof, The Netherlands, 255–264.

De Kreij C Janse J Van Goor B J and Van Doesburg J D J 1992. The incidence of calcium oxalate crystals in fruit walls of tomato (*Lycopersicon esculentum* Mill.) as affected by humidity phosphate and calcium supply. J. Hort. Sci. 67, 45–50.

De Kreij C 1996. Interactive effects of air humidity, calcium and phosphate on blossom-end rot, leaf deformation, production and nutrient contents of tomato. J. Plant Nutr. 19, 361–377.

De Kreij C Voogt W Van den Bos A L and Baas R 1997a. Voedingsoplossingen gesloten teelt systemen, Tomaten. Proefstation voor Bloemisterij en Glasgroente, Naaldwijk. Brochure VG 2, 21pp.

De Kreij C Voogt W Van den Bos A L and Baas R 1997b. Voedingsoplossingen gesloten teelt systemen, Komkommer. Proefstation voor Bloemisterij en Glasgroente, Naaldwijk. Brochure VG 3, 27pp.

De Kreij C Voogt W Van den Bos A L and Baas R 1997c. Voedingsoplossingen gesloten teelt systemen, Paprika. Proefstation voor Bloemisterij en Glasgroente, Naaldwijk. Brochure VG 1, 19pp.

De Kreij C Voogt W Van den Bos A L and Baas R 1997d. Voedingsoplossingen gesloten teelt systemen, Roos. Proefstation voor Bloemisterij en Glasgroente, Naaldwijk. Brochure VG 4, 35pp.

De Kreij C Voogt W Van den Bos A L and Baas R 1997e. Voedingsoplossingen gesloten teelt systemen, Anthrium. Proefstation voor Bloemisterij en Glasgroente, Naaldwijk. Brochure VG 16, 26pp.

De Kreij C Voogt W Van den Bos A L and Baas R 1997f. Voedingsoplossingen gesloten teelt systemen, Anjer. Proefstation voor Bloemisterij en Glasgroente, Naaldwijk. Brochure VG 6, 21pp.

De Kreij C Voogt W Van den Bos A L and Baas R 1997g. Voedingsoplossingen gesloten teelt systemen, Hippeastrum. Proefstation voor Bloemisterij en Glasgroente, Naaldwijk. Brochure VG 10, 21pp.

De Kreij C Voogt W Van den Bod A L and Baas R 1999. Bemestingsadviesbasis Substraten. Proefstation voor Bloemisterij en Glasgroente Naaldwijk The Netherlands, 145pp.

Ehret D L and Ho L C 1986. Translocation of calcium in relation to tomato fruit growth. Ann. Bot. 58, 679–688.

Gale J and Zeroni M 1984. Cultivation of plants in brackish water in controlled environment agriculture. *In*: Staples R C and Thoeniessen G H (eds), Salinity tolerance in plants. John Wiley and Sons, New York, 363–380.

Geraldson C M 1957. Factors affecting calcium nutrition of celery, tomato and pepper. Soil Sci. Soc. Am. Proc. 21, 621–625.

Greenway H and Munns R 1980. Mechanism of salt tolerance in nonhalophytes. Ann. Rev. Plant Physiol. 31, 149–190.

Gulzar S Khan M A and Ungar I A 2003. Effects of salinity on growth, ionic content, and plant-water status of *Aeluropus lagopoides*. Commun. Soil Sci. Plant Anal. 34, 1657–1668.

Hayward H E and Long E M 1940. Anatomical and physiological response of the tomato to varying concentrations of sodium chloride, sodium sulphate and nutrient solutions. Bot. Gaz. 102, 437–462.

Hilgard E W 1919. Soils. Macmillan & Co, New York, 593 pp.

Howell W and Bernhard R L 1961. Phosphorus response of some soybean varieties. Crop Sci. 1, 311–313.

Hsiao T C and Läuchli A 1986. Role of potassium in plant-water relationships. *In*: Tinker B and Läuchli T C (eds). Advances in Plant Nutrition, Vol. 2, 281–312. Praeger, New York.

Ingestad T 1970. A definition of optimum requirements in birch seedlings, I. Physiol. Plant. 23, 1127–1138.

Janse J 1985. Bemesting zeer bepalend voor kwaliteit van tomaten en komkommers. Bemesting en kwaliteit. Tuinderij 65 no 6, 14–16.

Janse J and De Kreij C 1989.Vooral stip bij hoog calciumgehalte in de vrucht. Groenten en Fruit 44 no 3, 40–41.

Lüttge U and Smith J A C 1984. Structural, biophysical, and biochemical aspects of the role of leaves in plant adaptation to salinity and water stress. *In*: Staples R C and Toenniessen G H (eds). Salinity Tolerance in Plants, John Wiley & Sons, New York, 125–150.

Maas E V and Hoffman G J 1977. Crop salt tolerance – Current assessment. J. Irr. Drainage Div. 103, 115–134.

Maas E V and Nieman R H 1978. Physiology of plant tolerance in salinity. *In*: Crop Tolerance to Suboptimal Land Conditions. ASA Special Publication no 32, OS Salinity Lab. ARS/USDA, Riverside, CA, 277–299.

Maas E V 1986. Salt tolerance of plants. Appl. Agric. Res. 1, 12–26.

Maaswinkel R H M and Welles G W H 1986. Factors influencing glassiness in lettuce. Neth. J. Agric. Sci. 34. 57–65.

Maaswinkel R H M 1988. Geringe verdamping beperkt kans op kelkverdroging. Groenten en Fruit 44 no 22, 39.

Magistad O C Ayers A D Wadleigh C H and Gausch H G 1943. Effect of concentration, kind of salt, and climate on plant growth in sand cultures. Plant Physiol. 18, 151–166.

Massey D and Winsor G W 1980. Some response of tomato to nitrogen in recirculation solutions. Acta Hort. 98, 127–137.

Meloni D A Oliva M A Ruiz H A and Martinez C A 2001. Contribution of proline and inorganic solutes to osmitic adjustment in cotton under salt stress. J. Plant Nutr. 24, 599–612.

Mizrahi Y 1982. Effect of salinity on tomato fruit ripening. Plant Physiol. 69, 966–970.

Mizrahi Y and Pasternak D 1985. Effect of salinity on quality of various agriculture crops. Plant Soil 89, 301–307.

Mizrahi Y Taleisnik E Kagan-Zur V Zohar Y Offenbach R Matan E and Golan A 1988. A saline irrigation regime for improving fruit quality without reducing yield. J. Am. Soc. Hort. Sci. 113, 202–205.

Nukaya A 1983. Salt tolerance studies in muskmelons and other vegetables. Technical Bulletin, no 8. Dep. Hort. Faculty, Shizuoka University, Japan. 97pp.

Nukaya A Voogt W and Sonneveld C 1991. Effects of NO_3, SO_4 and Cl ratios on tomato grown in recirculating system. Acta Hort. 294, 297–304.

Nukaya A and Hashimoto H 2000. Effects of nitrate, chloride and sulfate ratios and concentration in the nutrient solution on yield, growth and mineral uptake characteristics of tomato plants grown in rockwool system. Acta Hort. 511, 165–171.

Ohta K Ito N Hosoki T and Highashimura H 1991. Influence of the concentrations of nutrient solutions and salt supplement on quality and yield of cherry tomato grown hydroponically. J. Jpn. Soc. Hort. Sci. 60, 89–95.

Petersen K K Willumsen J and Kaack K 1998. Composition and taste of tomatoes as affected by increased salinity and different salinity sources. J. Hort. Sci. Biotechn. 73, 205–215.

Plaut Z Grava A Yehezkel Ch and Matan E, 2005. Osmotic adjustment of tomato fruits and leaves grown in sand and irrigated with saline water. Acta Hort. 697, 285–291.

Ploegman C 1976. Invloed waterkwaliteit bij kasrozen. Inst. Cultuurtechniek en Waterhuishouding, Wageningen, nota 913, 14pp.

Pluda D Rabinowitch H D and Kafkafi U 1993. Pepino dulce (*Solanum muricatum*Ait.) quality characteristics respond to nitrogen nutrition and salinity. J. Am. Soc. Hort. Sci. 118, 86–91.

Richards L A 1954. Diagnosis and improvement of saline and alkaline soils. Agric. Handbook no 60, US Dep. Agrc.

Roorda van Eysinga J P N L 1974. De opname van fluor door de wortel en de gevolgen daarvan voor diverse gewassen, in het bijzonder fresia. Thesis Wageningen Universiteit, 83pp.

Savvas D and Lenz F 1994. Influence of salinity on the incidence of physiological disorder "internal fruit rot" in hydoponical-grown eggplants. Angew. Bot. 68, 32–35.

Savvas D 2002. Nutrient solution recycling. *In*: Savvas D and Passam H (eds). Hydroponic Production of Vegetables and Ornamentals. Embryo Publications, Athens, 299–343.

Semer-Olsen A R and Gislerød H R 1980. Nutrient content of chrysanthemum grown in recirculating nutrient solution. Acta Hort. 98, 211–218.

Shear C B 1975. Calcium related disorders of fruit and vegetables. HortScience 10, 361–365.

Shimida N 1973. Excess injury of calcium and magnesium in the crops. Jpn. Agric. Res. Q. 7, 173–177.

Siddiqi M Y Kronzuker H J Britto D T and Glass A D M 1998. Growth of a tomatocrop at reduced nutrient concentrations as a strategy to limit eutrophication. J. Plant Nutr. 21, 1879–1895.

Sonneveld C and Van Beusekom J 1974a. The effect of saline irrigation water on some vegetables under glass. Acta Hort. 35, 75–85.

Sonneveld C and Van Beusekom J 1974b. De invloed van zout gietwater bij de teelt van peper en paprika onder glas. Landbouwk. Tijdschr. 86, 241–245.

Sonneveld C and Van den Ende J 1975. The effect of some salts on head weight and tipburn of lettuce and on fruit production and blossom-end rot of tomatoes. Neth. J. Agric. Sci. 23, 191–201.

Sonneveld C and Voogt S J 1978. Effects of saline irrigation water on glasshouse cucumbers. Plant Soil 49, 595–606.

Sonneveld C 1979. Effects of salinity on the growth and mineral composition of sweet pepper and egg plant grown under glass. Acta Hort. 89, 71–78.

Sonneveld C and Voogt S J 1980. The application of manganese in nutrient solutions for tomato grown in a recirculating system. Acta Hort. 98, 171–178.

Sonneveld C 1981. Specifieke zouteffecten bij chrysant. Proefstation voor Tuinbouw onder Glas Naaldwijk. Intern verslag 1981 nr 8, 18pp.

Sonneveld C and Voogt W 1983. Studies on the salt tolerance of some flower crops grown under glass. Plant Soil 74, 41–52.

Sonneveld C 1988. The salt tolerance of greenhouse crops. Neth. J Agric. Sci. 36, 63–73.

Sonneveld C and Welles G W H 1988. Yield and quality of rockwool grown-tomatoes as affected by variations in EC-value and climatic conditions. Plant Soil 111, 37–42.

Sonneveld C Van den Ende J and De Bes S S 1990. Estimating the chemical composition of soil solutions by obtaining saturation extracts or specific 1:2 by volume extracts. Plant Soil 122, 169–175.

Sonneveld C and Voogt W 1990. Response of tomatoes (*Lycopersicon esculentum*) to an unequal distribution of nutrients in the root environment. Plant Soil 124, 251–256.
Sonneveld C 1991. Rockwool as a substrate for greenhouse crops. *In*: Bajaj Y P S (ed) Biotechnology in Agriculture and Forestry 17, High-Tech and Micro-Propagation I, Springer-Verlag, Berlin, 285–312.
Sonneveld C De Kreij C and Van der Wees A 1991. Normen voor waterkwaliteit in de glastuinbouw. Proefstation voor Tuinbouw onder Glas. Series Voedingsoplossingen Glastuinbouw, no 11, 23pp.
Sonneveld C and Van der Burg A M M 1991. Sodium chloride salinity in fruit vegetable crops in soilless culture. Neth. J. Agric. Sci. 39, 115–122.
Sonneveld C and Van Elderen C W 1994. Chemical analysis of peaty growing media by means of water extraction. Commun. Soil Sci. Plant Anal. 25, 3199–3208.
Sonneveld C 1995. Fertigation in the greenhouse industry. *In*: Proc. of the Dahlia Greidinger Intern. Symposium on Fertigation. Technion – Israel Institute of Technology, Haifa, 25 March–1 April 1995, 121–140.
Sonneveld C and Van den Bos A L 1995. Effects of nutrient levels on growth and quality of radish (Raphanus sativis L.) grown on different substrates. J. Plant Nutr. 18, 501–513.
Sonneveld C and Heinen M 1997. Efficiënt gebruik van nutriënten in de glastuinbouw. In: Marcelis L F M en Haverkort A J (eds), Kwaliteit en milieu in de glastuinbouw: Stimulans tot vernieuwing. AB-DLO Thema's 4, 91–108.
Sonneveld C Baas R Nijssen H M C and De Hoog J 1999. Salt tolerance of flower crops grown in soilless culture. J. Plant Nutr. 22, 1033–1048.
Sonneveld C and De Kreij C 1999. Response of cucumber (*Cucumis sativis L.*) to an unequal distribution of salts in the root environment. Plant Soil 209, 47–56.
Sonneveld C 2000. Effects of salinity on substrate grown vegetables and ornamentals in greenhouse horticulture. Thesis Wageningen University, Netherlands, 151pp.
Sonneveld C and Voogt W 2001. Chemical analysis in substrate systems and hydroponics – use and interpretation. Acta Hort. 548, 247–259.
Sonneveld C 2002. Composition of nutrient solutions. *In*: Savvas D and Passam H (eds). Hydroponic Production of Vegetables and Ornamentals. Embryo Publications, Athens, 179–201.
Sonneveld C Van den Bos A L and Voogt W 2004. Modelling osmotic salinity effects on yield characteristics of substrate grown greenhouse crops. J. Plant Nutrition 27, 1931–1951.
Sonneveld C and Welles G W H 2005. Cation concentrations of plant tissues of fruit-vegetable crops as affected by the EC of the external nutrient solution and by humidity. Acta Hort. 697, 377–386.
Sonneveld C and Voogt W 2008. Nutrient concentrations of plant tissues of greenhouse crops as affected by the EC of the external nutrient solution. Acta Hort. 779, 313–320.
Ten Cate H R 1978. Ontijzeren van gietwater. Tuinderij 18 no 6, 52–55.
Udagawa Y 1995. Some response of dill (*Anethum graveolens*) and thyme (*Thymus vulgaris*) grown in hydroponic, to the concentration of nutrient solution. Acta Hort. 396, 302–210.
Urban I Brun R and Urban L 1995. Influence of electrical conductivity, relative humidity and seasonal variations on the behaviour of cut roses produced in soilless culture. Acta Hort. 408, 101–107.
Van Berkel N 1987. Bestrijding van rand bij Chinese kool. Groenten en Fruit 42 no 43, 28–29.
Van den Bos A 1993. Interne kwaliteit van groentegewassen. Proefstation voor Tuinbouw onder Glas Naaldwijk, Plantenvoeding in de Glastuinbouw Informatiereeks no 87, 69–74.
Van den Bos A L De Kreij C and Voogt W 1999. Bemestingsadviesbasis grond. Proefstation voor Bloemisterij en Glasgroente, Naaldwijk, Brochure PPO.171, 54pp.
Van den Ende J Koornneef P and Sonneveld C 1975. Osmotic pressure of the soil solution. Determination and effects on some glasshouse crops. Neth. J. Agric. Sci. 23, 1181–190.
Van den Ende J 1989. Estimating the composition of the soil solution of glasshouse soil. 1. Composition of soil solution and aqueous extracts. Neth. J. Agric. Sci. 37, 311–322.

Van Genugten M Th and Hoffman G J 1984. Analysis of crop salt tolerance dat. *In*: Shainberg I and Shalhevet J (eds). Ecological studies 51, Soil Salinity under irrigation, Processes and Management, Springer Verlag, Berlin, 258–271.

Van Uffelen J A M and Bakker J C 1987. EC-effecten bij paprika: Neusrot en zwelscheuren dwingen tot compromis. Groenten en Fruit 43 no 22, 30–31.

Veen B W and Kleinendorst A 1986. The role of nitrate in osmoregulation of Italian ryegrass. Plant Soil 91, 433–436.

Verberkt H 1997. Zomerklimaat cyclamen. Proefstation Bloemisterij Glasgroente, Aalsmeer. Rapport 79, 47pp.

Verkerke W Gielesen W and Engelaan R 1992. Langer houdbaar door steviger schil. Groenten en Fruit/Glasgroenten 2 no 7, 22–23.

Verkerke W De Kreij C and Janse J 1993 Keukenzout maakt zacht maar lekker. Groenten en Fruit/Glasgroenten 3 no 51, 14–15.

Verloo M G 1980. Peat as natural complexing agent for trace elements. Acta Hort. 99, 51–56.

Voogt W 1981. De zoutgevoeligheid van Anthurium andreanum. Proefstation voor de Tuinbouw onder Glas Naaldwijk. Intern verslag nr 2, 29pp.

Voogt W and Sonneveld C 1985. NH_4 application for sweet peppers in recirculating water. *In*: Glasshouse Crops Research and Experiment Station, Naaldwijk, The Netherlands, Annual Report, 23.

Voogt W and Sonneveld-Van Buchem 1989. Hoge fosfaatconcentraties hebben negatieve effecten. Groenten en Fruit 44 no 35, 38–39.

Voogt W and Sonneveld C 2004. Interactions between nitrate (NO_3) and chloride (Cl) in nutrient solutions for substrate grown tomato. Acta Hort. 644, 359–368.

Voogt W 2008. Normen voor de waterkwaliteit in de glastuinbouw, revised edition. Wageningen University Research, Greenhouse Horticulture, Wageningen The Netherlands, Report 256, 28 pp.

Wiebe H J 1967. Untersuchungen über den Blattrandbrand bei Kopfsalat. Gartenbauwissenschaft 32, 375–385.

Wild A Jones L H P and Macduff J H 1987. Uptake of mineral nutrients and crop growth: The use of flowing nutrient solutions. Adv. Agron. 41, 171–219.

Chapter 8
Crop Response to an Unequal Distribution of Ions in Space and Time

8.1 Introduction

Nutrient and salt ions often are unequally distributed in the root environment of plants and it will be expected that this strongly affect the plant reaction on the uptake of minerals and the osmotic potential. An unequal distribution of salts for example will be found with field grown crops in arid areas where the water supply is carried out by trickle irrigation (Meiri, 1984; Mmolawa and Or, 2000; Prichard et al., 1983). When under these conditions brackish water is used for irrigation, the salt accumulation at the soil surface of the dry areas between the emitters can become that strong that crystallization of salts occurs in the top layer whereby the surface is coloured white. Despite such tremendous local salt accumulations, crops often develop relatively quite well.

Also in greenhouses heavy salt accumulations and the accompanying differences in ionic composition will occur. This is shown in crops grown in the greenhouse border soil, but also with crops grown in substrate systems. Differences occur in the horizontal as well as in the vertical direction. In greenhouse soils for example great differences were measured with the use of trickle irrigation, where easily on distances of 10 cm ratios between concentrations with a factor of 10 are measured (Sonneveld et al., 1991). In substrate systems the differences in salt concentrations are most obviously for those measured between the nutrient solutions supplied and the nutrient solutions drained out (Sonneveld and Voogt, 2001). Also within the root environment of substrate systems great differences have been found; in rock wool (Van der Burg and Sonneveld, 1987) as well in peat substrates (Ondrasek et al., 2008). Ratios between highest and lowest EC value easily attained values between 3 and 5. Most striking differences are shown with ebb and flow systems for potted plants. Nutrient solutions supplied from the bottom strongly accumulate in the top layer by evaporation, as long as there is no irrigation from top. An example of such a situation is shown in Fig. 8.1, whereby the EC values are expressed as dS m^{-1} of the 1:1½ extract (De Kreij and Straver, 1988). The differences found with sub irrigation between top and bottom part are strongly dependent on the level of fertilization, as shown by Cox (2001).

C. Sonneveld, W. Voogt, *Plant Nutrition of Greenhouse Crops*,
DOI 10.1007/978-90-481-2532-6_8, © Springer Science+Business Media B.V. 2009

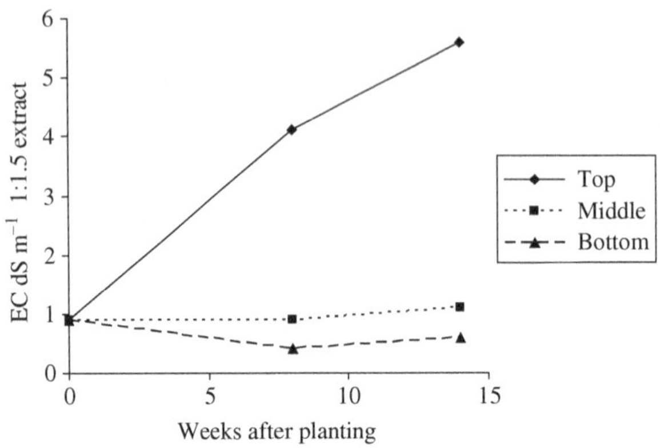

Fig. 8.1 The EC of the 1:1½ volume extract (dS m^{-1}) in three layers of substrate in pots with codiaeum grown on flooded benches in relation with the growing period. After De Kreij and Straver (1988). *Reprinted by permission of the International Society Horticultural Science*

Beside the spatial variation in ion concentrations as described before, variation in time is obvious. Such variation especially occur with the use of brackish irrigation water. During crop development the evaporation from the soil surface and the water use of the crop, together defined as the transpiration sum, increase and by this the accumulation of salts in the root environment. This especially occurs when leaching of salt during the cultivation period is restricted.

In studies on plant response to unequal distribution of salts, effects on nutrient (ion) absorption and effects on the osmotic potential should be distinguished. Plant reaction to unequal distribution of nutrients will be characterized on the one hand by preferential absorptions, thus, a seeking action of the plant to satisfy the nutritional needs. On the other hand, osmotic stress leads to actions of the plant due to escape from spots with high salt concentrations or to adjusted activities of roots present in spots with different salt concentrations. Thus, plants develop different activities to adjust for high salinity. The internal adjustments to resist high osmotic stress, as discussed in Section 7.4 and external adjustments to escape from places where the osmotic potential induces osmotic stress. In this chapter effects of seasonal as well spatial variations will be discussed. The adjustments to seasonal variation are met by plants by internal adjustments, while the spatial variations merely will be met by external adjustments.

8.2 Salinity Stress Response on Spatial Variations

In a study Meiri (1984) summarized effects of spatial variation of salinity, in which he proposed three models to estimate these effects. The first model was based on the average salinity in the root zone, the second on the water uptake from the compartment and its salinity and the third on presence of roots and its salinity. To test these models he used data of maize (Bingham and Garber, 1970) and data of alfalfa (Shalhevet and Bernstein, 1968). The estimations of the salinity based on the first and the third model gave best results, while the estimation of the salinity based on the second model showed lower correlation coefficients with the yield characteristics than the others. A disadvantage of the second and the third model is the difficulty in determining water uptake and root intensity in compartments under field conditions. So these models are less usefull as estimators for salinity response under practical conditions. Moreover, differences in water uptake from a compartment and measured root activity are results rather than causes of salinity. Thus, the causality between the variables in these models is under discussion.

In studies on crop reaction to salinity effects under spatial variation attention should be paid to the possibility of crops to escape from spots with high salt concentrations. Crops with a restricted root system can escape less easy from spots with high concentrations than crops with an extended root system. Furthermore, the growing conditions are important, like the root volume, the direction of the spatial variation and the length of the cultivation period. The direction of the spatial variation can be orientated horizontally, vertically and even in both directions. Escape from increasing salt concentration is time consuming because the plant has to adjust its root system. Thus, the length of the total growing period in relation to the time necessary for adjustment also will be important. Moreover, the adjustment of the plant roots at such seemed time consuming. In our experiments cucumbers were grown in split root systems in which 50% of the root volume got a high EC value and 50% got a standard EC value (Sonneveld and De Kreij, 1999). During a starting period of some weeks an equal salinity in the whole root system was maintained to ensure an equal root system in both halves. After the different EC values were realised the water uptake by the plant from the different root parts changed gradually. This is shown in Fig. 8.2, where the course of the water uptake in both root parts is shown in relation to the EC value realised. The water uptake by the crop in the low concentrated part gradually increased and in the high concentrated part it gradually decreased over a long period after the realisation of the different EC values in the root parts.

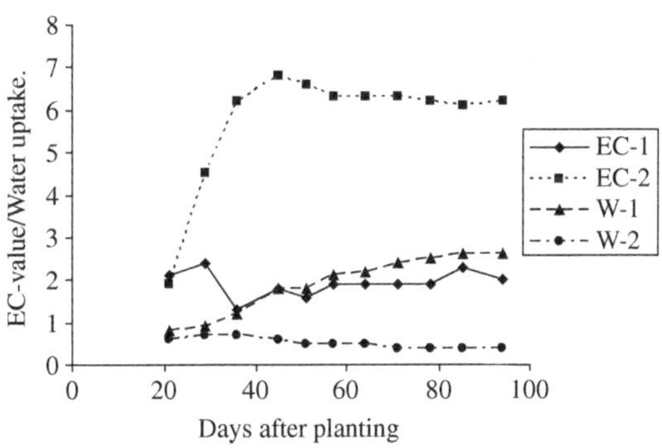

Fig. 8.2 The course of the EC value (dS m^{-1}) and the water uptake (l m^{-2}) of a cucumber crop grown in a split root system, of which 50% of the root volume was brought on a high EC value. The water uptake W-1 is realised by the root part with the standard (EC-1) value and the water uptake W-2 by the root part with the increased (EC-2) value

The different water uptakes from compartments with high and low EC values were generally found in experiments with spatial variations of salinity distribution (Bingham and Garber, 1970; Kirkham et al., 1969; Lunin and Gallatin, 1965; Shalhevet and Bernstein, 1968; Sonneveld and De Kreij, 1999). In all cases the water uptake was highest from the part with the highest osmotic potential (lowest EC). However, there were exceptions when the osmotic potential became very high in one of the compartments and the water uptake in such compartments became lower than in the higher concentrated part (Eaton, 1941; Sonneveld and De Kreij, 1999; Sonneveld and Voogt, 1990). Such effects apparently will be explained by an insufficient supply of certain essential nutrients at the root surface in the low concentrated root parts, combined with an insufficient redistribution by the plant of the nutrients absorbed by the roots grown in the higher concentrated part. The differences in water uptake as mentioned have been found by a horizontal as well as by a vertical unequal distribution of the osmotic potential in the root zone. However, with this comment, that in a root zone with an equally distributed osmotic potential commonly a vertical difference in water uptake will be found naturally. Such differences are characterized by a decreased of the uptake with increasing depth. Data of such differences are clearly shown by Bingham and Garber (1970) and summarized in Table 8.1. The water uptake in the treatment without any salinity in the root zone differs strongly from top to bottom. However, when the water uptake in one of the compartments is reduced by salinity, the uptake is compensated by the roots in one or both of the other non saline compartments, completely disturbing the original water uptake pattern.

The effects of spatial unequal osmotic potentials in the root zone on yield are not unequivocal. Sometimes, yields are related with the average salinity of the root zone, like calculated by Meiri (1984), while other data show not any effect (Bingham and Garber, 1970; Sonneveld and Voogt, 1990) or a reduced effect of the compartment(s) with the high osmotic potential. The last situation has been found in the experiments of Cerda and Roorda van Eysinga (1981), shown in Fig. 8.3. The yield of the treatments with a spatial variable EC in the growing medium is equal to or higher than the expected yield on basis of the average EC. It is not always clear from where

Table 8.1 Water uptake of corn (l per week per plant) as affected by vertically unequal distribution of the salinity in the root zone

Treatments salinity in root zone[1]			Water uptake			
Top	Centre	Bottom	Total	% Top	% Centre	% Bottom
0	0	0	4.800	46	35	19
S	0	0	3.674	10	42	48
0	S	0	4.460	60	7	33
0	0	S	5.009	51	44	5

[1] 0 – no salinity and S – saline part.
Data after Bingham and Garber (1970).

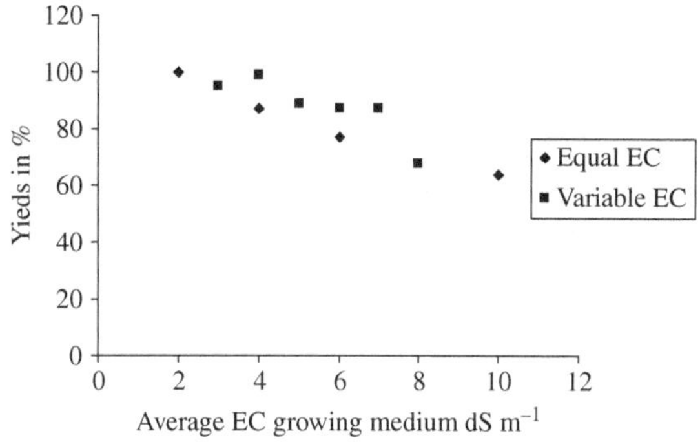

Fig. 8.3 Yield of tomato as affected by the average EC value (dS m^{-1}) in the growing media, with an equal EC in all root parts and with a variable EC in the root parts. After Cerda and Roorda van Eysinga (1981)

the different reaction on the spatial variation originate from. Sometimes, the different effects could be explained by the level of the high salinity part in combination with the sensitivity of the crop and the growing conditions (Sonneveld et al., 1991; Sonneveld and De Kreij, 1999). Sonneveld (2000) hypothesized an enlargement of the response model of Maas and Hoffman (1977) presented in Section 7.2 for crop reaction on unequal distributed salinity in the root zone, as shown in formula (8.1).

$$Y_r = 1 - SYD_l(c_l - c_{tl}) - SYD_h(c_h - c_{th}) \tag{8.1}$$

In which

Y_r= relative yield
SYD_l and SYD_h= salinity yield decrease values of the lowest and highest salinity levels in the root environment, respectively
c_l and c_h= lowest and highest EC in the root environment, respectively
c_{tl} and c_{th}= threshold values for lowest and highest salinity level in the root environment, respectively

Furthermore

$c_l > c_{tl}$
$c_h > c_{th}$
$SYD_l > SYD_h$

Data of experiments with an unequal distribution of the salinity in the root zone are listed in Table 8.2. The yield of the tomato crop grown in spring – summer (Sonneveld and Voogt, 1990) and the yield of the cucumber crop grown in autumn (Sonneveld and De Kreij, 1999) were not affected by a high EC value in 50% of the root environment. However, the yield of the cucumber grown in spring – summer (Sonneveld and De Kreij, 1999) was significantly reduced by a high EC value in part of the root environment. The parameters of formula (8.1) apparently are affected by

Table 8.2 Yield of tomato and cucumber grown in rock wool as affected by a horizontal unequally distributed EC in the root environment. The EC values are given in dS m^{-1} in two parts (x/y) both 50% of the root environment and the yield is given in kg m^{-2}

Tomato spring – summer		Cucumber spring – summer		Cucumber autumn	
EC	Yield	EC	Yield	EC	Yield
3.0/0.75	23.8	2.1/0.2	27.5 ab	2.2/0.8	4.6
3.0/3.0	24.0	2.0/1.2	28.1 a	2.3/1.4	4.6
3.0/5.0	25.1	2.0/2.0	26.5 bc	2.4/2.4	4.8
3.0/7.5	24.6	2.2/4.3	26.3 bc	2.5/4.3	4.7
3.0/10.0	23.6	1.9/6.2	25.5 cd	2.6/6.2	4.6
		2.2/7.7	24.2 d	2.4/7.8	4.6
LSD 0.05[1]	ns				ns

[1] ns – not significant; when significant values with the same letter do not differ significantly at P=0.05.

Data after Sonneveld and Voogt (1990) and Sonneveld and De Kreij (1999). *Modified by permission of Springer*

crop type and growing condition, just like the parameters in the Maas – Hoffman model.

High local salinity can strongly affect root development. Root growth in parts of the root environment with a high salinity is often strongly reduced in comparison with the root growth in non saline parts (Eaton, 1941; Lunin and Gallatin, 1965; Shani et al., 1993). Sometimes the reduced root development in the saline part is compensated by extra root growth in the non saline part (Kirkham et al., 1969) but such is not always the case (Bingham and Garber, 1970). Root development also can be hindered by too low concentrations of nutrients, as shown by Eaton (1941). The root growth of corn and tomato in compartments with distilled water was strongly reduced in comparison with compartments with nutrients. Comparable results with tomato have been found with the experiments presented by Sonneveld et al. (1991).

8.3 Stress Response to Variation of Salinity in Time

For long term salinity variations Meiri (1984) suggested in his review a linear response between the over time averaged osmotic potential in the root environment and the crop development. Such a response, however, cannot be expected under all conditions with long term variations. This is understandable, firstly, by the fact that under moderate and poor light conditions growth (yield) is more or less linearly related to light intensity and thus yield reductions under poor light conditions have relatively a lower contribution to the total yield depression than equal proportional yield reductions under ample light. Secondly, high EC values in the root environment at such are less detrimental under poor light conditions than under ample light (Maas and Hoffman, 1977; Magisted et al., 1943; Sonneveld and Welles, 1988). Thus, Meiris' (1984) response EC_t value calculated by multiplication of the length of the periods and the EC level maintained as given by Eq. (8.2) needs adjustment for light conditions.

$$EC_t = \frac{\Sigma d_i EC_i}{\Sigma d_i} \tag{8.2}$$

In which:

EC_t = response EC value (dS m^{-1}), calculated over time and EC value
d_i = a certain day i
EC_i = the EC (dS m^{-1}) on day i

Such an adjustment is presented by Sonneveld and Welles (1988) as a response EC value, based on the length of time, the EC value and the light intensity using Eq. (8.3).

$$EC_{tR} = \frac{\Sigma d_i R_i EC_i}{\Sigma d_i R_i} \tag{8.3}$$

In which:

EC_{tR}= response EC value (dSm^{-1}), calculated over time, light intensity and EC value
R_i= light intensity (joules day^{-1} cm^{-2}) on day i
and the other parameters as given at formula (8.2)

In experiments with tomato EC_{tR} showed a better response on yield than EC_t. The correlation coefficients for the relationship between yield and EC using EC_{tR} as independent variable were much higher than those using EC_t. The data are summarized in Table 8.3.

The results presented for tomato, were not fully confirmed by results for gerbera. An increase of the external EC value from 2 in summer to 4 in winter did not reduce yield, while an increase to 8 in winter seriously reduced yield at the same amount as found with such a high EC in summer (Van Os and De Kreij, 1987). However, high

Table 8.3 Regression equations and correlation coefficient for the relationship between yield of tomato (Y) and average EC values weighed over time (EC_t) and weighed over the product of time and radiation (EC_{tR})

Experiments	Regression equations	Correlation coefficients
A	$Y = -4.9\ EC_t + 112$	– 0.797
	$Y = -5.1\ EC_{tR} + 113$	– 0.944
B	$Y = -4.2\ EC_t + 112$	– 0.788
	$Y = -9.5\ EC_{tR} + 128$	– 0.950

After Sonneveld and Welles (1988). *Reprinted by Permission of Springer*

EC values of 6 dS m^{-1} with sweet pepper maintained from the start in winter up till 17 weeks after planting did not affect the yield in comparison with standard value of 1.5 dS $m^{-1}$from the beginning (Van Uffelen and Bakker, 1987). A high EC in the external solution of 6 dS m^{-1} with a spring crop eggplant from start until the beginning of harvest reduced growth and early yield, but such differences disappeared after 3–4 weeks when maintaining standard values (Savvas and Lenz,2000).

Short-term effects of changes of EC in the root environment will strongly differ from long term variation. A sudden increase of the EC in the root environment strongly will reduce the water absorption, because of the reduced difference between the osmotic potential of the internal and external solution. The water absorption recovers after such an osmotic shock by adjustment of the plant to the changed conditions in the root environment, mostly by an increased osmotic potential of the internal solution. In an experiment with tomato Van Ieperen (1996a) working with tomato noticed a strong decrease of the plant water content with a sudden decrease of the osmotic potential from –0.01 MPa (EC $\approx$ 0.3 dS m^{-1}) to –0.36 MPa (EC $\approx$ 10.8 dS m^{-1}). The switch from high to low was made at 10.00 a.m. and reduced the plant water uptake for about one hour before it recovered and became again in equilibrium with the transpiration. The reduction of the water uptake of the plant due to the reduced difference in osmotic potential between the internal and external solution as mentioned before includes also the most direct survival strategy of the plant. Because this reduced water uptake combined with a continued transpiration increases the solute concentrations in plants and recovers the necessary difference between the osmotic potential of the internal and external solution for water uptake.

In another experiment Van Ieperen (1996b) compared day/night EC regimes in the external solution with an average 24 hours day value of 5 dS m^{-1} with as well tomato as test crop. Total plant weights differed like shown in Table 8.4. Yields in the treatment with the continuous high EC value 9/9 day/night was surely reduced in comparison with the continuous 5/5 solution. The yield at the switching EC value, but with an equal average as the 5/5 regime, was improved when the high value was supplied during night time and decreased when the high value was supplied during day time. These results also show that high EC values are less detrimental when plants are under low stress conditions (night-time) than when they are under high stress conditions (day-time).

In other experiments with tomato day/night regimes of 1/8 and 2/8 were compared with a continuous 3.3 EC value in the external solution (Van Veen-Schotanus,

Table 8.4 Total plant weights (kg) of tomato as resulted from experiments with different day/night EC regimes, with an average 24 hours day value of 5 dS m^{-1} and one treatment with a continuously high EC value

	Experiments	
EC regime day/night	A	B
9/1	5.97	6.81
1/9	7.09	10.55
5/5	6.49	8.68
9/9	5.32	4.78

After Van Ieperen (1996a).

1999). No significant yield difference was found between the day/night regimes. The yield in the continuous 3.3 regime was 5% lower than in the other treatments. The lower yield can be explained by the fact that the EC value of 3.3 was above the salinity threshold value, which was estimated for tomato on 2.5 (Sonneveld and Van der Burg, 1991). Thus, the yield in this experiment seemed to be determined by the low day value and the high EC value during night is less or even not at all detrimental to crop development. The results of Nederhof (1997) with tomato also confirmed this effect, showing only a yield reduction of 8% with a 2/8 day/night EC regime in the external solution in comparison with the yield of a continuous EC of 2, while the yield reduction of the 8/2 day/night regime was 52% in comparison with the yield at the continuous EC of 2 coming close to the yield reduction of 58% of the continuous EC of 8. Short term salt shocks like carried out by Niedziela et al. (1993), twice a day for half an hour increasing the EC from about 2.2 to about 8.8 did not affect tomato yield.

With variation of salinity in time it is not always possible to conclude which conditions are responsible for differences in crop reaction, the growing stage of the crop or the climatic conditions. The results reported by Sonneveld and Welles (1988) for tomato give rise to the conclusions that the climatic conditions are more important than the growing stage. For other crops effects of climatic conditions also are shown to be evident (Maas and Hoffman, 1977; Maas and Hoffman, 1983).

8.4 Nutrient Uptake and Spatial Variations

With an unequal distribution of nutrients in the root environment plants are able to absorb sufficient nutrients from sites where these are available, when in other parts

Table 8.5 NO_3 and K uptake of cucumber grown in a rock wool substrate system with an unequal distribution of nutrients in the root environment. Both root parts x/y represented 50% of the root volume, in which the EC in all (x) parts was maintained on 2.0 dS m^{-1} and in the (y) parts varied from 0.1 till 7.7 dS m^{-1}. The differences of the EC values were realised by addition of all macro nutrients in equal ratios

NO_3				K				
In root env.[1]	Uptake[2]	Total uptake[3]	mmol kg^{-1} produce	In root env.[1]	Uptake[2]	Total uptake[3]	mmol kg^{-1} produce	Yield kg m^{-2}
10/1	35.5/ 1.4	2362	86	2.6/ 0.1	20.3/0.0	1299	47	27.5
10/6	19.8/16.1	2298	82	2.0/ 0.6	11.2/8.6	1267	45	28.1
11/11	17.0/16.9	2170	82	1.7/ 1.8	9.5/9.5	1216	46	26.5
13/32	29.2/ 6.4	2278	87	3.1/13.0	16.7/4.2	1338	51	26.3
10/49	31.2/ 6.9	2054	81	1.9/20.7	17.0/2.3	1235	48	25.5
12/63	33.4/–3.8	1894	78	2.7/29.0	19.7/–1.1	1190	49	24.2

[1]mmol l^{-1} in the substrate solution in the root environment;
[2]mmol m^{-2} day^{-1};
[3]mmol m^{-2}.

of the root environment the uptake is hindered. The results listed in Table 8.5 show data of an experiment with cucumbers where the plants were grown in a split-root system with different nutrient concentrations (Sonneveld and De Kreij, 1999). Both root sections represented 50% of the root volume. From the data it can be concluded that the uptake of NO_3 and K per kg produce of all treatments is more or less equal. An insufficient uptake of nutrients in one of the root sections caused by a too low supply or by a too high EC value is compensated in the other root part. Comparable results have been gained by tomato (Cerda and Roorda van Eysinga, 1981; Sonneveld and Voogt, 1990). The study of Jager (1985) with short duration experiments confirms this statement for very different crops. An example is presented in Table 8.6 where analytical data are listed of a tomato crop grown with an equal and an unequal distribution of nutrients in the root environment. With an equal EC value of the nutrient solution in the root environment the Na, Ca, Mg and Cl in the leaves were higher at a suboptimal supply of EC = 0.75 than at an optimal supply of EC = 2.5. The K and P on the contrary were lower. The differences with the suboptimal supply disappeared more or less completely when in part of the root environment, in this case 50%, was grown with an optimal supply of nutrients like in treatment 2.5/0.75. An insufficient uptake of nutrients, in this case K and P, at an EC of 0.75 was compensated by extra uptake in the higher concentrated root part. The plain effects in the young leaves were not evident for all elements in the fruits. In experiments, with tomato crops the uptake of nutrients from the high concentrated part was more prominent than with cucumber. With these results the strong effect of the metabolic control of the plant on the nutrient uptake is emphasized, which is experience to be most evident for the elements N and K.

Such a control, however, is not always operative under all condition and the uptake for all minerals. Spatial high EC values caused by NaCl will strongly reduce the uptake of nutrients from such parts (Flores et al., 2002; Sonneveld and De Kreij,

Table 8.6 Analytical data of a tomato crop grown in a root environment with an equal and an unequal distribution of nutrients. Both root parts x/y represented 50% of the root volume. The analytical data are expressed as mmol kg^{-1} dry matter

	EC values					
	0.75/0.75	2.5/2.5	2.5/0.75	0.75/0.75	2.5/2.5	2.5/0.75
Elements	Young laminae			Fruits		
Na	193	58	73	59	20	28
K	658	953	888	940	1116	1107
Ca	858	794	698	34	36	34
Mg	274	161	184	65	64	66
Cl	66	32	47	86	60	62
N	3340	3476	3561	1300	1298	1368
P	137	192	190	128	169	170
S	483	473	442	56	56	54

After Sonneveld and Voogt (1990). *Modified by permission of Springer*

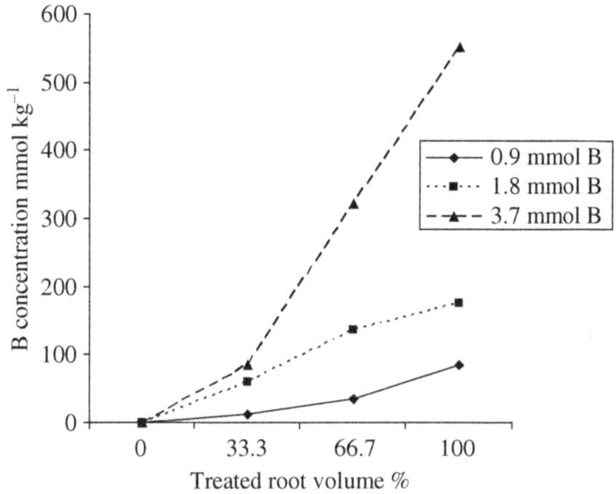

Fig. 8.4 The B concentration in leaves (mmol kg^{-1} dry matter) of corn as affected by the B concentration (mmol l^{-1}) in the soil solution and the percent of the root zone volume treated with B containing nutrient solution. After Bingham and Garber (1970). *Reprinted by permission of the Soil Science Society America*

1999). Bingham and Garber (1970) found a strong increase of the Na and Cl concentrations in corn leaves in relation to the percent of the root system that had a high salinity. They also found such an effect for B with this crop. The uptake of B was linearly related with the B concentration of the soil solution and the percent of the root system under treatment with B containing nutrient solution (Fig. 8.4).

Specific effects are found in a study of Tabatabaie et al. (2004) with the occurrence of blossom end rot in tomato. The plants were grown in rock wool and their roots were divided into two portions. Each portion was irrigated with water containing the same nutrient solution, but in different concentrations agreeing with EC values between 0 and 6 dS m^{-1}. High EC values in both portions strongly induced the occurrence of blossom end rot, up to 80% of the fruits. The disorder was reduced to about 15% when one rock wool portion was supplied with only water. The root xylem exudation measured on stem stumps of mature plants, was highly promoted in these treatments. Between the quantity of xylem exudation and the disorder a closely correlated negative linear relationship was found. Thus the increased xylem stream highly promoted the transport of Ca to the fruits, whereby the Ca concentration of the irrigation water of 0.65 mmol l^{-1} likely play an important part.

8.5 Problems and Possibilities

The unequal distribution of residual salts and nutrient elements in the root environment cause sometimes problems when it results to an unequal development of crops grown. In greenhouse cultivation this can occur when different crops are grown in succession, with a different planting design. In such cases it is likely that part of the plant will be placed on high concentrated and part on low concentrated spots.

Such easily induce an unequal start of the young plants. It especially will occur with succession of crop with a row planting design like tomatoes, sweet peppers and cucumbers by crops with a high planting density like lettuce, radish and many flower crops. The paths between the rows are at the end of the cropping period often high concentrated, while in the planting beds big differences occur when the crop was irrigated with drippers, like shown in Fig. 6.8. With such an unfavourable crop succession a thorough tillage and flooding of the soil is necessary to equalize the salt status of the whole area.

Another problem is the interpretation of the analytical data of soil and substrate samples. With an unequal distribution often the places where the sub-samples are gathered principally determines the results. Therefore, sufficient information about the salt distribution in the root environment and a thereupon focussed sampling technique is crucial for the interpretation, as discussed in Section 4.12.

An advantage of an unequal distribution is the possibility for plant to escape from high salt accumulations in the root environment and through that plants are less negatively affected by salinity, than will be expected on basis of the average salinity level, as is discussed in this chapter and in Chapter 7.

Sometimes, use is made of the plant reaction to an unequal distribution of nutrients by which such a distribution is intentionally brought in the root environment by specific placement of fertilizers. Such placements are applied to restrict nutrient losses by leaching or to realise specific spots in the root environment for an improved uptake of some elements, mostly micro nutrients. Conditions as mentioned are applied in soil grown as well in substrate systems. Geraldson (1963) presented a system for soil grown crops in which a banded fertilizer area in the top of the soil in combination with a shallow water table provided the plant from the top with nutrients and from the bottom with water. Later on also for substrate grown crops concepts were discussed (Benton Jones, 1999; Geraldson, 1990). Systems to improve micro nutrient uptake are presented by Kasten and Sommer (1990) and Sommer (1995) and are mainly focussed on the placement of concentrated NH_4 and possible other nutrients. The pH in the concentrated NH_4 spots will decrease gradually with the N uptake and NH_4 nitrification and the lowered pH will improve the uptake of many micro nutrients.

The fertilizer placement systems as described require that plants adjust their root system to the placement of the fertilizer. Such adjustments are time consuming and it is a drawback when crops suffer from nutrient stress before the root systems is adapted to the placement. This often is connected with growth reduction and an unequal plant development, which are unacceptable in greenhouse industry. Another objection can be the rigidity of such systems. The placement of the fertilizer offers less possibility for intermediate adjustment of the nutrient supply, while this is among others one of the advantages of the daily applications of water and fertilizers as common in fertigation systems. Surely, up to now the possibilities and consequences of an unequal distribution of nutrients and residual salts are insufficiently studied in greenhouse industry.

References

Benton Jones J Jr 1999. Advantages gained by controlling root growth in a newly-developed hydroponic growing system. Acta Hort. 481, 221–230.

Bingham F T and Garber M J 1970. Zonal salinization of the root system with NaCl and boron in relation to growth and water uptake of corn plants. Soil Sci. Soc. Am. Proc. 34, 122–126.

Cerda A and Roorda van Eysinga J P N L 1981. Tomato plant growth as affected by horizontally unequal osmotic concentrations in rockwool. Neth. J. Agric. Sci. 29, 189–197.

Cox D A 2001. Growth, nutrient content and growth medium electrical conductivity of poinsettia irrigated by subirrigation or from overhead. J. Plant Nutr. 24, 523–533.

De Kreij C and Straver N 1988. Flooded-bench irrigation: effect of irrigation frequency and type of potting soil on growth of codiaeum and on nutrient accumulation in the soil. Acta Hort. 221, 245–252.

Eaton F M 1941. Water uptake and root growth as influenced by inequalities in the concentration of the substrate. Plant Physiol. 16, 545–564.

Flores P Botella M A Martinez V and Cerda A 2002. Response to salinity of tomato seedlings with a split-root system: Nitrate uptake and reduction. J. Plant Nutr. 25, 177–187.

Geraldson C M 1963. Quantity and balance of nutrients required for best yields and quality of tomatoes. Proc. Florida State Hort. Soc. 76, 153–158.

Geraldson C M 1990. Conceptual evaluation of intensive production systems for tomatoes. *In*: Van Beusichem M L (ed) Plant Nutrition – Physiology and Applications. Kluwer Academic Publishers, Dordrecht, 539–544.

Jager A 1985. Response of plants to a localized nutrient supply. Thesis Botanical Laboratory, State University Utrecht, 137 pp.

Kasten P and Sommer K 1990. Cultivation of cut flowers with ammonium as nitrogen source. *In*: Van Beusichem M L (ed) Plant Nutrition – Physiology and Applications, Kluwer Academic Publishers, Dordrecht, 533–537.

Kirkham M B Gardner W R and Gerloff G C 1969. Leaf water potential of differentially salinized plants. Plant Physiol. 44, 1378–1382.

Lunin J and Gallatin M H 1965. Zonal salinization of the root system in relation to plant growth. Soil Sci. Soc. Am. Proc. 29, 608–612.

Maas E V and Hoffman G J 1977. Crop salt tolerance: Current assessment. J. Irr. Drainage Div. 103(IR2),115–134.

Maas E V and Hoffman G J 1983. Salt sensitivity of corn at various growth stages. California Agric. 37(7/8), 14–15.

Magisted O C Ayers A D Wadleigh C H and Gauch H G 1943. Effect of salt concentration, kind of salt, and climate on plant growth in sand cultures. Plant Physiol. 18, 151–166.

Meiri A 1984. Plant response to salinity: Experimental methodology and application to the field. *In*: Shainberg I and Shalhevet J (eds). Soil Salinity Under Irrigation. Process and Management. Springer Verlag. Berlin-Heidelberg, 284–297.

Mmolawa K and Or D 2000. Root zone solute dynamics under drip irrigation: A review. Plant Soil 222, 163–190.

Nederhof E 1997. Effects of different day/night conductivities on blossom-end rot, quality and production of greenhouse tomatoes. Acta Hort. 481, 495–502.

Niedziela C E Jr Nelson P V Willits D H and Peet M M 1993. Short-term salt-shock effects on production of tomato fruit quality, yield, and vegetative prediction of subsequent fruit quality. J. Amer. Soc. Hort. Sci. 118, 12–16.

Ondrasek G Romic D Romic M Tomic F and Mustac I 2008. Salt distribution in peat substrate grown with melon (Cucumis melo L). Acta Hort. 779, 307–312.

Prichard T L Meyer J L Hoffman G J Kogel F R and Roberts R 1983. Relationships of irrigation water salinity and soil water salinity. California Agric. 37(7/8), 11–14.

Savvas D and Lenz F 2000. Response of eggplant grown in recirculating nutrient solution to salinity imposed prior to the start of harvesting. J. Hort. Sci. Biotech. 75. 262–267.
Shalhevet J and Bernstein L 1968. Effects of vertically heterogeneous soil salinity on plant growth and water uptake. Soil Sci, 106, 85–93.
Shani U Waisel Y Eshel A Xue S and Ziv G 1993. Response to salinity of grapevine plants with split root systems. New Phytol. 124, 693–701.
Sommer K 1995. Ammonium and phosphate nutrition of plants growing in saturated solutions. Proc. Dahlia Greidinger Intern. Symp. Fertigation, Technion Inst. Technology Haifa, Israel, 155–164.
Sonneveld C and Welles G W H 1988. Yield and quality of rockwool grown-tomatoes as affected by variations in EC-value and climatic conditions. Plant Soil 111, 37–42.
Sonneveld C and Voogt W 1990. Response of tomatoes (*Lycopersicon esculentum*) to an unequal distribution of nutrients in the root environment. Plant Soil 124, 251–256.
Sonneveld C Van den Bos A L Van der Burg A M M and Voogt W 1991. Fertigation in the greenhouse industry in The Netherlands. *In*: Fertigation / Chemigation, FAO, Rome, 186–193.
Sonneveld C and Van der Burg A M M 1991. Sodium chloride salinity in fruit vegetable crops in soilless culture. Neth. J. Agric. Sci. 39, 115–122.
Sonneveld C and De Kreij C 1999. Response of cucumber (*Cucumis sativis L.*) to an unequal distribution of salts in the root environment. Plant Soil 209, 47–56.
Sonneveld C 2000. Effects of salinity on substrate grown vegetables and ornamentals in greenhouse horticulture. Thesis Wageningen University, Netherlands, 151 pp.
Sonneveld C and Voogt W 2001. Chemical analysis in substrate systems and hydroponics – use and interpretation. Acta Hort. 548, 247–259.
Tabatabaie S J Gregory P J and Hadley P 2004. Uneven distribution of nutrients in the root zone affects the incidence of blossom end rot and concentration of calcium and potassium in fruits of tomato. Plant Soil 258, 169–178.
Van der Burg A A M and Sonneveld C 1987. Variation of the EC in rockwool slabs. Glasshouse Crops Research Station Naaldwijk, The Netherlands. Annual Report, pp20.
Van Ieperen W 1996a. Dynamic effects of changes in electrical conductivity on transpiration and growth of greenhouse-grown tomato plants. J. Hort. Sci. 71, 481–496.
Van Ieperen W 1996b. Effect of different day and night salinity levels on vegetative growth, yield and quality of tomato. J Hort. Sci. 75, 99–111.
Van Os P C and De Kreij C 1987. Gerbera op steenwol. EC van 2 het beste voor produktie en kwaliteit. Vakblad Bloemisterij, 42 no 43, 56–57.
Van Uffelen J A M and Bakker J C 1987. EC effecten bij paprika: Neusrot en zwelscheuren dwingen tot compromis. Groenten Fruit 43 no 22, 30–31.
Van Veen-Schotanus L 1999. Productiewinst maar geen kwaliteitsverlies bij EC verlaging. Groenten Fruit, Glasgroenten 9 no 4, 16–17.

Chapter 9
Calcium Nutrition and Climatic Conditions

9.1 Introduction

The climatic conditions are one of the most striking differences between the growing conditions of field crops and those of protected crops, especially in the moderate climate zones. The increased temperature and the humidity in greenhouses are the dominating factors responsible for the differences. The radiation and the CO_2 level in greenhouses are lower, when not artificially adjusted Bakker (1991). Another striking difference between the cultivation under protected conditions in comparison with cultivation in the open field is the crop production under poor light conditions in moderate climate zones. Cultivation of most crops is impossible under field conditions in these climate zones in the period from late autumn until early spring, because of too low outside temperatures. However, under protected conditions crop production occurs year round in moderate zones, which includes production under winter conditions. Heating, and artificial lighting contribute to successful crop productions in winter, but the growing conditions differ strongly from those during summer. The low light intensity in combination with a high humidity and relatively high temperature stimulate the vegetative development of plants, which induces negative effects on the quality of the produce. This results in winter time to crops with a lush growth and high water contents (De Koning, 1994).

An optimal control on the climate as well as on the osmotic potential in relation to the addition of nutrients is important with respect to the production of high quality vegetables and flowers. The optimization of these factors sometimes requires rigorous interventions on the regulation of climate and plant nutrition, like lowering of the humidity and decrease of the osmotic potential in the root environment. Effects of the addition of nutrients with the regulation of the osmotic potential in relation to the plant development are discussed already in the Chapter 7 and will be discussed furthermore in Chapter 13. In the present chapter effects of the osmotic potential on the uptake and the internal transport of Ca will be considered in relation with climatic conditions. Both factors, the osmotic potential in the root environment and the climatic conditions play an important part in the regulation of the growth of crops in greenhouses and both factors strongly affect the uptake and transport of Ca in greenhouse crops and are strongly interacted. Therefore, precise regulation of these

C. Sonneveld, W. Voogt, *Plant Nutrition of Greenhouse Crops*,
DOI 10.1007/978-90-481-2532-6_9, © Springer Science+Business Media B.V. 2009

factors is necessary to get the effects aimed at. Miscalculations easily result to Ca disorders in the crop and by this a poor quality and often an unmarketable produce.

9.2 Ca Disorders in Greenhouse Crops

Ca disorders can occur as a result of a too high as well by a to low Ca status of plant tissues. Many symptoms of Ca deficiency have been described for long years as physiological disorders, with an indistinct origin. In the middle of the 20th century the first of such symptoms has been recognized as Ca deficiency (Shear, 1975). The long duration before it has been recognized as such can be explained by the fact that the unequal Ca distribution was the main cause and not the Ca concentration of the whole plant or even a whole plant organ. However, when the unequal distribution of Ca in plants and plant organs was recognized, the knowledge about Ca disorders quickly grew and Shear (1975) was able to publish a long list of Ca related disorders. The most well known symptoms in greenhouse crops will be discussed in this section, in relation to Ca concentrations and distribution in the tissues involved.

One of the most well known symptoms of Ca deficiency occurs in leafy vegetables with a head formation, enclosing the young leaves which in this way are excluded from substantial transpiration, like lettuce, Chinese cabbage and celery. The symptoms are called tip burn with lettuce (Kruger, 1966) and Chinese cabbage (Van Berkel, 1981) and blackheart with celery (Geraldson, 1957). The symptoms are characterized by necrosis on the tips of young leaves and black coloured, rotten hearts, respectively. The Ca concentrations in the leaves of such crops are characterized by a strong decrease from the outer leaves to the inner leaves and from the edge to the midrib. This is shown in Table 9.1, where Ca concentrations are listed of leaves from different positions and of different leaf parts of Ca disordered heads and of healthy heads (Sonneveld and Mook, 1983). Comparable results were gained with celery, the data of which are listed in Table 9.2 (Sonneveld and Van Beusekom, 1975). Tremendous differences are found between the enclosed and the outer leaves. It will be clear that the concentrations determined in the plant tissues mainly depend on the leaf parts sampled. Leaves enclosed in heads do not transpire and are thereby sparingly supplied with Ca which is nearly solely transported by the xylem bundles.

Table 9.1 Ca concentrations as has been found in lettuce leaves (mmol kg^{-1} dry matter) of healthy heads and of heads affected by tip burn. The data are derived from different leaf parts of different leaf positions

Plant part	Healthy head	Ca disordered head
Inner leaf midrib part	159	27
Inner leaf edge part	181	66
Outer leaf midrib part	672	82
Outer leaf edge part	663	163

Data after Sonneveld and Mook (1983).

Table 9.2 Ca concentrations (mmol kg^{-1} dry matter) of different parts of healthy celery plants and of plants affected by blackheart

Plant part	Healthy heads	Ca disordered heads
Inner part	214	50
Tips of outer leaves	822	478

After Sonneveld and Van Beusekom (1975).

Ca deficiency also can occur in young developing leaves of fruit bearing vegetables, especially when leaves in the beginning of the development easily are enclosed in the plant tops, like with cucumber (Bakker, 1985), tomato (Ho and Adams, 1989) and strawberry (Bradfield and Guttridge, 1984). Comparable symptoms are noticed with Ca deficiency in poinsettia, where the bracts can be affected by necrosis (Strømme et al., 1994) and the leaves by edge burn (Bierman et al., 1990). With cauliflower beside necrosis in young leaves, also curds can be affected by Ca deficiency (Krug et al., 1972). The curds show "glassy" spots with a diameter of 3–4 cm. In al cases the unequal distribution of Ca seem to play an important part (Chiu and Bould, 1976; Krug et al., 1972), just like shown with the leafy vegetables.

Other well known symptoms of Ca deficiency have been recognized in fruits of vegetable crops, as there are blossom-end rot in tomato (Spurr, 1959) and in pepper (Geraldson, 1957; Sonneveld, 1993), vitrescence in melon (Jean-Baptiste et al., 1999) and soft rot in eggplant (De Kreij, 1990). Ca deficiency in cucumber fruits seldom is reported under practical conditions, but still exist (Frost and Kretchman, 1989). It is characterized by the appearance of water-soaked lesions followed by necrotic lesions towards the blossom end of the fruits, like occur with many other Ca stressed fruits. The fact that it is scarcely occurs in cucumber fruits will be explained by the relatively high Ca concentration in the fruits of this crop (Adams and Ho, 1995), which possibly can be explained by an increased IAA synthesis during flowering and fruiting stage (Bengtsson and Jensén, 1983). The cause of Ca deficiency is often strongly dependent of the distribution of Ca within fruits (Ehret and Ho, 1986; Ho and Adams, 1989a). Generally, fruits show a strong decrease of the Ca concentration from the proximal end to the distal end, like also is shown for sweet pepper in Fig. 5.3.

Ca deficiency can occur also in the stems of flower crops like tulips (Van der Valk and Bruin, 1987) and with hippeastrum (Carow and Roeber, 1979). Stems of such crops locally become "glassy", lose their strength and bend.

Problems with too high Ca concentration in plant tissues are merely found in fruits. In tomato fruits gold speck, tiny yellowish spots on the fruit shoulder (De Kreij et al., 1992), and in sweet pepper fruits small green spots with a diameter of some mm on coloured fruits (Roorda van Eijsinga et al., 1973), have been recognized. Both disorders were related with high Ca concentrations in the fruit and accompanied with salt crystals in the cells of the fruit walls. The crystals consisted of a mixture of the mono- and di-hydrated form of calcium oxalate ($CaC_2O_4.H_2O$ and $CaC_2O_4.H_2O$) and thus, the spots on the fruits were considered as symptoms of

Ca excess (De Kreij et al., 1992). Other symptoms related with high Ca concentration in plant tissues are calyx browning on egg plant fruits (Maaswinkel, 1988) and chlorosis and necrosis on cucumber leaves (Abed, 1973; Shimida, 1973; Sonneveld and Voogt, 1978).

Picture 9.1 Necrosis at the edge of young tomato leaves caused by Ca deficiency.

Picture 9.2 Typical spherical shape of young unfolding cucumber leaves as a result of necrosis on the edge caused by Ca deficiency.

Picture 9.3 Ca deficiency in *Marantha* leaves.

Picture 9.4 Different stages of Ca deficiency in rose leaves.

Picture 9.5 Different stages of Ca deficiency in *Hippeastrum* leaves.

Picture 9.6 Ca deficiency in carnation.

Picture 9.7 Tip burn in lettuce is characterized by necrotic tips on the young leaves at the top of the head and is connected with low Ca concentrations in the inner leaves of the head.

Picture 9.8 Ca deficiency in celery appears in the young inner leaves and is called blackheart.

Picture 9.9 Tip burn in Chinese cabbage followed from a low Ca concentration of the inner leaves.

Picture 9.10 Ca deficiency in young unfolding strawberry leaves.

Picture 9.11 Ca deficiency in tomato fruits cause blossom-end rot (BER).This especially occurs with a low Ca concentration in the blossom-end of the fruit.

Picture 9.12 Gold specks on tomato fruit is connected with high Ca concentrations in the upper part of the fruit.

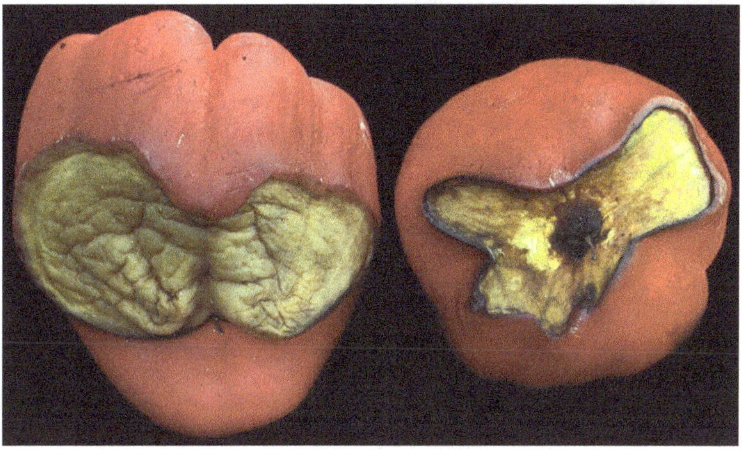

Picture 9.13 Blossom-end rot (BER) in sweet pepper fruits.

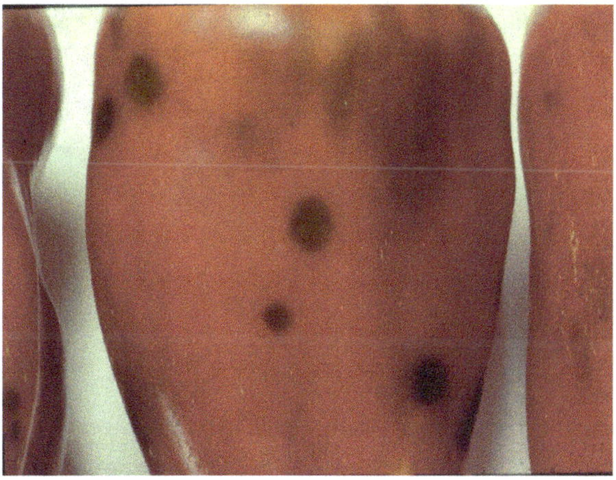

Picture 9.14 14 "Spot" on sweet pepper fruits is related with high Ca concentrations in the fruit and appear merely in the upper fruit part.

Picture 9.15 Ca deficiency in eggplant fruits is characterized by soft spots at the lower part of the fruits.

Picture 9.16 "Spot" on eggplant fruit is related with high Ca concentrations in the fruit and appear merely in the upper fruit part.

9.3 Required Ca Concentrations in Plant Tissues

An unambiguous system of critical concentrations of Ca in plant tissues, with which plainly deficient and healthy plant tissues are distinguished, cannot be presented. This conclusion will be based on different facts related with the occurrence of the phenomenon. Ca deficiency occurs in different organs of the plant, like in fruits and leaves of the same crop. Furthermore the distribution of Ca between plant organs and even within organs shows great variations, as discussed in the former section. Thus, the interpretation of analytical data of tissue tests will depend strongly on the organ sampled and even on the age and the part of that organ. This will be clear from the data of the experiment with lettuce shown in Table 9.1. The Ca disorder is related with the Ca concentration in the inner leaves as well as with those of the outer leaves. Nevertheless, the Ca concentration in part of the outer leaves of disordered plants is on the same level as those in part of in the inner leaves of healthy plants. The reasons for this phenomenon are firstly the fact that the disorder occur in specific plant parts with low Ca concentrations, in this case the edge part of the inner leaves and secondly that the Ca concentrations of the disordered plant parts are closely correlated with the Ca concentrations of the other plant parts, like shown with the data of Sonneveld and Mook (1983) in Fig. 9.1. Such relationships often occur and explain the strongly different critical values mentioned to explain Ca disorders. The Ca concentrations of the different leaf parts of lettuce heads in the experiment of Sonneveld and Mook (1983) were all suitable to explain the occurrence of Ca deficiency (tip burn). However, the tip burn occurred in the edge parts of younger leaves, thus where the Ca concentration was below about 100 mmol kg^{-1} dry matter. The concentration of 100 mmol kg^{-1} as has been found in the edge parts of the inner leaves is related with about 70, 290 and 360 mmol kg^{-1} respectively in the midrib parts of the inner leaves, the edge parts of the outer leaves and the midrib parts of the outer leaves. The results of the edge parts of the inner leaves showed a good agreement with the data of Barta and Tibbitts (1991), where in the edge part of lettuce leaves affected by tip burn also Ca concentrations were found

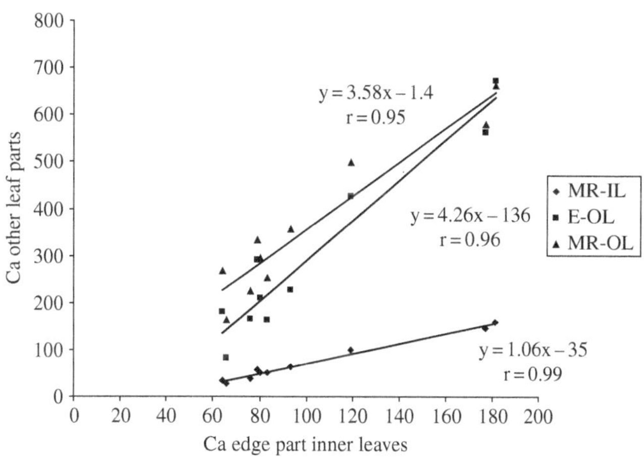

Fig. 9.1 Relationships between the Ca concentrations of lettuce tissues (mmol kg^{-1} dry matter) in relation to the Ca concentration of the edge part of the inner leaves. MR-IL = midrib parts of the inner leaves; E-OL = edge parts of the outer leaves; MR-OL = midrib parts of the outer leaves. After Sonneveld and Mook (1983)

below 100 mmol kg^{-1} dry matter. Thus, generally suitable interpretation values have to be related to the concentrations determined in the plant tissue where the disorder occurs.

Relationships between the Ca concentrations of different organs and parts of plant organs is not always unequivocal and differ in relation to the growing conditions, like shown in Fig. 9.2, derived from Adams (1990). In his experiment with tomato grown at different EC values in the nutrient solution young leaves and green fruits were sampled, which were divided into laminae and petioles and in distal end and remainder portions, respectively. The EC levels affect the Ca concentrations in the different organs upside down; an increase in the leaves and a decrease in the fruits. The Ca concentration within the organs also was differently affected: proportionate in the leaf parts and disproportionate in the fruit parts. Last effect can be explained by a specific hindrance of the Ca transport to the fruit end by a high osmotic potential in the root environment (Ehret and Ho, 1986a). Such data points to the conclusion that proportionate representative is not common for Ca distribution in plants and thus, sampling should be focussed on the part where the disorder is expected or occurs. In many fruits affected by Ca deficiency (blossom-end rot) the Ca concentration in the dry matter of the fruit is below 20 mmol kg^{-1} (Cerda et al., 1979; Chiu and Bould, 1976; Millikan et al., 1971; Morley et al., 1993; Wiersum, 1965). For eggplant fruits with internal rot also Ca concentrations below 20 mmol kg^{-1} dry matter had been found (De Kreij, 1990; Savvas and Lenz, 1994), while the distal end of affected sweet pepper fruits presented in Fig. 5.4 also shows comparable concentrations. In leaves much higher Ca concentrations are required than in fruits. This will be clear from following comparison. In an experiment (Holder and Cockshull, 1990) with greenhouse tomato grown at different humidity treatments leaves were seriously affected by Ca deficiency symptoms when the concentration of this element was below 250 mmol kg^{-1} dry

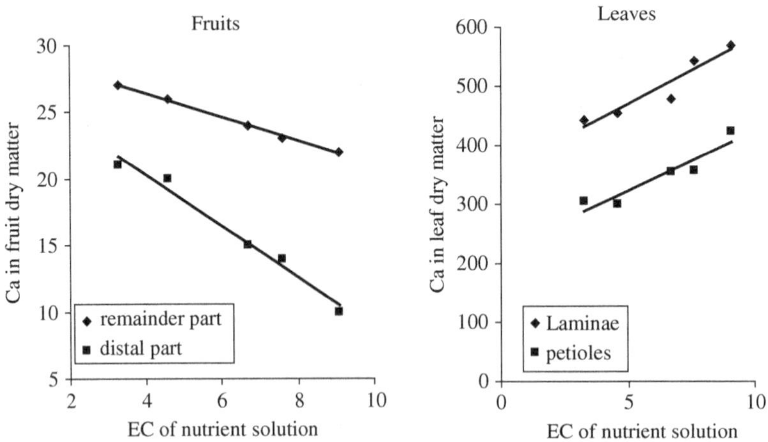

Fig. 9.2 Relationships between the EC of the nutrient solution and the Ca concentration in tomato tissues in mmol kg^{-1} dry matter. After Adams (1990). *Modified by permission of Springer*

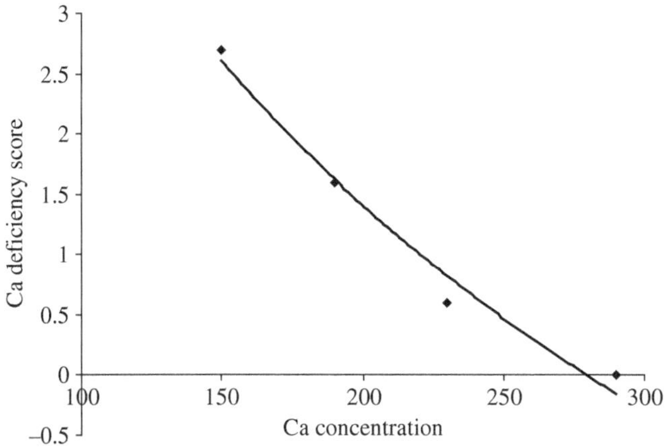

Fig. 9.3 Ca deficiency symptoms in the leaves of tomato as affected by the Ca concentrations (mmol kg^{-1} dry matter) in the leaves. After Holder and Cockshull (1990)

matter of (whole) leaves like shown in Fig. 9.3. The data discussed by De Kreij (1993) for fruit vegetable crops are in agreement with the already presented limits, when he stated that in fruits the Ca concentration should be 20–25, in growing points and in very young leaves about 70 and for old leaves 400–600 mmol kg^{-1} dry matter.

With flower crops comparable result are found with the translocation of Ca. With poinsettia the Ca concentrations of leaves and bracts presented by Strømme et al. (1994) showed great differences. On average for a series of treatments for the cv "Lilo" the Ca concentrations were 119 and 39 mmol kg^{-1}, respectively, while within the bracts also great differences were found with the marginal and the middle section, containing on average 28 and 74 mmol kg^{-1} dry matter, respectively. The bract necrosis disappeared at Ca concentration of about 35 mmol kg^{-1} dry matter in the marginal bract section (Strømme et al., 1994). Bierman et al. (1990) discovered a close relation between the Ca concentration in partly expanded leaves of poinsettia and the leaf edge burn of this crop. The leaf edge burn disappeared when the Ca concentration was higher than 175 mmol kg^{-1} dry matter.

Glassiness in the stems of tulips, connected with so called bending, occurred when the Ca concentration in the shoot was below 40 mmol kg^{-1} (Van der Valk and Bruin, 1987).

It is not always possible to relate the occurrence of Ca disorders to a distinguished concentration of this element in the plant tissue. This is already understandable because of the great variation of the Ca concentration in different parts of plants, as mentioned before. But misunderstandings are obvious, even when the sampling method and the plant part to be sampled are well defined. Besides the variations within plants a time delay between the occurrence of the disorder and sampling can seriously affect the analytical results. This for example, easily will happen in fruits, because the Ca concentration in fruits will change strongly with time. Shortly after anthesis the Ca concentration in fruits is relatively high and decreases quickly in the weeks following (Adams and Ho, 1985; Ehret and Ho, 1986; Gasim and Hurt (1986). This drop can be explained by the relatively rapid growth of the fruit in that period, while the Ca uptake is relatively slow. The weeks of the strong decrease of the Ca concentration is also the period that the fruit is most susceptible to Ca deficiency and the symptoms of blossom-end rot become visible. With increasing age the disordered fruits are strongly reduced in their development and remain small in comparison with healthy fruits growing to normal size. These different developments are responsible for a greater dilution of the Ca already available in fruits at anthesis in healthy fruits in comparison with those in the disordered fruits. In this way the difference in Ca concentrations between disordered and healthy fruits fade down or even invert and does not reflect any longer the concentrations at the moment that the disorder became visible. This for example explains the contradicting results sometimes found as shown by Millikan et al. (1971), presented in Table 9.3. The Ca concentration in the Ca deficient tomato fruits at normal Ca supply is higher than in the healthy fruits at low Ca supply. The explanation will be that fruits affected by Ca deficiency remain smaller than healthy fruits and that fruits received in the second growing period more Ca from the high Ca concentrated solution than from the low concentrated solution.

Table 9.3 Ca concentration (mmol kg^{-1} dry matter) as found by Millikan et al. (1971) in Ca disordered (BER) and healthy tomato fruits at low and normal Ca supply

	Low Ca supply		Normal Ca supply	
Fruit part	Healthy fruit	BER fruit	Healthy fruit	BER fruit
Stem end	9.5	8.2	21.4	18.0
Calyx end	7.0	7.0	14.2	11.2

9.4 Effects of Climate

Ca uptake by plants and the transport in plants is strongly affected by the climatic conditions, primarily by the humidity. Ca is easily transported in the plant by the xylem vessels and scarcely by the phloem vessels (Wiersum, 1979). The ratio between xylem and phloem transport and thus, thereby the distribution of Ca, is strongly determined by the humidity. In his research Wiersum grew ricinus plants in nutrient solutions with different K:Ca ratios and reported the concentrations of K and Ca in the leaves and in the phloem sap at four successive samplings during two months. Results summarized in Table 9.4 show that the concentrations in the plants change with time and with the composition of the external solution. However, on average the K/Ca ratio of 7 in the external solution is reflected in the plant with a K/Ca ratio of 16.2 in the leaves and with a K/Ca ratio of 824 in the phloem sap. For the K/Ca ratio in the external solution of 0.4 the K/Ca ratios in the leaves and in the phloem sap are 6.0 and 534, respectively. These ratios reflect the great difference

Table 9.4 Concentrations of K and Ca in leaves and phloem sap of ricinus plants as affected by the K:Ca ratios in the external solution. Results of four successive samplings

External solution		In leaves mmol kg^{-1} dm			In phloem sap mmol l^{-1}		
K/Ca mmol l^{-1}	Samplings	K	Ca	K/Ca	K	Ca	K/Ca
11.2/1.6	1	1664	77	21.6	97	0.11	882
	2	1399	80	17.5	115	0.14	821
	3	1297	95	13.6	103	0.12	858
	4	1156	95	12.2	103	0.14	736
3.3/7.6	1	1325	147	9.0	90	0.13	692
	2	1033	164	6.3	102	0.27	378
	3	893	162	5.5	96	0.19	505
	4	596	197	3.0	90	0.16	562

Derived from Wiersum (1979). *Modified by Permission of the Koninklijke Nederlandse Botanische Vereniging*

between the uptake of K and Ca and moreover the difference in the redistribution via the phloem sap. These results reflect that with actions to prevent Ca deficiency, the stimulation of xylem transport to plant parts generally poor in Ca will be very effective. This is possible with adjustments of climatic conditions in greenhouses.

An example of such an effect is shown in Fig. 9.4, where the relationship is shown between the occurrences of Ca deficiency in young leaves of cucumber, the humidity in the greenhouse and the Ca in the nutrient solution supplied (Bakker and Sonneveld, 1988). With a high supply of Ca no deficiency occurred. At a low Ca supply severe symptoms occurred at low VPD, which linearly decreased with increasing

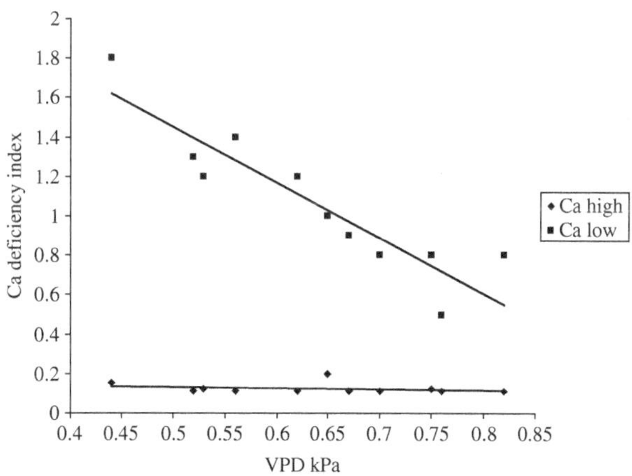

Fig. 9.4 The occurrence of Ca deficiency in cucumber leaves, as affected by the humidity (VPD) in the greenhouse and the Ca supply in the nutrient solution. High Ca 64% and low Ca 16% of C^+ covered by Ca. After Bakker and Sonneveld, (1988). *Modified by permission of the Journal Horticultural Science Biotechnology*

VPD. Herewith, the transpiration and by this the transport of Ca to the leaves was increased. The Ca concentration in the young leaves at the external solution with 16% Ca increased from 140 to 170 and at the external solution with 64% Ca from 205 to 245 mmol kg^{-1} dry matter from 0.4 to 0.8 kPa VPD, which is an increase of about 20% for both cases. Comparable data has been found by Holder and Cockshull (1990) with tomato. The Ca concentration in the leaves increased linearly from 150 to 300 mmol kg^{-1} by an increase of the VPD from 0.15 to 0.65 kPa. Simultaneously, the Ca deficiency symptoms determined following Holder and Cockshull (1988) disappeared completely by the increasing VPD, as shown in Fig. 9.5. A high humidity at night suppressed the Ca concentrations in tomato leaves somewhat stronger than a high humidity during day (Adams and Ho, 1995), but for cucumber no difference was found between the effects of day and night humidity. Tulips grown at different humidity during the forcing period showed an increased Ca in the shoot, closely related to the water uptake (Van der Valk and Bruin, 1987). The results presented, justify the general conclusion that with increasing transpiration the Ca concentration in leaves will increase.

However, the effect of humidity on the Ca transport in leaves can differ with the character and position of the leaves. For example, very young enclosed leaves can be supplied with extra Ca by very high, more or less saturated, humidity conditions during night, as reported for lettuce by Collier and Tibbitts (1984). Such effects possibly can be attributed to root pressure, which increase the xylem flow to plant parts with less transpiration. The strong reduction of tip burn in Chinese cabbage resulting from a cover of plastic film during night increased the Ca concentration of the inner leaves from 75 to 200 mmol kg^{-1} (Van Berkel, 1981) in one study and from 112 to 184 mmol kg^{-1} dry matter (Van Berkel, 1987) in a second study.

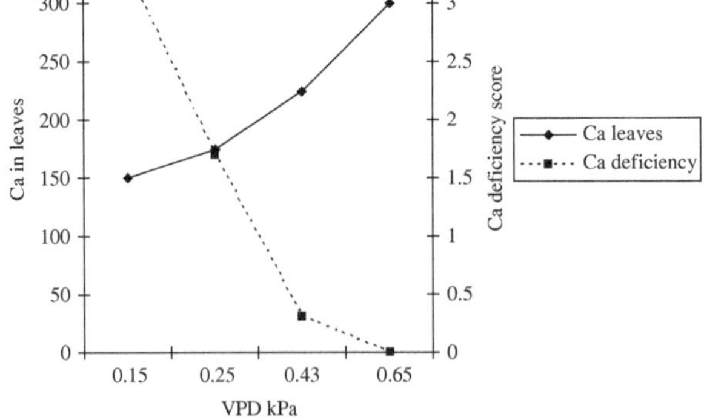

Fig. 9.5 Ca deficiency score (Holder and Cockshull, 1988) and Ca concentration of tomato leaves (mmol kg^{-1} dry matter) as affected by the humidity (VPD) in the greenhouse. After Holder and Cockshul (1990)

The effect of humidity on the Ca concentration in fruits is different from this effect in leaves. This different behaviour firstly is due to the fact that fruits merely receive the water via the phloem vessels, being water relatively poor in Ca. Secondly, surface area of fruits in relation to the total mass is always relatively small compared to leaves, and thus, the transpiration of fruits also is relatively small. A clear impression of the distribution of Ca in plants is given by Ho and Adams (1989) with their experiment with ^{45}Ca addition to a nutrient solution for tomato plants. Part of their data is shown in Fig. 9.6. The transport of Ca is mainly focussed on the high transpiring organs like leaves. The fruits contained no more than a few percentages of the total absorption, while the transport to the young leaves is surely reduced by the high humidity. Therefore, the transport of Ca in fruit bearing vegetables shows often a type of competition between leaves and fruits. Under high transpiring climatic conditions too much of the xylem stream is used by the leaves and the fruits are merely fed by the phloem and receive in this way less Ca. Upside down, when the transpiration is reduced too much, insufficient of the xylem stream will arrive in the young leaves, while the xylem stream to the fruits is improved. An example of such an effect is shown in Table 9.5 where the Ca concentrations of tomato fruits are shown in relation to the Ca supply in the nutrient solution and the humidity during day time (Adams and Holder, 1992). An example of the contradictory effect between fruits and leaves is listed in Table 9.6 (Adams and Ho, 1995). High humidity reduced the transport of Ca by a reduced transpiration. The leaves are much more affected by this reduction than the fruits, because the xylem stream in the plant is mainly utilized by the leaves. Thus, reduced transpiration offers possibilities to an improved xylem transport to the fruit and by this an improved Ca transport to these organs. This is evident for the high day humidity. The small reverse effect for the night humidity was not significant. See also the data of Adams and Holder (1990). When the high humidity specific is applied to the fruits, the Ca transport

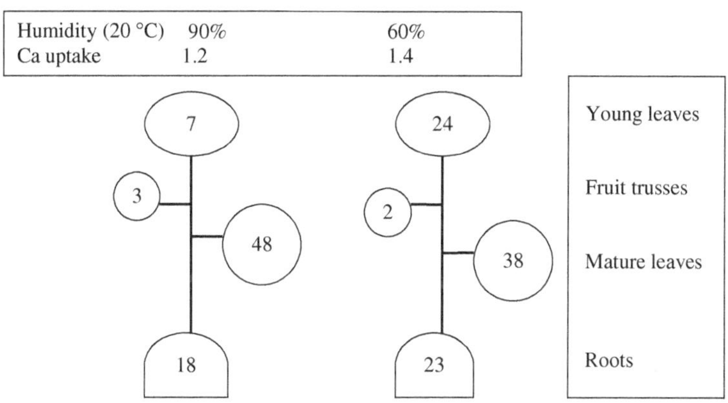

Fig. 9.6 Uptake (mmol kg^{-1} dry matter) and distribution (in %) of ^{45}Ca absorbed from nutrient solution by a tomato plant under different humidity conditions, grown in a nutrient solution with standard concentrations (EC 2 dS m^{-1}). After Ho and Adams (1989)

Table 9.5 Effects of day humidity and Ca concentration in the nutrient solution on the Ca concentration of tomato fruits (mmol kg^{-1} dry matter)

Day humidity VPD kPa	Ca in solution mmol l^{-1} 3.75	7.5	Average
0.65	10	12	11
0.15	14	18	16
Average	12	15	

After Adams and Holder (1992). *Modified by permission of the Journal Horticultural Science Biotechnology*

Table 9.6 Effects of humidity on the Ca concentration (mmol kg^{-1} dry matter) of tomato leaflets and fruits

	Day humidity					
	Leaflets			*Fruits*		
Night humidity	Low	High	Mean	Low	High	Mean
Low	279	229	254	18	21	20
High	177	140	158	16	20	18
Mean	228	184		17	21	

Adams and Ho (1995). *Modified by permission of the International Society Horticultural Science*

to fruits is strongly reduced. In this case the commonly low transpiration of fruits is further reduced and by this the Ca transport. Such effects occurred in experiments with fruits wrapped in plastic foil like shown for sweet pepper and bean by Mix and Marschner (1976) and for tomatoes by Paiva et al. (1998) and Wiersum (1965).

The effect of air temperature on the uptake of Ca is difficult to discriminate from the effect of the VPD, because in practice with increasing temperature VPD automatically rises and by this the transpiration of crops and the Ca absorption. Such effects are clearly shown by the data of an experiment with external mobile shading of a greenhouse in South of Spain (Lorenzo et al., 2003). The external shading was mainly used during the hottest periods of the days when the temperature went over 27°C. By this shading not only the temperature but also the radiation, as well the VPD were lowered. The experiment was carried out in combination with two EC values in the root environment. Results of yield, water use and the occurrence of blossom-end rot are listed in Table 9.7. The shading reduced strongly the water use with 32% by a reduced transpiration on the middle of the day and by this the occurrence of blossom-end rot was strongly reduced, especially in the treatments with a high EC in the root environment where the disorder was very serious. This means that the Ca transport to the fruit was much improved by reduction of the high water requirements on extreme temperature conditions, but cannot be ascribed as being a sole effect of the lower temperature.

Table 9.7 Yield (kg m^{-2}), water use (l m^{-2}) and % blossom-end rot (BER) of tomato as affected by EC in the root environment and by mobile shading of the greenhouse. The mobile shading was used at temperatures above 27°C

	Yield		Water use		% BER	
Shading	EC 3.1	EC 5.1	EC 3.1	EC 5.1	EC 3.1	EC 5.2
No	12.5	12.3	480	451	8.9	27.8
Yes	12.5	10.1	326	304	2.0	13.2

After Lorenzo et al. (2003).

With increasing root temperature the Ca uptake of tomato was seriously increased, as has been found by Adams (1988). However, beside the Ca also the water absorption was increased in the experiments by nearly 50% (Adams and Ho, 1993) and through that the uptake concentration of Ca was scarcely affected, as shown in Fig. 9.7. Thus, the increased Ca uptake will be due to the increased water uptake, which is understandable in view of the big difference in the water uptake and the small differences, only few percentages, within the Ca uptake concentrations. Benoit and Ceustermans (2001) reported an effect of cooling of the root environment on the appearance of blossom-end rot with rock wool grown sweet pepper. With ambient temperature in the root environment (23–33°C) more blossom-end rot occurred than with cooled down temperatures (17–22°C). The Ca concentration in the leaves was decreased, while it in the fruits was increased by the cooling.

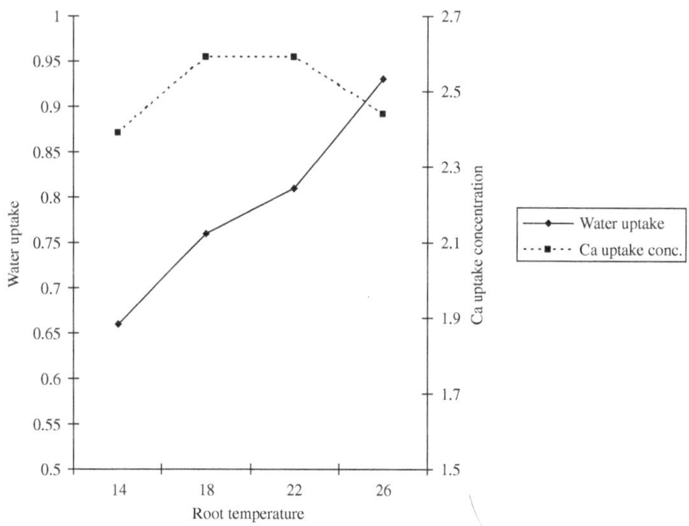

Fig. 9.7 The water absorption (l $plant^{-1}$ day^{-1}) and the Ca uptake concentration (mmol l^{-1}) of tomato as affected by the temperature in the root environment. After Adams and Ho (1993). *Modified by permission of Springer*

9.5 Effects of Cultivars

Cultivars show real differences with susceptibility to Ca deficiency, as has been found with different crops like lettuce (Collier et al., 1979), tomato (English and Barker, 1982) and sweet pepper (Morley et al., 1993). The susceptibility is not always directly related to the Ca concentration in the plant tissue (Collier and Tibbitts, 1982). With lettuce for example tip burn was not related with the Ca concentration in the tissue (Collier et al., 1977), but showed a relationship with the chlorogenic acid concentration (Collier et al., 1979). However, this concentration is less suitable as a guide line, because the relationship with the concentration of chlorogenic acid was weak and changed strongly with plant age. An extra handicap in the interpretation of analytical data of Ca in relation to the occurrence of tip burn are the different symptoms of tip burn that can be distinguished (Termohlen and Van der Hoeven, 1966). It is not sure that the different types of tip burn all are related to Ca deficiency. For tomato efficient and inefficient cultivars are distinguished (Giordano et al., 1982). The water-soluble Ca concentration in tissues of efficient cultivars was lower than those of inefficient plants (English and Barker, 1982). This seems in contradiction with the facts, but could be explained by ion competition in the plant. Therefore, in shoots different critical concentrations of total Ca for efficient and inefficient lines were established of 60 and 100 mmol kg^{-1} dry matter, respectively (English and Barker, 1987). However, the difference between susceptible and non susceptible cultivars mostly cannot be plainly attributed to a difference in Ca uptake (Adams and Ho, 1992). In their study with tomato they found an interaction between the susceptibility of the cultivar and the growing conditions. The susceptibility for blossom end rot in the cultivars changed with the osmotic potential and the weather conditions. They supposed that the cause of the differences between cultivars is a lack of co-ordination between cell enlargement in the distal placental tissue and the transport of Ca within the fruit.

9.6 Ca Sprays

In his review Shear (1975) stated that sprays of Ca salt solutions may supply sufficient added Ca to control disorders in leafy vegetables and that the Ca from sprays on fruits must move into these organs through their surface. In this way, he said, only very limited quantities can be supplied and that therefore most of the necessary Ca must be moved into the fruit by normal root nutrition. With this statement he directly pointed to the main problem arising with application of macro elements by sprays, being the relative small quantities of sprayed solution that get attached to leaves and surely to fruits. Calculations easily demonstrate that with repeated sprays only few percentages of the total need of crops will be covered. Moreover, it is not always possible to spray on the tissues where the deficiency is expected, like the growing-points and heart leaves of cabbage and lettuce. Therefore, best result will be expected with crops demonstrating the deficiency on leaves or tissues with sufficient expanded surface, like bracts of poinsettia (*Euphorbia pulcherrima*). The necrosis in these tissues can be successfully controlled by sprays of Ca salts.

Strømme et al. (1994) mentioned good results with weekly sprays of a solution of 9 mmol l^{-1} $CaCl_2$. Comparable results were obtained by and Harbaugh and Woltz (1989) with a solution of 11 mmol l^{-1} calcium nitrate. Meinken and Fischer (1991) recommend repeated sprays of $CaCl_2$ of 7 to 10 mmol l^{-1}. In their experiments the number and the time of the applications affected the reduction in the necrosis much more than an increased concentration from 5 to 10 mmol l^{-1} as was compared in one of the experiments. Bierman et al. (1990) mentioned favourable results with a solution of 12 mmol l^{-1}, with no difference between the sources: chelated Ca, $CaCl_2$ and $Ca(NO_3)_2$. The water used for the spray solution was tap water that contained 0.7 mmol l^{-1} Ca, the spray of which was also a treatment in the experiment. The low concentration in the tap water already reduced the leaf edge burn markedly. Thus, Ca sprays surely reduce the bract necrosis and further they increase the Ca concentration in the bracts, but did not clearly affect plant growth as such, as shown by the data summarized in Table 9.8 (Harbaugh and Woltz, 1989). Just the bract size was significant affected. Sprays with Ca solution also was effective in the prevention of the so called "upper leaf necrosis" in lily, which plainly was related by Ca deficiency (Chang and Miller, 2005). The sprays were effective for the cultivar Pirate (Berghoef and Elzinga, 1982), but were not effective for other cultivars of which the young leaves were insufficiently expanded. Concentrations up to 45 mmol l^{-1} were recommended, but higher concentration caused leaf scorch.

Spraying on leaves was also effective with Chinese cabbage when tip burn occurred in young plants and the heart is not yet closed. Sprays with 30 mmol l^{-1} three times a week reduced the disorder, but in this way it cannot always completely be prevented. An increased concentration up to 75 mmol l^{-1} was more effective (Van Berkel, 1987). Ca sprays were also effective to control blackheart in celery (Geraldson, 1957). Sprays with 40 mmol l^{-1} $CaCl_2$ prevented the disorder.

Sprays with Ca sometimes strongly reduce the occurrence of blossom-end rot in tomato, like published by Geraldson (1957) by sprays of 40 mmol l^{-1} $CaCl_2$. Borkowski (1984) obtained good results with $CaCl_2$ sprays every 3–5 days, with a concentration of 35 mmol l^{-1}. Sprayings on leaves had no any effect, but the sprays

Table 9.8 Effects of Ca sprays (11 mmol l^{-1} $Ca(NO_3)_2$) on bract necrosis of poinsettia

Plant characteristics	No sprays	Sprayed
Necrotic lesions < 1 cm	6.0	0.4
Necrotic lesions > 1 cm	17	1
Ca in bracts mmol kg^{-1}	62	100
Ca in leaves mmol kg^{-1}	521	556
Plant height cm	22	22
Plant width cm	42	42
Plant fresh weight g	142	148
Number of bracts ≥14 cm	25	28
Plant quality rating[a]	4.1	4.2

[a] subjective rating on scale 1–5; with 5 best.

Derived from Harbaugh and Woltz (1989). *Modified by permission of the American Society Horticultural Science*

on fruits reduced the percentage attached fruit from about 35% to about 7%. The Ca sprayed on leaves cannot be redistributed to other parts of the plant, as will be understandable from the foregoing information in Section 9.2. At the Experimental Station Noord Limburg in The Netherlands an experiment with sprays of $Ca(NO_3)_2$ was carried out to control blossom-end rot in tomato. The sprays were carried out twice a week with concentrations of 40 and 80 mmol l^{-1} and reduced the number of disordered fruit by 50%. Between both concentrations compared only minimal differences occurred in favour of the highest concentration. A problem of the sprays was the necrosis on leaves and fruits, caused by the high concentration of the spray solution (Proeftuin Noord Limburg, 1986). Therefore, sprays on fruits are recommended with sufficient low concentrations of mineral salts for example with $CaCl_2$ in a concentration of 30 mmol l^{-1} on only the very young fruits (Adams, 2002).

9.7 Reasons of Ca Deficiency

Many factors affect the Ca uptake and distribution. It is possible that Ca deficiency occurs, just by disturbance of one of these factors, when for example the supply of this element is far below the recommended limits. However, this rarely occurs in practise, because greenhouse crops are mostly grown under reasonable controlled conditions. Therefore, the occurrence of Ca deficiency generally is due to a combined action of different factors that negatively affect the uptake of Ca by the plant or the distribution within the plant. From the information presented in the foregoing sections it will be clear that there can be a contradiction between factors with respect to Ca deficiency in fruits and leaves. Conditions that lower the Ca concentration of fruits can increase that of leaves, and upside down. When such a factor plays a big part in the appearance of the disorder, a fine regulation of the growing conditions is required to control the disorder. Following factors are detected to promote the appearance of Ca deficiency in plants.

- Climatic conditions
  - High pressure vapour deficit reduces the transport of Ca to fruits and enclosed leaves.
  - Low pressure vapour deficit reduces the transpiration and by this the water uptake and the transport of Ca to expanding leaves.
  - High radiation increases the photosynthesis and by this the growth rate, which result in a stronger dilution of the Ca in the plant.
  - High air temperature increases growth rate and vapour pressure deficit, which result in dilution of the Ca in the plant and a reduced Ca transport to fruits and enclosed leaves.
  - CO_2 enrichment increases the photosynthesis and by this growth rate, which result in dilution of Ca in the plant.
- Conditions in the root zone
  - Low osmotic potential (high EC).
  - Low Ca supply.

  - Relatively low Ca concentration in relation to the concentrations of other cations.
  - Supply of NH_4 instead of NO_3 as N form.
  - Low P or low Cl supply.
- Susceptible crops
  - Susceptible cultivars.
  - Lush growth with strong leaf development (luxuriant crop).

The effects of the climatic conditions are discussed in the present chapter. The factors mentioned under the conditions in the root zone are for the effect of the osmotic potential discussed in Chapter 7, while the effects of the different anion and cations will be discussed at the appropriate place where the fertilization of the crops is discussed, like in the Chapters 12, 13, 14, 15 and 16. In a review Taylor and Locascio (2004) discussed factors affecting blossom-end rot in different crops. In this review much agreement exists with the experiences presented in this section.

9.8 Choose Between Two Evils

The control on the Ca nutrition of plants often is a choice between two evils, as can be extracted from the foregoing sections. A too low as well a too high Ca status of the plant can cause quality problems in for example vegetable fruits crops. This can be well demonstrated by the occurrences of blossom-end rot and gold speck in tomato as affected by the K/Ca ratio in the root environment, shown in Fig. 9.8 (Voogt, 2003). An increasing K/Ca ratio increases the occurrence of blossom-end rot and decreases the gold speck index. In the range of the K/Ca ratios presented, either blossom-end rot or gold speck occurred. From the figure can be concluded that a K/Ca ratio of about 1 both occurrences are reduced to an acceptable level. This is the ratio generally recommended to Dutch growers (Sonneveld and Straver, 1994). The choice will be adjusted dependent on different growing conditions, like for example a specific sensitivity for one of both disorders of cultivars grown.

When a strategy will be set up for the Ca nutrition of a crop, the effects of different factors on the Ca uptake and transport in the plant will be considered. The

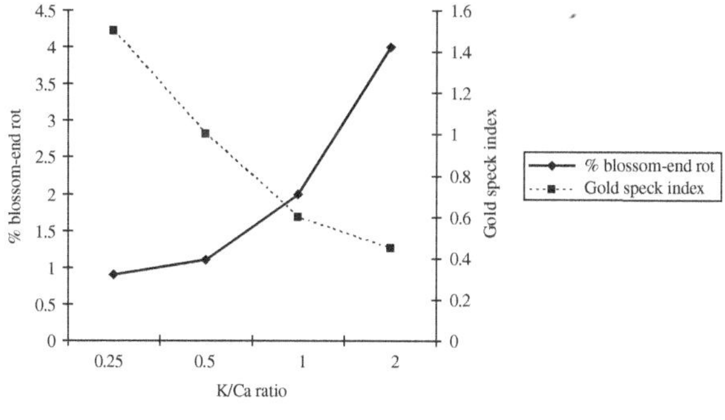

Fig. 9.8 The relationship between the K/Ca ratio and the occurrence of blossom-end rot and gold speck in tomato. Derived from Voogt (2003)

effect of a certain factor differs for crops and depends on growing conditions. In some experiments different factors affecting the Ca status of crops were compared, which gives an impression of the importance of such factors.

In a series of experiments Sonneveld and Welles (2005) compared different climatic conditions (humidity) and EC values in the root environment for fruit vegetable crops. Data are summarized in Table 9.9 and shows that the EC always affect the Ca status of the crops negatively. In young leaves the Ca concentration is decreased with 0.4 till 6.1% per unit EC value for autumn grown tomato and spring grown cucumber, respectively. In fruits the decrease per unit EC was from 3.3 till 8.8% for spring and autumn grown tomato, respectively. The Ca concentrations in fruits were relatively stronger affected than those in young leaves. The effect of the humidity on the Ca concentration in the plant tissues was not always unequivocal, but on average the concentrations were increased. An increase of the humidity by 0.2 VPD kPa affected the concentration in the leaves by a decrease from 11% till an increase of 9% for spring grown tomato and autumn grown sweet pepper, respectively, while the fruits were affected by a decrease from 3% till an increase of 34% for autumn grown tomato and autumn grown cucumber, respectively. Also the humidity affected the Ca concentration in the fruits relatively stronger than in the leaves.

Another example is an experiment of De Kreij (1996) with tomato grown in rock wool in a recirculation system. The humidity of the air and the supply of P and Ca in the nutrient solution were studied in this experiment. The humidity was compared in two treatments, ambient and high. In the high treatment the humidity was increased by a mist system. The average VPD day/night in the two treatments compared were 0.76/0/55 and 0.33/0.19 kPa. P was supplied in the added solution in

Table 9.9 Quotients for Ca concentrations in plant tissues of young leaves and fruits found with highest/lowest EC values and highest/standard humidity

			Spring		Autumn	
Crop	Factor	Tissue	Values compared	Quotient Ca conc.	Values compared	Quotient Ca conc.
Tomato	EC	Leaves	4.5/2.4	0.97	5.0/2.6	0.99
	dS m^{-1}	Fruits	4.5/2.4	0.93	5.0/2,6	0.79
	VPD	Leaves	0.53/0.73	0.89	0.54/0.72	0.97
	kPa	Fruits	0.53/0.73	1.17	0.54/0.72	0.97
Cucumber	EC	Leaves	4.4/1.6	0.83	5.7/2.0	0.95
	dS m^{-1}	Fruits	4.4/1.6	0.87	5.7/2.0	0.84
	VPD	Leaves	0.53/0.73	1.07	0.54/0.72	1.02
	kPa	Fruits	0.53/0.73	1.04	0.54/0.72	1.34
Sweet pepper	EC	Leaves	5.7/1.6	0.83	6.7/1.6	0.92
	dS m^{-1}	Fruits	5.7/1.6	0.74	6.7/1.6	nd
	VPD	Leaves	0.53/0.73	1.00	0.54/0.72	1.09
	kPa	Fruits	0.53/0.73	1.06	0.54/0.72	nd

After data of Sonneveld and Welles (2005).

Table 9.10 Ca concentrations in plant tops (mmol kg^{-1} dry matter) and blossom-end rot (BER) as affected by humidity and P and Ca supply in the nutrient solution in rock wool grown tomato in a circulation system

Factors	Values compared	Ca in plant tops	% BER
Humidity	Ambient	129	24
	High	138	19
P as developed in the root environment	0.8→ 0.2 mmol l^{-1}	88	40
	1.2→ 3.0	141	14
	1.8→ 6.0	172	11
Ca in root environment, relative of total cations	0.29 C^{+}	88	31
	0.72 C^{+}	180	13

After De Kreij (1996).

concentrations of 0.75, 1.25 and 1.75 mmol l^{-1} and were depleted or accumulated in the root environment to concentrations as mentioned in Table 9.10. The addition of Ca was adjusted to the concentration 4.0 and 11.2 measured in the drainage water. The total cation (C^{+}) concentrations were kept equal for both Ca treatments, thus the K and Mg concentration were adjusted. The results listed in Table 9.10 show big difference for the Ca concentrations in the top of the plant. They are in good agreement with the occurrence of blossom-end rot. Both characteristics are closely correlated following a linear relationship (Fig. 9.9). This relationship indicates that plants with a low Ca status will develop blossom-end rot, evidently irrespective of the cause of the deficiency. With Ca concentrations in the top above 200 mmol kg^{-1} the occurrence of blossom-end rot is minimal. The data in this figure again show the close correlation that can exist between the Ca concentration in any tissue and the disorder symptoms in a different tissue of the crop, as discussed in Section 9.3.

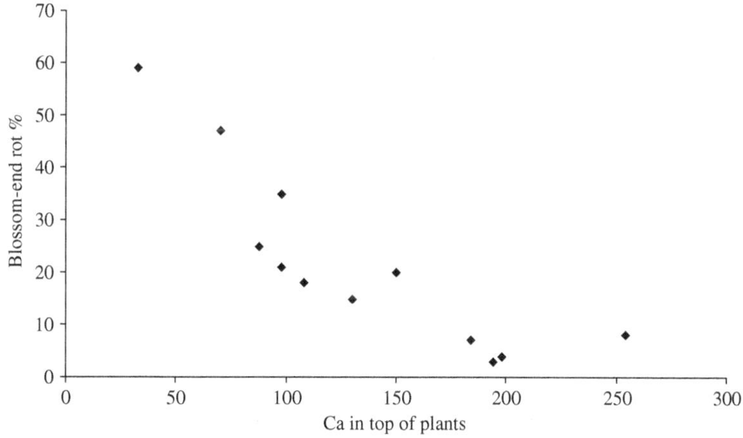

Fig. 9.9 The relationship between the Ca concentration in the top of tomato plants (mmol kg^{-1} dry matter) and the occurrence of blossom-end rot. Data from De Kreij (1996)

However, such a relationship is operative for the described situation and cannot be universally applied for different conditions.

Factors responsible for the Ca status of the plant have not always a linear effect on the Ca uptake, as for example is found for P in this experiment. The differences between the successive P additions is equal (0.5 mmol l^{-1}), but the Ca concentration in the tops increases with 53 and 31 mmol l^{-1} and the % blossom-end rot decreases with 26 and 3% for the first and second step, respectively. Non linear effects can be expected especially in sub-optimal supply of nutrients. Sub-optimal conditions easier occur in substrate grown than in soil grown crops. The root volume and by this the storage of nutrients in substrate growing is very restricted (see Section 3.4) and with insufficient control on the nutrient status depletion and by this a sub-optimal nutrient status in the root environment easily occur.

References

Adams P and Ho L C 1985. Tomatoes: two disorders but one cause. Grower 103(2), 17–25.

Adams P 1988. Some affects of root temperature on the growth and calcium status of tomatoes. Acta Hort. 222, 167–172.

Adams P 1990. Effect of salinity on the distribution of calcium in tomato (*Lycopersicon esculentum*) fruit and leaves. *In*: Van Beusichem M L (ed) Plant Nutrition Physiology and Applications. Kluwer Academic Publishers, Dordrecht, 473–476.

Adams P and Holder R 1990. Moves on Humidity. Grower 113(2), supplement: Horticulture Now HN, 9–11.

Adams P and Holder R 1992. Effects of humidity, Ca and salinity on the accumulation of dry matter and Ca by leaves and fruits of tomato (*Lycopersicon esculentum*). J. Hort. Sci. 67, 137–142.

Adams P and Ho L C 1992. The susceptibility of modern tomato cultivars to blossom-end rot in relation to salinity. J. Hort. Sci. 67, 827–839.

Adams P and Ho, L C 1993. Effects of environment on the uptake and distribution of calcium in tomato and the incidence of blossom-end rot. Plant Soil 154, 127–132.

Adams P and Ho L C 1995. Differential effects of salinity and humidity on growth and Ca status of tomato and cucumber grown in hydroponic culture. Acta Hort. 401, 357–363.

Adams P 2002. Nutritional control in hydroponics. *In*: Savvas D and Passam H (eds). Hydroponic Production of Vegetable and Ornamentals. Embryo Publications, Athens, 211–261.

Abed A H 1973. Einfluss von Salzkonzentration and Konzentrationsänderung in der Nährlösung auf Assimilation und Transpiration von Gurkenpflanzen. Dissertation Technischen Universität Hannover, 117 pp.

Bakker J C 1985. Physiological disorders in cucumber under high humidity conditions and low ventilation rates in greenhouses. Acta Hort. 156, 257–264.

Bakker J C and Sonneveld C 1988. Calcium deficiency of glasshouse cucumber as affected by environmental humidity and mineral nutrition. J. Hort. Sci. 63, 241–246.

Bakker J C 1991. Analysis of humidity effects on growth and production of glasshouse fruit vegetables. Dissertation Landbouwuniversiteit Wageningen, 155 pp.

Barta J D and Tibbitts T W 1991. Calcium localization in lettuce leaves with and without tipburn: Comparison of controlled-environment and field-grown plants. J. Amer. Soc. Hort. Sci. 116, 870–875.

Bengtsson B and Jensén P 1983. Uptake and distribution of calcium, magnesium and potassium in cucumber of different age. Physiol. Plant. 57, 428–434.

Benoit F and Ceustermans N 2001. Impact of cooling on blossom-end rot in soilless paprika. Acta Hort. 548, 319–325.

Berghoef J and Elzinga P 1982. Calciumchloride vermindert bladverbranding bij "Pirate". Vakblad Bloemisterij 37(11), 32–33.

Bierman P M Rosen C J and Wilkins H F 1990. Leaf edge burn and axillary shoot growth of vegetative poinsettia plants: influence of calcium, nitrogen form and molybdenum. J. Amer. Hort. Soc. 115, 73–78.

Borkowski J 1984. Study on the calcium uptake dynamic by tomato fruits and blossom-end rot control. Acta Hort. 145, 222–229.

Bradfield E G and Guttridge C G 1984. Effects of salt concentration and cation balance in soils on leaf tipburn and calcium content of strawberry leaves and fruits. Comm. Soil Sci. Plant Anal. 15, 681–693.

Carow B and Roeber R 1979. Calcium-Mangel verursacht Schäden an Hippeastrum. Deutscher Gartenbau 33, 2024–2025.

Cerda A Bingham F T and Labanauskas C K 1979. Blossom-end rot of tomato fruit as influenced by osmotic potential and phosphorus concentration of nutrient solution media. J. Amer. Soc. Hort. Sci. 104, 236–239.

Chang Y C and Miller W B 2005. The development of upper leaf necrosis in *Lilium* 'Star Gazer'. J. Amer. Soc. Hort. Sci. 130, 759–766.

Chiu T and Bould C 1976. Effects of shortage of calcium and other cations on Ca mobility, growth and nutritional disorders of tomato plants (*Lycopersicon esculentum*). J. Sci. Food Agric. 27, 969–977.

Collier G F Scaife M A and Huntington V C 1977. Nutritional aspects of physiological disorders. Nat. Veg. Res. Sta. Wellesburn, Ann. Report 1976, 43–44.

Collier G F Huntington V C and Cox E F 1979. A possible role for chlorogenic acid in calcium-related disorders of vegetable crops with particular reference to lettuce tipburn. Comm. Soil Sci. Plant Anal. 10, 481–490.

Collier G F and Tibbitts T W 1982. Tipburn of lettuce. Hort. Rev. 4, 49–65.

Collier C F and Tibbitts T W 1984. Effects of relative humidity and root temperature on calcium concentration and tipburn development in lettuce. J. Amer. Soc. Hort. Sci. 109, 128–131.

De Koning A N M 1994. Development and dry matter distribution in Glasshouse tomato: a quantitative approach. Thesis, Wageningen University, 240 pp.

De Kreij C 1990. Klappers veroorzaakt door gebrek aan calcium. Groenten en Fruit 46(9), 35.

De Kreij C Janse J Van Goor B J and Van Doesburg J D J 1992. The incidence of calcium oxalate crystals in fruit walls of tomato (*Lycopersicon esculentum Mill.*) as affected by humidity, phosphate and calcium supply. J. Hort. Sci. 67, 45–50.

De Kreij C 1993. Calcium transport het meest beïnvloed. *In*: Proefstation voor Tuinbouw onder Glas/Proefstation voor de Bloemisterij, Naaldwijk/Aalsmeer, The Netherlands. Luchtvochtigheid. Informatiereeks no 104, 36–37.

De Kreij C 1996. Interactive effects of air humidity, calcium and phosphate on blossom-end rot, leaf deformation, production and nutrient contents of tomato. J. Plant Nutr. 19, 361–377.

Ehret D and Ho L C 1986. Translocation of calcium in relation to tomato fruit growth. Ann. Bot. 58, 679–688.

Ehret D L and Ho L C 1986a. Effects of osmotic potential in nutrient solution on diurnal growth of tomato fruit. J. Exp. Bot. 37, 1294–1302.

English J E and Barker A V 1982. Water-soluble calcium in Ca-efficient and Ca-inefficient tomato strains. HortSci. 17, 929–931.

English J E and Barker A V 1987. Ion interactions in calcium-efficient and calcium-inefficient tomato lines. J. Plant Nutr. 10, 857–869.

Frost D J and Kretchman D W 1989. Calcium deficiency reduces cucumber fruit and seed quality. J. Amer. Soc. Hort. Sci. 114, 552–556.

Gasim A A and Hurd R G 1986. The root activity of fruiting tomato plant. Acta Hort 190, 267–265.

Geraldson C M 1957. Factors affecting calcium nutrition of celery, tomato and pepper. Soil Sci. Soc. Amer. Proc. 21, 621–625.

Giordano L de B Gabelman W H and Gerloff G C 1982. Inheritance of differences in calcium utilization by tomatoes under low-calcium stress. J. Amer. Soc. Hort. Sci. 107, 664–669.

Harbaugh B K and Woltz S S 1989. Fertilization practices and foliar-bract calcium sprays reduce incidence of marginal bract necrosis of poinsettia. Hort. Sci. 24, 465–468.

Ho L C and Adams P 1989. Calcium deficiency – a matter of inadequate transport to rapidly growing organs. Plants Today 2, 202–207.

Ho L C and Adams P 1989a. Effects of diurnal changes in the salinity of the nutrient solution on the accumulation of calcium by tomato fruit. Ann. Bot. 64, 373–382.

Holder R and Kockshull K E 1988. The effect of humidity and nutrition on the development of calcium deficiency in tomato leaves. *In*: Cockshull K E (ed). The effect of high humidity on plant growth in energy-saving greenhouses. Comm. E. C. Brussels, Dir. Gen. Agric. Report no. CEC-EUR-11261, 53–60.

Holder R and Cockshull K E 1990. Effects of humidity on the growth and yield of glasshouse tomatoes. J. Hort. Sci. 65, 31–39.

Jean-Baptiste I Morard P and Bernadac A 1999. Effects of temporary calcium deficiency on the incidence of a nutritional disorder in melon. Acta Hort 481, 417–423.

Krug H Wiebe H J and Jungk A 1972. Calciummangel an Blumenkohl unter konstanten Klimabedingungen. Z. Planzern. Bodenk. 133, Heft 3, 213–226.

Kruger N S 1966. Tip-burn of lettuce in relation to calcium nutrition. Queensland J. Agric. Sci. 23, 379–385.

Lorenzo P Sánchez-Guerrero M C Medrano E Garcia M L Caparrós I and Geménez M 2003. External mobile shading. *In*: Dirección General de Investigación y Formación Agraria de la Junta de Andalucía Spain, Improvement of Water Use Efficiency in Protected Crops, 197–218.

Maaswinkel R H M 1988. Geringe verdamping beperkt kans op verdroging. Groenten en Fruit 44(22), 39.

Meinken E and Fischer P 1991. Brakteen-Randnecrosen bei poinsettien. Deutscher Gartenbau 43, 2684–2685.

Millikan C R Bjarnason E N Osborn R K and Hanger B C 1971. Calcium concentration in tomato fruits in relation to the incidence of blossom-end rot. Austr. J. Exp. Agric. Anim. Husb. 11, 570–575.

Mix G P und Marschner H 1976. Einfluss exogener und endogener Factoren auf den Calciumgehalt von Paprika- und Bohnenfrüchten. Z. Pflanzenernaehr. Bodenk. 5, 551–563.

Morley P S Hardgrave M Bradley M and Pilbeam D J 1993. Susceptibility of sweet pepper (*Capsicum Annuum L.*) cultivars to the calcium deficiency disorder "Blossom-end rot". *In*: M A C Fragoso and Van Beusichem M L (eds). Optimization of Plant Nutrition. Kluwer Academic Publishers, Dordrecht, 561–567.

Paiva E A S Prieto Martinez H E Dias Casali V W and Padilha L 1998. Occurrence of blossom-end rot in tomato as a function of calcium dose in the nutrient solution and air humidity. J. Plant Nutr. 21, 2663–2670.

Proeftuin Noord Limburg 1986. Neusrotbestrijding. Jaarverslag 1986, 46–50.

Roorda van Eijsinga J P N L Rodenburg R and Van Uffelen L G 1973. Stip, een nieuw kwaliteitsprobleem bij paprikavruchten. Bedrijfsontwikkeling 4, 733–734.

Savvas D and Lenz F, 1994. Influence of salinity on the incidence of the physiological disorder 'internal fruit rot' in hydroponical-grown eggplants. Angew. Bot. 68, 32–35.

Shear C B 1975. Calcium-related disorders of fruits and vegetables. Hort. Sci. 10, 361–365.

Shimida 1973. Excess injury of calcium and magnesium in the crops. Japan. Agric. Res. Q. 7, 173–177.

Sonneveld C en Van Beusekom J 1975. Specifieke ion effecten bij tuinbouwgewassen. Proefstation voor de Groenten en Fruitteelt onder Glas, Naaldwijk, The Netherlands, Internal Report No 696, 32 pp.

Sonneveld C and Voogt S J 1978. Effects of saline irrigation water on glasshouse cucumbers. Plant Soil 49, 595–606.

Sonneveld C and Mook E 1983. Lettuce tipburn as related to the cation contents of different plant parts. Plant Soil 57, 29–40.

Sonneveld 1993. Gewasonderzoek. *In*: Proefstation Voor Tuinbouw Onder Glas Te Naaldwijk, The Netherlands, Informatiereeks No 87, Plantenvoeding in de Glastuinbouw (3th ed.), 62–68.

Sonneveld, C. and Straver, N., 1994. Nutrient solutions for Vegetables and Flowers grown in water or substrates. Research Station for Floriculture and Glasshouse Vegetables, Aalsmeer/Naaldwijk, The Netherlands, Series: Voedingsoplossingen Glastuinbouw, 8, 45 pp.

Sonneveld C and Welles G W H 2005. Cation concentrations of plant tissues of fruit-vegetable crops as affected by the EC of the external nutrient solution and by humidity. Acta Hort. 697, 377–386.

Spurr 1959. Anatomical aspects of blossom-end rot of tomatoes with a special reference to calcium nutrition. Hilgardia 28, 269–295.

Strømme A Selmer-Olsen A R Gislerød H R and Moe R 1994. Cultivar differences in nutrient absorption and susceptibility to bract necrosis in poinsettia (Euphorbia pulcherrima Wild. ex Klotzsch. Gartenbauwissenschaft 59, 6–12.

Taylor M D and Locascio S J 2004. Blossom-end rot: A calcium deficiency. J. Plant Nutr. 27, 123–139.

Termohlen G P and Van der Hoeven A P 1966. Tipburn symptoms in lettuce. Acta Hort. 4, 105–109.

Van Berkel N 1981. Three physiological disorders in glasshouse cultivation. Acta Hort. 119, 77–90.

Van Berkel N 1987. Bestrijding rand bij Chinese kool. Groenten Fruit 42(43), 28–29.

Van der Valk G G M and en Bruin P N A 1987. Voldoende transpiratie voorkomt kiepen. Bloembollencultuur 98(47), 12–14.

Voogt W 2003. Nutrient management in soil and soilless culture in The Netherlands: towards environmental goals. *In*: Proceedings 529 of the Dahlia Greidinger Symposium at Izmir, Turkey, The International Fertiliser Society, 2–27.

Wiersum L K 1965. Invloed van groei en verdamping der vruchten op het optreden van neusrot bij tomaten. Med. Dir. Tuinb. 28, 264–267.

Wiersum L K, 1979. Ca-content of the phloem sap in relation to Ca-status of the plant. Acta Bot. Neerl. 28, 221–224.

Chapter 10
Chemical Effects of Disinfestation

10.1 Introduction

In the greenhouse industry disinfestation is performed with soil as well with soilless cultures by various methods. The aim of this is the control of plant pathogens surviving in the root zone between successive crops. In soilless cultures it is carried out also with drainage water when it is reused in the growing system. The methods applied can be distinguished in chemical and physical methods and the choice which method will be applied depends on the pathogen and on the growing conditions. Chemical methods like fumigation with methyl bromide, the chemical method most recently applied on big scale is forbidden in most countries. However, when it legally can be used, it is not recommendable, because of the acute toxicity to people, the risk of environmental pollution and the Br residue in soil and substrate, which can be absorbed by plants to unacceptable levels. In this manner steam sterilisation survives as the method to sterilise soils and substrates.

The need for disinfestation was significantly reduced when substrate cultivation became popular in the greenhouse industry, because of the fact that for every cropping often a new substrate was used. However, when for successive crops in substrate growing no new substrate is applied, disinfestation is recommended to prevent the transmission of soil born diseases from one crop to another. Nevertheless disinfestations became an important issue in substrate cultivation since the reuse of drainage water was traced to be an important factor to prevent pollution of ground and surface water with nutrients. This reuse of drainage water includes a tremendous risk by spreading of infections possible present in such water. Thus, the need for sterilisation of drainage water of substrate systems was born. In the beginning heat sterilisation seemed to be the only applicable method. Later on also other methods came into being, like ultra violet radiation, dosage of residue free oxidants and still new methods are under research.

Besides the killing of plant pathogens, being the main item of sterilisation, all other types of microbial organism, the chemical and physical properties of soils, substrates or nutrient solutions treated can be affected. This chapter is focussed on the chemical aspects of the sterilisation methods used nowadays and will pass by the phytopathological aspects.

C. Sonneveld, W. Voogt, *Plant Nutrition of Greenhouse Crops*,
DOI 10.1007/978-90-481-2532-6_10, © Springer Science+Business Media B.V. 2009

10.2 General Effects of Steam Sterilisation

In a study on the influence of steam sterilisation of soils on different soil properties Jager et al. (1969) gave a review of aspects that occur due to steam sterilisation. In this review most properties of soils under review were differently affected. This will be explained by differences between soils under investigation and different heat treatments applied. However, three effects can be distinguished as being consistent in many studies, namely the release of soluble organic compounds, disturbance of the N chemistry and release of Mn. In later experiments it has been found that also Br is mobilised in greenhouse soils by steam sterilisation.

Picture 10.1 Steam sterilisation of greenhouse soils by pressure of steam under a sheet is the most widely used method

In an experiment of a research programme for greenhouse soils three different soils were analysed before (control) and after steam sterilisation at 100°C (Sonneveld, 1965a). Results of the analytical data listed in Table 10.1 show only small differences between the steamed soils and the control for pH, total soluble salts and water soluble mineral elements, being moreover not always consistent. The concentration exchangeable Mn is strongly increased by the steam sterilisation in all soils. These results are in good agreement with the study of Jager et al. (1969). The fact that the N concentrations of the soils were not consistently affected, have to be explained by the method of the determination. The total water soluble mineral N was determined and not the N sources separately. With steam sterilisation especially the N sources are affected, which will be discussed later on. Thus, with exception of Mn and N generally the concentrations of mineral elements are not that much affected that it will affect plant growth under prevalent greenhouse conditions.

Table 10.1 Chemical properties of soils determined before (control) and after steam sterilisation. pH was determined in a soil/water suspension; total soluble salts (TSS) are expressed as mg kg^{-1} dry soil and elements as mmol kg^{-1} dry soil

Soil type	Treatment	pH	TSS	Cl_{ws}[1]	N_{ws}	P_{ws}	K_{ws}	Mn_{exch}[2]
Humic sand	Steamed	7.1	1110	2.0	3.1	0.99	1.7	0.72
	Control	7.0	1110	1.4	4.1	1.24	1.2	0.17
Loamy clay	Steamed	7.2	3800	1.5	1.4	0.03	1.2	1.30
	Control	7.0	3400	1.0	1.0	0.03	1.0	0.56
Humic clay	Steamed	7.2	1600	3.1	3.3	0.06	4.0	1.21
	Control	7.0	1700	2.1	4.7	0.06	3.6	0.37

[1]ws-water soluble;
[2]exch-exchangeable, including the water soluble.
Sonneveld (1965a).

10.3 Steam Sterilisation and Mn Availability

Most soils contain much Mn subdivided in a great number of different compounds. The majority of these compounds consist of hydrated Mn oxides in which the Mn shows a valence between 2 and 4. Generally these compounds are divided into easily reducible and relatively inert compounds. Smaller quantities of Mn occur in the soil as exchangeable adsorbed on the clay and humus complex and as water soluble Mn available in the soil solution. In both last mentioned forms Mn occur in bivalent ionic form and is available to plants. The easily reducible Mn is not directly available for plant absorption, but can be considered as biological active, which mean that they can become available to plants by processes in the soil occurring under natural conditions. The trivalent and tetravalent compounds are readily reduced to bivalent Mn under such conditions. The relatively inert Mn cannot be easily converted to a plant available form under natural conditions. Therefore, these compounds are considered as biological inactive and are not important for plant nutrition. Apart from these Mn oxides and hydrated oxide compounds Mn salts occur in soils like Mn containing carbonates, silicates and phosphates. Furthermore, some organic Mn compounds are present. For horticultural purposes the Mn status of soils is characterized by the determination of water soluble, exchangeable and active forms. They are available to plants in the order given.

In the literature extended information is available about the chemistry of Mn in soils. For detailed information about the composition and the character of the compounds reference is made to Geering et al. (1969), De Groot (1963) and McKenzie (1972).

During steam sterilisation Mn compounds normally not available to plants are apparently converted into available compounds, shown by a strong increase of water soluble and exchangeable Mn concentrations as a result. Various processes may be responsible for it. In advance dehydration has been described as the most important one (Sherman, 1957). With this process complex Mn compounds are split by withdrawal of water. This process can be simply formulated as follows in formula (10.1).

$$(MnO)_x\ (MnO_v)_y\ (H_2O)_z - xMnO + yMnO_v + zH_2O \qquad (10.1)$$

The MnO released from the complex in this way is relatively readily soluble. The valence (v) of Mn in the MnO_v oxides complexes will vary between 2 and 4 and they are not available to plants. However, experiments indicated that other processes occur with the release of Mn during steam sterilisation of soils. This was shown with the results of an experiment in which dehydration and steam sterilisation were compared. Exchangeable Mn was determined in a series of soil samples after dehydration and after steam sterilisation. The dehydration was carried out by heating at 105°C during 36 h. Exchangeable Mn was determined with a Morgan's solution, a solution of Na acetate buffered at pH 4.8, and active Mn was determined with the same solution with hydroxylamine hydrochloride as reducing agent. The results are listed in Table 10.2, together with those of the untreated soil (Sonneveld and Voogt, 1975). On average 14, 23 and 70% of the active Mn has been found in exchangeable form, in the untreated soil, after dehydration and after steam sterilisation, respectively. Substantially more exchangeable Mn was found after steam sterilisation than after dehydration. It is likely that reduction is also an important factor during the steam sterilisation process. This process can be described as follows in formula (10.2).

$$2MnO_v \rightarrow 2MnO + (v-1)O_2 \qquad (10.2)$$

Besides dehydration and reduction, it is also possible that decomposition of Mn containing organic compounds with steam sterilisation plays a part in the release of Mn by steam sterilisation of soils (Page, 1964).

The average of 70% as mentioned before found after steam sterilisation agreed very well with the effect of heating under specific laboratory conditions at 100°C;

Table 10.2 The contents of active Mn of six different soils and the contents of exchangeable Mn of the same soils after dehydration and after steaming in comparison with the untreated soils. Mn contents are expressed as mmol kg^{-1} dry soil

		Exchangeable Mn[2]		
Soil type	Active Mn[1]	Untreated	After dehydrated	After steam sterilised
Sand	0.81	0.18	0.27	0.55
Loamy sand	1.14	0.27	0.46	0.82
Clayey loam	4.37	0.50	0.91	3.27
Clayey loam	6.28	0.46	0.64	2.55
Clayey peat	3.60	0.36	0.82	2.91
Clayey peat	3.55	0.32	0.91	3.01

[1] Including the exchangeable and water soluble;
[2] Including the water soluble

After Sonneveld and Voogt (1975). *Modified by permission of the International Society Horticultural Science*

Table 10.3 The effect of different temperatures during a heat treatment on the release of Mn from different soil types. Mn contents are expressed as mmol kg^{-1} dry soil

		Exchangeable Mn[1]		
Soil type	Active Mn[1]	Control	Heating at 70°C	Heating at 100°C
Dune sand	0.73	0.09	0.14	0.46
Loamy sand	1.09	0.41	0.36	0.91
River loam	7.15	0.41	0.64	3.37
Sea loam	6.69	0.73	0.82	2.09
Peaty clay	1.91	0.64	1.28	1.73
Sea loam	4.14	0.41	0.41	2.46
Clayey peat	3.55	0.23	0.23	2.46
Potting soil	0.55	0.23	0.14	0.18

[1]See notes at Table 10.2

After Sonneveld and Voogt (1973). *Modified by permission of Springer*

during 12 h (Sonneveld, 1979). Under practical conditions the release of Mn with steam sterilisation depends primarily on the Mn "storage" reflected as active Mn, the temperature realised with the heating and the duration of the heating. The effect of the heating temperature on the release of Mn on different soil types was clearly shown by the data presented by Sonneveld and Voogt (1973) listed in Table 10.3. At a temperature of 70°C on average nearly no increase of the exchangeable Mn is noticed. At 100°C huge concentrations of Mn are release, especially on soils with high concentrations of active Mn. The potting soil contained less active Mn and there is no Mn released by the heating. The reason of such an effect with potting soil will be the low clay content of this artificially prepared growing medium. Mn concentrations in soils are strongly related to the clay contents, on which it is mainly adsorbed. Furthermore, great differences are found among the origin of the clay, which means the origin of the sedimentation in which the clay is deposited (De Groot, 1963).

Mn released with steam sterilisation is very slowly fixed back in the active form. In greenhouse soils the fixation process often covers a period of 6–9 months, and even then sometimes the contents of exchangeable Mn are not yet arrived at the original level (Sonneveld, 1968, 1969a). In Fig. 10.1 the course of the exchangeable Mn content is shown as affected by steam sterilisation (day -1 and day1) and during a whole year after the steam sterilisation for two different greenhouse soils. The quantity of Mn released from the clayey soil is much higher than from the sandy soil, which is in agreement with the data presented in the Tables 10.1 and 10.2. In Fig. 10.1 the prompt increase of the Mn concentration directly after steam sterilisation is clearly shown as well the long duration before the exchangeable Mn is lowered to common values. This slow fixation of Mn is quite unusual for greenhouse soils with a reasonable pH value, like shown in Fig. 10.2. In this figure the results of a laboratory experiment are shown in which the course of the exchangeable Mn concentration in a loamy soil was followed after steam sterilisation and after addition

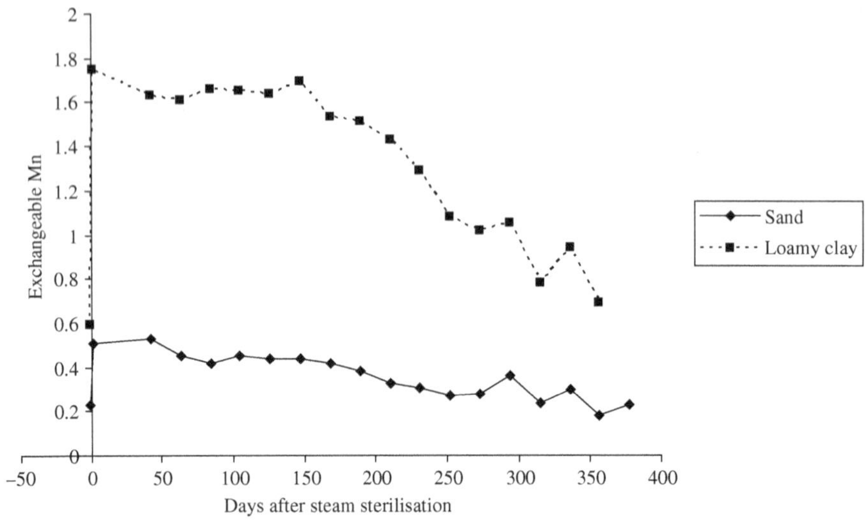

Fig. 10.1 The courses of the concentrations exchangeable Mn on two different greenhouse soils as affected by steam sterilisation. Mn was determined by Morgan's extraction and expressed as mmol kg^{-1} dry soil. Data of Sonneveld (1965)

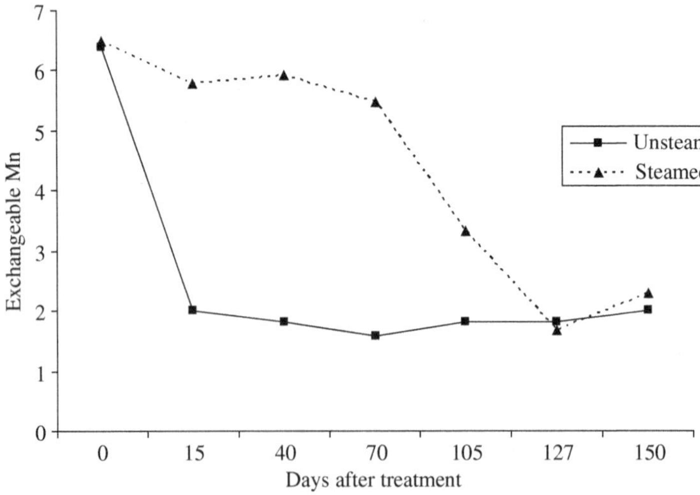

Fig. 10.2 The courses of the concentration of exchangeable Mn on greenhouse soils after steam sterilisation and after addition of $MnSO_4$ on the same but un-steamed soil. Mn was determined by Morgan's extraction and expressed as mmol kg^{-1} dry soil. After Sonneveld (1979). *Reprinted by permission of Elsevier*

of $MnSO_4$ to an equal level of exchangeable Mn on the same, but not sterilised soil. The exchangeable Mn content of the not steamed soil was decreased to a constant level within two weeks, while it required in the sterilised soil about 100 days before the Mn contents was decreased to this level.

The reason of this slow fixation will be caused by the lack of Mn oxidizing bacteria. It is well known that for a rapid oxidation specific species of bacteria are necessary (Bromfield and Skerman, 1950; Gerritsen, 1937; Leeper, 1947). Most bacteria are killed by steam sterilisation and it is likely that such also happens with the Mn oxidizing types. Therefore, research was carried out to the effect of inoculation of steam sterilised soils with Mn oxidizing bacteria on the course of the released Mn. The inoculation was carried out with the addition of a Mn oxidizing bacteria suspension or with addition of un-steamed soil (Sonneveld and Voogt, 1975a). The addition of un-steamed soil always resulted in a rapid Mn oxidation, while the addition of a suspension of Mn oxidizing bacteria was not always successful in this direction. It was supposed that cultured bacteria need time to adapt themselves to the soil environment. But it is also possible that not the right type of Mn oxidising bacteria were isolated or that the bacteria only can function in combination with other micro organisms. These subjects were not studied further on.

10.4 Mn Uptake and Steam Sterilisation

The Mn released at steam sterilisation strongly increases the Mn uptake by crops. In the relation between the Mn released by steam sterilisation and the uptake by crops the adsorption capacity of the soil play an important part. This is clear from the data presented in Fig. 10.3, where the relationship is shown between the exchangeable

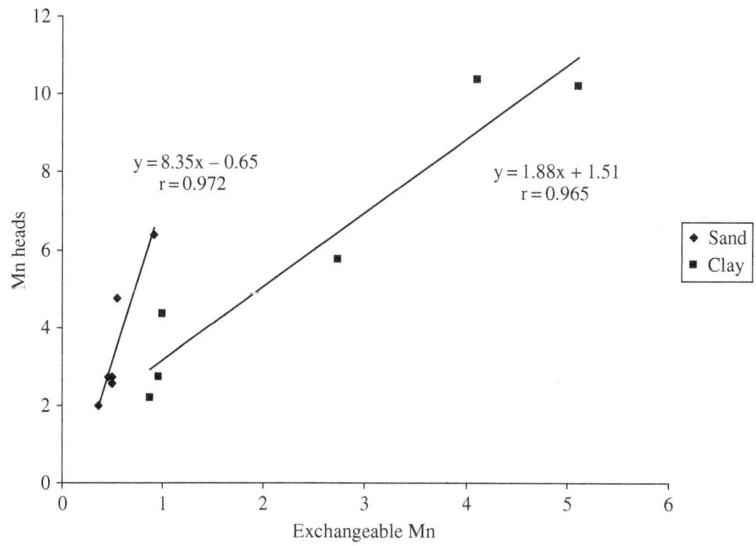

Fig. 10.3 The relationship between exchangeable Mn contents of the soil and the Mn concentrations of lettuce heads for a sandy and a clayey soil. The exchangeable Mn was determined by Morgan's extraction. The data are expressed as mmol kg^{-1} dry soil and mmol kg^{-1} dry matter of the lettuce heads. After Sonneveld and Voogt (1975a). *Modified by permission of Springer*

Mn content of the soil and the Mn concentration of a lettuce crop for a sandy soil and a clay soil (Sonneveld and Voogt, 1975a). The increase of the uptake by increasing exchangeable Mn concentrations of the soil is four times more in the sandy soil than on the clayey soil. This reflects the strong adsorption of Mn on the clay particles.

Another factor that strongly affects the Mn uptake is the pH of the soil (Roorda van Eysinga et al., 1978). Page (1962) showed that the contents of water soluble Mn was logarithmic related to the pH of the soil. In agreement with this finding Sonneveld (1968) calculated for un-steamed soils a logarithmic function between the pH of the soil and the Mn uptake of lettuce heads. This function can be converted to formula (10.3) as following.

$$\log y = -0.403pH + 2.711 \qquad \mathrm{r} = -0.867 \tag{10.3}$$

In which

y = Mn concentration of the lettuce heads in mmol kg^{-1} dry matter
pH = pH_{water} of the soil

However, for steam sterilised soils it was found that the uptake was mostly much higher than could be expected on basis of the pH (Sonneveld and Voogt, 1975a). In Fig. 10.4 the relationship for the un-steamed soil as given in formula (10.3) is given in comparison with a series of steamed soils. The Mn uptake from the steamed

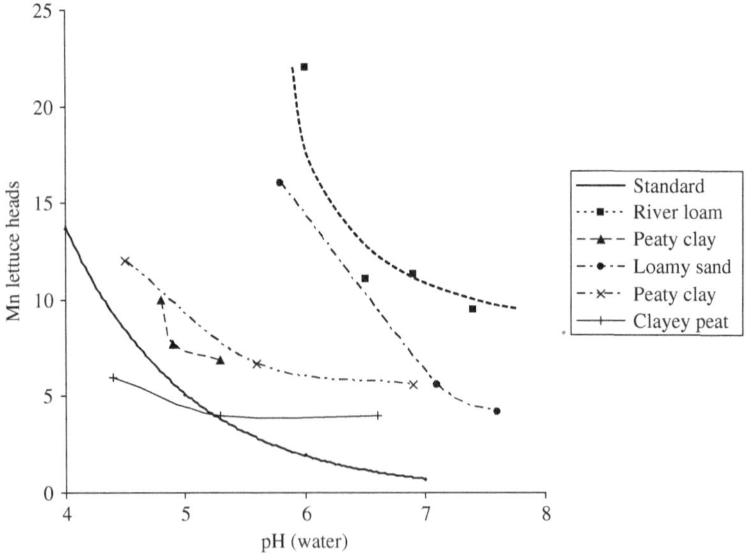

Fig. 10.4 The relationship between the pH and the Mn concentration (mmol kg^{-1} dry matter) of lettuce heads on un-steamed soils (standard) and on a series of steam sterilised soils. After Sonneveld and Voogt (1975a). *Modified by permission of Springer*

soils is mostly much higher than could be expected on basis of the prevailing pH in the root environment. This clearly reflects the strong disturbance of the Mn status of soils shortly after steam sterilisation. Thus, the first action on steam sterilised soils should be the control of the pH, when excessive quantities of available Mn are expected. It is experienced that the value should be at least 6.5 to restrict the uptake of Mn to a reasonable level. However, this often is insufficient to control the huge Mn uptake on steam sterilised soils. On such soils crops sensitive to Mn toxicity easily show toxicity symptoms after steam sterilisation, despite a sufficient high pH.

Crops show great differences in sensitivity for high Mn concentrations in the root environment and even among cultivars great differences occur. Foy (1973) has given a review of the sensitivity of numerous crops. The most sensitive crop in his review was oat already showing toxicity symptoms at about 1.5 mmol kg^{-1} dry matter in the crop and the most resistant crop was carrot only showing symptoms at concentrations of 125 mmol kg^{-1} dry matter. Among greenhouse crops lettuce (Messing, 1965; Sonneveld and Voogt, 1975), bean (Löhnis, 1960) and rose (Broer and Sonneveld, 1973) are indicated as susceptible, showing toxicity at levels below 7 mmol kg^{-1} dry matter. Cucumber (Messing, 1964; Sonneveld and De Bes, 1984) and carnation (Nilsson, 1967) are indicated as moderate susceptible, showing symptoms between 7 and 15 mmol kg^{-1} dry matter. Tomato (Dennis, 1968) is less susceptible showing symptoms at concentrations ever since 15 and 30 mmol kg^{-1} dry matter of the crop and sweet pepper (Gomi and Oyaki, 1972) is resistant showing symptoms only at higher concentrations.

The given indications for crops are very rough, because there are great differences in susceptibility among cultivars. Such differences cannot always be explained by the uptake of Mn. Such is shown for lettuce (Sonneveld and Voogt, 1975a), where big differences were found in the susceptibility of cultivars to Mn toxicity, while the differences in the manganese concentration of the crop were slight and not related to the degree of toxicity. See the data presented in Table 5.7. Another factor that will disturb the interpretation is the unequal distribution of Mn of plant organs and even within plant organs. Thus, the tissue sampled strongly determines the results. This factor is extensively discussed in Section 5.2. It is remarkable that in this case the resistance against high Mn is not related with the uptake from the external solution, but apparently with an "internal" resistance against the Mn absorbed. Differences between cultivars also have been found with rose. Two cultivars, namely Ilona and Baccara, were detected as severe sensitive to Mn toxicity (Broer and Sonneveld, 1973). For this crop it was not determined whether this could be related to differences in uptake, because there was no comparison with different cultivars. Besides the measures that could be carried out in the soil to reduce the concentration available Mn in the soil, breeding of resistant cultivars was quite effective in the prevention of Mn toxicity in sensitive crops.

It is possible to get informed on the risk of occurrence of Mn toxicity after steam sterilisation by soil testing. However, the Mn uptake mostly cannot be estimated by solely the determination of exchangeable Mn, when different soil types are under investigation. Different pH values and different adsorption capac-

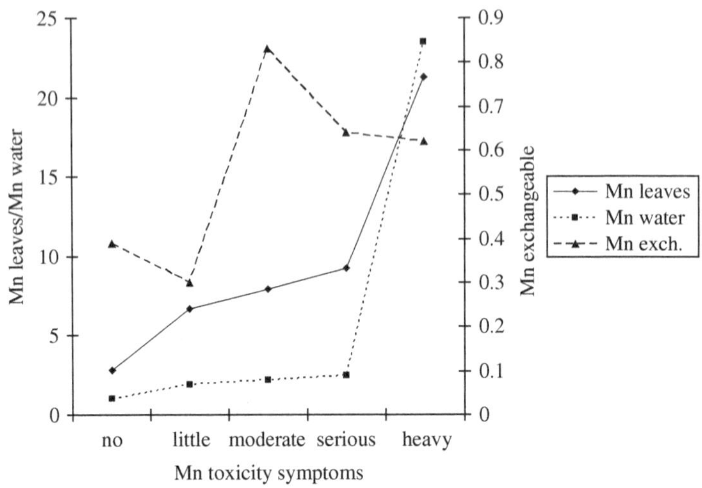

Fig. 10.5 The relationship between the occurrence of Mn toxicity in rose (cultivar Ilona) and the Mn concentrations in the leaves (mmol kg^{-1} dry matter) and in the soil. The water soluble Mn was determined by the 1:2 extract and expressed as μmol l^{-1} of extract and the exchangeable Mn determined by Morgan's extraction and expressed as mmol kg^{-1} dry soil. Derived from Broer and Sonneveld (1973)

ities of soils strongly affect this interpretation. The interpretation of water soluble Mn is less hindered by these factors. Therefore on steam sterilised soils the determination of water soluble Mn offers better results than the exchangeable concentrations (Sonneveld et al., 1977). A good example of this is shown in Fig. 10.5, where data of rose crops grown on different soil types are presented (Broer and Sonneveld, 1973). The toxicity symptoms are very well related with the Mn determined in the leaves and with the water soluble Mn concentration in the soil, while the relationship with the exchangeable Mn concentration in the soil is poor.

10.4.1 Soil Testing Methods

The soil testing methods used in this chapter are related to the common methods used in greenhouse industry and described in chapter 4. However, the data mentioned as exchangeable Mn are different, because they are obtained by extraction with a Morgan's solution instead of the method generally used with NH_4 acetate. The main difference is the pH on which this extraction solution was buffered, being 4.8 for Morgan's and in this case 7.0 for NH_4-Ac solution. The results of the determinations were closely correlated, as is clear from the relationship given in formula (10.4) (Sonneveld, 1975).

$$y = 0.69x - 0.24 \quad \mathrm{r} = 0.99 \tag{10.4}$$

In which:

x = Mn concentration in mmol kg^{-1} dry soil determined with Morgan's solution
y = Mn concentration in mmol kg^{-1} dry soil determined with NH_4 acetate

The equation was calculated for a loamy clay soil, with concentrations varying between 0.4 and 3 mmol kg^{-1} dry soil. With NH_4 acetate about 70% of the Mn is extracted of that extracted with a Morgan's solution.

10.5 Steam Sterilisation and N Status

In greenhouse soils the water soluble mineral N is usually present as NO_3. N applied in different forms, like NH_4 and urea are quickly converted to NO_3 by a high activity of micro organisms ensured by the continuous relatively high temperature and good aeration of the soil. This also is the case with N released from decomposition of organic matter, which at first is released as NH_4. However, steam sterilisation disturbs this process, because of a strong reduction of the microbiological activity. The microbial activity achieving the nitrification process can be separated into the conversion from NH_4 into NO_2 as given in formula (10.5) and the conversion of NO_2 into NO_3 as given in formula (10.6).

$$2NH_4 + 3O_2 \rightarrow 2HNO_2 + 2H_3O \qquad (10.5)$$

and

$$2HNO_2 + O_2 + 2H_2O \rightarrow 2NO_3 + 2H_3O \qquad (10.6)$$

The micro organisms involved for both processes are different and therefore, they can follow a different path with the re-colonization after steam sterilization. This includes a possible imbalance in the nitrification process and by this the possibility of the occurrence of NO_2 and NH_4 concentrations in the soil higher than usual.

Steam sterilisation at such directly affects the N status of soils by release of N as a result of decomposition of organic compounds. The N released in this process is found by a strong increase of NH_4 concentration and the quantity determined depends on the soil type, the temperature and the duration of the heating. The effects of these factors are shown by the data listed in Table 10.4. Another direct effect is the loss of mineral N, mainly NO_3. Some effects of steam sterilisation under greenhouse conditions on the mineral N status of soils are shown in Table 10.5, where the NO_3, NO_2 and NH_4 concentrations of six soils one day before and one day after steam sterilisation are shown. The soils were steam sterilised with sheets under greenhouse conditions. The contents of NO_2 and NH_4 of all soils were obvi-

Table 10.4 The effect of heating on the release of water soluble NH_4 from two soil types. Data of a laboratory experiment after Sonneveld (1969). Concentrations are expressed as mmol kg^{-1} dry soil

	Sandy loam			Clayey peat		
	Temperature			Temperature		
Heating time in hours	70	85	100	70	85	100
3	0.6	0.8	1.8	1.8	2.7	3.0
6	0.8	1.4	1.9	2.2	3.3	4.0
12	1.0	2.2	2.7	3.1	4.5	8.6
Control	0.4			1.8		

Table 10.5 Water soluble NO_3, NO_2 and NH_4 and exchangeable NH_4 as found on different greenhouse soils one day before and one day after steam sterilisation. The concentrations are expressed as mmol kg^{-1} dry soil

Soil type	NO_3 water		NO_2 water		NH_4 water		NH_4 exch[1]	
	Before	After	Before	After	Before	After	Before	After
Humic sand	7.1	3.4	0.00	0.05	0.00	1.43	0.00	3.43
Humic sand	7.6	1.6	0.00	0.07	0.14	1.14	0.00	3.21
Clayey loam	11.4	2.4	0.00	0.55	0.29	1.14	0.07	3.14
Loamy clay	2.1	1.1	0.00	0.15	0.21	1.29	0.00	2.64
Clayey peat	14.9	7.6	0.00	0.37	0.43	4.57	0.93	12.36
Clayey peat	11.8	7.1	0.00	0.03	0.21	2.93	0.29	6.00

[1]Exchangeable, the total in the extract after extraction with 1 M KCl.
After Sonneveld (1969a).

ously increased after steam sterilisation, while the contents of NO_3 are decreased. The increase of NH_4 can be explained by the decomposition of organic matter that occurs during steam sterilisation and the increase of NO_2 can be arisen from reduction of NO_3. However, this cannot explain the strong decrease of NO_3, which is much more than the increase of NO_2. Therefore, the decrease of NO_3 should be partly explained by leaching of NO_3 resulted from condensation of steam during the steam sterilisation treatment. However, there are indications that NO_3 totally can be got lost once reduced to NO_2. The NO_2 compound as being quite unstable can easily decompose to NO_x or N_2 and in this form it escapes to the atmosphere. This is supported by a laboratory experiment in which soils were heated at different temperatures and durations in closed pots. Especially with a long duration heating at 100°C the NO_3 contents were decreased. The results of the heating at 100°C during 12 h are listed in Table 10.6. The effect of the heat treatment differs strongly for the soils under investigation. At the two soils with original the lowest NO_3 concentrations small increases occurred. With two soils the NO_3 concentrations were more or less equal before and after the treatment. In most cases the NO_3 concentrations are decreased after steam sterilisation, especially in the peaty soils with the original

Table 10.6 NO_3 concentrations (mmol kg^{-1} dry soil) of different greenhouse soils before and after heat treatment in a laboratory experiment. The heating consisted of 100°C during 12 h and was carried out in closed pots

Soil type	Before heating	After heating	Soil type	Before heating	After heating
Sand	4.8	3.5	Clayey loam	1.2	2.8
Sand	3.4	2.5	Humic clay	4.3	4.4
Loamy sand	0.6	1.2	Humic clay	5.4	3.4
Loamy sand	5.4	5.6	Clayey peat	9.8	1.8
Clayey loam	3.6	2.7	Clayey peat	15.6	7.8

Experiment of Sonneveld (1967).

high concentrations. A decrease of NO_3 was also found after autoclaving at 105°C for half an hour (Jager et al., 1969). Bremner and Nelson (1968) in their study concluded that in sterilized soils NO_2 can decompose due to reaction with soil organic matter and by self decomposition of HNO_2. In the reaction with soil organic matter phenolic hydroxyl groups play an important part and have the ability to decompose NO_2 to N_2O and N_2. For the self decomposition of HNO_2 they claimed that the process best can be described as given in formula (10.7).

$$2HNO_2 \rightarrow NO + NO_2 + H_2O \tag{10.7}$$

With the determination of NO_2 in soils it is quite important that this determination is carried out in the fresh soil. During drying it is likely that important quantities of NO_2 decompose together with NH_4 following formula (10.8) (Ewing and Bauer, 1966). In our experiments we also got the experience that during drying at 50°C vast quantities of the NO_2 got lost from the soil.

$$NO_2 + NH_4 \rightarrow N_2 + 2H_2O \tag{10.8}$$

The release of NH_4 during steam sterilisation was also observed by Davies and Owen (1951) and Jager et al. (1969). Dawson et al. (1965) also found increases of NH_4 and NO_2 after heat treatments of potting compost. The effects were strongly related to the temperature of the heating and the addition of organic N compounds, in this case the addition of hoof and horn fertilizer. However, in their experiment the release of NH_4 occurred not during the heating treatment, but mainly in the weeks after the treatment when NO_2 exclusively resulted from the nitrification of NH_4. An increase of NO_2 concentrations as a result of the nitrification of NH_4 was also found with the greenhouse soils mentioned in Table 10.5. An example is shown in Fig. 10.6, where the courses of the concentrations of the different mineral N forms are presented of a clayey peat soil in relation with the time before and after the steam sterilisation. NO_3 is decreased by the steam treatment, possible by leaching as a result of steam condensation and by reduction and decomposition of NO_3 and NO_2, respectively, as mentioned before. The decrease of NO_3 during the following

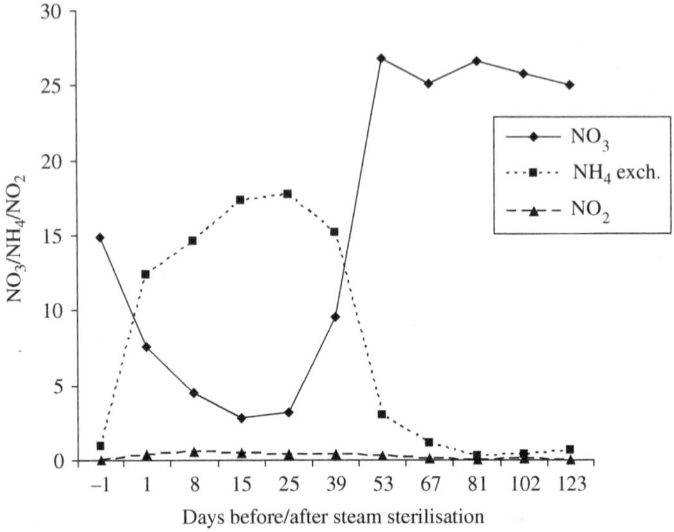

Fig. 10.6 The courses of the NO_3, exchangeable NH_4 and NO_2 concentrations in a clayey peat soil in relation to the time before and after steam sterilisation. The concentrations are expressed as mmol kg^{-1} dry soil. After Sonneveld (1969a)

weeks is caused by flooding and the strong increase afterwards is caused by nitrification of NH_4 and by the fertilizers applied. NH_4 is activated by steam sterilisation and the nitrification process starts about five weeks after the steam treatment. NO_2 is activated in periods when NH_4 is high, but the concentrations are low. The results presented in Fig. 10.6 are characteristic for the course of the different N forms in all six greenhouses under investigation. The NO_2 concentrations never went above 1 mmol kg^{-1} dry soil in one of the greenhouses and the concentrations exchangeable NH_4 never were higher than those found in the greenhouse soil presented in Fig. 10.6. Usually, the concentrations of NO_2 and of NH_4 arrive on the low levels common in greenhouse soils, within about 2 months after steam sterilisation.

10.6 Effects of N Form on Crop Development

Effects of NH_4 on the development of crops are studied for substrate growing and reviewed in Section 13.4. High concentrations of NH_4 in the root environment can become toxic to plants, like found for tomato (Maynard et al., 1966) and other crops (Barker and Mills, 1980). However, not all crops are sensitive for high NH_4 concentrations in the root environment. The growth reduction of crops by addition of solely NH_4 as N source differed strongly for the 8 different vegetable crops studied by Ikeda and Osawa (1982) in a water culture. With equal concentrations of NO_3 and NH_4 in the nutrient solution they got yields equal or even a little higher than with solely NO_3 fertilization. In experiments of Jager et al. (1970) with lettuce as a test crop the shape of the lettuce heads was negatively affected by the use of NH_4 in comparison with those when NO_3 was used. The use of NH_4 was connected with a poor heading, being an effect that also is experienced with winter grown lettuce grown on steam sterilised soils in The Netherlands.

In the greenhouses mentioned in Table 10.5 high NH_4 concentrations or high NH_4/N ratios after steam sterilisation merely occurred accidentally and when they occur it happened for short periods. This is shown in Table 10.7 where the highest NH_4 concentrations are mentioned in the six greenhouses together with the prevalent NO_3 concentrations. The highest NH_4 concentrations were found from one day until about 40 days after steam sterilisation. The plant mainly will react on the ratios of the concentrations in the soil solution. Because of the adsorption of NH_4 the ratio NH_4/N in the soil solution best can be estimated by the ratio (NH_4-water soluble)/(NO_3+NH_4-water soluble). This ratio varies between 0.14 and 0.38, with exception of the value calculated for the loamy clay where one day after steam sterilisation a ratio of 0.65 was found. This high ratio is more caused by the low NO_3 concentration than by a high NH_4 concentration. In view of these result can be concluded that plant development mostly will not be affected strongly by the NH_4 found in greenhouse soils after steam sterilisation.

NO_2 is much more toxic to plants than NH_4. This was shown in a pot experiment grown with lettuce. In this experiment $NaNO_2$ and $NaNO_3$ were compared as fertilizer in equivalent quantities on a loamy sand soil in steam sterilised and un-sterilized condition (Sonneveld, 1967a). The addition of the N was carried out as base dressing and as top dressing with the water added during the growing period with quantities of 7 mmol l^{-1} of soil and 14 mmol l^{-1} of water, respectively. The head weight of the lettuce was strongly reduced by the addition of NO_2 in comparison with NO_3, like shown in Fig. 10.7. The weight of the lettuce heads decreased linearly with increasing NO_2 concentrations in the drainage water. In this experiment the determination of NO_2 in the drainage water better reflected the addition of this compound than the determination by soil sampling. In the drainage water noticeable concentrations NO_2 were found at the end of the growing period, while in the soil only with the sampling some days after addition of the base dressing reasonable concentrations

Table 10.7 The day number after steam sterilisation when highest NH_4 concentrations were found in the greenhouse soils of Table 10.5, the prevalent water soluble NH_4, exchangeable NH_4 and NO_3 concentrations. All concentrations are expressed as mmol kg^{-1} dry soil

Soil type	Day nr	NH_4 water	NH_4 exch[1]	NO_3
Humic sand	15	2.4	4.0	5.8
Humic sand	37	3.7	9.2	7.9
Clayey loam	24	2.1	8.4	12.8
Loamy clay	1	1.1	2.6	0.6
Clayey peat	38	6.0	15.2	10.0
Clayey peat	39	3.8	8.4	12.0

[1]Exchangeable, including the water soluble

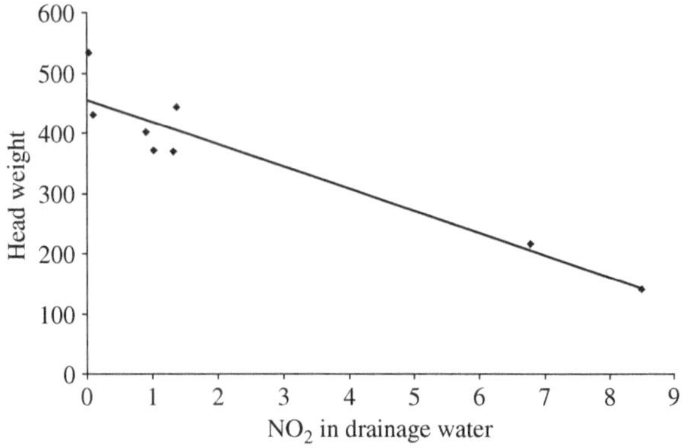

Fig. 10.7 The relationship between NO_2 concentrations in the drainage water (mmol l^{-1}) and the head weights of lettuce in g per head in a container experiment. After Sonneveld (1967a)

NO_2 were determined. At the end of the growing period no NO_2 was found in most soil samples from the treatments of the experiment, although the determination was carried out in the fresh soil. This again reflects the instability of the NO_2 compound in soils.

Highest NO_2 concentrations found in steam sterilised soils in a study on greenhouse holdings varied between 0.12 and 0.85 mmol kg^{-1} dry soil (Sonneveld, 1969a). Recalculated to the soil solution the concentrations varied between 0.26 and 1.12 mmol l^{-1}. Such concentrations only occurred for short periods and thus, in view of the results presented in Fig. 10.7, it is not to be expected that NO_2 strongly affects the growth of crops on steam sterilised soils.

10.7 Release of Br

Soil analysis before and after steam sterilisation learned that also Br is released in this process. This is clear from data of two greenhouse soils presented by Sonneveld (1993). Before the steam treatment the soils contained 6 and 12 μmol l^{-1} in the 1:2 volume extract and after the steam treatment 42 and 68 μmol l^{-1} were found, respectively. Most crops are not sensitive for somewhat high Br concentrations in soils, but some crops are very sensitive to Br among which carnation is best known. In pot experiments with this crop it has been found that irrigation with water containing 100 μmol l^{-1} Br damage can occur. By irrigation with such water in the 1:2 volume extract of the soil a concentration of 45 μmol l^{-1} was determined (Van den Bos and Zevenhoven, 1986). Thus, it is possible that for sensitive crops after steam sterilisation damage occurs as a result of too high Br concentrations. For edible crops, however, specific limits for Br are set up in view of human health. Leafy vegetables should not contain higher concentrations than 0.63 μmol kg^{-1} fresh weight and fruit crops not more than 0.38 μmol kg^{-1}. These limits can be crossed by the Br released by steam sterilisation, as is shown with the data listed in Table 10.8 (Sonneveld, 1993). In this table average and extreme values of Br concentrations of lettuce are shown, when grown on 10 different soils before and after steam sterilisation. Without steam sterilisation the concentrations are always below the limits, while after steam sterilisation the limits are crossed. Br can be leached from the soil quite easily as is clear from the data mentioned after flooding in Table 10.8. Thus, flooding of the soil is very effective in prevention of high Br concentrations in soil.

Table 10.8 Average and extreme Br concentrations of lettuce (μmol kg^{-1} fresh material) of ten different soils as determined before and after steam sterilisation and after steam sterilisation and flooding

Treatment	Average	Range
No steaming	0.14	0.03–0.34
Steaming	0.48	0.21–0.94
Steaming and flooding	0.09	0.02–0.18

Data from Sonneveld (1993).

10.8 Organic Matter

Besides effects on mineral elements steam sterilisation also affects the solubility of organic matter. Jager et al. (1969) determined after autoclaving of a humic clay soil an increase of the soluble organic matter from 280 to 860 mg kg^{-1} dry soil. The water holding capacity during the autoclaving had a strong effect on the release of the organic matter. With increasing water content of the soil from 33 until 100% of the moisture capacity, the quantity of soluble organic matter released with autoclaving increased strongly. Comparable results have been found by Sonneveld (1984). The chemical oxidation demand (COD) of the water drained from a clayey peat soil increased from about 100 mg l^{-1} for the un-steamed soil to 2500 mg l^{-1} O_2 for the steam sterilised soil. The effects of the higher concentrations of soluble organic matter in the soil solution after steam sterilisation are not yet studied in detail. In a simple experiment with cucumber grown in rock wool the addition of humic compounds to the substrate solution were compared with a control treatment (Sonneveld, 1984). The humic compounds were extracted with water from a steam sterilised well decomposed high moor peat. The nutrient solution in the growing system was circulated and the humic compounds were added regularly to keep the concentration in the solution on the proposed level. The cucumbers were grown during 5 months and in Table 10.9 the yield and analytical data of the leaves are listed. The yield in the treatment with the addition of soluble organic mater is not significantly affected in comparison with the control. The K and the Ca concentrations in the leaves are substantially decreased and increased, respectively by the addition of the organic matter. The concentrations of the other mineral elements in the leaves were slightly affected. In a study with herbs, oregano, thyme and basil, soluble organic matter extracted from white peat was added to the nutrient solution (De Kreij, 1995). The concentrations of metal micro nutrients, being Fe, Mn, Zn and

Table 10.9 Yield (kg m^{-2}) COD of the nutrient solution (mg O_2 l^{-1}) and analytical data of young fully grown leaves (mmol kg^{-1} dry matter) of a rock wool grown cucumber crop as affected by addition of soluble organic matter to the nutrient solution

Determination	Control	With organic matter
Yield	25.7	24.7
COD[1]	40	51
K	1101	730
Na	56	45
Ca	582	817
Mg	236	274
N	4552	4707
Cl	34	55
P	283	283
Fe	1.88	1.69
Mn	2.86	3.04
Zn	1.40	1.48
B	4.69	4.97
Cu	0.19	0.19

[1]Chemical oxidation demand of the nutrient solution in the root environment
Data of Sonneveld (1984).

Cu, in the shoots were reduced. This reduction occurred in peat as well in perlite substrate, but was mainly manifested at a pH value of 5.0 and not at a value of 6.5. The fresh weight of the herbs was significantly reduced by the addition of soluble organic matter, especially at the low pH. The yield reduction could not be explained by the reduced uptake of micro nutrients and was described as a toxic effect of the humic substances. Thus, soluble organic matter released by steam sterilisation of greenhouse soils will be able to affect nutrient uptake and crop development. However, the consequences under practical growing conditions are not yet studied sufficiently.

10.9 Pasteurization

From the data presented it is clear that the undesirable side effects of steam sterilisation of soils strongly depend on the time of heating and the temperature realised (Dawson et al., 1965; Dawson et al., 1967; Sonneveld, 1968, 1969). In addition it was proved that many harmful plant pathogens are killed at temperatures lower than 100°C (Bollen, 1969). Therefore, to escape from unfavourable side effects experiments were carried out with treatments of steam-air mixtures of 70°C (Sonneveld and Voogt, 1973). In a container experiment lettuce was grown in two series with in total 8 different soils, treated with steam (100°C) or with a steam-air mixture of 70°C. Compared with the not sterilised control the head weight of the lettuce grown in the soil treated with steam was on average 88% of the control with a variation from 23 till 109%. In the soils treated with the steam-air mixture the average head weight was 109% of the control with a variation from 102 till 130%.

The wide variation of the yields points to the fact that the undesirable side effects strongly varies with the soil type. In two field experiments sterilisation at 100°C was compared with steam/air treatment at 70°C. The head weight of the lettuce grown directly after the heat treatments was on average over two years highest at the 70°C treatment with a head weight 13 and 28% above the 100°C treatment for the two experiments. The tomato crop grown afterwards did not shown any significant difference between the treatments.

Despite the good results of the partial sterilisation of soils with steam-air mixtures of 70°C, the method is not widely applied. There is no other explanation for this slow development than the technical installation necessary to prepare the steam-air mixture, which is relatively extended compared with the equipment necessary for the widely applied simple steam treatment of the soil under sheets, see Picture 10.1. However, the equipment for preparation of steam-air mixture is not that complicated that it will block future developments in that direction (Dawson et al., 1967).

Picture 10.2 An installation used for sterilisation of drainage water from substrate systems by UV treatment

10.10 Substrate Growing

Sterilisation in substrate growing is important when substrates or drainage water are reused. Substrates are sterilised when the material is used for different crops successively grown to prevent infection from the one crop to another. Drainage water is sterilised when reused in the growing systems during crop cultivation to prevent transmission of a local infection of pathogens throughout the whole system. The method of sterilisation depends on the pathogen and the material involved. Runia (1995) and Wohanka (2002) in their reviews mentioned following disinfection systems.

- Heating
- UV radiation
- Chemical treatment, like O_3, Cl_2, ClO_2, I_2, H_2O_2 and Cu
- Filtration

Heating is used to sterilize substrates as well for water and nutrient solutions, while the other methods are solely applied for treatment of drainage water and possible irrigation water. When applied for sterilisation of water and nutrient solutions it is recommended to bring the pH of the solution at a value below 4, to prevent precipitation of Ca salt. This best can be done by nitric acid, which can be utilized later on by the crop as a nutrient, and it disturbs relatively scarcely the ratios of the nutrients. When heating is applied for sterilisation of substrates, it can have comparable effects as with steam sterilisation of soils. Release of Mn and NH_4 will be expected when available in forms comparable with those in soils. However, substrates are artificially composed and some materials are prepared in that way that the components have got a different structure and are differently affected by steam sterilisation. Such for example will be the case with rock wool. This material is prepared at very high temperature and the metals in the material are included that way that they are not easily released during steam sterilisation. However, Mn from the nutrient solution can be oxidised and will precipitate on the fibres during cultivation and will be released by a following steam sterilisation. Comparable effects of Mn can occur on all different types of substrate. However, when material rich in active Mn is mixed in the substrate substantial quantities will be released with heating. The addition of clay rich in Mn in peaty mixtures is an example.

UV radiation of nutrient solutions affects the availability of Fe by decomposition of the chelate (Daughtrey and Schippers, 1980; Himken, 1989; Runia and Boonstra, 2004; Stangellini et al., 1984). This should be taken into account by addition of extra fertilizer or possible by the choice of the type of Fe fertilizer. The Fe fertilizers used in substrate culture mostly consist of the chelate types EDDHA, DTPA and EDTA. The stability against the degradation by UV treatment was highest for EDDHA, while DTPA and EDTA showed little difference (Acher et al., 1997). The degradation by UV light was affected by the pH of the nutrient solution and was stronger at the extreme pH values of 3.5 and 7.5 tested in the experiment. At the values 4.5 and 6.0 a greater stability was found against UV radiation.

The chemical treatments are merely due to chemical oxidation, while Cu act as a growth inhibiter for microorganism. It will be expected that chemical oxidizers also affect Fe-chelates and Mn; Fe is decreased by oxidation of the chelate complex on which it is bound and Mn is decreased by precipitation following from a direct chemical oxidation. Such effects have been found with ozonization (Runia and Boonstra, 2004; Vanachter et al., 1988) and by addition of ClO_2 (Wohanka et al., 2005). Vanachter et al. (1988) in their experiments with ozonization found an interaction between the effect on different chelates and Mn. Fe was strongly decreased by ozonization when added as EDDHA, while Mn was not affected. When Fe was added as EDTA or DTPA the Fe concentration was hardly affected, but the Mn concentration surely decreased. Oxidation of DTPA was found also by addition of H_2O_2

(Runia and Boonstra, 2004). When compounds are used containing elements toxic to plants, like I and Cu, the concentration of such elements will be carefully controlled. Toxic concentrations of different elements are given by Chapman (1966).

References

Acher A Heuer B Rubinskaya E and Fischer E 1997. Use of ultraviolet-disinfected nutrient solution in greenhouses. J. Hort. Sci. 72, 117–123.

Barker A V and Mills H A 1980. Ammonium and nitrate nutrition of horticulture crops. Hort. Rev. 2, 395–423.

Bollen G J 1969. De invloed van stomen op de biologische eigenschappen van de grond. Tuibouwmededelingen 32, 475–479.

Bremner J M and Nelson D W 1968. Chemical decomposition of nitrite in soils. 9th Intern. Congress Soil Sci. Transactions, Adelaide, Australia II, 495–503.

Broer S and Sonneveld C 1973. Mangaanvergiftiging bij roos (var. Ilona). Vakblad Bloemisterij 28(48), 10–11.

Bromfield S M and Skerman V D B 1950. Biological oxidation of manganese in soils. Soil Sci. 69, 337–348.

Chapman H D 1966. Diagnostic criteria for plants and soils. University of California, Division of Agriculture Sciences, 793 pp.

Davies J N and Owen O 1951. Soil Sterilization. I Ammonia and nitrate production in some glasshouse soils following steam sterilisation. J. Sci. Food Agric. 2, 268–279.

Daughtrey M L and Schippers P A 1980. Root death and associated problems. Acta Hort. 98, 283–291.

Dawson J R Johnson R A H Adams P and last F T 1965. Influence of steam/air mixtures, when used for heating soil, on biological and chemical properties that affect seedling growth. Ann. Appl. Biol. 56, 243–251.

Dawson J R Kilby A A T Ebben M H and last F T 1967. The use of steam air mixtures for partially sterilizing soils infested with cucumber root pathogens. Ann. Appl. Biol. 60, 215–222.

De Groot A J 1963. Mangaantoestand van Nederlandse en Duitse Holocene sedimenten in verband met slibtransport en bodemgenese. Verslagen van Landbouwkundige Onderzoekingen, 69.7. Pudoc Wageningen, The Netherlands, 206 pp.

De Kreij C 1995. Effect of humic substances in nutrient film technique on nutrient uptake. J. Plant Nutr. 18, 793–802.

Dennis D J 1968. Manganese toxicity in tomato seedlings. New Zealand J. Agric. 117, 118–119.

Ewing G J and Bauer N 1966. An evaluation of nitrogen losses from the soil due to the reaction of ammonium ions with nitrous acid. Soil Sci. 102, 64–69.

Foy C D 1973. Manganese in plants. *In*: Lieben J (ed), Manganese. National Acdemy of Sciences, Washington, 51–76.

Geering H R Hodgson J F and Sdano C 1969. Micro nutrient cation complexes in soil solution: IV. The chemical state of manganese in soil solution. Soil Sci. Soc. Amer. Proc. 33, 81–85.

Gerritsen F C 1937. Manganese deficiency of oats and its relation to soil bacteria. Ann. Bot. 1, 207–230.

Gomi K and Oyaki T 1972. Studies on manganese nutrition of vegetable crops. I. Manganese deficiency and excess of vegetable crops. Bull. Faculty Agric. Miyazaki Univ. 19(2), 493–503.

Himken S 1989. Erprobung der Wirksamkeit einess UV-Gerätes zur Entkeimung gärtnerischer Nährlösungen. Diplomarbeit Research Center Geisenheim Germany,79 pp.

Ikeda H and Osawa T 1982. Effect of potassium and calcium levels and their accompanying anions in the nutrient solution on ammonium toxicity in vegetable crops. J. Japan. Soc. Hort. Sci. 51, 309–317.

Jager G Van der Boon J and Rauw G J G 1969. The influence of soil steaming on some properties of the soil and on the growth and heading of winter lettuce. I. Changes in chemical and physical properties. Neth. J. Agric. Sci. 17, 143–152.
Jager G Van der Boon J and Rauw G J G 1970. The influence of soil steaming on some properties of the soil and on the growth and heading of winter lettuce. III. The influence of nitrogen form, manganese level and shading studied in sand culture experiments with trickle irrigation. Netn. J. Agric. Sci. 18, 149–157.
Leeper G W 1947. The forms and reactions of manganese in the soil. Soil Sci. 63, 79–94.
Löhnis M P 1960. Effect of magnesium and calcium supply on the uptake of manganese by various crop plants. Plant Soil 12, 339–376.
Maynard D N Barker A V and Lachman W H 1966. Ammonium induced stem and leaf lesions of tomato plant. Proc. Amer. Soc. Hort. Sci. 88, 516–520.
Mckenzie R M 1972. The manganese oxides in soil – A review. Z. Pflanzenernährung Bodenk. 131, 221–242.
Messing J H L 1964. Manganese toxicity. Rep. Glasshouse Crops Res. Inst. 1963, 63–64.
Messing J H L 1965. The effects of lime and superphosphate on manganese toxicity in steam-sterilised soil. Plant Soil 23, 1–16.
Nilsson G 1967. Twee anjerproblemen. DCK Information 26, 1.
Page E R 1962. Studies in soil and plant manganese. II The relationship of soil pH to manganese availability. Plant Soil 16, 247–257.
Page E R 1964. The extractable manganese in soil. J. Soil Sci. 15, 93–102.
Roorda van Eysinga J P N L Van Dijk P A and De Bes S S 1978. The available manganese content of soils in The Netherlands determined by various methods. Comm. Soil Sci. Plant Anal. 9, 141–151.
Runia, W Th 1995. A review of possibilities for disinfection of recirculation water from soilless cultures. Acta Hort. 382, 221–229.
Runia W T and Boonstra S 2004. UV-oxidation technology for disinfection of recirculation water in protected cultivation. Acta Hort. 644, 549–555.
Sherman G D 1957. Manganese and soil fertility. The 1957 yearbook of agriculture (soil) USDA, 135–139.
Sonneveld C 1965. Onderzoek naar het verloop van grondanalysecijfers (1963–1964). Proefstation voor Tuinbouw onder Glas, Naaldwijk, The Netherlands. Internal Report, 7 pp.
Sonneveld C 1965a. De invloed van grondontsmetting op de stikstofgift bij verschillende grondsoorten. Proefstation voor Tuinbouw onder Glas Naaldwijk, The Netherlands. Internal Report 16 pp.
Sonneveld C 1967. De invloed van het verhitten van grond op de mangaan- en stikstofhuishouding. (laboratorium onderzoek 1966). Proefstation voor de Groenten- en Fruitteelt onder Glas te Naaldwijk, The Netherlands. Internal Report, 1967, 15 pp.
Sonneveld C 1967a. De invloed van nitrietstikstof bij sla. Proefstation voor de Groenten- en Fruitteelt onder Glas te Naaldwijk, The Netherlands. Internal Report, 1967, 15 pp.
Sonneveld C 1968. De mangaanhuishouding van de grond en de mangaanopname van sla. Meded. Dir.Tuinb. 31, 476–483.
Sonneveld C 1969. De invloed van het stomen op de stikstofhuishouding van de grond. Tuinbouwmededelingen 32, 197–203.
Sonneveld C 1969a. De invloed van het stomen van de grond op de stikstof- en de mangaanhuishouding (praktijkonderzoek 1967–1968). Proefstation voor de Groenten- en Fruitteelt onder Glas te Naaldwijk, The Netherlands. Internal Report 1969, 10 pp.
Sonneveld C and Voogt S 1973. The effects of soil sterilisation with steam-air mixtures on the development of some glasshouse crops. Plant Soil 38, 415–423.
Sonneveld C 1975. Het verloop van het gehalte uitwisselbaar- en water oplosbaar mangaan in gestoomde grond. Proefstation voor de Groenten- en Fruitteelt onder Glas, Naaldwijk. Internal Report No 688/1975, 13 pp.

Sonneveld C and Voogt S J 1975. The effect of steam sterilisation of soil on the manganese uptake by glasshouse crops. Acta Hort. 51, 311–319.

Sonneveld C and Voogt S J 1975a. Studies on the manganese uptake of lettuce on steam-sterilised glasshouse soil. Plant Soil 42, 49–64.

Sonneveld C Voogt S J and Van Dijk P A 1977. Methods for the determination of toxic levels of manganese in glasshouse soils. Plant Soil 46, 487–497.

Sonneveld C 1979. Changes in chemical properties of soil caused by steam sterilisation. *In*: Mulder D (ed) Soil Desinfestation, Elsevier, Amsterdam, 39–50.

Sonneveld C and De Bes S S 1984. Micro nutrient uptake of glasshouse cucumbers grown on rockwool. Comm. Soil Sci. Plant Anal. 15, 519–535.

Sonneveld C 1984. De invloed van organische stof toediening aan de voedingsoplossing bij komkommers. Proefstation voor Tuinbouw onder Glas te Naaldwijk The Netherlands. Internal Report 1984 no 10, 11 pp.

Sonneveld C 1993. Chemische veranderingen in de grond door stomen. Plantenvoeding in de Glastuinbouw. Proefstation voor Tuinbouw onder Glas. Informatiereeks no 87, 112–121.

Stangellini M E Stowell L J and Bates M L 1984. Control of root rot of spinach caused by *Pythium aphanidermatum* in a recirculating hydroponic system by ultraviolet irradiation. Plant Disease 68, 1075-1-76.

Vanachter A Thys L Van Wambeke E and Van Assche C 1988. Possible use of ozon for disinfestation of plant nutrient solutions. Acta Hort. 221, 295–302.

Van den Bos A L and Zevenhoven M 1986. Bromide injury in carnations. Glasshouse Crops Research Station Naaldwijk The Netherlands. Annual Report 1986, 12–13.

Wohanka W 2002. Nutrient solution disinfection. *In*: Savvas D and Passum H (eds), Hydroponic Production of Vegetables and Ornamentals. Embryo Publications, Athens, Greece, 345–372.

Wohanka W Molitor H-D und Weislein D 2005. Verhinderung der Krankheitsausbreitung bei Ebb/Flutbewaässerung mittels Wasserdesinfektion aurch Chlordioxid. Schlussbericht Furschungsanstalt Geisenheim Germany, 80 pp.

Chapter 11
Substrates: Chemical Characteristics and Preparation

11.1 Introduction

In this chapter the characteristics of substrates will be discussed with respect to their effects on plant nutrition. Therefore, the chemical composition will be taken into account in the first place, because the mineral elements present in the material can be directly available to plants or can become available to plants dependent on the growing conditions. Beside mineral elements also other chemical compounds can be available in the material, which affect the plant growth negatively as well positively. Furthermore, with the preparation of some substrates mineral fertilizers are added to supply the plants grown in it with sufficient nutrients at the start of the growing period. Such applications with the preparation depend on the objective for which the substrate is prepared. Requirements in this field differ for substrates, crops grown and growing conditions. Important factors with respect to the growing conditions are for example the length of the growing period of the plant – a short propagation or a long production period – the growing system aimed at, the irrigation system and in relation with the last the method of fertilization that will be applied. If for example a substrate is prepared for a growing system in which directly at the start a complete nutrient solution is supplied, the requirement for the addition of mineral nutrients is less in comparison when is started with irrigation of just pure water. Substrates with a high cation adsorption capacity (CEC) will be fertilized differently from substrates with a low CEC. In this chapter mainly characteristics of substrates that affect the uptake of mineral elements by plants will be presented, while physical characteristics not directly affecting the mineral composition of plants are outside the context of this book.

11.2 Physical Conditions

Generally, the uptake of macro nutrients by crops, will not be affected by the physical characteristics of the substrate. In an experiment with different crops a detailed study was made between the nutrient uptake of crops grown in two substrates with very different density and structure, being sand and rock wool fibre. In

C. Sonneveld, W. Voogt, *Plant Nutrition of Greenhouse Crops*,
DOI 10.1007/978-90-481-2532-6_11, © Springer Science+Business Media B.V. 2009

Table 11.1 Macro nutrient concentrations of plant tissues of different crops at optimal nutrient supply as affected by sand and by rock wool substrate. The contents are expressed as mmol kg^{-1} dry matter

Crops	Plant parts	Substrates	K	Ca	Mg	N	P	S
Chrysanthemum	Young leaves	Sand	1702	352	128	4134	158	97
		Rock wool	1684	380	143	4195	164	99
Kohlrabi	Leaves	Sand	1696	922	256	3426	141	400
		Rock wool	1818	818	216	3406	150	382
Lettuce	Heads	Sand	2495	244	168	3967	236	95
		Rock wool	2622	262	162	3995	238	94
Freesia	Young leaves	Sand	1567	194	108	2434	164	140
		Rock wool	1524	182	117	2338	124	144
Aster	Whole branches	Sand	821	210	54	1462	113	64
		Rock wool	756	202	54	1534	111	60
Hippeastrum	Leaves	Sand	1156	320	92	1468	154	99
		Rock wool	1252	308	82	1449	184	91
Lily	Whole branches	Sand	778	235	100	1528	88	54
		Rock wool	821	216	104	1502	84	48
Average		Sand	1459	354	129	2631	151	136
		Rock wool	1497	338	125	2631	151	131

Data Van den Bos (1994, 1994a, 1995, 1996, 1996a, b, 1997).

the experiment different concentrations of the nutrient solution in the root environment were compared with a series of crops (Van den Bos, 1994, 1994a, 1995, 1996, 1996a, b, 1997). In Table 11.1 just the data of the macro element concentrations of plant tissues at the optimal yield are shown. The average values show small differences for the cation uptake. A high density of the substrate reduces the uptake of K and aggravates that of Ca and Mg. This is in agreement with expectations that a high density of the substrate aggravate root branching and by this the uptake of ions, like Ca and Mg, which are preferably absorbed by newly formed roots (Scott Russell and Clarkson, 1976). However, the differences between the nutrient concentrations in the plant tissues are small and not systematic and therefore, must be mainly attributed to residual errors. Comparable conclusion can be drawn from data of Van Labeke and Dambre (1998) with a gerbera crop in coir and rock wool. The study of Imas and Bar-Yosef (1998) with lettuce in tuff, sand and rock wool confirm the conclusion for the P uptake. However, differences in the nutrient uptake between substrates are obvious, if the chemical composition of the substrate solutions differ (Maher and Prasad, 1995).

The data of the micro nutrient concentrations in the plant tissues in the already mentioned experiments are listed in Table 11.2. For lettuce no micro nutrients were determined. On average the concentrations of micro nutrients of the rock wool grown crops were a little lower than in the sand grown crops, which amount from 5 to 22%. However, the differences are not systematic. An explanation for this

Table 11.2 Micro nutrient concentrations of plant tissues of different crops at optimal nutrient supply as affected by sand and rock wool substrates. The contents are expressed as mmol kg^{-1} dry matter

Crops	Plant parts	Substrates	Fe	Mn	Zn	B	Cu
Chrysan-themum	Young leaves	Sand	1.98	0.52	0.92	2.76	nd[1]
		Rock wool	1.78	0.36	0.98	2.62	nd
Kohlrabi	Leaves	Sand	1.20	0.50	0.64	4.60	nd
		Rock wool	0.76	0.30	0.58	4.19	nd
Freesia	Young leaves	Sand	1.90	1.28	0.93	7.69	0.09
		Rock wool	2.03	1.11	0.88	9.08	0.10
Aster	Whole branches	Sand	0.61	0.50	0.60	3.84	0.08
		Rock wool	0.56	0.34	0.52	3.58	0.07
Hippeastrum	Leaves	Sand	0.73	0.20	0.34	5.33	0.06
		Rock wool	1.03	0.27	0.34	4.37	0.04
Lily	Whole branches	Sand	0.82	0.22	0.49	5.38	0.07
		Rock wool	0.72	0.16	0.45	3.60	0.05
Average		Sand	1.21	0.54	0.65	4.93	0.075
		Rock wool	1.15	0.42	0.62	4.57	0.065

[1] nd not determined
Data Van den Bos (1994, 1994a, 1995, 1996, 1996a, b, 1997).

phenomenon is possibly the same effect of root branching, like mentioned with the major elements. Such an effect for example was noticed in an experiment in which the Fe uptake of tomato in a substrate system and in a nutrient film technique system was compared at a low level of Fe supply. In advance, the uptake of Fe was more effective in the substrate, supposing that in substrate a more finely branched root system occurred, which promoted the Fe uptake (Sonneveld and Voogt, 1985). Later on, when the plants in the system without substrate also had formed a finely branched root system, the differences disappeared and the lower yield in advance changed in an even higher yield later on (Table 11.3). The fact whether the substrate contained Fe or not – rock wool or polyurethane, respectively – apparently had no effect on the Fe uptake. Thus, the uptake of extra micro nutrients as found in Table 11.2 merely should be explained by a more branched root system which can be expected in a denser substrate, in this case the sand versus the rock wool grown crops.

Table 11.3 The effect of some substrates on the appearance of Fe chlorosis in tomato grown in a circulation system. In the nutrient solution added 5 μmol l^{-1} Fe was supplied

	Substrates		
Characteristic	None	Rock wool	Polyurethane
Fe root environment μmol l^{-1}	9	10	11
Early chlorosis	severe	slight	slight
Early yield kg m^2	1.7	2.0	2.0
Total yield kg m^2	19.6	17.4	17.4

After Sonneveld and Voogt (1985). *Reprinted by permission of Springer*

11.3 Chemical Conditions

The uptake of mineral elements by crops will be affected by the chemical condition of substrates or substrate constituents. In the first place this will occur, when mineral elements are added with the preparation of the substrate. Secondly, substrates itself or substrate constituents may contain mineral elements available to plants. In the third place the fluctuations of the pH of the substrate in which the crop is grown can mobilize or immobilize mineral elements.

Addition of mineral elements to plant substrates merely occurs with the preparation of substrates with a high cation adsorption capacity, like the substrates composed from mainly natural organic materials. This subject will be discussed under Section 11.4.

The concentrations of nutrients in substrates or substrate constituents directly available to plants will be determined with water extraction, while the quantities determined as exchangeable nutrients only are gradually released during the cultivation. Therefore, the concentrations determined with water extraction best reflect the concentrations directly available to plants in the substrate solution. During crop growth the pH in the root environment will fluctuate, by which the availability of some elements to plants vary too. Therefore, determination of minerals in an exchangeable solution is advisable to get informed about minerals which can become available during crop growth, dependent on the development of the pH or fluctuations in the cation concentrations. Most minerals are better soluble at low values and therefore an extraction solution with a pH buffer of about 4.5 is most informative for those elements, being the lowest value occurring during crop cultivation. Some elements, however, are better soluble at high pH values, like Mo. When solubility at high pH values is expected, the determination with a solution buffered at pH 8 can be recommended, being the highest pH that can be expected in the root environment. The EN method of CAT presented in Section 4.9 is less suitable to this purpose, because of the strongly changing pH of this extract in relation with the pH and the adsorption capacity of the substrate under treatment. Extractions with NH_4Ac presented in Section 4.10 possible buffered at the two extreme pH values mentioned will better inform about potential availability of minerals. In Table 4.7 the differences between water extraction and CAT extraction of a series of substrates is already shown.

In Table 11.4 the analytical data of different unfertilized substrates or substrate constituents are listed as found by water extraction (Kipp et al., 2000). The analyses are carried out by the 1:1½ volume extract or by "three times substrate moisture content". Last method is related to the 1:1½ volume extract (Sonneveld and Van Elderen, 1994) used for peaty substrates realizing a three times dilution of the soil solution. Thus, for interpretation the results of the "three times moisture extraction" can be compared with those of the 1:1½ extract.

The data show that rock wool fibres is the most inert substrate, showing just a little B the concentration of which will be ignored. The data indicate that coir can

Table 11.4 Analytical data of unfertilized substrates and substrate constituents. The analyses are carried out by the 1:1½ volume extract or "three times substrate moisture content". The EC is expressed as dS m^{-1} and the mineral elements are expressed as mmol and μmol l^{-1} on the extract

Characteristics	Rock wool	Peat	Coir chips	Pumice	Perlite	Wood fibre
pH	6.2	3.9	5.7	6.3	6.3	4.8
EC dS m^{-1}	0.0	0.2	0.5	0.2	0.1	0.2
K mmol l^{-1}	0.0	0.0	1.3	0.7	0.0	0.9
Na	0.0	0.6	2.8	0.5	0.1	0.1
Ca	0.0	0.1	0.1	0.1	0.0	0.4
Mg	0.0	0.1	0.3	0.1	0.0	0.3
NO_3	0.0	0.3	0.0	0.4	0.1	< 0.1
Cl	0.0	0.4	2.5	0.2	0.1	0.1
SO_4	0.0	0.1	0.0	0.2	0.0	0.1
P	0.0	0.0	0.1	0.2	0.0	0.1
Si	0.0	–.–	0.0	0.2	0.1	0.1
Fe μmol l^{-1}	0.1	1.1	3.9	1.9	0.1	5.2
Mn	0.0	0.5	0.3	11.3	0.1	14.1
Zn	0.0	0.3	1.5	1.3	0.1	4.3
B	2.1	2.5	5.0	15.8	11.0	27.0
Cu	0.0	0.1	2.3	2.3	0.0	0.3

Data derived from Kipp et al. (2000).

contain noticeable concentrations of K, Na and Cl, These concentrations fluctuate strongly (Prasad, 1997) and often the concentrations of Na and Cl are on a level that leaching is necessary (Verhagen and Van Schie, 1996). The leaching of Na is promoted by the addition of Ca ions to the water with which the coir is washed out. $Ca(NO_3)_2$ is suitable to this purpose. It will be noted that the coir material after this treatment will contain higher concentrations Ca and NO_3, which will be taken into account with possible addition of fertilizers. The treatment of the coir also will lower the K concentrations seriously. It is recommended that the concentrations of Na and Cl after leaching are below 2 mmol l^{-1} in the 1:1½ extract (Wever, 1995). The results of pumice and wood fibre indicate serious concentrations B and Mn. This also can occur with rice hulls and bark (Bos et al., 2002; López-Cuadrado et al., 2008). Materials like pumice, coir and wood fibre are from natural origin and won under varying conditions and such materials often show a great variation around the average value. The ranges of the data for wood fibre were 0–36 and 0–70 for Mn and B, respectively, while for the pumice these ranges were 0–56 and 0–42 respectively. The data on the high side of the ranges includes the possibility of toxic concentrations in the substrate solution. Such is less dangerous for B than for Mn, because the B is more or less completely extracted by water and will be easily washed out. For Mn the water extraction only indicate the activity of this element, while tremendous quantities can be in storage, which can become available depending of the use of the substrate. Beside the Mn soluble in water, very big quantities are adsorbed or in storage as Mn oxides. These oxides will easily become soluble during cultivation with lowering of the pH, which can easily induce Mn toxicity.

Table 11.5 Mn uptake of a rock wool grown gerbera crop in relation to different pH values in the root environment. The Mn concentration in the root environment (substrate solution) is expressed as μmol l^{-1} and the concentrations in the plant as mmol kg^{-1} dry matter

pH in root environment	Mn in root environment	Mn in young leaves	Mn in old leaves	Chlorosis index*
6.7	0.7	0.27	0.39	4.8
6.4	0.9	0.49	0.94	2.4
5.6	2.0	0.71	1.07	1.6

*0 – no chlorosis en 9 – serious chlorosis.
After Sonneveld and Voogt (1997).

Mn release, for example is clear shown by the data of an experiment in which the addition of Mn in a nutrient solution was studied at different pH levels in rock wool grown gerberas (Sonneveld and Voogt, 1997). In the treatments without addition of Mn in the nutrient solution, the gerberas were able to absorb sufficient Mn when the pH was sufficient low, as shown in Table 11.5. Despite that the Mn in the rock wool is not available by extraction with water, during cultivation at low pH continuously low concentrations are released and absorbed by plants.

Therefore, it is important to know which elements and chemical compounds can be on hand in the substrates and substrate constituents used, to get informed what can be released. A total analysis of substrates and substrate constituents is important and will be useful. However, the standard package of determinations is restricted and cannot cover all elements and compounds possible present in materials used for substrate preparation. It is always possible that elements and compounds are present in the substrate or substrate constituent different from the standard package. Therefore, it is important that substrate producers know the composition, the safety, the origin and the possible chemical reactions of their materials that possible occur during the production process. In this way they can be aware of the risks of their substrates and constituents used. See also the discussion in Section 11.5. In Table 11.6 the chemical composition of a number of mineral substrates are listed. The elements are expressed as oxides, but this does not mean that these elements always really occur as oxides. This depends on the origin and a possible fabrication

Table 11.6 Chemical compositions of some mineral substrates, determined by total destruction. Elements are expressed as oxides in percentages of the dry matter

Materials	SiO_2	Al_2O_3	CaO	MgO	Fe_2O_3	Na_2O	K_2O	MnO	TiO
Rock wool[1]	47	14	16	10	8	2	1	1	1
Perlite[2,3]	73–74	13–15	0.8	0.1–02	0.8–1.0	3.3–3.6	4.3–4.5	0.05	0.1
Vermiculite[4]	20–25	5–10		35–40	32–35				
Pumice[4]	70–75	12–14	1–3	0.1–0.6	0.8–2.0	3–6	4–5		
Zeolite[5]	64–68	9–13	0.8–4.4	0.0–1.2	0.0–1.8	0.0–3.0	0.6–0.3		

[1]Verwer, 1974; [2]Olympios 1992; [3]Grillas et al., 2001; [4]Raviv et al., 2002; [5]Stamatakis et al., 2001.

Table 11.7 Analytical data of expanded clay granules extracted by different methods. Average values of 4 samples Lamstedt hydro granules. The extractions were carried out in a 1:5 v/v ratio, with water, ammonium acetate (NH_4Ac) and CAT as described in Chapter 4 and are expressed as $\mu mol\ l^{-1}$ substrate

Extraction method	pH extract	Fe	Mn
Water	6.0	23.0	0.7
NH_4Ac (pH 4.6)	4.9	56.0	9.8
CAT	2.8	136.5	11.0

Derived from Sonneveld and Wever (2005).

process. Therefore, the sum of the percentages of the oxides is not always 100. Zeolite for example contains a lot of crystallization water, which is not mentioned in the composition. Furthermore, the materials mostly contain more elements and compounds than analysed. True enough mostly lower contents than those listed, but this does not mean that from such compounds no toxic concentrations can be released under specific conditions, like a low pH. This appears for example from the analytical data of different extraction methods of expanded clay granules as shown in Table 11.7. In the exchangeable extracts much more Fe and Mn is available than with water extraction. Besides release by exchange, a higher solubility at a lower pH and surely for Fe complex formation by DTPA will play an important role.

Some organic substrate constituents contain high concentrations of nutrients and sometimes also high concentrations of residual salts, which make such constituents less suitable for substrate preparation. Examples of such constituents are different composts types. Green compost is the most widely used type, which is abundantly available and cheap, but its utility for substrates is limited. The limitations mainly rest on a high salinity and high contents of some nutrients. Another handicap is the great variability of the composition. In a study of De Kreij and Van der Gaag (2003) 20 samples from different European countries were analysed by the 1:1½ extract from which the EC for example varied between 1.5 and 9 dS m^{-1}. The K varied between 1.5 and 18 mmol l^{-1} in the extract, such values correspond with about 3–36 mmol l^{-1} of substrate. Addition of peat, 80% by volume, lowered the EC and K mostly sufficiently and by specific adjusted addition of nutrients most levels could be brought more or less within an acceptable range. In Table 11.8 an

Table 11.8 Analytical data of the 1:1½ extract of green waste compost, peat and a mixture (v/v 20/80) of both materials after adjustment with nutrients

Materials	pH	EC	NH_4	K	Na	Ca	Mg	NO_3	Cl	SO_4	P	HCO_3
Compost	7.8	2.6	<0.1	13.9	2.9	1.3	0.8	5.8	10.0	0.6	0.09	1.6
Peat	3.5	0.2	0.2	0.2	0.2	<0.1	0.1	0.3	0.2	0.1	0.21	<0.1
Mixture	5.5	1.0	0.6	3.1	1.0	1.3	0.8	4.0	2.4	0.3	0.1	0.96
Target values	5.5	0.9	1.0	1.6	<1.7	1.2	0.5	4.0	<1.7	0.8	0.50	<0.1

Data derived from De Kreij and Van der Gaag (2003).

example of the analytical data of one of the samples is given. Surrage and Carlile (2008) showed comparable K concentrations in green waste composts in the UK, as those mentioned by De Kreij and Van de Gaag (2003). However, some composted green waste products contain much higher salt and K concentrations, like mentioned by Carrión, et al. (2005). In agricultural vegetable waste compost, they mentioned K concentrations up to 350 mmol l^{-1} of compost. The use of such material as a substantial constituent of substrates is nearly impossible. Leaching of such material hardly offers possibilities to lower the salt content to an acceptable level and creates an environmental problem by the drainage water which possibly will be transported to ground- and surface water (Carrión et al., 2005). Addition of small quantities of up to 20% of the volume offers possibilities, but the effects on plant development will depend on the type of the compost, the plants grown and the right adjustments of the fertilization (Fischer and Schmitz, 1997; Jespersen and Willumsen, 1993).

Composts prepared from green waste can absorb N in advance, because of insufficient stability. This also can be the case with bark, wood fibre products and coir (Raviv et al., 2002). The immobilization of N can be compensated by addition of extra N. The immobilization of N can be determined by the so called nitrogen fixation index (NFI) (Blok et al., 2008). When insufficient stabilized material is used for the preparation of growing media addition of extra N during cultivation can be necessary to compensate the immobilisation. However, this merely has to be applied as an emergency.

11.4 Preparation of Substrates

Substrates sometimes consist of one type of material and are in other cases composed by a mix of different materials. The materials often are of natural origin but can be also artificially produced in a factory. Pumice, for example, is a material of natural origin purely used as a substrate, with possibly no other treatments than sieving and washing. Looked away from the adjustment of the pH, many types of peat are suitable to be used directly as a substrate too, but are often mixed with other materials. The most common reason for mixing materials in substrate production is improvement of the physical conditions, like the water to air ratio of the ultimate product. Other reasons can be a special preference for the use of certain materials in the substrate or the reduction of other materials by the substrate producer or by the grower. This choice can be determined by the availability of the materials, by technical reasons, or by economic reasons. However, also political considerations can play a part. An example of the last reason is the movement to peat free or peat reduced mixtures, which is strongly supported by environmental organisations (Carlile, 2008). From these environmental lobbies often the use of composted waste materials is pleaded. However, the use of composted green material is strongly bound on limits, because of the strong growth reductions that can occur with the use of such material. For some plants apparently the quantity of 20% by volume mentioned in Section 11.3 reduces plant growth, and the growth reductions often increase with increasing green waste compost additions (Frangi et al., 2008; Surrage and Carlile, 2008). Such growth reductions will vary for the type green compost used and cannot always be explained by changes in the salt and nutrient status of the substrate mixtures. It is thinkable that the growth reduction as occurs with green waste compost has an advantage for crops with which a growth reduction is desired. For bedding plants for example studies are carried out to restrict a lush growth by reduction of the P addition in the substrate (Warmenhoven et al., 2008). This growth reduction possible also can be achieved by the use of compost and can act as a favourable side effect. Another favourable side effect of the use of compost can be the suppression of soil borne diseases that sometimes occurred (Raviv, 2008; Van der Gaag et al., 2007). The study to the effects of green waste composts has been started since some years and surely need a further intensity before it safely can be used in growing media under different growing conditions.

In this section the composition of substrates with respect to their physical characteristics will not be discussed, because the effect on the fertilization is small, which is discussed in Section 11.2. Attention will be paid on the chemical aspects of substrate production. In the production following situations will be distinguished.

- The pH of the substrate is too low or too high and needs adjustment.
- The substrate has a low cation buffer capacity and will be fertilized by a nutrient solution.
- The substrate has a substantial cation buffer capacity, with which two situations can be occur:
  - Addition of nutrients is desired.
  - Addition of nutrients is not desired.

Picture 11.1 Storages of base materials due to the preparation of peaty substrates.

Picture 11.2 Rock wool substrate slabs placed in a greenhouse for cucumber production.

11.4.1 pH Adjustment

The pH among substrates differs strongly, but also the differences within the same type of substrate are substantial. This is shown by the data of Kipp et al. (2000) from which a summary is listed in Table 11.9. The pH measured in the substrate is of secondary importance if there is only a small buffering in the material, like with rock wool and expanded clay granules. This is shown by the results of pH measurements of expanded clay granules with water and with a nutrient solution. Both measurements were carried out following routine handlings, using either water or a standardized nutrient solution with an EC of 2.0 dS m^{-1} adjusted to a pH of 5.5 (Wever, 2005). The buffer in the nutrient solution consisted of the required P, added

Table 11.9 The pH of different substrates and substrate constituents as given by Kipp et al. (2000)

Materials	Average	Range
Wood fibre	4.8	3.8–5.4
Expanded clay granules	8.1	7.7–8.6
Coir chips	5.7	5.4–6.1
Coir dust	6.2	6.0–6.7
Perlite	6.3	5.2–7.7
PU-foam	6.6	4.7–8.9
Pumice	6.3	4.7–7.6
Rock wool	6.2	5.2–7.8
Peat	3.9	3.4–4.4

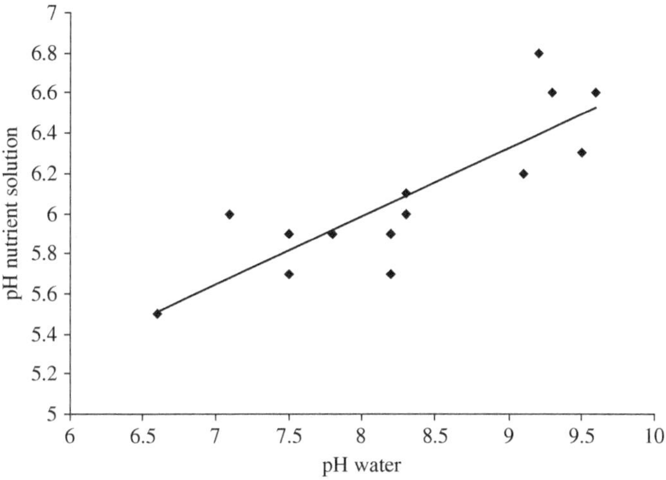

Fig. 11.1 The relationship between the pH of expanded clay granules determined with water and with a nutrient solution of an EC of 2.0 dS M^{-1}. Data after Sonneveld and Wever (2005)

as H_2PO_4 in the solution, which is quite a weak one. The results are presented in Fig. 11.1. The pH in the water suspension varied between 6.6 and 9.6 and in the nutrient solution suspension between 5.5 and 6.8, whereby is shown that the pH is strongly affected by the weak buffer of the nutrient solution. The very high values measured in water sometimes need no adjustment beforehand, but a close control during crop cultivation and the adjustment of the right NH_4 concentrations in the nutrient solution used are much more important to keep the pH within the required limits than the adjustment before use (see Section 13.4).

Substrates containing constituents like peat and clay have a high cation exchange capacity (CEC). At low pH the adsorption complex is heavily saturated with H_3O ions which will be neutralized by addition of OH, HCO_3 or CO_3 addition. A special behaviour has been found for some types of poly phenol foam. This substrate has no cation buffer, but the pH is sometimes buffered at a low level (Voogt, 1983).

Generally, substrates with a too low substantial buffer pH values are adjusted for that parameter in the factory, because adjustment during cultivation is problematic. The quantities of neutralizing chemicals needed for optimization of the pH is substantial and are too bulky for addition with the nutrient solution. Therefore, the adjustment is carried out with the preparation of the substrate by addition of fertilizers commonly used to this purpose, like $Ca(OH)_2$, $CaCO_3$ and mixtures of $CaCO_3$ and $MgCO_3$. The quantities required depend on the material and the pH difference to be bridged. In this process the CEC of the material used is a main factor.

Cation exchange capacities for different materials were gathered and published by Bunt (1988). Some of his values are listed in Table 11.10 together with values published elsewhere. The CEC between materials show great differences, but also the variation within materials is substantial. This variation can be explained by factors like the method of determination applied, the weathering stage (Raviv

Table 11.10 Cation exchange capacities (CEC) of some substrates and substrate constituents expressed as C^+ mmol kg^{-1} dry matter

Material	CEC	References
Humus	2000	Bunt, 1988
Peat	560–1580	Lamaire, 1995; Lamaire 1998; Puustjärvi,1977
Coir dust	350–600	Evans et al.,1996; Verhagen, 1999
Sawdust (fresh)[1]	100	Jokova et al., 1997
Compost (fresh)	270–1080	Jokova et al., 1997; Lamaire,1998
Compost (stable)	640–1810	Chen et al.,1989; Jokova, et al.,1997; Lamaire, 1998
Clay[1]	100–300	RHP, 2007
Vermiculite[1]	390–530	Van der Mark, 2008
Perlite	10–70	Bunt, 1988; Gizas et al., 2001; Lamaire, 1995
Zeolite	400–1200	Stamatakis, 2001; Maloupa, 2002
Tuff[2]	70–600	Silber et al., 1994
Pumice	60–80	Gizas et al., 2001
Rockwool	0	Lamaire, 1995

[1]See text; [2]Within pH 4.0–7.0.

et al., 2002); the particle size (Bunt, 1988; Gizas et al., 2001) and the pH (Silber et al., 1994). The degree of decomposition also has a huge effect, like shown for compost in Table 11.10. This also has been found with the sawdust mentioned by Jokova et al. (1997). The value mentioned concerns fresh material, like often used as a growing medium, but after one year the CEC was about 830. For peat it sometimes also is mentioned that the dark types, being materials generally older and more decomposed, have a higher CEC than the light types. However, Puustjärvi (1977) showed that such is not usually the case. The common experience that higher alkali applications are necessary in dark than in light peat to increase the pH, merely will be explained by the higher bulk density of the dark material. Fertilizers are usually added on volume basis. The values noticed for vermiculite are debatable and the data presented concern the quality on the Dutch market. Very different values are mentioned elsewhere, from 20 up to 1500 mmol kg^{-1} (Bunt 1988; Lamaire, 1995; Maloupa, 2002). Presumably, the very high values sometimes mentioned for this material are associated with the original clay mineral. For clay also great difference are mentioned, dependent on the particle size distribution. The data mentioned in the table concern the quality recommended as a constituent in growing media by the Dutch certification organisation RHP. CEC values published for substrate constituents sometimes show great differences and thus such also will be the case for reviews of these values. So, some of our values substantially differ from those published in a review of Silber (2008).

Lime requirements of substrates mostly are based on volume basis. When expressed in this way, beside of the factors mentioned, lime requirements will depend on the bulk density too. In Fig. 11.2 the relationship is shown between the addition of limestone and the pH of two high moor peat substrates; a low decomposed type and a frozen black peat. The relationships are somewhat curvilinear. The frozen black type also requires a higher application of lime stone than the low

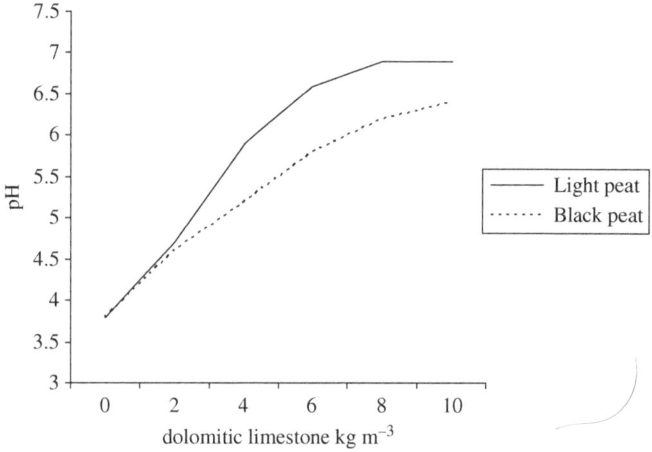

Fig. 11.2 The effect of the addition of lime on the pH of high more peat. After Van Schie (1989)

decomposed type, to increase the pH value. From Fig. 11.2 will be concluded that for a pH raise of one unit 2 till 3 kg limestone per m^3 is required in the range up to pH 6.0. Bunt (1988) concluded that for such a pH raise 1 till 2 kg m^{-3} lime is required for peat mixed with 25% by volume with sand. Differences will be caused by the characteristics of the substrate, the type of lime used, the range in which the pH raise is realised and the duration of the time between the addition of the lime and the sampling. The effect of the character of the mixture is already discussed. The lime best can be characterized by the carbonate content. In the situations presented, in both cases a mixture of $CaCO_3$ and $MgCO_3$ was used. These compounds are equal effective on basis of molecular weights, which means that 100.1 g $CaCO_3$ is equivalent with 84.3 g $MgCO_3$. With respect to the range in which the raise is realised, there is a tendency that the raise is not linear over the whole range presented, as mentioned already from Fig. 11.2. Such has been presented for different peat and sand mixtures by Bunt (1988) too, as shown in Fig. 11.3 for a mixture of Irish peat and sand. The required lime by Bunt (1988) was calculated over the linear part of his curves. The effect of the measuring time can be explained by the slow release of the carbonates from the lime added. After addition of lime it will take some weeks until the raise of the pH is fully realised (De Kreij, 1993).When the peat is mixed with other constituents of higher pH the quantities of lime will be reduced, or even omitted, when the pH is sufficiently increased by the constituent added (Fischer and Schmitz, 1997). Fisher et al. (2006) and Nemati and Fortin (2008) developed an algorithm for the lime requirement of growing media, based on different constituents of the growing media and the to those belonging pH values. Beside the characteristics of the substrate Fisher et al. (2006) demonstrated that also the characteristics of the fertilizer affect the pH rise in the substrate. A course granulation of the fertilizer slackens the reaction and the residual fertilizer act as a buffer on the long run.

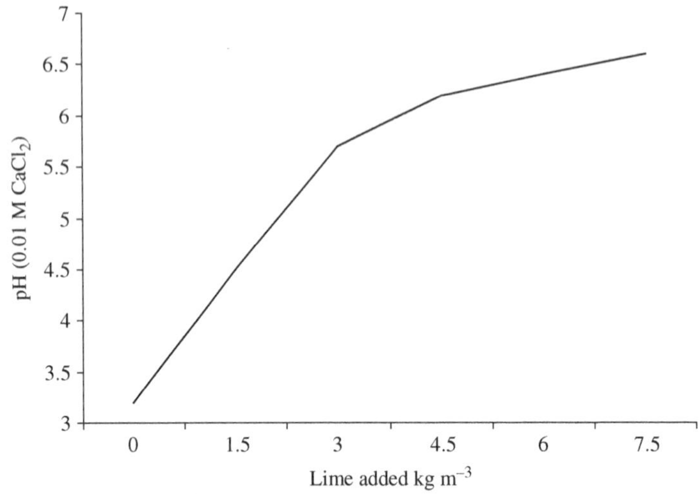

Fig. 11.3 The relationship between the addition of lime and the pH of a mixture of 75% Irish sphagnum peat and 25% by volume of sand. Derived from Bunt (1988)

The optimum value of the pH depends on crop and varies for most crops between 5 and 6 (Bos et al., 2002), but will be adjusted to specific crop requirements. However, the value realised at the start easily changes during cultivation (Willumsen, 2001). Such changes result from factors like crop growth, irrigation water quality (Maher and Prasad, 2004) and fertilization. These factors are discussed further in the Chapters 13 and 14.

11.4.2 Macro Elements

Addition of macro elements with substrate preparation mostly occurs for substrates with low nutrient concentrations and a substantial cation adsorption capacity. Substrates with a restricted cation adsorption capacity do not affect the composition of the nutrient solution with which the substrate is saturated at the start of the cropping period and thus, for such substrates there are no reasons for the addition of fertilizers beforehand. In such cases, the composition of the nutrient solution used for the saturation is equal to those of the substrate solution in the root environment desired at the start of the growing period. The composition of the substrate solution in the root environment at the start mostly differs from the standard solution and adjustment to the target values is recommended as will be discussed in Section 12.3. This policy is suitable for cultivations in rather inert substrates like rock wool and foam (Sonneveld, 1995). Also substrates with a restricted cation adsorption capacity and a reasonable pH hardly affect the composition of an added nutrient solution, like shown by Gizas et al. (2001) for perlite and pumice. The quantities of adsorbed cations were marginally changed, after saturation with nutrient solution. Substrates with a substantial adsorption capacity, like several peat types and clay, will adsorb cations strongly, especially when they are acid. It is advisable to bring the substrate of such material on a reasonable nutrient status by addition of sufficient fertilizer. It can be expected that clayey material will adsorb specific K while organic material merely adsorb Ca and Mg. When acid organic material is provided with $CaCO_3$ and $MgCO_3$, important parts of the adsorption capacity is occupied with the cations released from these compounds.

With fertilization of substrates, effects of cation adsorption should be taken into account. The quantities of Ca and Mg adsorbed by organic matter can substantially exceed the quantities soluble in the substrate solution. The adsorption for K by organic matter is much lower. One thing and another is shown in Fig. 11.4, where the relationship is shown between the concentrations of cations in the 1:1½ volume extract and in the concentrations extra released (exchanged) in this extract by addition of 0.1 M $BaCl_2$. The data are derived from a series of peaty substrates, composed from different constituents (De Zeeuw, 2003). The quantities K and Na released by exchange show a clear relationship with the quantities soluble in water. The release of K from the adsorption complex is equal or less than those found as water soluble, which is in agreement with findings for soils with a substantial exchange capacity, like loamy soils (Németh et al., 1970). The exchangeable Na is less than half of the Na found as water soluble. The single dot that shows an exceptional high value for exchangeable K originates from a mixture with 15% clayey material and could be explained by specific adsorption. Ca and Mg are adsorbed quite strongly and showed no reasonable relationship with the water soluble concentrations. Average values of concentrations are listed in Table 11.11. The adsorption of the different cations can be put in the following series, Ca, Mg, K, and Na. The difference as has been found between both bivalent cations Ca and Mg are reasonable in agreement with the findings for peat (Puustjärvi, 1977) and for mineral soils (Németh et al., 1970). However, the relationship between soluble and exchangeable concentrations for both bivalent cations from our material is different from those

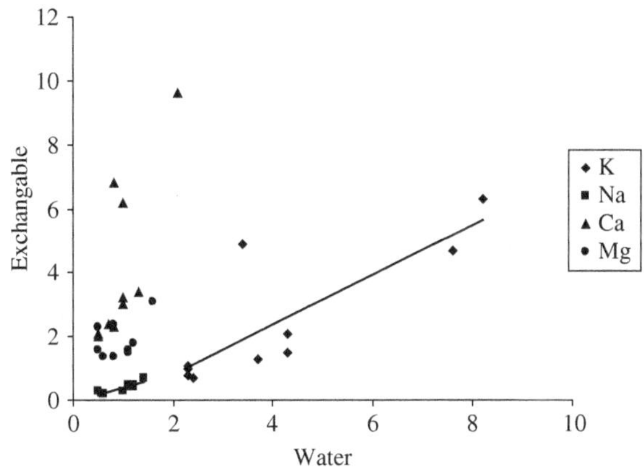

Fig. 11.4 The relationship between cation concentrations in the 1:1½ volume water extract and the concentrations released (exchangeable) by addition of 0.1 mol l^{-1} $BaCl_2$ to the 1:1½ volume extract. After De Zeeuw (2003)

Table 11.11 Average concentrations cations in the 1:1½ volume water extract and the extra cations released (exchangeable) by addition of 0.1 mol l^{-1} $BaCl_2$

Cations	Water soluble	Exchangeable	Ratio
K	4.08	2.44	0.60
Na	0.93	0.37	0.40
Ca	0.97	4.10	4.23
Mg	0.87	1.94	2.23

Data of different substrates after De Zeeuw (2003).

found for mineral soils by last authors. This possibly will be explained by the great variations in the organic materials as used in organic substrates. NH_4 was not determined and the behaviour will be comparable with K. With coir dust a comparison between water soluble and exchangeable cations was made by Verhagen (1999) and the results show comparable characteristics. However, the concentrations released in the extract as exchangeable K and Na were relatively higher than those found with the peaty substrates by De Zeeuw (2003), while the Ca and Mg concentrations of the coir dust were much lower than those in the peaty substrates. Thus, beside the character of the material as suggested by Prasad (2001), the occupation of the adsorption complex also can be a factor. Furthermore, the concentration of the different cations can play a part. Prasad (1997) showed a curve linear relationship for water soluble and exchangeable K. The curve linearity was caused by a relative high concentration exchangeable K in the low range.

With respect to exchangeable cations it will be concluded that in those substrates, the composition of which is based on natural organic and clay material, substantial quantities cations are adsorbed. From the data presented can be concluded that in substrates with reasonable concentrations water soluble nutrients, for K and NH_4 the ratio between exchangeable and water soluble quantities are easily between 0.5 and 1.0. For Ca and Mg the ratio between exchangeable and water soluble quantities varies strongly, but under mentioned conditions for Ca the ratio easily will reach values between 2 and 4; for Mg somewhat lower ratios will be expected, values between 1.5 and 3.0 frequently occur.

Guide values for analytical data of peaty substrates in the 1:1½ extract are given by Kipp et al. (2000) and listed in Table 11.12. The values supplied are not crop specific, but relate to existing standards (Straver et al., 1999). On basis of the data in Table 11.12 calculations for the fertilizer addition to organic substrates are possible. The data are expressed on the concentrations in the 1:1½ extract. The extraction is carried out with fresh substrate with a moisture content equal with the moisture condition at a pressure head –3.2 kPa (Sonneveld and Van Elderen, 1994). At this moisture condition the water content is about 50% of the volume (Kipp and Wever, 1993). Together with the 1½ volume of water added with the extraction, the water to substrate ratio in the suspension is 1:2 v/v. Thus, for the expression of water soluble elements on the substrate volume, the results of the 1:1½ extract should be multiplied by about 2.

With the aid of this knowledge, the quantity of fertilizer required to realize a beforehand fixed concentration in the 1:1½ extract can be easily calculated.

Table 11.12 Guide values for analytical data of macro nutrients for peaty substrates on basis of the 1:1½ volume extract

Determination	Level of fertilization		
	Lightly	Moderate	Heavily
EC Ds m^{-1}	0.5	1.0	1.5
NO_3 mmol l^{-1}	2.0	3.2	4.0
NH_4	1.0	1.8	2.5
K	1.1	1.8	2.5
Ca[1]	0.7	1.0	1.8
Mg	0.7	1.0	1.8
SO_4	0.9	1.2	1.5
P	0.5	0.75	1.0

[1]Ca is relatively low, because of the strong adsorption of this element and the slow release in the beginning from the $CaCO_3$ generally added. Later on more Ca will be released. After Kipp et al. (2000). *Modified by permission of Elsevier*

Naturally, the calculation is operative for organic substrates and can be carried out by following formula.

$$F_{app} = 2Nu_{(1:1\frac{1}{2})}\alpha_{BD}(1 + C_r)A_rQ^{-1} \tag{11.1}$$

In which

F_{app} = fertilizer application in g m^{-3}
$Nu_{(1:1½)}$ = required increase of the concentration of the nutrients in the 1:1½ extract
α_{BD} = ratio between the density of the substrate at application of the fertilizer and the density at preparation of the extract
C_r = adsorption relative to water soluble elements
A_r = atomic weight of the nutrient element
Q = relative mass of the element in the fertilizer used

The value for α_{BD} depends on the material in charge and the pressures applied (Wever and Van Leeuwen, 1995) and with usual pressures the value of α_{BD} varies between 1.1 and 0.9. The values of C_r depend on the material, the element and the concentration in charge, rough information about it is given before in this section. When $\alpha_{BD} = 1$, $C_r = 0$ and $Q = 1$, the formula can be compared with the results calculated by De Kreij (1993). For an increase of 1 unit of each element in the 1:1½ water extract, quantities are mentioned of 25–30, 70 and 100 g m^{-3} for N, P and K, respectively. With our formula quantities are calculated of 28, 62 and 78 g m^{-3}, respectively. For N and P there is a good agreement. However, De Kreij gave a somewhat higher quantity for K, perhaps he has taken some adsorption into account. When an adsorption of 0.60 to the water soluble is taken into account, in agreement with the average listed in Table 11.12, the quantity calculated with formula (11.1) increases to 125 g m^{-3}.

The behaviour of P can differ, as has been found by Sonneveld et al. (1974). In this study big differences were found between the solubility of P for concentrations below about 0.15 mmol l^{-1} in the substrate solution and for higher values. At the higher concentrations the behaviour of P in the dilution system studied was in agreement with that of other elements, while at low concentrations only with strong dilutions much P became soluble. All substrates in the study with a deviant behaviour of P in the low range could be related either to mixtures containing clay or to unfertilized substrates. In mixtures with clay, the solubility of P is affected by Fe and Al, and the solubility products of these elements with P compounds are much lower than most solubility products of P compounds with Ca. In most substrates the solubility of P is related to Ca and the pH around 5.5. At this pH value the bulk of P occurs as H_2PO_4, which ion is well soluble in combination with Ca. Therefore, in many substrate mixtures P will behave like NO_3. However, when constituents are added containing Fe and Al, like clay, some composts and some volcanic materials, the solubility of P is also affected by Fe and Al. Besides precipitation of P as mineral salts the availability in such mixtures can also be affected by adsorption (Silber, 2008). Effects of the addition of clay are clearly shown by the data of Verhagen (2004). In an Irish peat substrate the P concentration in the 1:1½ volume extract decreased from 0.72 mmol l^{-1} extract to values between 0.10 and 0.20 mmol l^{-1} when clay, 15% by volume, was mixed in the substrate. In substrates without Fe and Al precipitation of P can be expected at pH values above 6.5, because under this condition important parts of the P occur as HPO_4, and this ion also has a low solubility in combination with Ca.

When materials rich in nutrients are mixed in the substrate, like compost, substantial parts of the nutrients are supplied with such a constituent, like discussed in Section 11.3. The residual nutrients will be added in such cases by specific fertilizer applications based on substrate analysis (De Kreij and Van der Gaag, 2003). In other situations often special prepared compound fertilizers are used, like the so called PG-mixtures, the composition of which are focussed on the fertilization of substrates poor in all nutrients. The ratios between N, P_2O_5 and K_2O in these fertilizers, like 14+ 16+18, 15+10+20 and 12+14+24, are determined in consideration of some adsorption and precipitation. The nutrient levels recommended from lightly until heavily fertilized peaty substrates, as summarized in Table 11.12, can be realized by addition of 1 to 2 kg m^{-3} of PG-mix, when the substrate itself does not contain substantial quantities of nutrients. For some potted plants lower nutrient levels are mentioned as being optimal (Straver et al., 1999), than the lowest level mentioned by Kipp et al. (2000) and will be realised by adjusted applications of fertilizers.

The use of slow release fertilizers is not a common practice in substrates for greenhouse cultivation. This only is meaningful when during cultivation insufficient possibilities exist for effective top dressings. Therefore, slow release fertilizers are applied to substrates for some specific potted plants, sometimes for bedding plants and more frequently in the production of container grown nursery stock. See further the remarks about this type of fertilizers in Section 2.2.

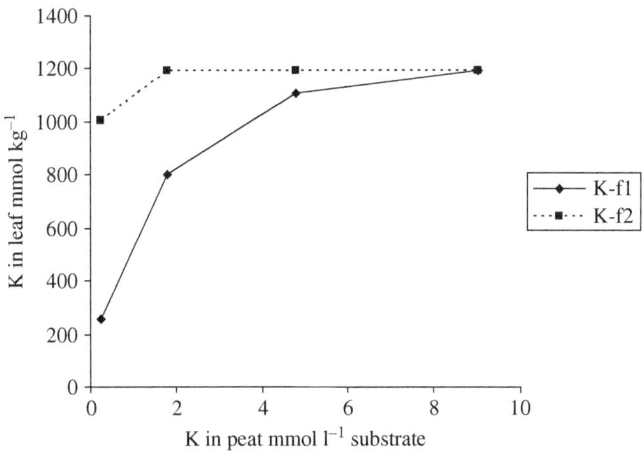

Fig. 11.5 The relationship between the K added to the peat substrate before planting and the K concentration in tomato leaves 8 weeks after planting at two levels of top dressing (Adams 2002). The levels of the top dressing were K-f1 = 2.6 mmol l^{-1} and K-f2 = 7.7 mmol l^{-1}. After Adams (2002). *Modified after permission of Embryo Publications.*

A clear interaction can be expected between the level of fertilizer added to the substrate at preparation and the required level of the top dressing in the beginning of the cultivation period. This for example is shown by Adams (2002) for the potassium supply to tomato grown in peat substrate. With a low K supply in the substrate a high K in the top dressing is required to attain an optimal uptake in due time, as shown in Fig. 11.5.

11.4.3 Micro Nutrients

The application of micro nutrients to substrates differ from the application of macro nutrients because of the much smaller quantities required for an optimal plant development and the dominant role of the pH with the uptake. The small quantities required by the crop easily achieve that the quantities on hand in the substrate or added with preparation of the substrate are of importance for the crop supply and easily exceed the quantities required by the crop (Sonneveld, 2002). The effect of the pH in the root environment on the uptake of micro nutrients is shown in Table 11.13 for a rose crop grown in rock wool at different pH values with standard applications of all micro nutrients (Sonneveld and Voogt, 2001). Despite an equal application, the uptake of Fe, Mn, Zn and Cu was strongly increased by a decreased pH value. For Mo, an upside down effect was noticed. B was not clearly affected by the pH in this experiment. However, B uptake mostly is clearly affected by the pH level, like found by Alt and Rosen (1987) for tomato grown in peat substrate and shown in Fig. 11.6. The uptake of B at the low pH of 4.6 is much more effective than at the

Table 11.13 Micronutrient concentrations (mmol kg^{-1} dry matter) in young fully grown rose leaves in an experiment with different pH values in rock wool slabs

pH	Fe	Mn	Zn	B	Cu	Mo
7.4	0.53	0.14	0.48	3.4	0.022	0.010
7.1	0.61	0.28	0.30	2.5	0.024	0.007
6.9	0.85	1.01	0.47	2.5	0.034	0.007
6.6	1.22	1.14	0.63	2.5	0.053	0.006
6.1	1.53	1.23	0.75	2.4	0.074	0.006
6.0	1.03	1.68	0.73	2.5	0.054	0.005

After Sonneveld and Voogt (2001). *Reprinted by permission of the International Society Horticultural Science*

high pH value of 6.8. The well known pattern of availability of nutrients in relation with the pH published for organic field soils (Lucas and Davis, 1961) fits well with these findings. Depending on the composition of the substrate gradual variations are possible in this relationship, but the main structure of the model remains (Peterson, 1982). The dominant role of the pH with the uptake of micro nutrients often involves that sometimes the uptake is more affected by the level of the pH than by the concentrations of the element applied. Such especially can be the case with Mn. The availability (Page, 1962) and uptake (Sonneveld and Voogt, 1975) of this element are logarithmic related with the pH in the root environment. In substrate grown crops the pH in the root environment can easily be affected by the management of the fertilization programme and the best equilibrium for the uptake of the different micro elements will be gained with values between 5 and 6, these values are mentioned already in this section as being optimal for most crops.

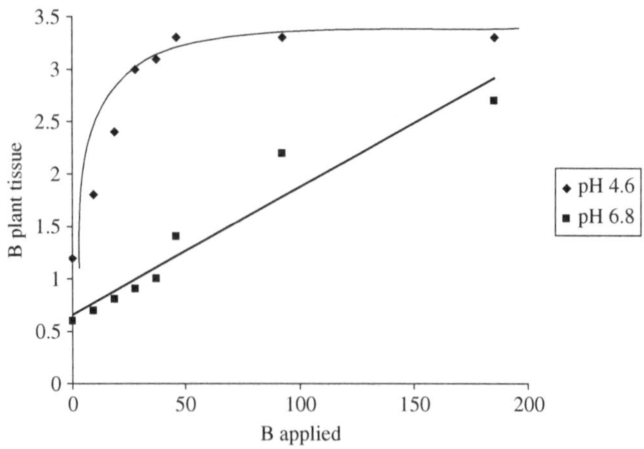

Fig. 11.6 The relationship between the B application to a peaty substrate and the B concentration in the plant tissue of young tomato plants as affected by the pH of the substrate. B application is expressed as μmol l^{-1}substrate and B in plant tissue as mmol kg^{-1} dry matter. Data after Alt and Rosen (1987)

As well as with macro nutrients addition of micro nutrients makes sense for substrates with a substantial adsorption capacity, because in such substrates part of the metal micro nutrients will be fixed on the adsorption complex. The quantities available as water soluble and as NH_4Ac exchangeable (Table 11.7) differ substantially. However, the quantities extracted by the DTPA/$CaCl_2$ (CAT) solution (see Section 4.9) as developed by CEN exceed the quantities of exchangeable enormously for some micro nutrients, as shown by the data of Sonneveld and De Kreij (1995). In this study they compared the 1:1½ v/v substrate to extraction solution as developed by Sonneveld and Van Elderen (1994) water, NH_4Ac and CAT as extraction solution for micro nutrients. The average values are listed in Table 11.14. The results show clearly that with CAT by complex formation much more cation micro nutrients are dissolved than present as exchangeable with NH_4Ac. This is understandable from the character of DTPA. For Fe, Zn and Cu the ratio exchangeable (released by NH_4Ac addition) and water soluble ranged between 0.5 and 2, which is reasonable in agreement with the factors found with macro element cations. For Mn a much higher ratio was found, which will be caused by the lower pH of the NH_4Ac extract, being buffered at a value of 4.6. The availability of Mn is strongly increased by this low pH.

In substrates Fe is generally added as a chelate, while the other micro nutrients are added as mineral salts. Results of research on the application of micro nutrients to peaty substrates from different institutes in The Netherlands were summarized by De Kreij (1993) and listed in Table 11.15. In this table also the quantities of micro nutrients applied with the usual application of 1¼ kg m^{-3} PG-mix (Klapwijk and Mostert, 1992) and the mutual ratios of the micro nutrients in nutrient solutions (Sonneveld, 2002) are listed. The application with PG-mix was calculated from the average concentrations of different PG-mix fertilizers as listed in Table 2.7, being 14+16+18, 15+10+20 and 12+14+24. The mutual ratios in nutrient solutions are reasonable in agreement with the uptake as calculated by Sonneveld (2002), except for Mo. This element is relative to the other micro nutrients abundantly available in nutrient solutions. This is even more the case for the addition of Mo in peaty substrates. However, beside Mo in substrates also Cu is added in relatively huge quantities. This is understandable, for Cu is fixed by organic compounds in peat substrates (Verloo, 1980) and is less available to plants. The quantities as summa-

Table 11.14 Average values of analytical data of water soluble and NH_4Ac and $CaCl_2$/DTPA extractable micro nutrients determined in peaty substrates. All data are determined in a 1:1½ v/v substrate to extraction solution ratio and expressed as μmol l^{-1} of the extract

Extract	Fe	Mn	Zn	Cu
Water	4.2	2.1	1.8	0.43
NH_4Ac[1]	6.6	19.1	5.4	0.91
DTPA/$CaCl_2$[1]	98.1	24.4	17.5	5.58

[1]Total in the extract, together with water soluble.
After Sonneveld and De Kreij (1995).

Table 11.15 Application of micro nutrients to peaty substrates (De Kreij, 1993), approximate concentrations applied to nutrient solutions (Sonneveld, 2002), uptake by a high yielding crop (Sonneveld, 2002) and application with 1¼ kg m^{-3} PG-mix per m^3

	Application in mmol m^{-3}		In nutrient solutions	Uptake	mmol in 1¼
Elements	Minimum	Maximum	μmol l^{-1}	mmol m^{-2}	kg PG-mix
Fe	54	107	15–25	14	20.1
Mn	55	109	5–10	7	36.4
Zn	9	18	3–5	2	7.6
B	37	74	20–30	14	34.7
Cu	31	63	0.50–0.75	0.7	27.5
Mo	21	42[1]	0.50	0.01	26.0

[1] for cauliflower 63

rized by De Kreij (1993) are determined as being sufficient for a long growing period of the crop. The additions with standard applications of PG-mix fertilizers are somewhat lower than those mentioned by De Kreij, and more in agreement with the fact that with long crops usually also micro nutrients are added by top dressings with fertilizers. However, also by the addition of the PG-mix fertilizers the relatively abundant applications of Cu and Mo remain. It can be discussed whether the high application in organic substrates are always required. In view of the modern fertigation methods addition of micro nutrients during the growing period is normal practice. Thus, there is no reason for huge additions of micro nutrients from start, to cover the need for long growing periods. A better tuning between applications at start and top dressings during cultivation is advisable. The application of Cu and Mo to substrates is not yet completely crystallized and need further studies (Sonneveld and Voogt, 2009). In these studies it will be considered that the need of the high addition of Mo to substrates merely are empirically demonstrated for young plants of specific crops, which not necessarily include that these high additions are desired for the whole growing period. Beside the well known example of the high requirements for young cauliflower plants, lettuce was determined as being sensitive for Mo deficiency during plant raising. Boertje (1969) found serious Mo deficiency in lettuce plants grown in white peat. The pH had a strong effect on the disorder and without Mo addition the symptoms virtually disappeared at a pH of 6.5. With lower pH values an addition of 17 μmol l^{-1} was sufficient for a healthy plant production.

Addition of Fe in the form of chelates also can affect the availability of other micro nutrients. This is clear from data of De Kreij (1993), listed in Table 11.16. In an experiment with a peaty substrate different quantities of Fe as Fe-EDTA and Cu as $CuSO_4.5H_2O$ were added at different pH values of the substrate. At the low pH the EDTA compound is completely occupied by Fe. Such follows from the Fe concentration of 55 μmol l^{-1} in the extract, which must be doubled when expressed on the substrate, thus, which is reasonably in agreement with the addition of 95 μmol per litre of substrate. With higher pH values Fe is released from the chelate complex and will precipitate, while the chelate is occupied by other cations like Mn, Zn, Cu

Table 11.16 Analytical data of the determination of water soluble micro nutrients by the 1:1½ volume extract after addition of 95 μmol FeEDTA and 55 μmol $CuSO_4.5H_2O$ per litre of a peaty substrate. The data, expressed as μmol l^{-1} of extract, are derived from De Kreij (1993)

pH	Fe	Mn	Zn	Cu
4	51	0.2	0.2	2.0
6	32	1.7	4.3	9.6
7	9	3.9	5.1	6.2

and possible also major nutrient cations. This is in agreement with the characteristics of the EDTA compound (Lindsay et al., 1966; Lindsay and Norvell, 1969). Thus, the results of the determination of water soluble micro nutrients in such cases strongly depend on the pH of the substrate and moreover, this effect depends on the type of chelate used. Mostly, these factors are insufficiently taken into account with the application of chelates to substrates and the estimations of the thereby expected effects on the chemical analysis. Besides, an extra complicated factor in such assessments is the fact that the plant uptake of micro nutrients as determined by chemical analysis sometimes interacts with the pH and the type of chelate (Sonneveld and Voogt, 2001; Voogt and Sonneveld, 2009).

In Table 11.17 general guide values are listed for the 1: 1½ extract, as derived from data of Kipp et al. (2000) and Wever et al. (2005). Data of Boertje (1982) and Van der Wees (1993) offered generally higher values, but the recent data are more obvious. Beside the guide values, limits for maximum values are given for crops sensitive and tolerant for one of the elements. Values higher than the limits given can induce toxic effects to plants. With these limits it should be remembered, that a specific sensitivity of a crop mostly holds for one element. For Mo no reliable limits are available.

The applications as mentioned in Table 11.15 are not directly in agreement with the guide values. However, the dynamics of adsorption and fixation as described have to be taken into account. Therefore, guide value and application are more or less in equilibrium just for B, which element is not affected by such processes. The guide values can be reasonably compared with the guide values for the composition

Table 11.17 Guide values and maximum limits for analytical data of micro nutrients for peaty substrates on basis of the 1:1½ volume extract (Wever et al., 2005; Kipp et al., 2000); expressed as μmol l^{-1} of extract. The limits are given for crops specific sensitive and tolerant to a certain element. Composition of substrate solution as given by Kipp et al. (2000)

		Limits		Substrate
Determination	Guide values	Sensitive	Tolerant	solution
Fe	8	<20	<40	25
Mn	2	<3	<10	5
Zn	2	<6	<10	5
B	15	<10	<40	30
Cu	0.7	<3	<3	1

of substrate solutions as given in the last column of Table 11.17. The dilution factor of the 1: 1½ extraction in comparison with the substrate solution is three, thus, the concentrations of the 1: 1½ extract is about ⅓ as found by Sonneveld and Van Elderen (1994) for macro elements. Recent data show (Sonneveld and Voogt, 2009) that for micro nutrients a somewhat lower factor will be taken into account. This is reasonably in agreement, with exception of Cu, because of the organic Cu complexes soluble in the extract, determined on the laboratory, but hardly available to plant.

11.5 Safety

The safety of substrates to plants is a continuous care of the grower and cannot be completely ensured by a check on the EC, pH, salt and nutrient status. These characteristics are terms for plant growth and do not include the determination on the presence of harmful substances, like neither harmful concentrations of elements other than the likely nutrients and salts, any chemical compound that negatively affect plant growth nor micro organism pathogenic to plants grown in the substrate. The discussion about this subject is more or less outside of the contents of this book, but is closely connected with the preparation of substrates. There are no any chemical or biological laboratory tests that exclude all possible risks in this field. The number of chemical compounds and through that the number of possible toxic compounds that can occur is endless. Therefore, chemical laboratory tests in this direction are meaningful as a check on expected harmful compounds. The same can be mentioned for tests on plant pathogens or other components which can be important in the practical use of substrates, like the presence of weed seeds. Checks on pathogens are suitable only on known species and is carried out when the presence can be supposed.

The best security for producer and grower to a safe substrate is a well developed information system about the origin of the substrate and by this the origin of components and constituents and the precision of the production process of the substrate. Therefore, the safety of substrates will be merely allotted to the substrate producers. A strict control on the different production factors is important and the basis of the inspections by the so called certification of substrates by institutes, like KIWA (2003) and RHP (2007) in The Netherlands. Besides tests on physical, chemical and biological characteristics of terms to plant growth, holders of a certificate are obliged to keep a precise administration to their used raw materials in detail and to the characteristics of their production process. Moreover, the use of raw materials and all other constituents for substrate preparation under certification is restricted to a list of materials admitted.

Specific regulations for growing media can be expected for the concentrations of heavy metals in substrates. The uptake of these elements from substrates is not yet studied in detail and cannot be simply compared with results of soil grown crops as will be discussed in Section 16.6. The great variation in materials used for substrate production induces a strong variation in the availability of total concentrations on hand. Furthermore, the great difference between the rooting volume of soil and substrate growing ensure a great difference in the storages disposed to the crop at equal concentrations in soil and substrate. However, tentative limits have been set by RHP (2007) for certification of substrate constituents and by Eco Label (2007) for certification of growing media. In Table 11.18 a review of these limits is listed. The limits of RHP for organic constituents are derived from the limits set by LNV, (undated) for peaty soils. The limits accepted for mineral constituents, like rock wool, clay, pumice and perlite, are generally much higher than for agricultural purposes. The concentrations determined as totals in such materials are merely poorly available for plant uptake. Therefore, the limits for mineral substrates are higher and are partly determined with in view of the acceptance for building materials, in consideration that the residual products of such substrates often are reused in that industry.

A rough check to determine toxic concentrations of substances not covered by the routine testing methods is presented by the so called plant growth tests, (Kipp et al., 2000; KIWA, 2003a). Such tests offer information about plant growth capability of a substrate compared to a standard substrate during a short growing period. The results of the tests show great differences between crops (Wever, 2004). Drawbacks of such tests are numerous. The growing period in the test is short, while in practice crops stay in the substrate for the whole growing period varying from some months until some years. Not all toxic effects become visible by growth reduction or toxicity symptoms within one or two weeks, the common duration of such tests. The optimal growing conditions for various substrates are different. When the growing conditions within a test is accommodated to the requirements of the standard substrate, the growth of the crop in the substrate tested will easily be reduced; not by toxic concentration of a substance in the substrate, but as a result of an interaction

Table 11.18 Maximum acceptable concentration of heavy metals as given for RHP (2007) certification and as voluntary proposed by ECO Label (2007). The concentrations are expressed on the dry material

	ECO Label[1]		RHP organic material		RHP mineral materials	
Elements	μmol kg^{-1}	mg kg^{-1}	μmol kg^{-1}	mg kg^{-1}	μmol kg^{-1}	mg kg^{-1}
Cd	8.9	1	6.4	0.72	32.0	3.6
Cr	1923	100	962	50	4808	250
Cu	1572	100	377	24	1887	120
Hg	5.0	1	1.1	0.23	5.7	1.15
Mo	20.9	2		–		–
Ni	852	50	170	10	3407	200
Pb	483	100	314	65	1569	325
Se	19.0	1.5		–		–
Zn	4587	300	1116	73	5581	365
As	134	10	67	5	1402	105
F	10526	200		–		–

[1]For Mo, Se, As and F relating to the presence of these elements are needed only for products containing material from industrial processes.

between substrate and growing conditions. Furthermore, it is well known that there are interactions between crops and the character and the concentration of the toxic compound. The sensitivity of crops for different toxic substances varies greatly and not all crops are tested in the standard procedures of the tests. One thing and another makes clear that plant tests will give information about growth capabilities of substrates, but cannot completely ensure the safety. Therefore, application of really new components, constituents and production processes need an intensive testing programme. Such a programme will be characterized by tests with sufficient variation of crops in long duration experiments.

Recently work by CEN (2007 and 2008) is started to develop a plant response test due to get a European agreement about this subject. However, the tests developed will have equal drawbacks as mentioned and the work of CEN only is fruitful in standardization of the tests existing in the various countries.

References

Adams P 2002. Nutritional control in hydroponics. *In*: Savvas D and Passam H (ed) Hydroponic Production of Vegetables and Ornamentals. Embryo Publications, Athens, 211–261.

Alt D and Rosen A 1987. Bor-Ernährung von Tomaten-Jungpflanzen. Gemüse 25, 168–173.

Blok C De Kreij C Baas R and Wever G 2008. Analytical methods used in soilless cultivation. *In*: Raviv M and Lieth J H (eds). Soilless Culture: Theory and Practice. Academic Press San Diego, 245–290.

Boertje G A 1969. Molybdenum and lettuce. Glasshouse Crops Research and Experimental Station Naaldwijk, The Netherlands. Annual Report 1969, 29–32.

Boertje G A 1982. Spoorelementen bij tomaten op veen. Tuinderij 62 nr 13, 46–49.

Bos E J F Keijzer R A W Van Schie W Verhagen J B G M and Zevenhoven M A 2002. Potgronden en Subtraten, Stichting RHP Naaldwijk, The Netherlands, 90pp.

Bunt A C 1988. Media and mixes for container grown plants. Unwin Heyman London.

Carlile W R 2008. The use of composted materials in growing media. Acta Hort. 779, 321–332.

Carrión C Abad M Maquieira A and Pchades R 2005. Leaching of compost from agricultural waste to prepare nursery potting media. Acta Hort. 697, 117–124.

CEN 2007. Soil improvers and growing media – Determination of plant response. CEN TC 223 WG4 N 139, 17pp.

CEN 2008. Soil improvers and growing media – Determination of plant response Part 2: Petri dish test. CEN 223 WG4 N 144 rev2, 8pp.

Chen Y Inbar Y Hadar Y and Malcolm R L 1989. Chemical properties and solid-state CPMAS ^{13}C-NMR of composted organic matter. Sci. Total Environ. 81/82, 201–208.

De Kreij C 1993.Bemesten van teelten in veensubstraat. *In*: Plantenvoeding in de Glastuinbouw. Proefstation voor Tuinbouw onder Glas te Naaldwijk. Informatiereeks no 87, 137–143.

De Kreij C and Van der Gaag D J 2003. Optimizing fertilizer supply to peat-based growing media containing composted green waste. *In*: Schmilewski (ed) Peat in Horticulture, Proceedings of the International Peat Symposium, Amsterdam, 31–37.

De Zeeuw J 2003. Bezetting adsorptioncomplex mengsels. RHP, Naaldwijk, The Netherlands. Reaearch report, 23pp.

Eco Label 2007. Revised ecological criteria and the related assessment and verification for the award of the Community Eco-label to growing media. Official J. European Union L32/137, 7pp.

Evans M R Konduru S and Stamps R H 1996. Source variation in physical and chemical properties of coconut coir dust. HortSc. 31, 965–967.

Fischer P and Schmitz H J 1997. Composts from residues of anaerobically treated household waste and their suitability in growing media. Acta Hort. 450, 229–235.

Fisher P R Huang J and Agro W R 2006. Modelling lime reaction in peat-based substrates. Acta Hort. 718, 461–468.

Frangi P Castelnuovo M Pozzi A Valagussa M Crippa L and Genevini P L 2008. A comparison of methods for the analysis of compost-based growing media. Acta Hort. 779, 113–120.

Gizas G Savvas D and Mitsios I 2001. Availability of macro cations in perlite and pumice as influenced by the application of nutrient solutions having different cation concentration ratios. Acta Hort. 548, 277–284.

Grillas S Lucas M Bardopoulou E Sarafopoulos S and Voulgari M 2001. Perlite based soilless culture systems: current commercial applications and prospects. Acta Hort. 548, 105–113.

Imas P and Bar-Yosef B 1998. Response of lettuce plants grown on different substrates to phosphorus fertigation. Acta Hort. 458, 171–178.

Jespersen L M and Willumsen J 1993. Production of compost in a heat composting plant and test of compost mixtures as growing media for greenhouse cultures. Acta Hort. 342, 127–142.

Jokova M Kostov O and Van Cleemput O 1997. Cation exchange and reducing capacities as criteria for compost quality. Biol. Agric. Hort. 14, 187–197.

Kipp J A and Wever G 1993. Wortelmedia. Proefstation voor Tuinbouw onder Glas te Naaldwijk, The Netherlands, Series Informatiereeks no 103, 48pp.

Kipp J A Wever G and De Kreij C 2000. International Substrate manual. Elsevier International Business Information Doetinchem, The Netherlands, 94pp.

KIWA, 2003. Assessment Guideline for the Product Certificate for Growing Media. BRL K10001, 43pp.

KIWA, 2003a. Test of phytotoxicity. Assessment Guideline for the Product Certificate for Growing Media. Appendices on Analytical Methods, 46–50.

Klapwijk D en Mostert J 1992. Richtlijnen voor de productie van potgronden en substraten. Proefstation voor tuinbouw onder Glas te Naaldwijk, The Netherlands. Informatiereeks no 73, 65pp.

Lemaire F 1995. Physical, chemical and biological properties of growing medium. Acta Hort. 396, 273–284.

Lamaire F 1998. Consequences of organic matter biodegradability on the physical, chemical parameters of substrates. Acta Hort. 469, 129–138.

Lindsay W L Hodgson J F and Norvell W A 1966. The physico-chemical equilibrium of metal chelates in soils and their influence on the availability of micro nutrient cations. Intern. Soc. Soil Sci. (Scotland 1966) Trans. Comm. II and IV, 305–316.

Lindsay W L and Norvell W A 1969. Equilibrium relationships of Zn^{2+}, Fe^{3+}, Ca^{2+}, and H^{+} with EDTA and DTPA in soils. Soil Sci. Soc. Amer. Proc. 33, 62–68.

LNV undated. Zuiveringsslib, compost en zwarte grond; kwaliteit en gebruik in de landbouw. Ministerie Landbouw, Natuurbeheer Visserij, 27pp.

López-Cuadrado M C Ruiz-Fernández J Masaguer A and Moliner A 2008. Utilization of different organic wastes from Madrid as growth media for *Pelargonium zonale*. Acta Hort. 2008, 623–629.

Lucas R E and Davis J F 1961. Relationship between pH values of organic soils and availabilities of 12 plant nutrients. Soil Sci. 92, 177–182.

Maher M J and Prasad M 1995. Comparison of substrates, including fractioned peat, for the production of greenhouse cucumbers. Acta Hort. 401, 225–233.

Maher M J and Prasad M 2004. The effect of peat type and lime on growing medium pH and structure and on growth of *Hebe pinguifolia* "Sutherlandii". Acta Hort. 644, 131–137.

Maloupa E, 2002. Hydroponic systems. *In*: Savvas D and Passam H (eds) Hydroponic Production of Vegetables and Ornamentals. Embryo Publications, Athens, 143–178.

Nemati R and Fortin J P 2008. development of models for predicting the pH equilibrium of organic substrates. Acta Hort. 779, 105–112.

Németh K Mengel K and Grimme H 1970. The concentration of K, Ca and Mg in the saturation extract in relation to exchangeable K, Ca and Mg. Soil Sci. 109, 179–185.

Olympios C M 1992. Soilless media under protected cultivation: rock wool, peat, perlite and other substrates. Acta Hort. 323, 215–234.

Page E R 1962. Studies in soil and plant manganese II. The relationship of soil pH to manganese availability. Plant Soil 16, 247–257.

Peterson J C 1982. Effects of pH upon nutrient availability in a commercial soilless root medium utilized for floral crop production. Ohio Agric. Res. Dev. Centre, Research Circular 268, 16–19.

Prasad M 1997. Physical, chemical and biological properties of coir dust. Acta Hort. 450, 21–29.

Prasad M 2001. Relationship between the 1:1.5 water extract and $CaCl_2$/DTPA (CAT) extract in various growing substrates. Acta Hort. 548, 269–275.

Puustjärvi V 1977. Peat and its use in horticulture. Turveteollisuusliitto Ry, Helsinki. Publication 3, 160pp.

Raviv M Wallach R Silber A and bar-Tal A 2002. Substrate and their analysis. *In*: Savvas D and Passam H (eds) Hydroponic Production of Vegetables and Ornamentals. Embryo Publications, Athens, 25–101.

Raviv M 2008. The use of compost in growing media as suppressive agent against soil-borne diseases. Acta Hort. 779, 39–49.

RHP, 2007. Productcertificatieschema. Stichting RHP 's Gravenzande, The Netherlands, Looseleaf Publication.

Scott Russell R and Clarkson D T 1976. Ion transport in root systems. *In*: Sudderland N (ed) Perspectives in Experimental Biology, 2 Botany, Pergamon Press, Oxford, 401–411.

Silber A Bar-Yosef B Singer A and Chen Y 1994. Mineralogical and chemical composition of three tuffs from northern Israel. Geoderma 63, 123–144.

Silber A 2008. Chemical characteristics of soilless media. *In*: Raviv M and Lieth J H (eds) Soilless Culture – Theory and Practice. Elsevier, Amsterdam, 209–244.

Sonneveld C Van den Ende J and Van Dijk P A 1974. Analysis of growing media by means of $1{:}1\frac{1}{2}$ volume extract. Comm. Soil Sci. Plant Anal. 5, 183–202.

Sonneveld C and Voogt S J 1975. Studies on the manganese uptake of lettuce on steam-sterilised glasshouse soil. Plant Soil 42, 49–64.

Sonneveld C and Voogt W 1985. Studies on the application of iron to some glasshouse vegetables grown in soilless culture. Plant Soil 85, 55–64.

Sonneveld C and Van Elderen C W 1994. Chemical analysis of peaty growing media by means of water extraction. Comm. Soil Sci. Plant Anal. 25, 3199–3208.

Sonneveld C 1995. Fertigation in the greenhouse industry. *In*: Proc. of the Dahlia Greidinger Intern. Symposium on Fertigation. Technion – Israel Institute of Technology, Haifa, 25 March – 1 April 1995, 121–140.

Sonneveld C and De Kreij C 1995. Standardisation of chemical analysis of growing media. Acta Hort. 401, 569–574.

Sonneveld C and Voogt W 1997. Effects of pH value and Mn application on yield and nutrient absorption with rockwool grown gerbera. Acta Hort. 450, 139–147.

Sonneveld C and Voogt W 2001. Chemical analysis in substrate systems and hydroponics – use and interpretation. Acta Hort. 548, 247–259.

Sonneveld C 2002. Composition of Nutrient solutions. *In*: Savvas D and Passam H (ed) Hydroponic Production of Vegetables and Ornamentals. Embryo Publications, Athens, 179–210.

Sonneveld C and Wever G 2005. Eigenschappen van kleikorrels. PPO Naaldwijk, The Netherlands. Report Projectnummer 41680033, 27pp.

Sonneveld C and Voogt W 2009. Determination of micro nutrients in substrates by water extraction and interpretation of the analytical data. Acta Hort. 819, 87–98.

Stamatakis M Koukouzas N Vassilatos Ch Kamenou E and Samatouras K 2001. The zeolites from Evros region, Northern Greece: a potential use as cultivation substrate in hydroponics. Acta Hort. 548, 93–103.

Straver N De Kreij C and Verberkt H 1999. Bemestingsadviesbasis potplanten. Proefstation voor Bloemisterij en Glasgroente, Naaldwijk, 55pp.

Surrage V A and Carlile W R 2008. Variation in quality of composted green wastes of UK origin and their suitability for inclusion in growing media. Acta Hort. 779, 631–636.

Van den Bos, A L 1994. EC in relatie tot het type substraat bij chrysanten in een gesloten teeltsysteem. Proefstation voor Tuinbouw onder Glas te Naaldwijk, The Netherlands. Intern verslag 22, 13pp.

Van den Bos, A L 1994a. EC in relatie tot het type substraat bij koolrabi in een gesloten teeltsysteem. Proefstation voor Tuinbouw onder Glas te Naaldwijk, The Netherlands. Intern verslag 27, 15pp.

Van den Bos, A L 1995. EC in relatie tot het type substraat bij sla in een gesloten teeltsysteem. Proefstation voor Bloemisterij en Glasgroente te Naaldwijk, The Netherlands. Intern verslag 4, 22pp.

Van den Bos, A L 1996. EC in relatie tot het type substraat bij freesia in een gesloten teeltsysteem. Proefstation voor Bloemisterij en Glasgroente te Naaldwijk, The Netherlands. Rapport 45, 34pp.

Van den Bos, A L 1996a. EC in relatie tot het type substraat bij asters in een gesloten teeltsysteem. Proefstation voor Bloemisterij en Glasgroente te Naaldwijk, The Netherlands. Rapport 50, 29pp.

Van den Bos, A L 1996b. EC in relatie tot het type substraat bij amaryllis in een gesloten teeltsysteem. Proefstation voor Bloemisterij en Glasgroente te Naaldwijk, The Netherlands. Rapport 55, 30pp.

Van den Bos, A L 1997. EC in relatie tot het type substraat bij lelie in een gesloten teeltsysteem. Proefstation voor Bloemisterij en Glasgroente te Naaldwijk, The Netherlands. Rapport 81, 34pp.

Van der Gaag D J Van Noort F R Stapel-Cuijpers L H M De Kreij C Termorshuizen A J Van Rijn E Zmora-Nahum S and Chen Y 2007. The use of green compost in peat-based potting mixtures: Fertilization and suppressiveness against soilborne diseases. Sci. Hort. 114, 289–297.

Van der Mark R 2008. Pull B. V. Rhenen. The Netherlands. Personal communication

Van der Wees A 1993. Betekenis analysecijfers bij teelten in substraat. *In*: Plantenvoeding in de Glastuinbouw. Proefstation voor Tuinbouw onder Glas te Naaldwijk. Informatiereeks no 87, 127–136.

Van Labeke M C and Dambre P 1998. Gerbera cultivation on coir with recirculation of the nutrient solution: a comparison with rockwool culture. Acta Hort. 458, 357–362.

Van Schie W, 1989. Kalk verhoogt pH potgrond. Vakblad Bloemisterij 44(14), 24–25.

Verhagen H and Van Schie W 1996. Keurmerk voor de beste kokos. Groenten en Fruit/Glasgroenten 6(44), 32–33.

Verhagen J B G M 1999. CEC and the saturation of the adsorption complex of coir dust. Acta Hort. 481, 151–155.

Verhagen J B G M 2004. Effectiveness of clay in peat based growing media. Acta Hort. 644, 115–122.

Verloo M G 1980.Peat as a natural complexing agent for trace elements. Acta Hort. 99, 51–56.

Verwer F L J W A 1974. Teelt en opkweek in kunstmatige media. Instituut voor Mechanisatie, Arbeid en Gebouwen, Publicatie 2, 28pp.

Voogt W 1983. pH problemen bij cymbidium in kunstschuim. Vakblad Bloemisterij, 38(48), 38–39.

Voogt W and Sonneveld C 2009. The effects of Fe-chelate type and pH on substrate grown roses, Acta Hort. 819, 411–417.

Warmenhoven M G Van Noort F Dijkstra T and Verberkt H 2008. Effect van fosfaatreductie op de andere (hoofd)elementen bij de teelt van eenjarige zomerbloeiers. Wageningen UR Glastuinbouw Bleiswijk, Report Project 3242015900, 20pp.

Wever G and Van Leeuwen A A 1995. Measuring mechanical properties of growing media and the influence of cucmber cultivation on these properties. Acta Hort. 401, 27–24.

Wever G 1995. Kokos vraagt andere aanpak. Groenten en Fruit/Glasgroenten 5(28), 21–22.

Wever 2004. Comparison of different bioassays for testing phytotoxicity of growing media. Acta Hort. 644, 473–477.

Wever G 2005. Bijlagenboek bepalingsmethoden, BRL K 10001 Substraatmaterialen. KIWA Rijswijk, The Netherlands, 81pp.

Wever G Verhagen J B G M Baas R and Straver N 2005. Bemestingsadviedbasis Potplanten voor Europese EN 1:5 volume methode. PPO Naaldwijk The Netherlands, Publicatie 593, 60pp.

Willumsen J 2001. Effect of substrate liming and fertilizer supply on cabbage seedlings grown organically in plugs. Acta Hort. 548, 335–342.

Chapter 12
Nutrient Solutions for Soilless Cultures

12.1 Introduction

Nutrient solutions intended for plant growth are already used from the middle of the 19th century, when the importance of mineral elements for plant growth was made clear by Justus von Liebig. In advance, the nutrient solutions used to grow plants in so called "water cultures" had a simple composition and consisted of salts like KNO_3, $Ca(NO_3)_2$, $KHPO_4$, $MgSO_4$, and a little Fe-compound (Hoagland and Arnon, 1950), thus, containing all the major elements and some Fe. The relative success with these solutions will be due to the not knowingly supplied micro nutrients from the impurities of the chemicals and fertilizers used to compose the solutions. It can be supposed that the impurities contained sufficient micronutrients, to prevent the crops grown from serious nutrient disorders. Knowledge about the necessity of micro nutrients for plant growth was mainly gathered in the first half of the 20th century (Marschner, 1997), when the purification of fertilizers and chemicals were improved. The first systematic description for the preparation of nutrient solutions was given by Hoagland and Arnon (1950) and since then in many publications reference is given to them, when one or another nutrient solution is used to grow plants in soilless cultivation systems.

Hydroponics systems are used for many years for experimental purposes, to study effects of nutrient supply on agriculture crops. Despite, that many experiments were carried out with hydroponics or hydroponics related systems and these systems in advance got an enthusiast reception, it never came to a big scale professional development and practical application until the seventies of the 20th century. In that time, the development with soilless growing started with the production of tomato, cucumber and some cut flowers in peat bales, mostly called sausages, on the Channel Islands (Dally, 1974; States of Guernsey Horticultural Advisory Service, 1974). These sausages were long and slim sacks of about 80 cm length and 20 cm diameter, filled with 25 l of peat. In the same period growing in rock wool was initiated in the Scandinavian countries and soon developed in The Netherlands as a high technological system (Verwer, 1974; Sonneveld, 1980), while in the UK nutrient film technique (NFT) has been developed (Graves, 1983). In all these systems the development of nutrient solutions played a key role, because the quantity of nutrients available at any moment in hydroponics (NFT) or in related systems, like rock

C. Sonneveld, W. Voogt, *Plant Nutrition of Greenhouse Crops*,
DOI 10.1007/978-90-481-2532-6_12,

wool, peat and perlite, is restricted and cover only few percentages of the total crop requirements on minerals (Sonneveld, 1981; Voogt, 2002). For some substrates, like peat, the quantities of micro nutrients applied in advance are considerable in relation to the total absorption (Sonneveld, 2002), but it cannot be exactly predicted whether these elements will be sufficiently available for the whole growing period. Leaching and inactivation of these elements will be considerable during long growing periods. Thus, even in such cases a regular check on the availability of micro nutrients will be recommended to control the application.

When soilless culture in the greenhouse industry started on a large scale with the growing systems mentioned soon it appears that the development of nutrient solutions specific to crops and growing conditions was necessary. The use of the Hoagland solutions mentioned and the so called universal nutrient solutions (Steiner, 1968 and 1984) are only suitable in small scale experiments and if a regular replacement of the solution is carried out. Therefore, in line with the development of soilless cultivation in the seventies of the 20th century nutrient solutions for specific crops in relation to growing conditions have been developed (Sonneveld and Straver, 1994). Important factors to be considered with the development of nutrient solutions were for example: the specific uptake of the crops, the characteristics of the substrate, the chemical composition of the irrigation water used, the growth stage of the crop, the climatic conditions, and most of all whether or not reuse of the drainage water.

In the present chapter the preparation of nutrient solutions for specific crops will be explained in relation to the different factors mentioned. Fertilizers suitable to the purpose will be discussed and an algorithm for the required calculations presented.

12.2 Characteristics of Nutrient Solutions

Modern nutrient solutions for soilless culture mostly contain more or less all nutrients necessary for plant growth. This despite that substrates used sometimes contain nutrient elements that can become available to plants. The availability of those nutrients is quite often uncertain and depends strongly on the pH, which is not a stable factor in substrate growing. The risk of nutrient disorders is economical not acceptable in the high technical greenhouse industry, because the costs of the damage in such cases has no relationship to possible extra fertilizer costs. Therefore, usually all plant nutrients are added to nutrient solutions. Substantial quantities of nutrients present in the irrigation water or possibly released from the substrate will be taken into account in advance or will be controlled during the cropping period. Such adjustments will be based on the data of chemical analysis of substrates or solutions from the root environment.

The elements added to nutrient solutions are following:

Major nutrients N, P, S, K, Ca, Mg
Micro nutrients Fe, Mn, Zn, B, Cu, Mo

Elements being essential, but deliberately not added are Cl and Ni, while Na and Co are essential for some plants and beneficial for others. Si is beneficial for some plants, but not essential. Cl and Na are usually abundantly available in the irrigation water used to cover the essential requirements. An exception is the addition of Cl in nutrient solutions for tomato to promote the uptake of Ca as discussed further on in this section. For Co knowledge is still restricted and it is to be expected that in many cases this element is sufficiently available from impurities.

The necessity of the addition of Ni to nutrient solutions for substrate systems is not yet clear. For soil grown crops there is no clear evidence of Ni deficiency, but for pot experiments with soil effects have been found (Marschner, 1997). Thus, for substrate grown crops effects can be expected also, which mainly will depend on the composition of the substrate. Insufficient research is carried out for the determination of required concentrations. Addition on basis of rough estimations is dangerous, because the optimum and toxic concentrations for sensitive crops are close together. Rahman et al. (2005) found for barley grown in de-ionized water optimal plant weights with a concentration of 1 μmol l^{-1}. Balaguer et al. (1998) working with acid washed volcanic material found a substantial improvement of the growth of tomato at a concentration of 85 μmol l^{-1} in the nutrient solution added, compared with no addition. However, this concentration can be considerably above the optimum, because Marschner (1997) mentioned 5 μmol l^{-1} in the root environment as a maximum acceptable level for sensitive crops. In the experiment of Balaguer (1998) the level of 85 μmol l^{-1} was the lowest application rate, therefore this experiment was merely suitable for conclusions about toxic levels. The addition of Ni affected strongly the uptake and transport of other metal micro nutrients Rahman et al. (2005).

Si is identified as beneficial for a number of plants, but estimated as being not essential. The uptake differs strongly between plant species. It is abundantly available in soils and a number of substrates. For soilless cultures it is valuable to pay attention to the application of Si to nutrient solutions, dependent on the type of substrate used, the Si content of the primary water and the expected reaction of the crops (Voogt and Sonneveld, 2001). Si has an effect on the suppression of powdery mildew mainly with cucurbits and roses (Bélanger et al., 1995) and will increase the yield of some crops, (Miyake and Takahashi, 1983; Voogt and Sonneveld, 2001).

Besides uptake of the elements mentioned, being essential for the development of crops, plants also absorb many other elements present in the root environment. Effects of application of most of these elements to nutrient solutions for soilless culture are not yet studied and therefore, not yet applied to nutrient solutions as long as the application has not been proved to be relevant.

Sometimes, an extra addition or a reduction on the application of an element aggravate or reduces the uptake of a different element, which sometimes shows a beneficial effect on the development of plants. This well known antagonism and synergism effects, respectively do not only exists mutually among anions and cations, but also between anions and cations. In Fig. 12.1 an example is shown with replacement of NO_3 by Cl at high EC values in the root environment for a tomato crop

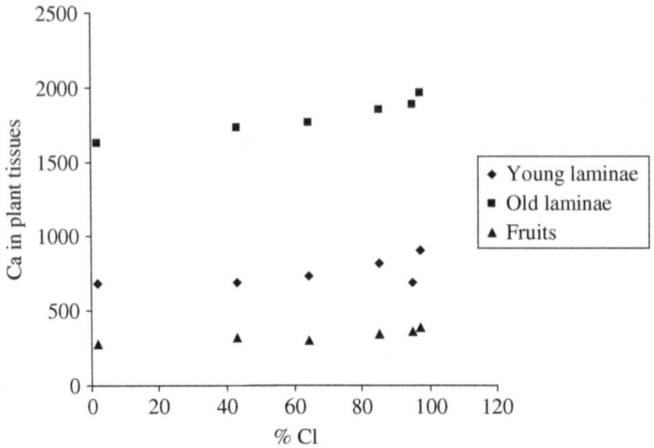

Fig. 12.1 Relationships between the % Cl ions of the sum (NO_3+ Cl) in the root environment and the Ca concentrations of plant tissues of rock wool grown tomatoes (mmol kg^{-1} dry matter, for fruits multiplied by 10). The sum of NO_3 + Cl was 22.5 mmol l^{-1}. Data after Voogt and Sonneveld (2004)

(Voogt and Sonneveld, 2004). This replacement strongly increases the uptake of Ca and thereby reduces the appearance of blossom-end rot, which can be a serious problem for example at high EC values, as has been discussed in Chapter 9.

Beside the addition of the elements mentioned the osmotic potential and the pH are important characteristics of nutrient solutions. The osmotic potential of nutrient solutions is mostly measured by the electrical conductivity (EC) and is build up in nutrient solutions by mineral salts. Therefore, a linear relationship between the osmotic potential and the EC will be supposed (Sonneveld et al., 1966; Sonneveld and Van den Ende 1967; Van den Ende, 1968). The pH in the root environment is important, because very low pH values in the root environment are toxic to plants. However, in practice the pH is mainly important because of its effects on the availability of many plant nutrients, especially micro nutrients, in the root environment (Lucas and Davis, 1961; Peterson, 1982).

The nutrient elements are mostly supplied in the form of fertilizers, mineral salts, acids and bases. N can be given in the form of NH_4 as well NO_3 and even as NH_2. The choice is mainly determined by the pH in the root environment. Addition of NH_4 lowers the pH in the root environment, because of an activation of the cation (NH_4) uptake and a reduction of the anion (NO_3) uptake. Standard quantities of NH_4 added to nutrient solutions for soilless culture are between 5 and 10% of the total N supply and seldom will exceed 15%. The tuning of the NH_4 addition merely occurs during crop growth in relation with the pH development in the root environment and will be discussed in detail in Section 13.4. High concentrations of NH_4 can be toxic to plants (Barker and Mills, 1980), especially at high pH values (Bennet, 1974).

Fe in mineral salts is very unstable and easily precipitates in solutions. Therefore, this element is supplied in the form of organic complexes, best known as chelates.

With the addition of major elements to nutrient solutions by fertilizers or other mineral compounds one always should be aware that besides the nutrient ion in view one or more other ions are added. Such accompanying ions should be taken into account too. For micro nutrients the accompanying ions are mostly Cl, NO_3 or SO_4, being major nutrient ions. The quantities of major ions added by this application are insignificant and can be neglected, because they are very low in comparison with the quantities of major elements required.

12.3 Nutrient Solutions for Different Crops and Development Stage

The nutrient uptake differs strongly between crops, not only with respect to the quantity of nutrients absorbed, but also the ratios between the nutrients are different. This is illustrated by the data listed in Table 12.1 where the yearly uptakes of major elements of high yielding tomato, chrysanthemum and cymbidium crops are compared. The data of tomato and chrysanthemum are derived from regression equations presented by Sonneveld (1997) with which the relationship between yield and nutrient uptake of two specific crops are presented. The data of cymbidium are derived from a study on different nurseries (Voogt et al., 2005). The yearly uptake of N and K of high yielding tomato and chrysanthemum crops are of the same order and ratio, but the uptake and the ratio of the other elements differ strongly. For cymbidium the quantities as well the ratios for most elements differ strongly from those of both other crops.

The variation in the uptake over the year differs strongly and is related with the growth rate of the crop. The driving force behind the growth rate is the radiation and therefore, the daily uptake varies strongly with the radiation input. Thus, the daily uptake is much higher in summer time than in winter time, especially in areas where the global radiation input in the winter is low. It has been found that the water use of crops, mainly due to transpiration, is also strongly related to the radiation input

Table 12.1 Yearly uptakes of major elements (kg ha^{-1}) by high yielding tomato (round and beefsteak types), chrysanthemum and cymbidium (cv Yonina) crops; being 60 kg of fruit fresh weight and 22.5 and 2.3 kg flower fresh weight yr^{-1} m^{-2} respectively

	Tomato		Chrysanthemum		Cymbidium	
Nutrients	kg ha^{-1}	Ratio N = 100	kg ha^{-1}	Ratio N = 100	kg ha^{-1}	Ratio N = 100
N	1185	100	903	100	110	100
P	284	24	129	14	18	16
S	290	24	61	7	37	34
K	2044	172	1567	174	163	148
Ca	863	73	259	29	111	101
Mg	208	18	82	9	10	9

(De Graaf and Esmeijer, 1998). This implies that the ratio between nutrient uptake and water uptake will be more or less constant. This ratio, often called uptake concentration, has no physiological basis, because the uptake of water and nutrients are independent processes (Sonneveld and Voogt, 1990). Nevertheless, the uptake concentration appears to be not altogether stable (Kläring et al., 1997) and increases with low and decreases with high radiation input. This can be explained by the fact that the photosynthesis of crops is curve linearly related with radiation input, while transpiration is merely linearly related with radiation input (Sonneveld, 2002). Experimental data showed that the fluctuations of the uptake concentrations are less than those of the daily uptake (Savvas and Lenz, 1995). Therefore, the uptake concentration is often used as a rough basis for the addition of fertilizers to the irrigation water in substrate systems.

For an accurate nutrient application the development of standard nutrient solutions is important (Sonneveld and Straver, 1994). Such solutions will be defined as nutrient solutions due to addition in soilless cultures to keep the nutrient status in the substrate optimal for the crop under the current growing conditions. These growing conditions will be defined and mainly concern the climatic conditions, the irrigation patterns and the development stages of the crop. With the climatic conditions, the light intensity and the transpiration rate are detected as being important. With the irrigation, the fraction drain to waste, the reuse of drain water and the circulation rate are main factors. With the development stages, the change from the vegetative to the reproductive phase is important. Mainly the K:Ca ratio of the uptake increases in the reproductive phase as has been described for tomato (Voogt, 1993) and other fruit vegetables (Voogt, 2002). It also occurs with head formation of lettuce crops and the shoot up of a flower flush with cut flowers like roses and carnations. The explanation is the relative much higher K:Ca ratio in the reproductive organs, like fruits and flower stems than in the vegetative organs, mainly leaves. (Van Goor et al., 1988; Voogt, 1988; Voogt and Sonneveld, 1997; Ward, 1967). The cation distribution of a tomato and a cucumber crop are shown in Table 12.2. The K:Ca ratios of fruits are much higher than those of leaves, especially with tomato. Therefore, adjustment of the solution added to the growing system is necessary during crop development, like shown in Table 12.3 (Voogt, 1993).

Table 12.2 Distribution of dry matter and K, Ca and Mg over different plant parts in % of the total aboveground uptake for long term tomato and cucumber crops

	Tomato				Cucumber		
	Leaf	Shoots[1]	Stem	Fruit	Leaf	Stem	Fruit
Dry matter	20	4	14	62	28	14	58
K	18	5	11	66	22	10	67
Ca	76	4	15	5	56	23	20
Mg	50	5	15	30	44	19	37

[1]Young shoots removed with pruning
Data derived from Roorda van Eijsinga and Haeff (1964); Voogt (1993) and Voogt (2002).

Table 12.3 Guidelines for the adjustment of the nutrient solution supplied to tomatoes in a closed growing system in relation to the growth stage

	mmol.l^{-1}						
Growth stages	NH_4	K	Ca	Mg	NO_3	SO_4	H_2PO_4
Standard solution	1.0	6.5	2.75	1.0	10.75	1.5	1.25
Saturation root medium	−0.5	−2.5	+1.0	+0.5	[2]		
Start till 1st truss[1]		−1.2	+0.3	+0.3			
1st till 3rd truss							
3rd till 5th truss		+1.0	−0.25	−0.25			
5th till 10th truss		+3.5	−1.25	−0.50			
10th till 12th truss		+1.0	−0.25	−0.25			
After the 12th truss							

[1]Adjustments are related to the truss number at anthesis
[2]Where no value is given, standard concentration is recommended
After Voogt (1993). *Reprinted by permission of the International Society Horticultural Science.*

Effects of climatic conditions are shown by the results of Bakker and Sonneveld (1988) with cucumber. With a high humidity the uptake of Ca is strongly reduced and thus, a high Ca supply is necessary to prevent Ca deficiency, as shown in Fig. 9.4. Effects of the irrigation practises are shown with the nutrient solutions used in a free drainage system and in a closed system to realise equal concentrations in the substrate solution in the root environment with a tomato crop, as listed in Table 12.4. In the free drainage system a leaching fraction of about 30% is taken into account. The mutual ratios of the nutrient concentrations in the solution supplied in a free drainage system and those supplied in a closed system differ substantially. The bivalent ions are relatively lower in the solution for closed systems. This is linked up with the usually desired strong accumulation of the bivalent nutrients in the root environment, because of the relatively low uptake of these ions like discussed in Section 3.5.

Table 12.4 Composition of the nutrient solution in the root environment recommended for tomato and the composition of the supplied solution recommended for a closed system and for a free drainage system

Parameters	Root environment	Supplied closed system	Supplied free drainage	Ratio free/closed
EC dS m^{-1}	4.0	1.6	2.6	1.6
NH_4mmol l^{-1}	< 0.5	1.0	1.2	–
K	8.0	6.5	9.5	1.5
Ca	10.0	2.75	5.4	2.0
Mg	4.5	1.0	2.4	2.4
NO_3	23.0	10.75	16.0	1.5
SO_4	6.75	1.5	4.4	2.9
H_2PO_4	1.0	1.25	1.5	1.2

After Sonneveld (2002). *Reprinted after permission of Embryo Publications.*

12.4 Water Quality

The quality of the primary water affects the requirements for the composition of nutrient solutions in different ways. Preferably, water is used with low concentrations of mineral elements, that means concentrations equal or lower than the uptake concentrations. When such is the case for all elements, the drainage water can be completely reused. This situation mainly exists with the use of rain water or desalinated water. Natural ground waters or surface waters commonly contain substantial concentrations of mineral elements often beyond those of the uptake concentrations. When such is the case, the water is less suitable for closed growing systems. Different effects of the presence of mineral elements in the irrigation water on the preparation of nutrient solutions will be discussed successively.

In the first place the situation that the water contains minerals acting as plant nutrients and residual salts, the concentrations of which are equal or below the uptake concentration. The adjustment exists in a reduction on the addition of the required nutrient(s).

Secondly, the concentration of a nutrient element or a residual salt in the primary water is higher than the uptake concentration of the crop grown. The addition of the nutrient element in question is eliminated from the addition. Nevertheless, the residual concentration of the nutrient or the residual salt will accumulate in the root environment. The maximum acceptable accumulation levels, vary for crops and growing conditions. Under certain conditions the accumulation fits in the composition of the nutrient solution in the root environment, before leaching is necessary (Sonneveld, 2000). Guidelines for adjustments in relation to the acceptable accumulations of residual salts are discussed in Section 7.7.

Thirdly, when concentrations of minerals in the primary water are much higher than the uptake concentrations and the accumulation of such minerals easily exceeds the maximum acceptable level, a systematic drainage to waste is necessary. With this drainage water, beside the mineral(s) in question, also other minerals (nutrients) are removed from the root environment and by this the addition should be adjusted to a systematic leaching requirement, as discussed in Section 12.3.

Fourthly, adjustments are necessary when the primary water contains (bi)carbonate. When substantial concentrations of these ions are present, the pH of the nutrient solution easily becomes too high, because of the alkaline buffering capacity of the carbonate. This problem can be solved by addition of equivalent concentrations of free acids in the nutrient addition. HNO_3 and possible H_3PO_4 or H_2SO_4 are most suitable to the purpose. Following reaction occurs with the acid addition.

$$HCO_3 + H_3OA \rightarrow 2H_2O + A + CO_2 \uparrow \tag{12.1}$$

In this equation A is placed for the anion of the acid used and can be in the current situation a NO_3, H_2PO_4, or SO_4 ion.

Finally, Fe available in natural waters cannot be taken into account, because it easily precipitates, like discussed in Section 6.4. Cu in such waters sometimes can be affected by complex formation in combination with soluble organic compounds present in the water. Hereby, the metal can be bound that strongly in the complex that it is scarcely available to plants. See also the discussion about this effect in Section 13.4. Therefore, it is experienced that Cu as determined by chemical analysis cannot always completely taken into account in natural waters.

12.5 Adjustments

The intention of the application of nutrients during crop cultivation is the maintenance of a nutrient level in the root environment for an optimal crop development, which holds a maximum production of the required quality. To this purpose as much as possible information about nutrient use during crop cultivation should be outlined in a systematic scheme, as discussed in Section 12.3. Important processes that determine the equilibrium between the addition to and the extraction from the nutrient solution in the root environment are the uptake by the crop and the immobilisation processes. In case of free drainage systems also the leaching of nutrients is an important factor. Often insufficient information is available about these processes, which results in disturbance of the equilibrium and an unsettlement of the chemical composition of the nutrient solution in the root environment. Therefore, a regular check on the chemical composition is necessary. The sampling frequency varies between once a month and once a week dependent on risks and chances to be expected. The analysis is mostly carried out by professional laboratories. Besides the determination of chemical composition on the laboratory, frequent measurements of EC and pH are recommended with the aid of portable instruments on the greenhouse holdings itself.

The results of the measurements will be compared with the guide values developed for crop and prevailing growing conditions. These values are different from those of the nutrient solution added as given by Sonneveld and Straver (1994); see also the examples in Table 12.4. On basis of the comparison conclusions will be drawn about the necessity of adjustments on the nutrient solution added. The extent of the adjustments depend on the deviation of the current values from the standardized guide values. For major nutrients with a high absorption rate like N, P, and K deviations – reduction or increase – of 30% on the addition are applied, while for Ca, Mg, S and micro nutrients deviations up to 50% are thinkable. Seldom results lead to exclusion of the application of a nutrient element. Judgement of the analytical data of P and Mn need extra attention, since the results of these elements will be judged in relation to the pH and the duration of the cropping period. The availability of both elements is strongly affected by the pH. At pH values >6.5 the concentration can be low independent of the concentration added with the irrigation water. P is less soluble at high pH values. Mn at such pH values can become very low after some months by the development of Mn oxidising micro organisms in the root environment, as will be discussed in detail in Section 13.4. Sometimes the ratio between elements is more important than the actual concentration. This for example is the case for the K:Ca ratio for fruit vegetables, as will be explained in Section 13.4.

Adjustments with regard to pH and EC mostly occur on basis of the frequent measurements by the portable instruments. The pH will be regulated mainly by adjustments of the NH_4 application since the pH in the root environment can be hardly affected by acid or base applications. Only in systems with a high circulation rate, addition of acids also offers possibilities for pH control (Voogt, 2009). At the moment research is going to the development of *in situ* measurements of different nutrients in solutions (Gieling, 2001) under growing conditions, but these systems are not yet operative. In future they offer possibilities for a full automation of nutrient supply.

For interpretation of the analytical data the sampling method, the sampling place and the extraction method on the laboratory are important factors too to be taken into account with the interpretation. See therefore the Chapters 4 and 8.

12.6 Fertilizers

Not all fertilizers used in horticultural practise are suitable for the preparation of nutrient solutions for soilless culture. The fertilizers, but also the other chemicals used to that purpose, should be rapidly and completely soluble in water and should not contain insoluble residues, contaminants of heavy metals or other components leading to concentrations in nutrient solutions toxic to plants or, after absorption by plants, to human beings. Insoluble residues, although not directly harmful to plants, are undesirable because such residues easily block the narrow canals of the irrigation systems commonly used. In Table 12.5 common fertilizers used for the preparation of nutrient solutions in greenhouse cultivation are listed.

Table 12.5 Fertilizers and acids used in the greenhouse industry to compose nutrient solutions, chemical compositions and molecular weights

Fertilizer	Chemical composition	mol weight
Calcium nitrate	$5[Ca(NO_3)_2.2H_2O]NH_4NO_3$	1080.5
Ammonium nitrate	NH_4NO_3	80
Potassium nitrate	KNO_3	101.1
Magnesium nitrate	$Mg(NO_3)_2.6H_2O$	256.3
Nitric acid 100%	HNO_3	63
Mono potassium phosphate	KH_2PO_4	136.1
Phosphoric acid 100%	H_3PO_4	98
Potassium sulphate	K_2SO_4	174.3
Magnesium sulphate	$MgSO_4.7H_2O$	246.3
Manganese sulphate	$MnSO_4.H_2O$	169
Zinc sulphate	$ZnSO_4.7H_2O$	287.5
Copper sulphate	$CuSO_4.5H_2O$	249.7
Borax	$Na_2B_4O_7.10H_2O$	381.2
Sodium molybdate	$Na_2MoO_4.2H_2O$	241.9
Iron chelate EDTA 13% Fe	Fe-EDTA	430
Iron chelate DTPA 6% Fe	Fe-DTPA	932
Iron chelate EDDHA 5% Fe	Fe-EDDHA	1118

Picture 12.1 Installation for the preparation of nutrient solutions. At the back the reservoirs for concentrated fertilizer solutions are placed

The formula given for calcium nitrate is a little complicated, but necessary to use in the calculations, to get the right NH_4 concentration in the solution. With the fertilizer of this source, often 50% or more of the required NH_4 is brought in. For the acids molecular weights of 100% pure chemicals are given. However, such acid solutions do not exist. The strength of the solutions in trade is different. The molecular weights of such solutions will be calculated following formula (12.2).

$$mol\ weight\ current\ solution = \frac{mol\ weight\ 100\%\ solution}{0.01 \times current\ \%} \tag{12.2}$$

For micro nutrients most common types are given, but also other types are suitable, like the chloride or nitrate salts of Mn, Zn and Cu. Furthermore, the water bound in the salt crystals can differ, which affects also the molecular weight. For Fe DTPA and EDTA chelates can be used and are suitable up to a pH of 6.5 (Lindsay et al., 1967). When the pH will arise above this value Fe-EDDHA is often used, because the compound is stable at all pH values and thus suitable under all pH conditions. However, the applicability is not unequivocal due to side effects, see Section 13.4.4. Moreover, the price of the latter compound is mostly much higher than those of the other compounds and therefore less frequently used. All Fe-chelates are available in different qualities (% of chelated Fe) and with the deviation from the percentages mentioned the dosing will be adjusted accordingly.

Besides the group of common fertilizers and chemicals mentioned in Table 12.5, the Dutch fertilizer industry has manufactured special groups of chemical compounds or mixtures of such compounds tuned to substrate growing (Sonneveld and

Voogt, 1994). Such special groups merely consist of highly concentrated solutions of mineral salts, fertilizers, acids and alkalines. The acids and alkalines are due to chemical reactions when put together in the stock solution at the greenhouse holdings, with mineral salts as a result. Following reaction occur when a base is used in combination with an acid

$$H_3OA + MOH \rightarrow 2H_2O + A + M \tag{12.3}$$

And when a carbonate is used

$$H_3OA + MHCO_3 \rightarrow 2H_2O + CO_2^{\uparrow} + A + M \tag{12.4}$$

In which is

A = an anion
M = a cation

Nowadays several methods are available for the preparation of nutrient solutions. Traditionally fertilizers and chemicals were brought together in two concentrated stock solutions in separate reservoirs mostly called A and B. Compounds containing Ca and Fe should be added to the A reservoir and compounds containing SO_4 and H_2PO_4 and the bulk of the acids to the B reservoir. In this way precipitation of Ca-phosphates ($CaHPO_4$ and $Ca_3(PO_4)_2$ or $CaSO_4$ will be prevented as well as decomposition of Fe-chelates, which easily occurs in a strong acid solutions.

The A and B stock solutions are usually 100 times higher concentrated than the nutrient solution to be supplied. This is possible until an EC value in the supplied solution of about 2.0, for the commonly used compositions. When higher concentrated solutions are required, the stock solutions become over saturated with precipitation of some salts and an unsettlement of the composition as a result.

Addition of the stock solutions to the water is carried out by devices, automatic or semi- automatic types. The semi- automatic types inject the stock solution in the water flow or in a mixing tank on a preset proportional basis without a feedback of an EC measurement. The automatic types inject the stock solution in the water flow or into the mixing tank, whereby the final concentration of the nutrient solution supplied is regulated by means of continuous EC measurements. In last case, also a continuous measurement of the pH is carried out and possible adjustments are performed by addition of small quantities from separate either acid or alkaline stock solutions. For the preparation of the acid stock solution HNO_3 is preferred to H_3PO_4, because the small applications make relatively more sense to the P than to the N application. Besides, long duration of P overdoses especially with reuse of drainage water will easily induce high P levels in the root environment, which can become toxic to some plants (Howell and Bernhard, 1961). Alternatively, the use of H_2SO_4 to the purpose is possible. The alkaline stock solution best can be prepared with KOH or $KHCO_3$.

More advanced fertilizer supply is carried out by individual and direct injection of the fertilizers in the water flow or onto a mixing tank. In this case all the fertilizers have to be in liquid form or have to be dissolved in separate concentrated stock solutions. These systems are furnished with the same number of injectors as the number of fertilizer stock solutions that will be added. Such systems are fully computerized and have the advantage that the recipe of the nutrient solution can directly be brought in the managing computer and that the recipe thus can instantly be adjusted.

The application of Si to nutrient solutions is complicated because the Si compounds applicable for plant nutrition purposes are not stable in diluted, ready to use, nutrient solutions at pH <9 and in particular not in concentrated stock solutions. Precipitation of Si can become a serious problem in irrigation systems, especially with respect to blockings in trickle irrigation systems. For the addition of Si to nutrient solutions highly alkaline concentrated Si compounds, containing Si, K and OH in the mol ratios of 1:2:2, respectively are recommended. These compounds are stable in concentrated form, but not in concentrated nutrient stock solutions. Therefore, it should be separately added from the other fertilizers in the ready (diluted) nutrient solutions. Beforehand, the fertilizer recipe will be adjusted for the K and alkaline input of the Si compound, being an equivalent reduction of K and addition of H_3O. After the Si addition the alkalinity is instantly neutralized by the H_3O in the adjusted fertilizer recipe. In the diluted form at "neutral" pH values a maximum concentration of about 1.5 mmol l^{-1} Si will be soluble in the monomer ($Si(OH)_4$ form, likely the only form absorbed by plants (Iler, 1979). Optimal effects of Si are derived by a relatively low concentration, as is clear from the data shown in Fig. 12.2. Higher applications increase the uptake of Si like shown in Fig. 12.3, but did not increase yield. It only increases the risk on blocking of the irrigation system. Thus, if there is no Si available from the substrate or from the irrigation water a concentration of about 0.75 mmol. l^{-1} in the nutrient solution is sufficient for an optimal response for most crops that react on Si application (Voogt and Litjens, 1991; Voogt 1992).

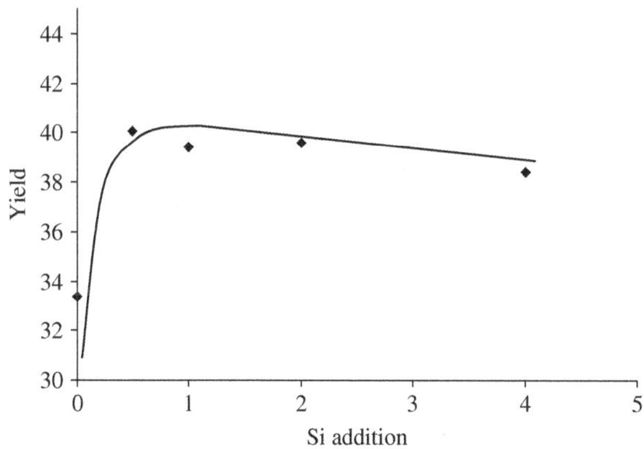

Fig. 12.2 Relationship between the addition of Si to the nutrient solution (mmol l^{-1}) and the yield of cucumbers (kg m^{-2}). Data derived from Voogt and Sonneveld (2001)

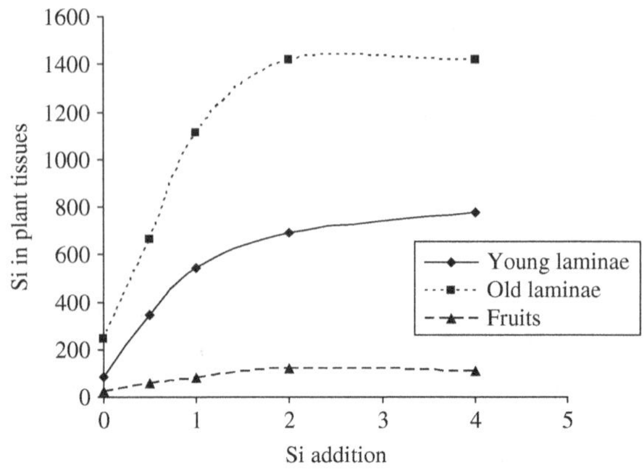

Fig. 12.3 Relationships between the addition of Si (mmol l^{-1}) to the nutrient solution and the Si uptake (mmol kg^{-1} dry matter) in different organs of cucumbers grown in rock wool. Data derived from Voogt and Sonneveld (2001)

12.7 Algorithm

For the calculation of nutrient solutions a universally suitable algorithm is developed (Sonneveld et al., 1999). With this algorithm the fertilizer composition of nutrient solutions can be calculated based on standard recipes (Sonneveld and Straver, 1994) with adjustments on water quality, pH, EC and recommendations by chemical analysis of the nutrient solution in the root environment. Last adjustments can be formulated as single anions and/or cations, because the algorithm equalizes the possible difference between the sum of anions and of cations relatively over the anions and/or cations. In this way the mutual ratios of anions and the mutual ratios of cations are kept constant.

Following steps in the calculations of the algorithm should be carried out in the order given.

1. Selection of the standard nutrient solution depending on crop, growing system and possible growth stage and growing conditions.
2. Possible adjustments of nutrient elements, for example recommendations based on the chemical composition of the nutrient solution in the root environment.
3. Equalizing of differences between anion and cation sums, keeping the EC equal to those of the standard solution. NH_4, P, and micronutrient are excluded from this equalization.
4. Adjustment on the EC required in the supplied solution, from which components are excluded as in step 3.
5. Adjustment on the quality of the primary water
6. Calculation of fertilizer composition.

In step 3 and 4 NH_4 is excluded from the equalization, because the application of the ion is primarily connected with the pH regulation. P is excluded from it because an over supply of this element easily disturb the uptake of some other nutrients and P can be toxic, when the concentration in the root environment become too

Picture 12.2 Rooting system of chrysanthemum grown in a hydroponics system

high, as already mentioned in Section 12.6. Micro nutrients are excluded from these corrections, because the EC is not notably affected by these elements.

The standard nutrient solution for step 1 can be selected from the available sources, or drawn up by the grower or by an advisor. The adjustments under step 2 are mostly based on chemical analysis of a sample of the nutrient solution from the root environment. For the calculations necessary in the other steps following formulae are developed (Sonneveld et al., 1999; Sonneveld, 2002).

For step 3 calculations of cations:

$$c_{s(x)}' = \frac{\{c_{st(x)} + c_{adj(x)}\} \times \{C_{st}^{+} - c_{st(NH_4)} - c_{adj(NH_4)}\}}{\{C_{st}^{+} - c_{st(NH_4)} + \sum[V_{(x)} \times c_{adj(x)}]\}} \tag{12.5}$$

For step 3 calculations of anions:

$$c_{s(x)}' = \frac{\{c_{st(x)} + c_{adj(x)}\} \times \{A_{st}^{-} - c_{st(H_2PO_4)} - c_{adj(H_2PO_4)}\}}{\{A_{st}^{-} - c_{st(H_2PO_4)} + \sum[V_x \times c_{adj(x)}]\}} \tag{12.6}$$

In which

$c_{s(x)}'$ = calculated concentration in the water supplied after adjustment of the mutual ratios of the macro nutrients in mmol l^{-1}
$c_{st(x)}$ = concentration of any ion x in the standard nutrient solution in mmol l^{-1}
$c_{adj(x)}$ = concentration adjustment of any ion x except P and NH_4 in mmol l^{-1}
A_{st}^{-} = sum of electrons of the standard solution in mmol l^{-1}
C_{st}^{+} = sum of protons of the standard solution in mmol l^{-1}
$V_{(x)}$ = valence of ion x

It should be noticed that for the concentrations NH_4 and P the following is operative: $c_{s(NH4)'} = c_{st(NH4)} + c_{adj(NH4)}$ and $c_{s(H2PO4)'} = c_{st(H2PO4)} + c_{adj(H2PO4)}$, respectively

If the desired EC of the nutrient solution supplied differs from the EC of the standard nutrient solution, Step 4 will be carried out.

Step 4 calculation of cations:

$$c_{s(x)} = \frac{10EC_s - c_{s(NH_4)'}}{10EC_{st} - c_{s(NH_4)'}} \quad (12.7)$$

Step 4 calculation of anions:

$$c_{s(x)} = \frac{10EC_s - c_{s(H_2PO_4)'}}{10EC_{st} - c_{s(H_2PO_4)'}} \quad (12.8)$$

In which

$c_{s(x)}$ = concentration after adjustment on the EC in the nutrient solution supplied

The ultimate concentrations that will be supplied by addition of fertilizers are derived after correction on the nutrients available in the primary water. These calculations are carried out by steps 5 and 6, the adjustments of anions and cations on the composition of the primary water. The composition of which will be given by chemical analysis.

$$c_{f(x)} = c_{s(x)} - c_{w(x)} \quad (12.9)$$

In which

$c_{f(x)}$ = the concentration of ion x added by fertilizer application in mmol l^{-1}
$c_{w(x)}$ = the concentration of any ion x in the primary water in mmol l^{-1}

The HCO_3 will be neutralized by addition of equivalent quantities of acid, as given in Section 12.4. An example of the calculations is given in Table 12.6.

An example of the calculation of the fertilizer composition is listed in Table 12.7. In this calculations it should be noted that 1 mol of a salt in a fertilizer after dissociation often contain more than one mole of the element concerned. This is for example quite clear with K_2SO_4, but more complicated with the fertilizer calcium nitrate. After dissociation 1 mol contains following nutrient compounds.

$$5[Ca(NO_3)_2.2H_2O]NH_4NO_3 \rightarrow 5Ca + 11NO_3 + NH_4$$

Beside the calcium nitrate fertilizer, there are high concentrated calcium nitrate solutions on the market. These solutions are often free from NH_4. There are much

Table 12.6 An example of the calculation of a nutrient solution worked out for a cucumber crop with reuse of drainage water

	Steps of the calculations					
Elements	1 (c_{st})	2 (c_{adj})	3 (c_s')[1]	4 (c_s)[2]	5 (c_w)	6 (c_f)[3]
EC dS m^{-1}	1.5		1.5	1.9		
NH_4 mmol l^{-1}	1.0	+0.25	1.25	1.25		1.25
K	6.5		6.88	8.88		8.88
Ca	2.75	−0.5	2.38	3.07	0.6	2.47
Mg	1.0		1.06	1.37	0.3	1.07
NO_3	11.75		11.54	14.96		14.96
SO_4	1.0		0.98	1.27	0.4	0.87
H_2PO_4	1.25	+0.25	1.50	1.50		1.50
HCO_3					1.0	
H_3O						1.0
Fe μmol l^{-1}	15		15	15		15
Mn	10		10	10		10
Zn	5		5	5		5
B	25		25	25	10	15
Cu	0.75		0.75	0.75		0.75
Mo	0.5		0.5	0.5		0.5

[1] Using formulae (12.5) and (12.6);
[2] using formulae (12.7) and (12.8);
[3] using formula (12.9)
The standard solution is derived from Sonneveld and Straver (1994).

Table 12.7 Calculations of the quantities of fertilizer necessary for the composition of the nutrient solution given in the last column of Table 12.6

			Elements added in mmol l^{-1}							
Fertilizers	mmol l^{-1}	mg l^{-1}	NH_4	K	Ca	Mg	H_3O	NO_3	SO_4	H_2PO_4
Calcium nitrate[1]	2.47/5	534	0.49		2.47			5.43		
Ammonium Nitrate	0.76	61	0.76					0.76		
Mono pot. phosphate	1.50	204		1.50						1.50
Magnesium nitrate	1.07	274				1.07		2.14		
Potassium sulfate	0.87	152		1.74					0.87	
Nitric acid 38%	1.00	166					1.00	1.00		
Potassium nitrate	5.64	570		5.64				5.64		
Sums			1.25	8.88	2.47	1.07	1.00	14.97	0.87	1.50

[1] fertilizer

more concentrated fertiliser solutions on the market containing more than two nutrient elements varying in concentrations and ratios between the nutrients. It is obvious that the calculations become more complicated with such compound fertilizers than with fertilizers containing a single salt. Furthermore it should be noticed that 1 mol Borax ($Na_2B_4O_7.10H_2O$) ≡ 4 mol B, but when for example boron acid (H_3BO_3) is used that 1 mol H_3BO_3 ≡ 1 mol B.

References

Bakker J C and Sonneveld C 1988. Calcium deficiency of glasshouse cucumbers as affected by environmental humidity and mineral nutrition. J. Hort. Sci. 63, 241–246.

Balaguer J Almendro M B Gómez I Navarro Pedreño J and Mataix J 1998. Tomato growth and yield affected by nickel presented in the nutrient solution. Acta Hort. 58, 269–272.

Barker A V and Mills H A 1980. Ammonium and nitrate nutrition of horticulture crops. Hort. Rev. 2, 395–423.

Bélanger R R Bowen P A Ehret D L Menzies J G 1995. Soluble silicon, its role in crop and disease management of greenhouse crops. Plant Disease 79, 329–336.

Bennet A C 1974. Toxic effects of aqueous ammonia, copper, zinc, lead, boron and manganese on root growth. *In*: Carson E W (ed) The Plant Root and its Environment, University of Virginia, Charlottesville, 669–683.

Dally H 1974. The cash advantage of growing tomatoes in peat modules. The Grower 82, 239–242.

De Graaf R and Esmeijer M H 1998. Comparing calculated and measured water consumption in a study of the (minimal) transpiration of cucumbers grown in rockwool. Acta Hort. 458, 103–110.

Gieling Th G 2001. Control of water supply and specific nutrient application in closed growing systems. Thesis Wageningen University, 176 pp.

Graves 1983. The nutrient film technique. Hort. Rev. 5, 1–44.

Hoagland D R and Arnon D I 1950. The water culture method for growing plants without soil. Agric. Exp. Sta. Berkely, Calif., Circular 347, 39 pp.

Howell W and Bernhard R L 1961. Phosphorus response of soybean varieties. Crop Sci. 1, 311–313.

Iler R K 1979. The chemistry of silica. Wiley Interscience, New York.

Kläring H P Schwarz D and Heissner A 1997. Control of nutrient solution concentration in tomato using models of photosynthesis and transpiration: a simulation study. Acta Hort. 450, 329–334.

Lindsay W L Hodgson J F and Norvell W A 1967. The physico-chemical equilibrium of metal chelates in soils and their influence on the availability of micro nutrient cations. Intern. Soc. Soil Sci. (Scotland 1966) Trans. Comm. II and IV, 305–316.

Lucas R E and Davis J F 1961. Relationship between pH values of organic soils and availabilities of 12 plant nutrients. Soil Sci. 92, 177–182.

Marschner H 1997. Mineral Nutrition of Plants, Second edition. Harcourt Brace & Company, Publishers, London.

Miyake Y and E Takahashi 1983. Effect of silicon on the growth of solution cultured cucumber plant. Soil Sci. Plant Nutr. 29, 71–83.

Peterson J C 1982. Effects of pH upon nutrient availability in a commercial soilless root medium utilized for floral crop production. Ohio Agric. Res. Dev. Center, Circular 268, 16–19.

Rahman H Sabreen S Alam S and Kawai S 2005. Effects of nickel on growth and composition of metal micronutrients in barley plants grown in nutrient solution. J. Plant Nutr. 28, 393–404.

Roorda van Eijsinga J P N L and Haeff J N M 1964. Onttrekking van voedingselementen aan de grond door komkommers. Glasshouse Crops Research Station Naaldwijk The Netherlands, Annual Report 1964, 35–38.

Savvas D and Lenz F 1995. Nährstoffaufnahme von Aubergine (Solanum melongena L.) in Hydrokultur. Gartenbauwissenschaft 60, 29–33.

Sonneveld C Koornneef P and Van den Ende J 1966. De osmotische druk en het electrische geleidingsvermogen van enkele zoutoplossingen. Meded. Dir. Tuinb. 29, 471–474.

Sonneveld C and Van den Ende J 1967. De samenstelling van de zouten in het oppervlaktewater in het Zuidhollands glasdistrict. Meded. Dir. Tuinb. 30, 411–416.

Sonneveld C 1980. Growing cucumbers and tomatoes in rockwool. Proc. fifth Intern. Congr. Soilless Culture, Wageningen, 1980, 253–262.

Sonneveld C 1981. Items for application of macro-elements in soilless cultures. Acta Hort. 126, 187–195.

Sonneveld C and Voogt W 1990. Response of tomatoes (Lycopersicon esculentum) to an unequal distribution of nutrients in the root environment. Plant Soil 124, 251–256.

Sonneveld C and Straver N 1994. Nutrient solutions for Vegetables and Flowers grown in water or substrates. Research Station for Floriculture and Glasshouse Vegetables, Aalsmeer/Naaldwijk, The Netherlands, Series: Voedingsoplossingen Glastuinbouw, 8, 45 pp.

Sonneveld C and Voogt W 1994. Het berekenen van voedingsoplossingen voor planteteelt zonder aarde. Proefstation voor Tuinbouw onder Glas, Series Voedingsoplossingen Glastuinbouw, 10 negende druk, 21 pp.

Sonneveld C 1997. Mineralenopname bij teelten onder glas. Proefstation voor Bloemisterij en Glasgroente. Intern verslag 81, 35 pp.

Sonneveld C Voogt W and Spaans L 1999. A universal algorithm for calculation of nutrient solutions. Acta Hort. 481, 331–339.

Sonneveld C 2000. Effect of salinity on substrate grown vegetables and ornamentals in greenhouse horticulture. Thesis Wageningen University, Netherlands, 151 pp.

Sonneveld C 2002. Composition of Nutrient solutions. *In*: Savvas D and Passam H (eds) Hydroponic Production of Vegetables and Ornamentals. Embryo Publications, Athens, 179–210.

States of Guernsey Horticultural Advisory Service 1974. Peat Culture. Bulletin A 3, 9 pp.

Steiner A A 1968. Soilless culture. Proc. 6th Coll. Intern. Potash Inst. Florence/Italy. Intern. Potash Inst., Berne. Switzerland, 324–341.

Steiner A A 1984. The universal nutrient solution. Proc. Sixth Intern. Congress on Soilless Culture, Lunteren 1984. Intern. Soc. Soilless Culture, Wageningen/Netherlands, 633–649.

Van den Ende J 1968. Analysis of greenhouse soils by means of aqueous extracts. Proc. 6th Coll. Intern. Potash Inst. Florence, 246–255.

Van Goor B J De Jager A and Voogt W 1988. Nutrient uptake by some horticultural crops during the growing period. Seventh Intern. Congress on Soilless Culture, Flevohof 1988, Proceedings, 163–176.

Verwer F L J A W 1974. Teelt en opkweek in kunstmatige media. IMAG Wageningen, The Netherlands, Publication 2, 28 pp.

Voogt 1988. The growth of beefsteak tomato as affected by K/Ca ratios in the nutrient solution. Acta Hort. 222, 155–165.

Voogt W and Litjens H 1991. Teveel silicium geeft problemen. Groenten en Fruit/Glasgroenten 1(1), 28–31.

Voogt W 1992. Silicium zinvol bij roos in steenwol. Vakblad voor de Bloemisterij 47(33), 28–29.

Voogt W 1993. Nutrient uptake of year round tomato crops. Acta Hort. 339, 99–112.

Voogt W and Sonneveld C 1997. Nutrient management in closed growing systems for greenhouse production. *In*: Goto E et al. (eds) Plant Production in Closed Ecosystems, 83–102. Kluwer Academic Publishers, Dordrecht, The Netherlands.

Voogt W and Sonneveld C 2001. Silicon in horticulture crops grown in soilless culture. *In*: Datnoff L E Snijder G H and Konadörfer G H (eds) Silicon in Agriculture. Elsevier Science B.V., 115–131.

Voogt W 2002. Potassium management of vegetables under intensive growth conditions. *In*: Pasricha N S and Bansal S K (eds) Potassium for Sustainable Crop Production. Proc. Int. Symp. on role of potassium in nutrient management for sustainable crop production in India. Int. Potash Inst. Bern, 347–362.

Voogt W and Sonneveld C 2004. Interactions between nitrate (NO_3) and chloride (Cl) in nutrient solutions for substrate grown tomato. Acta Hort. 644, 359–368.

Voogt W Kromwijk A and Lagas P 2005. De N and P opname van cymbidium. Wageningen uUniversity, Praktijkonderzoek Plant & Omgeving B.V. Business unit Glastuinbouw. Internal Report june 2005, 26 pp.

Voogt W 2009. Lage pH positief, verhoging van NH_4 kan beperkt helpen. Onder Glas 6(1), 62–63.

Ward G M 1967. Growth and nutrient absorption in greenhouse tomato and cucumber. Amer. Soc. Hort. Sci. Proc. 90, 335–341.

Chapter 13
Nutrient Management in Substrate Systems

13.1 Introduction

Speaking about nutrient solutions in soilless cultivation, different solutions can be discerned. Originally, in soilless culture only one nutrient solution was taken into account, being the solution in the containers in which the plants were grown. Such solutions were intensively moved by air bubbling and thus, the composition of the solution in the whole root environment was equal. The root environment was restricted to the container in which the plants were grown and thus, the whole root system of the plant was surrounded by the same nutrient solution. However, this is not the case for hydroponics and substrate systems under practical growing conditions, where great differences occur in time and place within the root environment. The main reason for these differences of salt concentrations between spots within the root environment are the inequality of water supply and water uptake by the crop as discussed in Section 6.3, at the one hand and the lack of movement of the solution within the root environment to equalize them on the other. In Chapter 8 some examples were shown of the inequality of the distribution of nutrients and salts within the root environment of substrate grown plants and the consequences of it on plant development.

Beside the nutrient solution in the root environment, in substrate cultivation other solutions that will be distinguished are: the nutrient solution supplied to the system, the drainage water and the circulating nutrient solution. Last solution exists only in growing systems where drainage water is reused and consists of a mix of the fresh nutrient solution supplied to the system and the drainage water reused in the system.

Clear relationships often exist between the compositions of the different solutions. These relationships depend on factors like the growing system used, the crop grown and the growing conditions. With the management of the nutrient status in substrate systems information about these relationships are valuable to realise optimum conditions for plant development. These optimum conditions will not be focussed solely on maximum production, but also on the quality of the produce. In this chapter the nutrient management for systems used to grow vegetables and cut flowers will be discussed, while those for potted plants and bedding plants will be presented in Chapter 14. Cut flowers and vegetables are mainly grown in bags

C. Sonneveld, W. Voogt, *Plant Nutrition of Greenhouse Crops*,
DOI 10.1007/978-90-481-2532-6_13,

or basins mostly filled with inert material like mineral wool, perlite, expanded clay granules or artificially produced foam, and are watered by drip irrigation. Sometimes, vegetables and cut flowers are grown in hydroponics, with which systems without any substrate are meant. In such systems the nutrient solution in which the plants are grown is circulated with a high speed, like with nutrient film technique (NFT) or intensively bubbled by air, like with deep flow culture (DFT) (Graves, 1983; Jensen and Collins, 1985; Maloupa, 2002; Sakamoto et al., 2001). Potted plants and bedding plants generally are grown in containers mostly filled with an organic substrate of natural origin and mainly irrigated by flooding.

13.2 Technical Equipment of the Growing Systems

The growing systems of which the nutrient management will be discussed in this chapter mostly consist of bags, containers or troughs filled with an inert substrate like defined in the introduction or consist of pre-shaped slabs of substrate wrapped in plastic sheet. The shape of the substrate volume will be very important with respect to the water holding capacity and the air space in the substrate, but is of minor importance with respect to the management of the plant nutrition. With few exceptions, the substrate itself is of secondary importance for the management of the plant nutrition, because the contribution of most substrates to the nutrient supply for the crops grown is negligible. Despite that most substrates are not really inert, especially with respect to micro nutrients, the contributions are mostly restricted in view of the concentrations of these elements required for plant nutrition. Possible contributions are already discussed in Section 11.3.

Picture 13.1 Tomato growing in rock wool slabs

The irrigation of the systems under discussion are generally characterized by a top down stream of the nutrient solution and the difference between the composition of the solution supplied and these drained out is often substantial, especially if the rate of the nutrient solution in the substrate is low. Such differences are important for the management of the nutrient status in the system. The strong movement of the nutrient solution in the hydroponics ensures a more equal distribution of nutrients and residual salts in the root environment. However, in situations of a restrictive nutrient supply even with a high circulation rate of the nutrient solution, great differences can occur between inlet and outlet of the circulating solution. This is shown by an experiment with kohlrabi grown in gullies with circulating water, shown in Table 13.1. At the low concentration of 1.0 dS m^{-1} in the circulating solution the plants strongly suffered from nutrient disorders, from which the N deficiency was most striking. Yield, leaf growth and the NO_3 concentration decreased dramatically with the distance from the inlet. True enough, from the viewpoint of the nutrient management, hydroponics are easier to control than substrate systems, especially when the concentrations of the solution and the flow rate are sufficient to ensure an equal distribution. The control of the nutrition of both systems is based on the composition of the solution in the root environment and thus, there are strong relationships in the management of the mineral nutrition of both systems. Differences between the management in substrate systems and hydroponics mainly are related to the distribution of salts and nutrients as mentioned before, which is mostly more equal in hydroponics than in substrate systems. The consequences of such differences are discussed in Chapter 8 and will be explained further on in the proper sections of present chapter. The discussion in this chapter is mainly focussed on substrate systems, because these systems are most widely used and offers specific opportunities in the management.

The management of the nutrition in substrate systems in which the drainage water flows to waste should be clearly distinguished from that in which the drainage water partly or completely is reused. In Fig. 13.1 both systems are schematically shown; for more detailed information reference is made to Voogt and Sonneveld (1997).

Table 13.1 Fresh bulb and top weights of kohlrabi (g $plant^{-1}$) and the NO_3 concentrations of the tops (mmol kg^{-1} dry matter) grown in a circulation nutrient solution in relation to the distance from the inlet. With the standard supply the EC of the nutrient solution was maintained at 1.5 and at the low supply at 1.0 dS m^{-1}

Distance from inlet in m	Level of supply circulating solution					
	EC 1.5	EC1.0	EC 1.5	EC1.0	EC 1.5	EC1.0
	Bulb weight		*Top weight*		*NO_3 top*	
1.25	320	284	105	91	1298	942
3.75	349	237	104	73	1042	602
6.25	403	170	118	53	982	124
8.75	382	121	113	38	880	28

Data Van den Bos (1997).

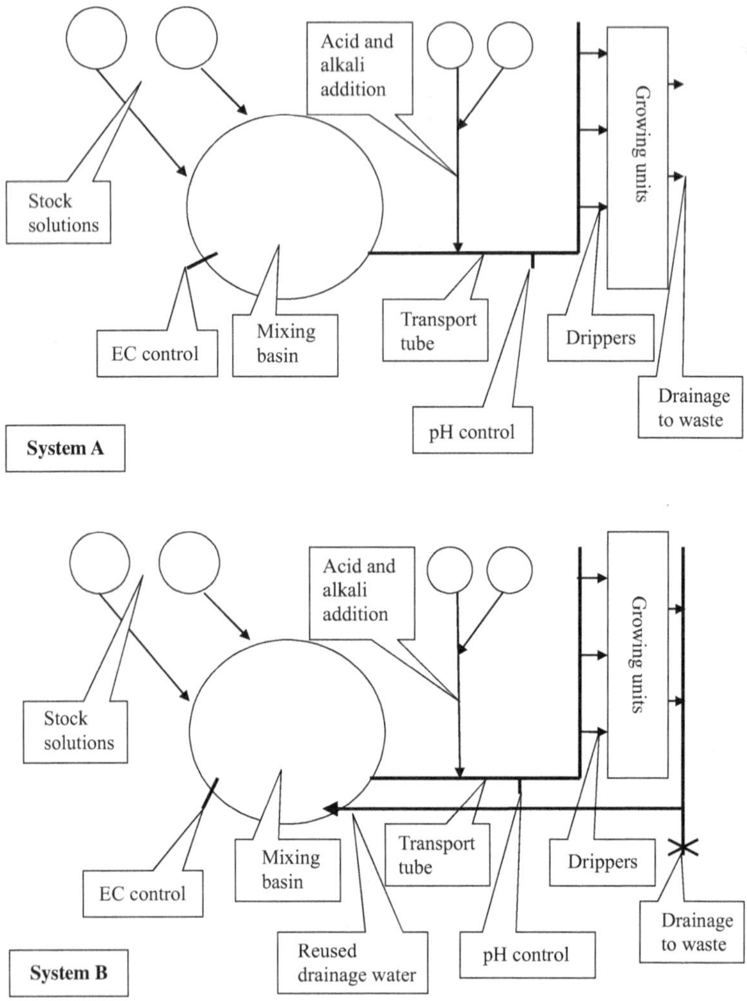

Fig. 13.1 Schemes of growing systems (**A**) without and (**B**) with reuse of drainage water

Different modifications on the outlines presented are possible, but different systems will be suitable under conditions that the systems are able to produce a warranted nutrient solution in the root environment of an exact ionic composition based on a required EC and pH value, in the same degree to every plant grown in the system. Traditional installations for example are equipped with four stock solutions and injectors; two for the nutrients and two for pH control. Modern installations, however, are provided with a series of stock solutions and injectors, for separate injection of all the different fertilizers and chemicals necessary, like shown by Van Os et al. (2002). Such installations are provided with a computer able to calculate the composition of nutrient solutions on every relevant adjustment at the proper time.

Picture 13.2 Rose production in coir bags, in a mobile growing system

The results of these calculations directly guide the injectors to produce the nutrient solution required.

Following subscripts will be used for the EC and the ion concentrations of different solutions operative in this chapter. All concentrations are expressed as mmol l^{-1} and μmol l^{-1} for macro and micro nutrients, respectively.

- *ss*, soil (substrate) solution. Being the concentration in the solution in the root environment. When not specified further on it is the average value of possible different occurring values in the root environment.
- *ad*, addition. The concentrations added to the system, including the nutrients and minerals in the primary water from origin. Not those reused with the drainage water.
- *su*, supply. The concentrations supplied by the irrigation system to the plants in the substrate system. Including those from the reused drainage water.

Picture 13.3 *Hydrangea* cut flowers, grown in containers with perlite

- *dr*, drainage. The concentrations in the drainage water.
- *rw*, raw water. The concentrations in the primary water. Thus before any nutrient is added.
- *up*, uptake. The uptake concentration, being the ratio between a nutrient and the water absorbed by plants.

The preparation of the stock solutions is discussed in Section 12.6. The distribution of the fertilizers in the A and B stock solutions is summarized in Table 13.2. Principally, there is no difference between the results of the traditional system preparing an A and B stock solution and the more advanced system working with

Table 13.2 Fertilizer distribution in two stock solutions, mostly denoted as A and B. In the columns is given which fertilizer is suitable to mix either in one or in both stock solutions

Stock solution A	Stock solution B
Calcium nitrate	Potassium sulphate
Ammonium nitrate	Potassium nitrate
Potassium nitrate	Mono potassium phosphate
	Mono ammonium phosphate
Magnesium nitrate	Magnesium sulphate
Nitric acid	Magnesium nitrate
Iron chelates	Nitric acid
	Phosphoric acid
	Micro nutrients other then iron

separate injectors for all fertilizers. However, with the modern installations adjustments can be instantly realised, while with the traditional systems mostly a quantity of stock solution is in storage in the reservoirs, which have to be used before possible adjustments can be realised.

The difference between both systems in Fig. 13.1 is the possibility that the drainage water in system B, is transported to the mixing reservoir and thus, is reused. In this way a closed system is accomplished. This is preferred with respect to environmental pollution. When the primary water contain too much residual salts, at least a partial drainage to waste is required. This drainage to waste will be restricted as much as possible from viewpoint of environmental pollution. Guidelines to this depend on the crop grown, the water quality and the growing conditions and will be discussed later on.

The equipment used for irrigation depends on the requirements of the growing system. In hydroponics, often the nutrient solution is supplied from one or more inlets per gully or basin and pumped round with a high flow rate. In substrate systems the nutrient solution is supplied with a drip irrigation system or a sprinkler system. Sprinkler systems are used when a crop is grown with such a high plant density that drip irrigation is not an option. When drip irrigation is used every plant need to be supplied with a dripper, to ensure an equal supply of water and nutrients to all plants and to prevent differences that will occur with pH values in the root environment near plants with and without dripper, like shown in Fig. 13.2 (Sonneveld and Voogt, 2001). Such differences especially will occur if the nutrient solution supplied contains NH_4. This ion is absorbed by plants preferentially, when supplied in relatively small quantities of the total N supply. Thus, most NH_4 is absorbed by the plant near the dripper, whereby the pH is lowered in the surrounding root environment of this plant. The other plants received the nutrient solution from which most of the NH_4 is removed, which results in a higher pH in the surrounding root environment of the plants not directly supplied with nutrient solution from the dripper. Perhaps the problem will be merely met, when the dripper is placed such that all plants have an equal position to the dripper. Nevertheless, even when every plant is supplied with a dripper, spatial differences of pH values occur (Voogt et al., 1989). This will be related to site specific uptake of nutrients, especially NH_4, but also to other processes, like root respiration and microbiological activities. See

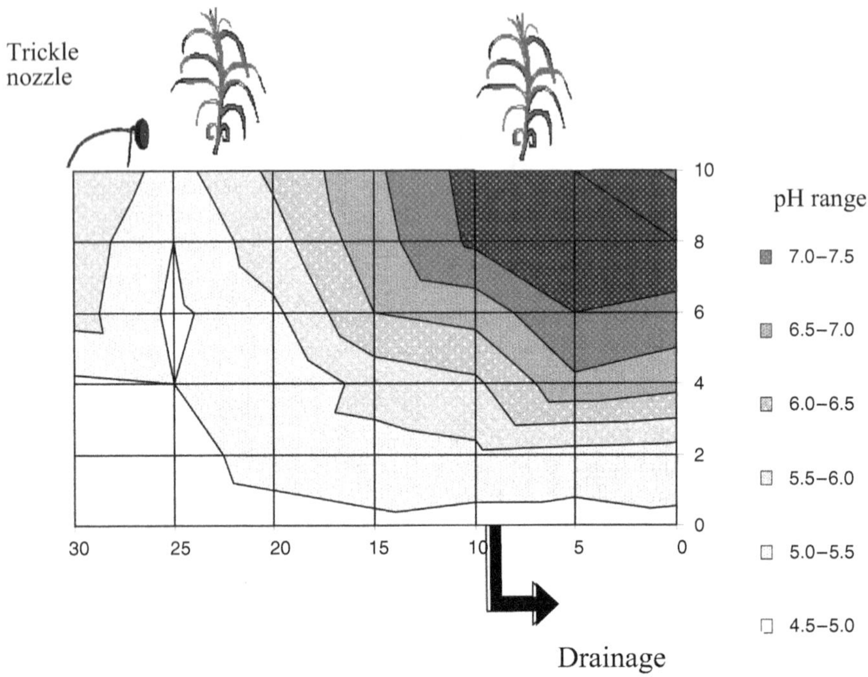

Fig. 13.2 The variation of the pH value in rock wool slabs grown with two rows of carnations. One row was supplied with drip irrigation and the other row got the nutrient solution by water movement only caused by differences in total head. After Sonneveld and Voogt (2001). *Reprinted by permission of the International Society Horticultural Science*

also the discussion in Section 13.4. However, when every plant is supplied with a dripper such differences in the root environment does not cause an unequal plant development.

The velocity of the nutrient solution in the root system of hydroponics differs strongly from this in a substrate system. A high velocity is necessary for the oxygen supply, because merely all the oxygen in such systems will be supplied by the aeration of the nutrient solution. In a NFT system for example the flow rate, the relation between the inlet and the water use, is about 40, while the flow rate in substrate systems varies mostly between $1\frac{1}{4}$ and 2. The majority of the oxygen in the root environment of a substrate system is supplied by air diffusion. Sometimes a high flow rate in substrate systems is unwanted, because of a too high water content in the root environment, which reduces the oxygen availability. Another reason to restrict the flow rate in substrate systems is the environmental pollution when the drainage water is wasted and when the disinfection costs become too high if the water is reused in the system. For many crops the drainage water is disinfected before it is reused to prevent an eventual outbreak of root diseases (Runia, 1995).

A high velocity of the nutrient solution in hydroponics results in an equal distribution of salts and nutrients in the root environment. To a certain extent this is also operative in a substrate system, which can be an advantage from the viewpoint of an equal crop development. However, it can be a disadvantage from viewpoint of water and nutrient efficiency. Partial accumulation nearly does not occur in the root environment in substrate systems with a high flow rate of the nutrient solution. Partial high concentrations of residual salts in substrate systems offers possibilities for high concentrations of residual salts in drainage to waste, which restricts the volume drained off and by this the quantity of nutrients discharged to the environment. With such a partial accumulation the plant reacts due to an osmotic escape, merely on the low concentrated spots in the root environment, being the added solution, see Chapter 8. With a high velocity of the solution in the root environment the accumulation takes place over the whole space, which offers no osmotic escape to the crop when it will be drained off with the same high concentration of the drainage water.

13.3 Fertilization at Start

The composition of the nutrient solution in the root environment recommended at the start of the growing period generally differs distinctly from those recommended for the standard nutrient solution added. The differences can exist of those generally existing between the compositions of the standard solution added and those of solution in the root environment during cultivation (Sonneveld and Straver, 1994), but for some crops specific adjustments are recommended at the start. These differences vary and depend on crops and growing conditions and often exist of a high EC_{ss} value at the start, like shown in Fig. 12.3, but also can concern the ratios between the elements. The most extreme situation occurs for tomato growing on rock wool slabs in The Netherlands, especially under poor light conditions during winter. The high EC_{ss} value in the beginning can reduce the Ca uptake and transport in the plant. Therefore, adjustments of the cation ratios are recommended too at the start, like shown in Table 12.3 and mainly result in a relative increase of the bivalent cation concentrations.

The EC_{ss} maintained in the different growing periods is mainly related to crop growth and fruit quality and is strongly related to the climatic conditions. Starting under poor light conditions, like is the case during winter in Northwest Europe, many crops easily show a lush growth, which often is connected with a thin leaf structure, a poor fruit setting and an insufficient formation of tubers, like found for radish (Sonneveld and Van den Bos, 1995). To escape from this negative growth effects under poor light conditions the osmotic potential in the root environment is lowered by addition of mineral salts. A total concentration of nutrients agreeing with an EC_{ss} between 1.5 and 2.0 is sufficient for optimal plant nutrition of most crops (Sonneveld, 2000). However, this is too low for a good plant condition for many crops at start, especially under poor light. For the start of a tomato crop in The Netherlands in winter time for example EC_{ss} values are maintained between 4 and 8, while with other fruit vegetable crops mostly values between 3 and 6, are practised. But also in summer when the light conditions are optimal, increased EC_{ss} values are applied as well to prevent luxurious vegetative development. True enough, less extreme than in winter, but EC values up to 4 are a normal practise for crops that need growth correction at start.

The level on which and the duration that the high EC_{ss} values are maintained varies depending on the development of the crop and the climatic conditions. Growers often take decisions on that by every day measurements of the EC_{ss} in relation to the crop development. For example, a guide line for the start of a tomato crop under poor light conditions is an EC_{ss} of 8 during the first 4 till 6 weeks after planting. This value is gradually lowered in the following 4 till 6 weeks to a value of about 4. In Fig. 13.3 an example is given of the course of EC_{su} and EC_{dr} in a recirculation system with tomato, from the start in December until October. The drainage water was reused in the system. The average concentration in the root environment will be between those found in the supply and in the drainage water and varies from about 9 in the beginning at winter until 4 in the summer.

High EC values are always gradually realised. A strong instantaneously increase at once is dangerous for the plants, because a reverse osmotic effect can occur which will seriously damages plants. A plant need time to adjust to low osmotic potentials and thus an increase of the EC_{ss} should be restricted to one or two EC units per day.

If drip irrigation is used, the dripper will be placed near the stem of the plant to prevent salt accumulation around the stem. Especially in periods when high EC values are maintained and huge salt accumulations will occur at the top of the substrate. It is experienced that such extreme salt accumulations will damage the stem on the interface of substrate and air and easily promotes stem rot by fungi like pythium or phytophthora.

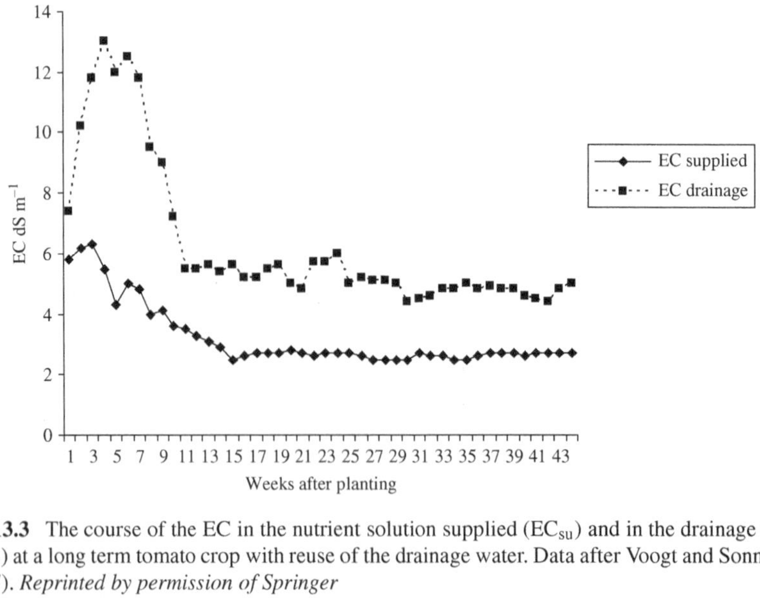

Fig. 13.3 The course of the EC in the nutrient solution supplied (EC_{su}) and in the drainage water (EC_{dr}) at a long term tomato crop with reuse of the drainage water. Data after Voogt and Sonneveld (1997). *Reprinted by permission of Springer*

Restriction of the root volume is helpful with the control of plant development. Therefore, sometimes at the start the root volume is restricted to the propagating cube for a number of weeks. In this case, with planting out the cube is placed on top of the substrate unit on the wrapping material. After a couple of weeks, when the plants are in the right condition, they are placed on their final places on the substrate material. The small volume of the propagating cube in which the plants grow during these first weeks offers excellent possibilities for a precise daily adjustable control on the growth, by a quick adjustment of the water availability and of the level of the EC_{su} based on the results of frequent measurements of EC_{ss}.

When plants are directly sown in the substrate, like done sometimes in propagating cubes, it is not recommended to maintain high EC_{ss} values during germination, because this process easily can be hindered by low osmotic potentials. In such cases the high EC_{ss} values must be realised gradually after the germination stage.

The increased EC_{ss} values can be realised by addition of extra nutrients, while eventual residual salts in the primary water also can be taken into account. With increased EC_{ss} values the uptake or the transport of Ca can be reduced (Sonneveld and Welles, 2005), which can promote Ca deficiency in leaves or fruits (Adams and El-Gizawy, 1986; Ho and Adams, 1994; Savvas and Lenz, 1994; Sonneveld and Van der Burg, 1991). Therefore, a relative high Ca concentration is advisable with such conditions.

The EC_{ss} values mentioned are tailored to values measured in the substrate solution and are operative in case that the substrate solution is analysed. For natural organic substrates often samples of the substrate are analysed by the 1:1½ extract. The values measured in this extract are 35–40% of the values in the soil solution (Sonneveld and Van Elderen, 1994) as a result of the dilution applied with the extraction on the laboratory. The interpretation of the data of such analysis can be adjusted to this dilution by calculating the concentrations in the substrate solution, see Table 4.4. In natural organic substrates a high EC_{ss} at start is realised sometimes by high fertilizer application as a base dressing mixed with the substrate at preparation. However, very high values of EC_{ss} will be realised gradually, to prevent a possible osmotic shock, as already discussed in this section.

13.4 Nutrient Management During Crop Growth

The nutrient management during crop growth will be divided in four sections, covering following subjects: pH control, EC regulation, mutual ratios of the major elements and addition of minor elements.

13.4.1 pH Control

The pH of substrates itself is of minor importance for inert materials. However, substrates with a substantial buffer capacity have a notably contribution to development

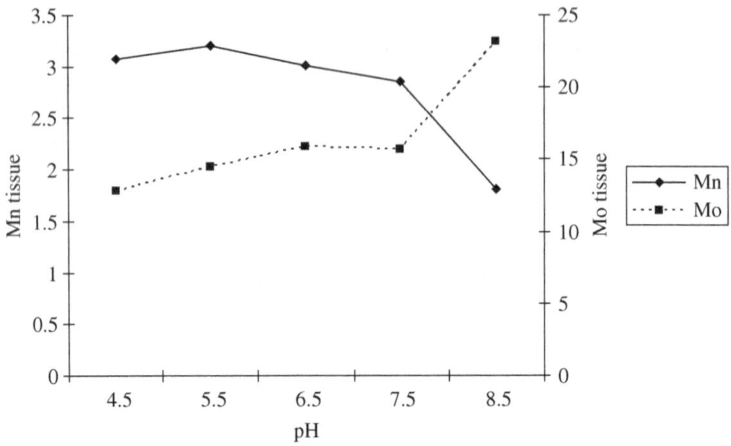

Fig. 13.4 Relationship between pH and the concentrations of Mn (mmol kg^{-1} dry matter) and Mo (μmol kg^{-1} dry matter) of young carnations leaves. Data after Voogt (1995)

of the pH_{ss} during cultivation. For last substrates, a right tuning of the pH to the requirements is important with the preparation, as discussed in Section 11.4. The pH_{ss} affects various factors and need to be regulated within distinct limits. So, it is an important factor in the uptake of micro nutrients, like shown for carnation in Fig. 13.4. The uptake of most micro nutrients is aggravated by a low pH, like shown for Mn. However, the uptake of Mo is decreased by a lower pH. Optimal pH_{ss} values are between 5 and 6 for more or less all crops, but there is a great difference among the sensitivity of crops for high pH values. Gerbera, cucumber and rose for example easily show chlorosis and yield reduction when the pH_{ss} rises above 6. Sometimes with these crops, the micro nutrient concentrations in the plant were reduced, which could be traced as being the cause for these effects. However, this supposition was not always fully confirmed by the data of the tissue analysis (Sonneveld and Voogt, 1994). Nevertheless, a positive yield response was found when crops were grown at pH_{ss} values of about 5.5, in comparison with values of about 6.5. Such effects are clearly detected for cucumber (Sonneveld and Voogt, 1994), gerbera (Sonneveld and Voogt, 1997), carnation (Voogt, 1995) and rose (Voogt and Sonneveld, 2009). Many other crops can survive quite well with values between 6 and 7. Values below 4.5 are not advisable, for either root damage or an increased risk of root diseases. The occurrence of corcky roots with tomato was aggravated by pH values decreasing from a value of 5.5 (Voogt and Paternotte, 1997).

The pH_{ss} in a substrate system varies strongly during cultivation. The small volume of the substrate allows such variations, especially when a substrate is used with a low buffer capacity. Therefore, the variations will be biggest in inert substrates, like rock wool and perlite, while substrates composed with materials like peat and coir material are more stable to pH changes. The buffering capacity of the nutrient solutions used is very small and more or less solely determined by the P concentration (Sonneveld, 2002).

The main reason for the changes of the pH during cultivation is the variation of the difference between the uptake of anions and cations by the crop. Uptake of anions is connected with release of HCO_3 or OH, while uptake of cations is connected with release of H_3O by plants. The exchange of cations and anions is an electrochemical process to keep the total electric charge between plant and external solution unchanged. The exchange occurs in equivalent valences, thus, when the uptake of cations exceeds those of the anions the pH in the root environment decreases and the other way round, the pH will increase. When the N is supplied solely as NO_3, it appears for most crops, that the sum of valences of the absorbed anions is higher than those of the cations and thus, the pH during cultivation generally shows a leaning to increase. As the anion uptake is dominated by the N absorption and given that N can be supplied as anion (NO_3) and as cation (NH_4), the N form plays a prominent part in the regulation of the pH_{ss} value. When a crop is supplied with a nutrient solution in which the N is partly supplied as NH_4, the uptake of cations is stimulated in comparison with a nutrient solution containing the N solely as NO_3 and thus, in the first case pH_{ss} will become lower. Since NH_4 at low concentrations is preferential absorbed by crops, playing around with the addition of NH_4 to nutrient solutions is very effective in the control of the pH_{ss} value.

Logically pH adjustments can be carried out by addition of alkalis and acids. The quantities that can be added are restricted to avoid extreme low or high pH_{su} values. Adjustment of the pH during cultivation by addition of these chemicals only is possible in hydroponics systems when the nutrient solution has a high flow rate. Because of this high flow rate the pH of the circulating solution does not deregulate too much before it passes the control place and it is possible to add frequently small quantities of alkalis or acids for adjustment. However, in substrate systems the rate of the nutrient solution in the root environment is that low, that the pH becomes too high or too low, without possibility to a direct correction by addition of sufficient alkalis or acids. The quantities of these chemicals that can be added to the solution supplied are restricted, because extreme low or high values of pH_{su} will be prevented. Values below 4.5 have risks of damage on roots as mentioned before, but also the matrix of some substrates are easily damaged by such low pH values.

The control of the pH_{ss} during cultivation is carried out by adjustment of the composition of the nutrient solution added, whereby the quantity of NH_4 is regulated in relation to the results of the measurements of the pH_{ss} values. The total N supply is usually not affected, but NO_3 is partly replaced by NH_4. The effect of NH_4 addition on the course of the pH in the root environment is shown in Fig. 13.5 for a rock wool grown cucumber crop. Without NH_4 supply the pH went up to values around 7, while with 15% NH_4-N the pH was lowered below 5.0. With 8% NH_4-N the pH mostly varied between 5.5 and 6.0, which is within the optimal range. However, dependent on the growing conditions the pH will switch during the growing period. This for example will occur when the crop switches from the vegetative to the reproductive stage, like shown for rose in Fig. 13.6. In this example a period of low yield corresponded with the development of young shoots and involved a relative high N uptake. In this period a high addition of H_3O was required to control

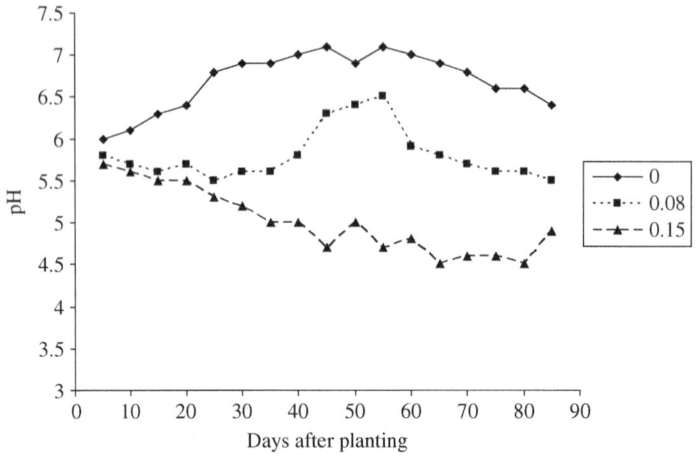

Fig. 13.5 Effect of different NH_4/N ratios in the nutrient solution on the pH in the root environment of a rock wool grown cucumber crop. After Sonneveld (1991). *Reprinted by permission of Springer*

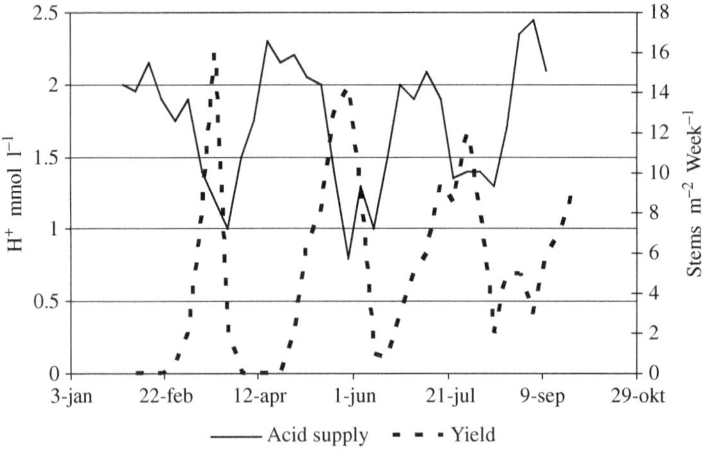

Fig. 13.6 Relationship between the harvested number of stems per week and the H_3O addition to the growing system to control the pH in the root environment. Derived from additional data of Voogt (1994)

the pH_{ss}. The period of high yield corresponds with a strong stem elongation and a relative high K uptake, which is characterized by the requirement of low additions of H_3O.

The quantity of NH_4 necessary to keep the pH on the required level mostly varies between 5 and 10% of the total N uptake. For rose it tends to 25% during the vegetative stage, whilst for example for melon it tends to 0% during fruit development (Sonneveld, 1988).

High concentrations of NH_4 can be toxic to plants. The tolerance depends on crop, cultivar and growing conditions (Barker and Mills, 1980). The concentrations toxic to plants are much higher than those mentioned presently to control the pH and exceed mostly 25% of the total N (Ingestad, 1972; Ikeda and Osawa, 1983; Feigin et al., 1984).

With the concentrations mentioned to control the pH_{ss} often yield increases has been found, as mentioned before. Data from the Glasshouse Crops Research Station at Naaldwijk in The Netherlands showed yield increases for different vegetable and flower crops by addition of NH_4 to nutrient solution used in rock wool cultures up to about 100% (Sonneveld, 2002). It is often suggested that the positive effect of NH_4 application is caused by an improved uptake of micro nutrients (Ikeda and Osawa, 1980) or by a lower energy use in the protein metabolism (Mengel and Kirkby, 1987). However, in a series of trials with cucumber in which different pH_{ss} values were compared realised either by NH_4 addition or by addition of acids or hydroxides showed that the effects on yield and quality as well the effect on nutrient uptake could merely attributed to the realised pH_{ss} (Voogt, 1996). Three successive crops were grown in glass wool and the nutrient solution was circulated frequently to equalize the pH within the root environment. Results of the total yields of these experiments are listed in Table 13.3. Highest yields were gained with the addition of 1.5 mmol l^{-1} NH_4. Equal yield was obtained without NH_4 addition at the lowest pH and by addition of 3 mmol NH_4 at a pH value of 5.5. The lower yield at the other treatments can be explained by possible too high pH values in the rhizosphere of the roots for the combinations 0 NH_4 and pH 5.5 and 6.5, while the combination 3 NH_4 pH 4.5 possible had too low pH in the rhizosphere. The low yield at the combination 3 NH_4 pH 6.5 possibly can be explained by NH_3 toxicity, what easily occur at pH values of about 7.0. The chlorosis was solely affected by the pH level.

Addition of NH_4 as a replacement of NO_3, can reduce the uptake of other cations, like K, Ca and Mg, which can be explained by cation competition of NH_4 and these cations. The proportion of these effects depend on different factors, like crop, growing conditions and the adjustments made in the ionic balance of the nutrients

Table 13.3 Effects of the pH on the total yield of three successive cucumber crops grown at different pH values in the root environment. The pH values were realised by either NH_4 addition mmol l^{-1} or by addition of acids and hydroxides

Treatments		Yield		
pH	NH_4	Fruits m^{-2}	kg m^{-2}	Chlorosis
4.5	0	137 a	51.1 a	5.3
5.5	0	129 b,c	47.3 b,c,d	5.4
6.5	0	126 c	45.8 d	6.0
4.5	1.5	131 a,b,c	48.8 a,b	5.0
5.5	1.5	126 c	49.6 a,b	5.0
6.5	1.5	135 a,b	49.4 a,b	5.7
4.5	3	130 a,b,c	47.5 b,c,d	5.6
5.5	3	124 c,d	48.5 a,b,c	6.0
6.5	3	118 d	45.5 d	6.1

Data of Voogt (1996).

(Conover and Poole, 1986; Feigin et al., 1986; Marti and Mills, 1991; Masui et al., 1982; Sonneveld and Voogt, 1994; Voogt, 1995; Zozorna et al., 1987). Therefore, a careful use of NH_4 is recommended for crops sensitive to Ca deficiency. This especially is true when such crops are grown under climatic conditions that reduce the Ca transport to fruits. Good examples of this are the production of tomato and sweet pepper under dry and hot conditions. Both crops are sensitive to blossom-end rot, caused by Ca deficiency in the fruit, which is stimulated by a hot and dry climate (Adams and Ho, 1993; De Kreij, 1996; Ho and Adams, 1989; Savvas, 2008). Under such conditions every reduction in Ca uptake can becomes dangerous and thus, the use of NH_4 too.

The use of NH_4 in substrate systems with a low circulation rate, like most substrate systems with a low drainage fraction, often lead to local great pH differences. When NH_4 is present in the nutrient solution supplied, lowest pH will be found under irrigation spots, as shown in Fig. 13.7 (Voogt, et al., 1989). This is mainly caused by preferential NH_4 absorption, as discussed in Section 13.2. The NH_4 is relatively strongly absorbed around the irrigation point, which decreases the pH on the spot. The result of the pH measurement in a randomized gathered sample mostly tends to the higher values of the root environment. This can be explained by the accumulation of HCO_3 in the substrate solution. This HCO_3 build-up is part of the CO_2/carbonate equilibrium and becomes substantially with increasing pH values from 5.5. In mixed samples the HCO_3 concentration in part of the sub-samples with a high pH nullifies the H_3O concentration in the other part of the sub-samples with a low pH. In the situation described in Fig. 13.7, average pH values of 4.8, 6.8 and 6.4 were found for the samples taken under drippers, between drippers and the randomized sample, respectively. The high average value in the randomized sample

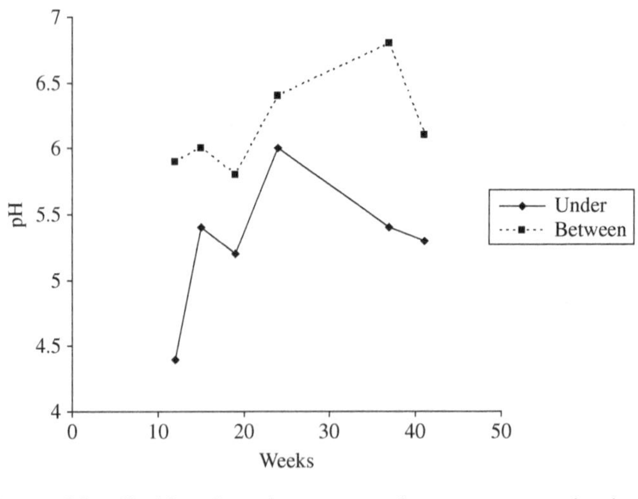

Fig. 13.7 Course of the pH with rock wool grown cucumber crops, measured under and between the drippers of the irrigation system. The results until week 28 are derived from spring grown crops and the others from fall grown crops. (After Voogt et al., 1989)

is more or less misleading. Therefore, separate measurements from spots with high and low pH_{ss} values will be instructive and reflect better the real situation in the root environment. More equal distribution of the pH values can be derived by more movement of the substrate solution by extra water supply. This, however, has also restrictions, as discussed in Section 13.2.

When for pH control acids are used which are not incorporated in the calculations of the nutrient solution, nitric acid is preferred above phosphoric acid, because the effect of such "uncontrolled" addition is relatively much higher for the P balance than for the NO_3 balance, see also Section 12.6.

13.4.2 Control of the EC

The EC_{ss} plays a prominent role in the equilibrium between yield and quality of the harvested produce of many crops grown in substrate and hydroponics. Plants grown in such systems mostly have the disposal over plenty of water at a low matrix tension. This together with the fact that mostly crops in such systems are grown in protected cultivation, which means crop production at a temperature and a humidity higher than in the field. Such growing conditions stimulate a lush growth. In addition, sometimes such cultivations are carried out under poor light conditions, like in the North-West European greenhouse industry in winter. One thing and another promotes the development of crops with a vegetative development and possible a poor quality of the produce.

It has been found, that with a decreasing osmotic potential in the root environment, equivalent with increasing EC_{ss}, growth and quality can be controlled. In this way the management of EC_{ss} has different aspects.

- In the first place it is a measure for the availability of nutrients. When the substrate solution does not contain high concentrations of residual salts, a minimum EC_{ss} is required to supply sufficient nutrients for optimal productions. From this point of view, the EC_{ss} for most crops will be at least between 1 and 4 dS m^{-1} dependent on crop and growing conditions. An exception is orchid which requires a value of 0.6.
- In the second place EC_{ss} is increased above values necessary for maximum productions to control growth and produce quality. Under conditions that plants develop insufficient generative parts a generative development will be stimulated by addition of extra nutrients or by accumulation of residual salts. To control growth and produce quality EC_{ss} values are required between 2 and 10 dS m^{-1}.
- In the third place by the use of saline water or by an unbalanced supply of nutrients EC_{ss} is increased by accumulation of residual salts, which reduces growth and production unnecessary and which can be harmful for the quality when excessive high or low concentrations of nutrients occur. In such cases the measurement of EC_{ss} offers insufficient information and additional information about the nutrient status is urgently required.

Summarized, a systematic measurement of the EC_{ss} during crop production is of great importance to realize a combination of high productions and optimum quality.

The EC required for optimum productions depend on crop and growing conditions. In an extended research (Sonneveld et al., 2004) with 8 different vegetable and flower crops it was found that optimum productions were obtained by EC_{ss} values between 0.5 and 3.0 dS m^{-1}. For a winter grown radish crop even a value of 4.6 was required to get sufficient tuber formation. Under poor light conditions EC_{ss} values higher than required for maximum production to improve the quality of plants of the harvested produce are normal practice for Dutch growers (Sonneveld et al., 1991) and are incorporated in the standard recommendations (De Kreij et al., 1999). But also under normal light conditions improvement of the quality of fruits of cucumber, pepino, strawberry and tomato are reported. A better colour and longer shelf life higher sugar and acid contents and improved taste have been found as mentioned in Section 7.3. Therefore, EC_{ss} values are not always maintained on the level that is connected with the highest growth rate and yield, because plant condition and produce quality also are main items in protected cultivations. High EC_{ss} values are not always favourable for the produce quality. Flower crops for example often show negative quality effects by the reduction of the length and thickness of stems and flower size (Section 7.3). The recommendations for EC_{ss} given by the Glasshouse Crops Research Centre at Naaldwijk to Dutch growers are focussed on the demand on the market considering the relations between yield and quality for the different crops grown. In this consideration sometimes EC_{ss} values are recommended higher than the salinity threshold values, to apply for a better quality. Which mean that small yield reductions have to be accepted to get a required quality of the harvested produces. A summary of EC_{ss} values for important greenhouse crops as recommended to Dutch growers is listed in Table 13.4 (Sonneveld and Straver, 1994). The

Table 13.4 EC_{ss} values as recommended to Dutch growers, after Sonneveld and Straver (1994). The values mentioned under special are maintained under poor light conditions, when a lush growth is expected

Crops	EC_{ss}		Growing medium mainly used
	Standard	Special	
Strawberry	2.4		Peaty substrate
Eggplant	3.0	8.0	Rock wool
Cucumber	3.0	6.0	Rock wool
Sweet pepper	3.0	5.0	Rock wool
Lettuce	2.5		Circulating water
Tomato	4.0	8.0	Rock wool, circulating water
Carnation	2.5		Rock wool
Anthurium	1.0		Rock wool-, foam-, course organic granules
Cymbidum	0.8		Rock wool-, foam-, course organic granules
Gerbera	2.2		Rock wool
Rose	2.2		Rock wool, expanded clay

values mentioned under "Standard" are often maintained during the main growing period. Those mentioned under "Special" are practiced at start or under poor light conditions or a combination of both factors. Beside the average EC_{SS} values as mentioned in Table 13.4, interpretations on basis of high and low values in the root environment according Chapter 8 will be considered.

When EC_{SS} is increased above the values necessary for plant nutrition and no residual salts are accumulated, often nutrient elements are used to that purpose. Mostly, all nutrients, except P and NH_4, are increased with equal ratios. In this way the absorption of nutrients by the crops are hardly affected (Sonneveld and Welles, 2005; Sonneveld and Voogt, 2008). The application of extra P will be excluded from the EC increase by addition of nutrients, because some plants have insufficient control for P absorption and absorb toxic quantities (Howell and Bernhard, 1961). High P also can induce micro nutrient deficiencies, especially Zn (Robson and Pitman, 1983). NH_4 will be excluded also from this increase, because of its effect on the pH_{SS} value like explained in section 13.4.1. Micro nutrients do not affect the osmotic potential, and thus extra application with increasing EC_{SS} has no sense.

For crops sensitive to Ca deficiency, like some vegetable fruit crops, a relatively higher Ca concentration will be supplied with high EC_{SS} values. These crops showed a reduced Ca uptake with increasing concentrations, while the K uptake is aggravated (Sonneveld and Welles, 2005).

When the irrigation water contains sufficient residual salts to increase the EC, these salts will be utilized to increase EC_{SS} above the level of nutrients necessary for optimum plant nutrition. When crops do not show specific salinity effects, residual salts have comparable osmotic effects as the addition of extra nutrients (Sonneveld and Van der Burg, 1991; Sonneveld et al., 1999). See also the discussion about this subject in Section 7.7.

13.4.3 Mutual Ratios of Major Elements

From former paragraphs it will be clear that not only the absolute concentration of a nutrient determines the uptake by the crop, because the mutual ratios of nutrients in the root environment are mostly more important than the absolute concentrations. This is shown by the data presented in Table 13.5 (Sonneveld and Welles, 2005) in which is shown that the uptake of cations is much more affected by their mutual ratios than by the absolute concentrations. The cation concentration in young leaves of cucumber and sweet pepper are slightly affected by increasing EC_{SS}, despite strongly increasing concentrations of the cations in the root environment. The Ca and Mg concentrations in the dry matter of the plant even can be reduced with increasing EC_{SS}, while those of K only was slightly increased in comparison with changes that occur when the mutual ratios of the cations in the root environment were changed at equal EC_{SS} values.

The effects of an increased uptake of K and a suppressed uptake of Ca and Mg with increasing EC_{SS} by nutrients like found with vegetable fruit crops is not evidently for all crops. In a series of experiments with other crops such an effect has

Table 13.5 Ratios between highest and lowest K_{ss} Ca_{ss} and Mg_{ss} concentrations (mmol l^{-1}) and in the young fully grown leaves (mmol kg^{-1} dry matter) of rock wool grown cucumber and sweet pepper crops, for results of experiments with an equal EC, but different cation ratios (Sonneveld and Voogt, 1985) and those of experiments with different EC values, but equal cation ratios (Sonneveld and Welles, 2005). *Modified by permission of the International Society Horticultural Science*

	Cucumber			Sweet pepper		
Highest/lowest concentrations	K	Ca	Mg	K	Ca	Mg
	Experiments with equal EC, but different cation ratios					
Ratios in slabs	2.98	1.90	3.36	2.21	1.50	2.22
Ratios in leaves	1.50	1.06	2.08	1.17	1.10	2.22
	Experiments with different EC, but equal cation ratios					
Ratios in slabs (spring crop)	6.15	3.53	3.42	8.19	4.61	5.00
Ratios in leaves	1.26	0.83	0.85	1.10	0.83	0.86
Ratios in slabs (fall crop)	4.31	3.17	2.94	6.48	3.31	3.47
Ratios in leaves	1.07	0.95	1.19	1.11	0.92	0.95

not been found (Sonneveld and Voogt, 2008). Indeed, for lily extra K absorption was found with increasing EC_{SS}. However, this effect could be traced as an osmotic adjustment by this crop and did not affect the uptake of Ca and Mg.

In Table 13.6 an example is given for a lettuce crop (Van den Bos, 1995). The K concentrations are relatively suppressed in the substrate solution and the Ca and Mg concentrations are relatively increased in comparison with the addition, while this situation in the plant is upside down. This accentuates the preference of plants for the uptake of K. Such effects also have been found with many other crops too (Voogt and Sonneveld, 1997; Sonneveld, 2002). The K/Ca ratio is quite important because of the quality of the produce harvested. A high K/Ca ratio aggravates blossom-end rot, caused by Ca deficiency in fruits like tomato, sweet pepper and eggplant (Voogt, 2002; Bakker et al., 1989). Therefore, the nutrient management

Table 13.6 Cation concentrations in the dry matter (mmol kg^{-1}) of lettuce as affected by the cation concentrations in the substrate solution (mmol l^{-1}). The cation concentrations in the nutrient solution added were increased in equal ratios; the ratios of the valences of the cations were 51:38:11. The figures in brackets reflect the percentages of the valences of $C^{+(1)}$

K		Ca		Mg	
Root environment	Plant	Root environment	Plant	Root environment	Plant
2.7 (27)	2142 (74)	2.7 (53)	246 (17)	1.0 (20)	134 (9)
5.9 (30)	2400 (74)	4.6 (47)	261 (16)	2.3 (23)	160 (10)
10.1 (32)	2410 (73)	7.2 (45)	271 (16)	3.6 (23)	169 (10)
14.0 (32)	2416 (72)	10.0 (46)	286 (17)	4.8 (22)	176 (11)
21.3 (37)	2394 (71)	12.4 (43)	298 (18)	6.0 (21)	185 (11)
23.7 (36)	2405 (71)	14.4 (44)	312 (18)	6.4 (20)	183 (11)

(1) Sum of valences: K + 2Ca + 2 Mg.
Data Van den Bos (1995).

will be focussed on relative high Ca_{ss} concentrations in relation to the uptake, and opposite for those of K_{SS}. However, too high Ca concentrations in the root environment also can be unfavourable for the quality, it aggravate symptoms like gold speck in tomato, which is connected with high Ca in fruits (De Kreij et al., 1992; Voogt, 2002). The recommendation for the K/Ca mol/mol ratio in the nutrient solution in the root environment for many crops is about 1.0, while Mg on average is applied at a ratio of 0.5 (Sonneveld and Straver, 1994; Sonneveld and Welles, 2005). Crops sensitive to Ca deficiency are supplied by ratio with some extra Ca and crops sensitive to Mg deficiency with some extra Mg.

It has been found that also the concentrations of anions can affect the uptake of cations. The uptake of Ca by tomato was strongly reduced at low P and low Cl (De Kreij, 1996; Voogt and Sonneveld, 2004). Last authors suggested that relatively high P and Cl stimulate the Ca uptake, while high NO_3 and SO_4 reduce it. This, however, was observed with tomato, but has not been proved for other crops.

The uptake of anions is dominated by N, for this element covers 80 till 90% of the total anion uptake with most crops. In substrate systems, N is supplied for the greater part as NO_3, while only small percentages are supplied as NH_4 to stabilize the pH as discussed in section 13.4.1. Plants will absorb sufficient N and P at relatively low concentrations, but there is no sense to maintain such low concentrations in substrate systems, since the available quantity is limited in small root volumes as practiced in substrate cultivation. For optimum productions apparently a certain EC_{ss} is required and there is no reason to maintain a relatively low N or P concentration in the ionic balance at this EC_{ss} value. It only should have advantages with a drain to waste to restrict the environmental pollution. In Fig. 13.8 the effect of increasing P_{ss} concentrations on the yield of cucumbers is shown. Maximum yield

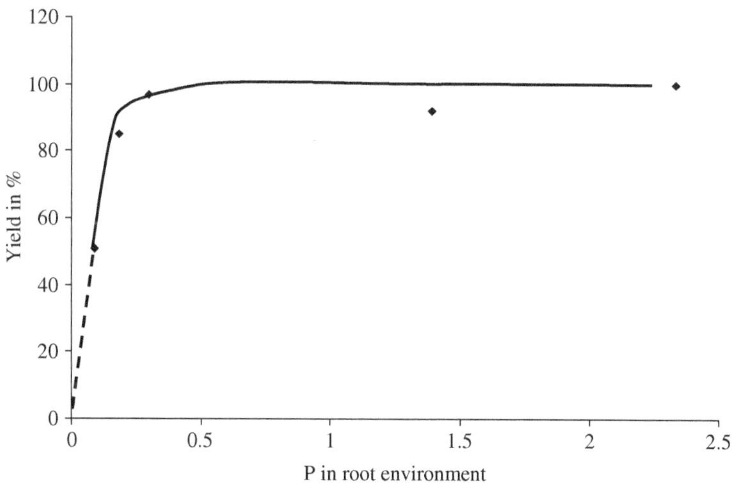

Fig. 13.8 Relationship between the P concentrations in the root environment (mmol l^{-1}) and relative yield (%) of cucumber grown in rock wool. After Sonneveld (1991). *Modified by permission of Springer*

were already obtained at P_{ss} concentrations below 0.5 mmol l^{-1}, which is lower than the uptake concentration of this crop, which is 1.0 mmol l^{-1} (Sonneveld and Voogt, 2001). Thus the concentration in the solution supplied must be much higher, than those necessary in the root environment to provide the crop with sufficient P. The same is true for N, maximum yield will be gained with N concentrations in the solution in the root environment lower than the uptake concentration (Voogt and Sonneveld, 2004).

The NO_3 concentration maintained in the root environment is much higher than those of P and SO_4. In the recommendations to growers about 70% of the valences of anions are occupied by NO_3 (Sonneveld and Straver, 1994). Only with some crops favourably reacting on SO_4, like tomato, relatively somewhat higher concentrations of this ion are maintained. With salinity when Cl accumulates in the root environment lower NO_3 concentrations will be accepted to restrict the rise of EC_{ss}, without negative effects on the N uptake and yield (Voogt and Sonneveld, 2004). Comparable results were gained with rose (Voogt et al., 2006).

With the management of the application of major nutrients in substrate systems, the required EC_{ss} is the leading parameter. Low EC_{ss} easily lead to an insufficient nutrient uptake, while high values reduce growth by too low an osmotic potential. In this context also the accumulation of residual salts has to be taken into account. In Table 13.7 interpretations for analytical data are given with a tomato crop as an example in connection with expected yield and produce quality requirements on the one hand and water quality and growing conditions on the other hand. The interpretations are given for the composition of a substrate solution.

Following considerations are taken into account for the situations listed in Table 13.7. Focussed on maximum yield, an EC_{ss} should be maintained sufficiently below the salinity threshold value, but high enough for optimal nutrient uptake of the crop. The salinity threshold value is estimated on 2.5 (Sonneveld and Van der Burg, 1991), and a value of 2.0 is sufficient to ensure an optimal nutrient uptake. This is

Table 13.7 Interpretations of analytical data in connection with quality requirements and growing conditions. The data are based on concentrations in the substrate solution for tomato. EC_{ss} is expressed as dS m^{-1} and the ion concentration as mmol l^{-1} substrate solution

EC_{ss}	K_{ss}	Na_{ss}	Ca_{ss}	Mg_{ss}	$NO_{3(ss)}$	Cl_{ss}	P_{ss}	$SO_{4(ss)}$
Maximum yield								
2.0	5	0	5	2½	15	0	1	2
Quality production								
4.0	10	0	10	5	30	0	1	4½
Quality production and saline irrigation water								
4.0	5	17	5	2½	15	17	1	2
Quality production, saline irrigation water under hot and dry growing conditions								
4.0[1]	4	17	6	2	15	17	1	2
3.0[2]	4	9	6	2	15	9	1	2

[1]and [2]two options, for explanation see text.

in agreement with the first interpretation option of Table 13.7. However, high quality production requires a higher EC_{SS}=4.0, which is in agreement with the second option. When saline irrigation water with high NaCl concentrations is used, part of the decreased osmotic potential can be realised by accumulation of NaCl, which introduces the third interpretation. Finally, when the crop is grown under hot and dry conditions, easily blossom-end rot can occur. This suggests a relative increase of Ca_{SS} and a maximum acceptable reduction of other nutrient cations. Another option in such case is a reduction of EC_{SS}, which also will reduce the risk on the occurrence of blossom-end rot. With this option the salt accumulation is reduced and the drain to waste increases followed by extra environmental pollution of nutrients in the water drained out to waste.

Data for interpretations as standardized for different crops are presented in Appendix C. The interpretation is focussed on prevailing climate conditions in North-West Europe and on the use of primary water with low salt contents. For adjustments to different growing conditions reference is made to the present section.

13.4.4 Micro Nutrient Concentrations

The application of micro nutrients in substrate systems is on the one hand less complicated than those of the macro nutrients. This is due to the fact that with the calculations of the fertilisers only one element will be taken into account. Generally, the accompanying ion of the micro element in question is the ion of a macro element, the concentration of which can be ignored on the total concentration of that major element in the nutrient solution (Sonneveld, 2002). On the other hand it is more complicated, because of the great number of external factors that affect the availability of micro elements in the root environment and thus on the uptake by crops.

Following factors are mentioned as a being most important with substrate growing (Sonneveld, 2002).

- The quantities and the form in which the micro nutrients are supplied.
- The quantities and the form in which they are present in the substrate and the capability of the substrate to absorb micro nutrients.
- The ability of plants to absorb the micro elements and to adjust the conditions in the rhizosphere to improve the uptake.
- The pH in the substrate and in the rhizosphere.
- Organic compounds available in the root environment.
- Antagonistic effects of other elements.

In handbooks of plant nutrition, a lot of examples will be found about these effects on soil grown crops. Many of these effects are operative in substrate systems too. However, in protected cultivation specific applications of micro nutrients on soils in situ are scarce, because a lot of soils naturally contain sufficient of these elements or they are supplied with irrigation water, manures or fertilizers containing

Table 13.8 Mn uptake of a tomato crop as affected by the pH in the root environment and the Mn application in the nutrient solution added. The tomato was grown in a NFT system and the Mn contents of the laminae are expressed on the dry matter

	pH 7.0			pH 6.0	
Mn added μmol l^{-1}	18	27	36	18	36
Mn in root environment μmol l^{-1}	3	5	5	10	37
Mn in young laminae mmol kg^{-1}	5.4	5.1	6.5	11.6	22.5

After Sonneveld and Voogt (1980).

casually micro elements. This is different from the fertilization programmes for substrates, with which micro nutrient application is common practise as well with substrate preparations for a base dressing as with nutrient solutions for top dressings.

In substrate systems, the uptake of a micro element can be affected more by the growing conditions than by the available quantity of the element at such. This for example is the case with the effect of the pH on the Mn uptake, like shown in Table 13.8. The Mn_{ss} concentration is much more determined by the pH than by the concentration supplied by the nutrient solution added. For Mn this striking effect of the pH is probably due to the presence of Mn oxidising bacteria developing rapidly in the nutrient solution, at high pH (Bromfield, 1978). Comparable data has been found in an experiment with gerbera (Sonneveld and Voogt, 1997) as shown in Fig. 13.9. It is well known that the uptake of micro nutrients mostly is increased by decreasing pH, except Mo the uptake of which is improved by increasing pH (Lucas and Davis, 1961), like already shown in Fig. 13.5. The small rooting volume,

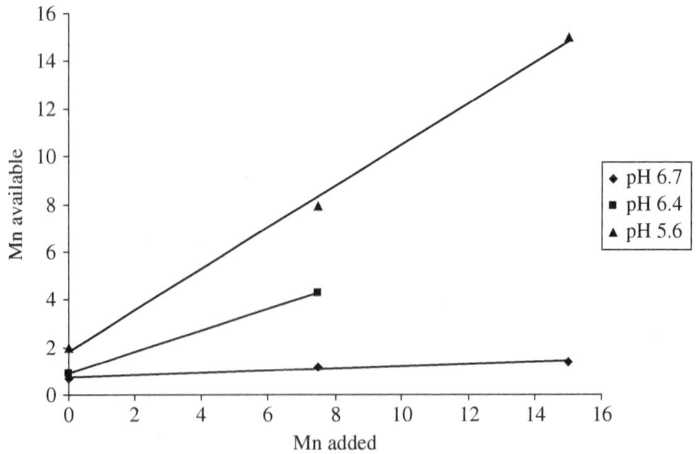

Fig. 13.9 Relationships between the Mn concentration (μmol l^{-1}) in the nutrient solution added and the Mn concentration in the nutrient solution of rock wool slabs at different pH values, during gerbera cultivation (after Sonneveld and Voogt, 1997)

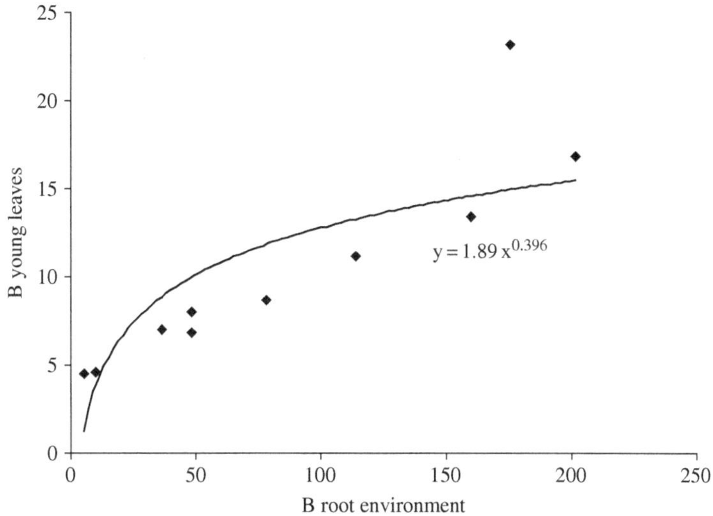

Fig. 13.10 Relationship between the B_{ss} concentrations in rock wool slabs (μmol l^{-1}) and those in the young leaves of cucumber (mmol kg^{-1}). After Sonneveld and De Bes (1984)

intensive fertiliser applications and possibilities to control the pH_{ss} promote strong fluctuation of micro nutrient concentrations despite equal additions (Sonneveld, 1993; Sonneveld and Voogt, 2001) on the one hand, but possibilities for a rapid adjustment on the other hand.

The relationships found between the concentrations of micro nutrients in the root environment and the absorptions by crops is often curve linear, like shown for B with cucumber in Fig. 13.10. For cucumber exponential functions fitted best and exponents below 1 were found, which means that the uptake relatively decreases with increasing concentration (Sonneveld and De Bes, 1984). Thus, cucumber has some control over the uptake, which gives plants some protection from too high uptakes and by this from toxicity, but it surely does not ensure plants free from risks of toxicity. Reactions on high concentration of micro elements can differ among plant types, cultivars and even plant parts, like found with tomatoes, where in laminae over a wide range of application a more or less stable Zn uptake was found. Thus, the exponent found for the function adjusted was low. For the petioles, however, an exponent above 1 was found (Sonneveld et al., 1986). This indicates a preferential accumulation of Zn in the petioles of tomato leaves with increasing concentrations in the root environment, like shown in Fig. 13.11. Thus, too high applications or too strong accumulations of micro nutrients in some plant tissues surely will induce toxicity in plants by high uptakes and therefore a well controlled application is necessary in connection with sampling and analysis of the solution in the root environment to check unwanted accumulations and depletions.

Fe in substrate systems is applied solely in the form of artificial produced Fe chelates. Different types are available in the horticulture industry and the availability of the Fe depends on the pH and the type of the chelate on which the Fe is

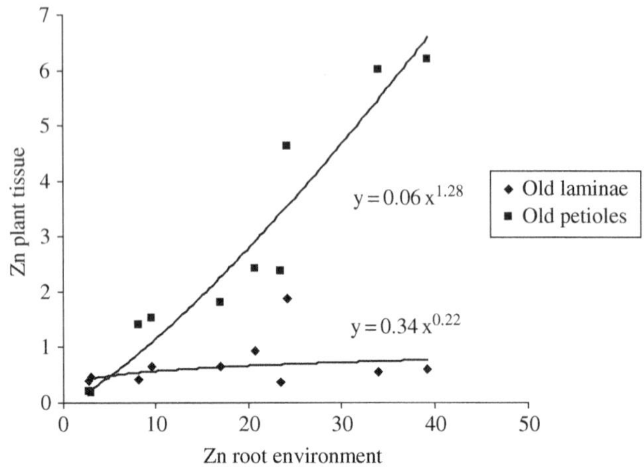

Fig. 13.11 Relationships between the Zn concentration in the root environment (μmol l^{-1}) of rock wool slabs and the Zn concentration of different tissues parts of tomato leaves (mmol kg^{-1} dry matter). After Sonneveld et al. (1986)

bound, like shown in Fig. 13.12 (Lindsay et al., 1967). EDTA and DTPA are suitable until a pH value in the root environment of 6.5 and 7.5 respectively, while EDDHA is suitable until much higher pH values. The measurements as shown in Fig. 13.12 are carried out in the presence of 2.5 mmol l^{-1} Ca. However, with the use of DTPA above a pH value of 6.5 and sufficient Zn in the solution the Fe on the chelate can be replaced by Zn (Lindsay and Norvell, 1969). Under these conditions the Zn is such strongly bound by the DTPA complex, that plants are not able to absorb it (Sonneveld and Voogt, 2001). Cu more or less showed the same behaviour with DTPA at high pH. Thus, a combination of DTPA, high pH and high Zn in

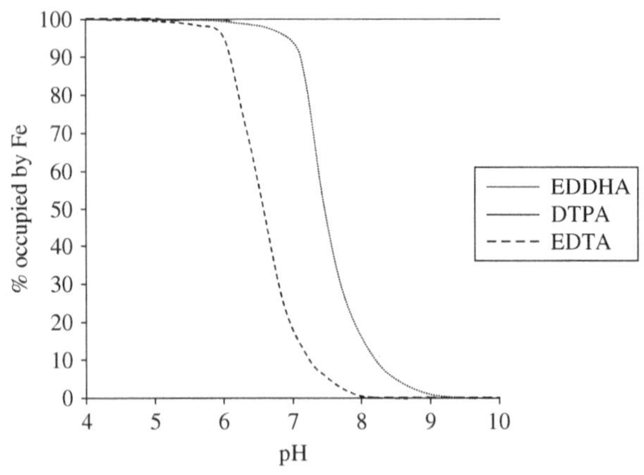

Fig. 13.12 The percent of different chelates occupied by Fe in relation to the pH in a solution of 2.5 mmol l^{-1} Ca. After Lindsay et al. (1967). *Modified by permission of International Union of Soil Science*

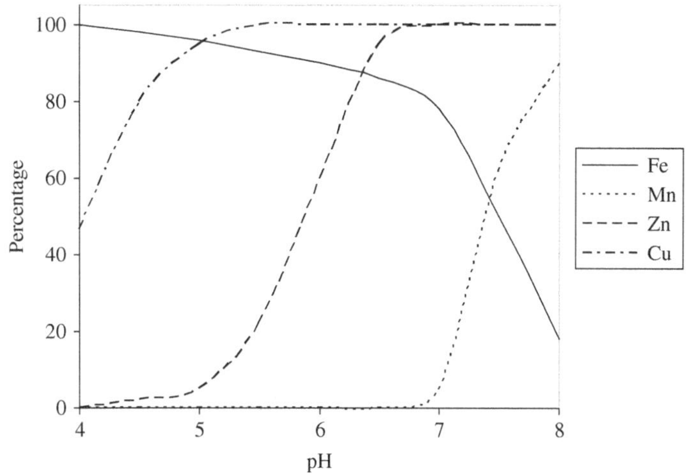

Fig. 13.13 The percentage metal chelated by DTPA in a mixed nutrient solution. After Voogt and Sonneveld (2009)

the solution can induce Fe deficiency, because of an insufficient availability of Fe. Moreover, with the use of DTPA at high pH, Zn or Cu deficiency in plants can occur, while sufficient concentrations of both elements are measured in the substrate solution, because the Zn-DTPA and Cu-DTPA complexes are soluble, but not available (Voogt and Sonneveld, 2009). In Fig. 13.13 the behaviour of different micro nutrient metals are shown in a mixed nutrient solution in relation to the pH, as calculated by Reichwein (2007) using the MINEQL programme of AKZO (Voogt and Sonneveld, 2009).

Thus, during cultivation the Fe_{ss}, Zn_{ss} and Cu_{ss} concentrations will be related to the chelate complex in relation with the pH to estimate the uptake. Below a pH value of 6.0 in the root environment all chelate complexes mentioned are suitable. In case that high pH values become too high in the root environment and cannot be sufficiently controlled, often EDDHA is used. This statement is merely based on widespread common practice than it is proved by research (Voogt and Sonneveld, 2009). The Fe uptake is enhanced by the use of EDDHA when compared with DTPA, However, a sufficient uptake of other micro nutrients is not ensured. Thus, a well controlled pH is preferable. Moreover, Fe EDDHA is more expensive than both other chelates. Furthermore, it should be noticed that EDDHA has three isomers, from which the ortho-ortho form is sufficiently stable to be used in nutrient solutions (Garcia-Mina et al., 2003).

Micro nutrients can be present in the substrate or are sometimes supplied as a base dressing to the substrate in much greater quantities than adsorbed by plants during the growing period (Sonneveld, 2002). However, these nutrients are not always available to plants. Micro nutrients on the one hand can be released from substrates, but on the other hand a strong immobilisation of applied micro nutrients is also possible. Cu, for example, can be strongly bound by organic compounds of peat substrates (Verloo, 1980). In these substrates Cu deficiency can occur in plants, despite

Table 13.9 Approximate concentrations of micro nutrients in nutrient solutions supplied in substrate systems with inert substrates. The concentrations are expressed as $\mu mol\ l^{-1}$

Elements	Range of application
Fe	15–25
Mn	5–10
Zn	3–5
B	20–30
Cu	0.5–0.75
Mo	0.5

After Sonneveld (2002). *Reprinted after permission of Embryo Publications*

high concentrations analysed in the material. The Cu is such strongly bound on the organic compounds that plants cannot adsorb it and extra supply of Cu is necessary. Firstly, as a base dressing to the substrate and later on during cultivation in the nutrient solution supplied (Verhagen, 1992; Roelofs and Van Emmerik, 1992). In an experiment with cucumbers grown in peat substrate concentrations up to 15 $\mu mol\ l^{-1}$ in the substrate solution were required to supply the crop sufficiently with Cu (De Kreij et al., 1993). Boertje (1982) concluded that the Cu concentration in the 1:1½ extract for tomato grown in peat substrate will be 1–2 $\mu mol\ l^{-1}$, which is comparable with 3.5–7.0 $\mu mol\ l^{-1}$ in the substrate solution. The great differences found in the Cu requirements for substrate grown crops insists to further studies, with special attention to effects of organic compounds in relation to crops requirements (Sonneveld and Voogt, 2009).

In more or less inert substrates in which the pH is maintained between roughly 5 and 6 the addition of micro nutrients to the nutrient solution supplied varies within the values listed in Table 13.9 (Sonneveld, 2002). Within the given range the supply will be adjusted on crop and growing conditions.

13.5 Interpretation of Analytical Data

During crop growth the uptake of nutrients is not constant, but the quantity and the mutual ratios fluctuate in relation to climatic conditions and growth stage of the plant. Therefore, the composition of the nutrient solution in the root environment will fluctuate. The grower needs to be informed about the changes in the root environment to control them. Partly the measurements can be carried out by the grower himself with the aid of portable apparatus, like for the measurements of pH_{ss} and EC_{ss}. Several times a week a sample will be gathered and by the results of the measurements in these samples, the grower will adjust the pH_{su} and EC_{su}, including the NH_4 application, when the adjustment of the pH_{su} is insufficient to keep the pH_{ss} within the required values. The pH_{su} will not exceed values between 5.0 and 6.0, to prevent damages by too low values and precipitations of minerals by too high values.

Furthermore, during the growing period samples are gathered from the nutrient solution in root environment for determination of the nutrient status. Mostly these determinations are carried out on agricultural laboratories. The frequency of the sampling to this purpose depends on the growing system, the crop grown, the growing conditions and the growth stage. The nutrient status in the root environment is easier unbalanced in closed systems than in a system with drain to waste. With last system eventual discrepancies between supply and uptake of nutrients more or less are washed out. In practice, the frequency varies from weekly to once in four weeks, in a closed system with a fast growing crop and in a drain to waste system in winter season, respectively.

With interpretation of the analytical data it is not realistic to focus on a full agreement with the guide values recommended (De Kreij et al., 1999; Sonneveld and Straver, 1994). Plants accept certain deviations before any serious problem will occur by nutrient disorders. An example of the deviations accepted under Dutch growing conditions is shown in Table 13.10 (De Kreij et al., 1997; 1999). As shown, relatively big deviations are acceptable. This is understandable, because when the analytical data are within the limits, standard nutrient concentrations are continuously supplied with the irrigation water and available to the crop. When the analytical data of the nutrient solution from the root environment are beyond the limits, adjustments will be considered. Seldom, these adjustments lead to an omission of nutrient elements in the supplied solution, which easily induce a depletion of nutrients in the root environment. Especially, for elements with which the ratio between the concentration in the root environment and the uptake is narrow, like N, K and P. In Table 13.11 limits for the adjustments on basis of the standard nutrient solution are listed (Sonneveld, 2004).

Adjustments during cultivation can be related to growth stages distinguished in plant development, like heavy fruit bearing or a massive development of the stems of cut flowers. When such periods can be well defined, adjustments of the nutrient solution applied can be carried out in anticipation of the crop development, see the

Table 13.10 Guide values and limits for nutrients in the root environment for a tomato crop grown in rock wool

Determination	Guide values	Limits
EC dS m^{-1}	4.0	
pH	5.5	5.0–6.0
NH_4 mmol l^{-1}	<0.1	0.0–0.5
K	8	6.5–10.0
Ca	10	8–12
Mg	4.5	2.7–6.5
NO_3	23	17–28
P	1.0	0.7–2.0
SO_4	6.5	4–9
Fe μmol l^{-1}	15	9–25
Mn	5	3–10
Zn	7	5–10
B	50	35–65
Cu	0.75	0.5–1.5
Mo	0.5	0.3–0.8

After De Kreij et al. (1997).

Table 13.11 Maximum adjustments recommended for standard nutrient solutions when too low or too high values are traced in the root environment

Elements	Limits for adjustment
EC	With too high or too low values in the root environment reduce or increase the EC of the supplied solution, respectively. See also considerations in text
pH and NH_4	pH too low, reduce NH_4, minimum supply 0 mmol l^{-1} pH too high, increase NH_4, maximum until 15% of total N, for short periods until 25%
NO_3, K and P	With too high and too low concentration in the root environment maximum reduction or increase respectively both 30%
Ca, Mg, SO_4 and micro nutrients	With too high and too low concentration in the root environment maximum reduction or increase respectively both 50%

After De Kreij et al. (1999) and Sonneveld (2002).

example in Table 12.3. Another well known example of adjustment in anticipation is the vegetative and generative development of cymbidium. The shoot development in the vegetative growing period requires relatively high nutrient concentrations, while the generative development is promoted by a low EC_{ss} or more in particular by low concentration of N, the addition of which is sometimes completely omitted (De Kreij and Van den Berg, 1990; Van Os 1991). Thus, the EC_{su} can be focussed on such developments.

Special attention will be paid to EC_{ss}, because this parameter determines the yield and the quality. The value can be strongly affected by accumulations of residual salts from the primary water. Well known are accumulations of Na and Cl, being ions often abundantly present in primary water, but scarcely absorbed by the crop. However, also other salts can occur in the primary water in concentrations higher than the uptake concentrations, with accumulation in the root environment as a result. In Chapter 7 indications for interpretation are given, as well the relation with the nutrient status. When a high EC_{ss} is recommended with respect to fruit quality, part of the EC_{ss} can be realised by accumulation of residual salts. However, there are restrictions with respect to specific salinity effects and to minimum required nutrient concentrations to ensure an optimum nutrient supply to the crops. In case of crops specific sensitive to a specific ion, for example Na or Cl, the accumulation of that ion must be restricted to concentrations that initiate specific symptoms. The minimum EC_{ss} value brought about by nutrients is determined in the first place by the easily absorbable nutrients, like N, K and P, because the minimum required concentrations of these nutrients are relative lowest in relation to the uptake. It has been found that plants can absorb sufficient nutrients at very low concentrations (Clement et al., 1978; Ingestad, 1970; Massey and Winsor, 1980; Sidiqi et al., 1998; Voogt and Sonneveld, 2004; Wild et al., 1987). With low nutrient concentrations, however, a high flow rate and intensive monitoring of the chemical composition of the nutrient solution in the root environment are necessary to ensure continuously the availability of sufficient nutrients for optimum productions. Utmost, it happens that low nutrient concentrations in advance does not affect yield, while long term experiments did (Voogt and Sonneveld, 2004). In a long term experiment with a series of different flower and vegetable crops optimum values for EC_{ss} of about 2.5 were found for most crops (Sonneveld et al., 2004). Only some bulb crops showed optimum growth at EC_{ss} of about 1.0 dS m^{-1}. For a radish crop grown in winter a value of 4.5 dS m^{-1} seemed to be optimal. On basis of these results it will be concluded, that generally it is not wise to strive for lower concentrations of $NO_{3(ss)}$, K_{ss} and P_{ss} of about 5, 4 and 0.5 mmol l^{-1}, respectively. Nutrient solutions with a K concentration of 4 mmol l^{-1} and by ratio adjusted Ca and Mg concentrations easily tend to EC_{ss} values of 1.5 dS m^{-1}. This value has been suggested also as minimum concentration to get maximum crops productions (Sonneveld, 2000; Sonneveld and Welles, 2005). Mostly, values for EC_{ss} recommended to growers (Table 13.4) are higher. See section 13.4.2 and Chapter 7.

When the primary water contains more residual salts than are absorbed by crops, EC_{ss} easily exceed the values recommended by accumulation of these residuals in the root environment. The levels of residual salts that will be accepted depend on the sensitivity of the crop and the possible yield reduction accepted by the grower when EC_{ss} exceed the salinity threshold value. Replacement of nutrient levels by residual salts as discussed in Sections 7.7 and 7.8 will be considered as well as the effects of spatial distribution of salts presented in Chapter 8. In the spatial salt distribution the circulation speed of the nutrient solution will be taken into account too, because the circulation rate affects this distribution. When the crop is grown in a pure hydroponics system, the circulation speed is very high and differences in the spatial distribution can be ignored. On thing and another will be made clear by following examples.

The first example is listed in the top of Table 13.12 and concerns cucumber growing in a substrate system. The parameters on which the calculations are based should be defined by the grower and depend on his considerations on yield, quality and the acceptance of environmental pollution as a result of the drainage to waste. The example in Table 13.12 is based on following requirements: $EC_{ss} = 3$ and $EC_{dr} = 4$. The decision to maintain higher values for EC_{ss} than necessary for optimal production can be made in view of quality requirements. The question in the current situation is the leaching fraction required when the NaCl concentration of the primary water is 4 mmol l^{-1} and no drainage water is reused.

Following formulae are available (McNeal et al., 1979; Sonneveld et al., 1966; Sonneveld, 2000).

$$EC_{ss} \approx \frac{EC_{ad} + EC_{dr}}{2} \tag{13.1}$$

$$EC_{dr} = \frac{EC_{ad} - (1 - LF)EC_{up}}{LF} \tag{13.2}$$

$$EC_{NaCl} = 0.115c_{NaCl} \tag{13.3}$$

In which LF = leaching fraction

EC_{NaCl} = EC caused by the NaCl concentration
c_{NaCl} = the NaCl concentration in mmol l^{-1}
Other denotations as given before (Section 13.2)
The formulae (13.1) and (13.2) can be used also for calculations of all specific ions that occur in substrate solutions

To supply the crop with sufficient nutrients, the EC of the nutrient solution added should be at least 1.5 as pointed before. The EC_{up} nutrients together with Na + Cl will be 1.7 (Sonneveld and Voogt, 2001). The calculations show that under these conditions a leaching fraction is required of 0.13. However, it can happen that a limit is set for the NaCl concentration in the root environment, because the crop is specific sensitive to NaCl. For example, the same crop is grown under the same conditions, but with the restriction $NaCl_{ss}$ = 10 mmol l^{-1}. The results of the calculations are listed in the second part of Table 13.12.

With the calculations presented attention have to be paid on the minimum required nutrient level as discussed before. This level depends on crop and growing conditions, but the required level is not always exactly known. When insufficient information is available, a nutrient level corresponding with EC_{ss} = 1.5 dS m^{-1} mostly is safe for an optimum supply of nutrients to get a maximum production. However, it is experienced that EC_{up} can exceed this concentration, which requires an $EC_{su} > 1.5$. In such cases a frequent measurement of EC_{ss} is necessary. Moreover, when residual salts are accumulated in the root environment additional measurements are required to get a right impression of the nutrient status.

Interpretation of EC_{ss} can be carried out on basis of either the average or highest and lowest value (Chapter 8). On basis of the first method, for a summer grown cucumber crop a salinity threshold value of 2.3 dS m^{-1} and the SYD value of 5.8 has been found (Sonneveld and Van der Burg, 1991).This mean that with an EC value of 3.0 a yield reduction will be expected of 4.1%. Interpretation on basis of

Table 13.12 Calculation of the leaching fraction in a cucumber crop grown in a free drainage system with irrigation water (ir) containing 4 mmol l^{-1} NaCl, $EC_{ss} = 3.0$ and $EC_{dr} = 4.0$

Parameter to calculate	Formulae used	Result
Cucumber $NaCl_{ir}$ = 4 mmol l^{-1}, EC_{ss} = 3.0, EC_{dr} = 4.0, EC_{up} = 1.7, $NaCl_{up}$ =2.0 mmol l^{-1}		
EC_{ad}	13.1	2.0
LF	13.2	0.13
$NaCl_{dr}$	13.2	17.4
Cucumber $NaCl_{ir}$ = 4 mmol l^{-1}, EC_{ss} = 3.0, EC_{dr} = 4.0, EC_{up} = 1.7, $NaCl_{up}$ =2.0 mmol $l^{-1}$and $NaCl_{ss}$ = 10 mmol l^{-1}		
$NaCl_{dr}$	13.1	16 mmol l^{-1}
LF	13.2	0.14
EC_{ss} (nutrients)	13.3	1.8
EC_{ad} (nutrients)	13.1 and 13.2	1.6
EC_{dr} (nutrients)	13.1	2.2

the highest and lowest value requires estimation of the yield reduction on basis of 2.0 dS m^{-1} as lowest value and 4.0 as highest value. Results of Sonneveld and De Kreij (1999) on basis of an unequal distribution of salts indicate no yield reduction with cucumber for such values.

References

Adams P and El-Gizawy A M 1986. Effect of salinity and watering level on the calcium content of tomato fruit. Acta Hort. 190, 253–259.

Adams P and Ho L C 1993. Effects of the environment on the uptake and distribution of calcium in tomato and on the incidence of blossom-end rot. Plant Soil 154, 127–132.

Bakker S Welles G W H Janse J De Kreij C and Buitelaar K 1989. The effect of air humidity in combination with the composition of the nutrient solution on yield, the incidence of blossom end rot and keeping quality of aubergine in an autumn crop. Glasshouse crops Research Station Naaldwijk, The Netherlands, Annual Report 1989, 42.

Barker A V and Mills H A 1980. Ammonium and nitrate nutrition of horticultural crops. Hort. Rev. 2, 395–423.

Boertje G A 1982. Spoorelementen bij tomaten op veen. Tuinderij 62(13), 46–49.

Bromfield S M 1978. The effect of manganese oxidising bacteria and pH on the availability of manganous ions and manganese oxides to oats in nutrient solutions. Plant Soil 49, 23–39.

Clement C R Hopper M J and Jones L H P 1978. The uptake of nitrate by *Lolium perenne* from flowing nutrient solution. J. Exp. Bot. 29, 453–464.

Conover C A and Poole R T 1986. Nitrogen source effects on growth and tissue content of selected foliage plants. HortSci. 21, 1008–1009.

De Kreij C and Van den Berg 1990. Effect of electrical conductivity of the nutrient solution and fertilization regime on spike production and quality of *Cymbidium*. Scientia Hort. 44, 293–300.

De Kreij C Janse J Van Goor B J and Van Doesburg J D J 1992. The incidence of calcium oxalate crystals in fruit walls of tomato (*Lycopersicon esculentum Mill.*) as affected by humidity, phosphate and calcium supply. J. Hort. Sci. 67, 45–50.

De Kreij C Martignon G and Van Elderen C W 1993. Comparison of water, DTPA, and nitric acid as extractants to assess the availability of copper in peat substrates. Comm. Soil Sci. Plant Anal. 24, 227–236.

De Kreij C 1996. Interactive effects of air humidity, calcium and phosphate on blossom-end rot, leaf deformation, productivity and nutrient contents of tomato. J. Plant Nutr. 19, 361–377.

De Kreij C Voogt W Van den Bos A L and Baas R 1997. Voedingsoplossingen voor de teelt van tomaat in gesloten teeltsystemen. Proefstation voor Tuinbouw onder Glas te Naaldwijk, The Netherlands, Brochure VG 2, 21 pp.

De Kreij C Voogt W Van den Bos A L and Baas R 1999. Bemestingsadviesbasis Substraten. Proefstation voor Bloemisterij en Glasgroente Naaldwijk The Netherlands, 145 pp.

Feigin A Ginzburg C Ackerman A and Gilead S 1984. Response of roses growing in volcanic rock substrate to different NH_4/NO_3 ratios in the nutrient solution. *In*: Proc. 6th Internat. Congr. Soilless Culture, Lunteren, 1984 ISOSC, Wageningen, The Netherlands, 207–214.

Feigin A Ginzburg C Gilead S and Ackerman A 1986. Effect of NH_4/NO_3 ratio in nutrient solution on growth and yield of greenhouse roses. Acta Hort. 189, 127–135.

Garcia-Mina J Cantera RG Zamarreno A 2003. Interaction of different iron chelates with an alkaline and calcareous soil: a complementary methodology to evaluate the performance of iron compounds in the correction of iron chlorosis. J. Pl. Nutr. 26, 1943–1954.

Graves C J 1983. The nutrient film technique. Hort. Rev. 5: 1–44.

Ho L C and Adams P 1989. Calcium deficiency – a matter of inadequate transport to rapidly growing organs. Plants Today 2, 202–207.

Ho L C and Adams P 1994. Regulation of partitioning of dry matter and calcium in relation to fruit growth and salinity. Ann. Bot. 539–545.

Howell W and Bernhard R L 1961. Phosphorus response of soybean varieties. Crop Sci. 1, 311–313.

Ingestad T 1970. A definition of optimum nutrient requirements in birch seedling I. Physiol. Plant. 23, 1127–1138.

Ingestad T 1972. Mineral nutrition requirements of cucumber seedlings. Plant Physiol. 52, 332–338.

Ikeda H and Osawa T 1980. Comparison of adaptability to nitrogen source among vegetable crops. II. Growth response and accumulation of ammonium and nitrate nitrogen by leafy vegetables cultured in nutrient solution containing nitrate, ammonium and nitrite as nitrogen source. J. Japan. Soc. Hort. Sci. 48, 435–442

Ikeda H and Osawa T 1983. Effects of ratios of NO_3 to NH_4 and concentrations of N source in the nutrient solution on growth and leaf N constituents of vegetable crops and solution pH. J. Japan. Soc. Hort. Sci. 52, 159–166.

Jensen M H and Collins W L 1985. Hydroponic vegetable production. Hort. Rev. 7, 483–558.

Lindsay W L Hodgson J F and Norvell W A 1967. The physico-chemical equilibrium of metal chelates in soils and their influence on the availability of micro nutrient cations. Intern. Soc. Soil Sci. (Scotland 1966) Trans. Comm. II and IV, 305–316.

Lindsay W L and Norvell W A 1969. Equilibrium relationships of Zn^{2+}, Fe^{3+}, Ca^{2+}, and H^+ with EDTA and DTPA in soils. Soil Sci. Soc. Amer. Proc. 33, 62–68.

Lucas R E and Davis J F 1961. Relationship between pH values of organic soils and availabilities of 12 plant nutrients. Soil Sci. 92, 177–182.

Maloupa E 2002. Hydroponic systems. *In*: Savvas D and Passam H (eds) Hydroponic Production of Vegetables and Ornamentals. Embryo Publications, Athens, 143–176.

Marti H R and Mills H A 1991. Nutrient uptake and yield of sweet peppers as affected by stage of development and N form. J. Planr Nutr. 14, 1165–1175.

Masui M Nukaya A and Ishida A 1982. Effects of nitrogen form on growth of muskmelons. J. Japan. Soc. Hort. Sci. 50, 475–480.

Massey D and Winsor G W 1980. Some response of tomato to nitrogen in recirculation solutions. Acta Hort. 98, 127–137.

McNeal B L Oster J D and Hatcher J T 1979. Calculation of the electrical conductivity from solution composition data as an aid to in situ estimation of soil salinity. Soil Sci. 110, 405–414.

Mengel K and Kirkby E A 1987. Principles of plant nutrition. Int. Potash Inst. Bern, 4th edition 687 pp.

Reichwein AM 2007. Model calculations micro element stability in relation to chelate type and environmental conditions, using MINEQL+. (personal communication)

Roelofs Th and Van Emmerik P 1992. Teelt op veen vraagt extra aandacht voor element koper. Vakblad Bloemisterij 47(29), 55.

Robson A D and Pitman M G 1983. Interactions between nutrients in higher plants. *In*: Läuchli A and Bieleski R L (eds), Inorganis Plant Nutrition, Encyclopedia of Plant Physiol. (New series) 15a, Springer-Verlag, Berlin, 147–180.

Runia W 1995. A review of possibilities for disinfection of recirculation water from soilless cultures. Acta Hort. 382, 221–229.

Sakamoto Y Watanabe S Okano K 2001. Growth and Quality of chrysanthemum (Dendranthema grandiflora) grown in wet sheet culture and deep flow technique. Acta Hort. 548, 459–467.

Savvas D and Lenz F 1994. Influence of salinity on the incidence of the physiological disorder "internal rot" in hydroponical-grown eggplants. Angew. Bot. 68, 32–35.

Savvas D Ntatsi G Passam H C 2008. Plant nutrition and physiological disorders in greenhouse grown tomato, sweet pepper and eggplant. Europ. J. Plant Sci. Biotechn. 2, 45–61.

Sidiqi M Y Kronzucker H J Britto D T and Glass A D M 1998. Growth of a tomato crop at reduced nutrient concentrations as a strategy to limit eutrophication. J. Plant Nutr. 21, 1879–1895.

Sonneveld C Koornneef P and Van den Ende J 1966. De osmotische druk en het electrische geleidingsvermogen van enkele zoutoplossingen. Meded. Dir. Tuinb. 29, 471–474.

Sonneveld C and Voogt S J 1980. The application of manganese in nutrient solutions for tomato grown in a recirculating system. Acta Hort. 98, 171–178.

Sonneveld C and De Bes S S 1984. Micro nutrient uptake of glasshouse cucumbers grown on rockwool. Comm. Soil Sci. Plant Anal. 15, 519–535.

Sonneveld C and Voogt W 1985. Growth and cation absorption of some fruit-vegetable crops grown on rockwool as affected by different cation ratios in the nutrient solution. J. Plant Nutr. 8, 585–602.

Sonneveld C De Bes S S and Voogt W 1986. Zinc uptake and distribution in tomatoes grown in rockwool. Soilless Culture 2 no 2, 49–60.

Sonneveld C 1988. Kali-Calciumverhoudingen bij meloen in steenwol. Proefstation voor Tuinbouw onder Glas Naaldwijk, Intern Rapport no 18, 13 pp.

Sonneveld C 1991. Rockwool as a substrate for greenhouse crops. *In*: Bajaj Y P S (ed) Biotechnology in Agriculture and Forestry 17, High-Tech and Micro-propagation I, Springer-Verlag, Berlin, 285–312.

Sonneveld C and Van der Burg A M M 1991. Sodium chloride salinity in fruit vegetable crops in soilless culture. Neth. J. Agric. Sci. 39, 115–122.

Sonneveld C Van den Bos A L Van der Burg A M M and Voogt W 1991. Fertigation in the greenhouse industry in The Netherlands. *In*: Fertigation/Chemigation, FAO, Rome 1991, 186–193.

Sonneveld C 1993. An overview of nutrition in hydrponics. *In*: Hanger B and Laffer B, Hydroponics and the Environment, Proc. Australian Hydroponic Conference Monash University Melbourne, Australia 1993, 21–36.

Sonneveld C and Straver N 1994. Nutrient solutions for Vegetables and Flowers grown in water or substrates. Research Station for Floriculture and Glasshouse Vegetables, Aalsmeer/Naaldwijk, The Netherlands, Series: Voedingsoplossingen Glastuinbouw 8, 45 pp.

Sonneveld C and Van Elderen C W 1994. Chemical analysis of peaty growing media by means of water extraction. Comm. Soil Sci. Plant Anal. 25, 3199–3208.

Sonneveld C and Voogt W 1994. Effects of calcium and ammonium on the appearance of secondary shoot chlorosis in rockwool grown cucumbers. Proc. Sino Intern. Coll. Soilless Culture, Hangzou 1994, 88–94.

Sonneveld C and Van den Bos A L 1995. Effects of nutrient levels on growth and quality of radish (Raphanus sativis L.) grown on different substrates. J. Plant Nutr. 18, 501–513.

Sonneveld C and Voogt W 1997. Effects of pH value and Mn application on yield and nutrient absorption with rockwool grown gerbera. Acta Hort. 450, 139–147.

Sonneveld C and De Kreij C 1999. Response of cucumber (*Cucumis sativis L.*) to an unequal distribution of salts in the root environment. Plant Soil 209, 47–56.

Sonneveld C Baas R Nijssen H M C and De Hoog J 1999. Salt tolerance of flower crops grown in soilless culture. J. Plant Nutr. 22, 1033–1048.

Sonneveld C 2000. Effects of salinity on substrate grown vegetables and ornamentals in greenhouse horticulture. Thesis Wageningen University, Netherlands, 151 pp.

Sonneveld C and Voogt W 2001. Chemical analysis in substrate systems and hydroponics – use and interpretation. Acta Hort. 548, 247–259.

Sonneveld C 2002. Composition of Nutrient solutions. *In*: Savvas D and Passam H (ed) Hydroponic Production of Vegetables and Ornamentals. Embryo Publications, Athens, 179–210.

Sonneveld C Van den Bos A L and Voogt W 2004, Modelling osmotic salinity effects on yield characteristics of substrate grown greenhouse crops. J. Plant Nutrition 27, 1931–1951.

Sonneveld C 2004. Nutrient solutions in substrate culture – composition and use. I Congreso Internacional de Horticultura Intensiva. 2 y 3 deciembre 2004, Centro de la Fundación Ruralcaja Valencia, 82–88.

Sonneveld C and Welles G W H 2005. Cation concentrations of plant tissues of fruit-vegetable crops as affected by the EC of the external nutrient solution and by humidity. Acta Hort. 697, 377–386.

Sonneveld C and Voogt W 2008. Nutrient concentrations of plant tissues of greenhouse crops as affected by the EC of the external nutrient solution. Acta Hort. 779, 313–320.

Sonneveld C and Voogt W 2009. Determination of micro nutrients in substrates by water extraction and interpretation of analytical data. Acta Hort. 819, 87–98.

Van den Bos A L, 1995. EC in relatie tot het type substraat bij de teelt van sla in een gesloten teeltsysteem. Proefstation voor Bloemisterij en Glasgroente Naaldwijk The Netherlands, Intern verslag 4, 22 pp.

Van den Bos A L, 1997. Stikstofaanbod en NO_3-gehalte bij koolrabi. Proefstation voor Bloemisterij en Glasgroente Naaldwijk The Netherlands, Intern verslag 90, 25 pp.

Van Os P C 1991. Lage EC geeft goede kwaliteit en hoge productie. Vakblad Bloemisterij 46(25), 52–53.

Van Os E C Gieling Th H and Ruijs M N A 2002. Equipment for hydroponic installations. *In*: Savvas D and Passam H (eds) Hydroponic Production of Vegetables and Ornamentals. Embryo Publications, Athens, 103–141.

Verhagen J B G M 1992. Ervaringen met de teelt van chrysant op veensubstraat in een gesloten teeltsysteem. Proefstation voor Tuinbouw onder Glas Naaldwijk, The Netherlands, Intern Verslag 67, 16 pp.

Verloo M G 1980. Peat as natural complexing agent for trace elements. Acta Hort. 99, 51–56.

Voogt W Blok C Smulders P and De Visch D 1989. Grote pH verschillen in de steenwolmat. Groenten en Fruit 44(49), 38–39.

Voogt W 1994. pH belangrijke factor productieverhoging bij roos. Vakblad Bloemisterij 49(43), 32–33.

Voogt W 1995. Effect of the pH on rockwool grown carnation (*Dianthus Caryophyllus*) Acta Hort. 401, 327–336.

Voogt W 1996. Komkommer: lage pH positief, effect NH_4 beperkt. Groenten en Fruit/Glasgroenten 6(15), 16–17.

Voogt W and Sonneveld C 1997. Nutrient management in closed growing systems for greenhouse production. *In*: E. Goto et al. (eds) Plant Production in Closed Ecosystems. Kluwer Academic Publishers, Dordrecht, The Netherlands, 83–102.

Voogt W and Paternotte P 1997. Tomaat : lage pH veroorzaakt wortelverkurking. Groenten en Fruit Vakdeel Glasgroenten 7(49), 12–13.

Voogt W 2002. Potassium management of vegetables under intensive growth conditions. *In*: Pasricha N S and Basal S K (eds) Potassium for sustainable crop production. Proc. Int. Symp. on role of potassium in Nutrient Management for sustainable crop production in India. Int. Potash Inst. Bern. 347–362.

Voogt W and Sonneveld C 2004. Interactions between nitrate (NO_3) and chloride (Cl) in nutrient solutions for substrate grown tomato. Acta Hort. 644, 359–368.

Voogt W Garcia N Straver N Van den Burg N 2006. Onderzoek naar de mogelijkheden om rozen te telen met een permanent dan wel tijdelijk lagere N concentratie in het wortelmilieu om de N emissie te verminderen. Wageningen UR Glastuinbouw, Rapport 4161607, 38 pp.

Voogt W and Sonneveld C 2009. The effect of Fe-chelate type and pH on the development of roses. Acta Hort. 819, 411–417.

Wild A Jones L H P and Macduff J H 1987. Uptake of mineral nutrients and crop growth: the use of flowing nutrient solutions. Adv. Agron. 41, 171–219.

Zozorna P Caselles J and Carpena O 1987. Response of pepper plants to NO_3:NH_4 ratio and light intensity. J. Plant Nutr. 10, 773–782.

Chapter 14
Fertigation Management of Potted Plants

14.1 Introduction

The horticultural crops considered in this chapter are characterised by the fact that the plants are grown in a restricted volume, like pots, containers, plastic trays or compressed peat blocks. In the market these crops are recognized as potted plants, bedding plants and container grown nursery stock, mostly for ornamental purposes. Another group is the raising of young vegetable and cut flower plants, due to production holdings. Although extremely diverse, all these plants are grown as single units and this makes the water supply of the plants complicated. Trickle irrigation is generally unsuitable because it is laborious, due to rapid changing crops and plant densities and besides expensive because of the large number of units per area. Exceptions are the plants grown in large containers, like balcony and patio plants, some nursery stock and subtropical trees with a long growing period and a low plant density. Overhead sprinklers induce excessive losses of irrigation water and nutrients, since a considerable amount of the irrigated water falls alongside the pots or drips from the leaf canopy mostly also alongside the pots. Moreover, the water supplied is absorbed per individual plant or plant batch, which strongly enhances the variation in the water and nutrient status of such units. To avoid these problems, potted plants already have been grown for many years on concrete floors or on container benches with sub-irrigation. Nowadays, in all modern greenhouses this is the common growing system. In contrast with vegetable and cut flower nurseries, potted plant nurseries are not always specialised in one crop, but in different plant species. Even if they are specialised in one plant type, different plant stages are present at the same time, like with *Kalanchoë and Chrysanthemum*. A variety of crops creates diversity in nutrient demands and other parameters for the root environment. This complicates a crop-specific nutrient management in greenhouses grown with potted plants, which can also be aggravated by the relative short growing period of some of these crops. Moreover, unlike vegetables and cut-flowers, the research on nutrient demands for many potted plants is limited and the number of species and cultivars grown change rapidly with the constant introduction of novelties. Therefore, the nutrient management is partly based on experience and the complexity of the management induces the need for a general and robust approach.

C. Sonneveld, W. Voogt, *Plant Nutrition of Greenhouse Crops*,
DOI 10.1007/978-90-481-2532-6_14,

Picture 14.1 Potted plant production in a modern greenhouse

In virtually all crops peat or peat based mixes are used as growing medium. The chemical characteristics of potting media and the analytical methods used are extensively discussed in the Chapters 11 and 4 respectively. In this chapter the 1:1½ by volume extract is meant in all cases of nutrient analysis in the potting medium, unless stated otherwise. The rooting volume is very restricted for this category of plants, especially in plant propagation at which it varies from only less then 50 ml per plant for pressed peat blocks to 250 ml for fruit vegetable plants in rock wool blocks. Also for bedding plants the root volume is restricted to less than 100 ml per plant.

Hydroculture plants form a specific group and are mostly grown in expanded clay granules. For the nutrient management these plants are treated as hydroponics. A nutrient solution as generally used for this type of cultivation is added to Appendix C.

Potted plants are typically marketed in the growing medium. Therefore, the nutrient management must not only be focussed on the growing process in the greenhouse but also on the post-harvest phase. The plant must be prepared for the transport, the shopping phase, as well as the stay at the consumer.

The conditions for the propagation of vegetable and cut flower plants are for a great part equal to those of potted- and bedding plants. These nurseries nowadays are often very large and specialised in crop type and virtually all of them grow the plants on concrete floors with ebb and flood irrigation.

Picture 14.2 *Beaucarnea* grown on a concrete floor with ebb and flood irrigation

14.2 Classification

As mentioned, the number of different potted plant species grown is huge and as a consequence the requirements for the nutrient supply are divers and will be met with a classification to groups with reasonable agreement. The species originate from different climatic zones and from places with very different soil or growing medium, which makes a classification on basis of such parameters more or less impossible. Classification by family or genus is also inadequate, since differences between species can be larger within one genus than between genera. For the Dutch situation a simple system is chosen which is applicable in practice and based on a classification in groups with more or less equal demands for the root environment. The system is characterized by three parameters, viz. the nutrient status, the EC level and pH requirements (Straver et al., 1999).

The first parameter concerns the nutrient concentrations and the different nutrient ratios in the root environment. The differences of last item mainly exist in the K:Ca and the K:N ratios. In total 11 classes are defined, each with its typical nutrient solution and accompanying target values for the nutrients in the root environment, as determined in the 1:1½ extract. The micro element concentrations are kept equal for all groups. All known potting- and bedding plants are assigned to one of these 11 nutrient classes.

The second parameter of the system is aiming at the salt sensitivity of the crops. Three levels are defined by the total EC in the root environment (Table 14.1).

Table 14.1 Parameters for salinity and pH used for the classification of potted and bedding plants. The values for salinity represent the maximum acceptable values per class determined in the 1:1½ extract. The values for the pH indicate the range for each class, the boundary value indicates when additional NH_4 should be supplied

Class	Salt sensitivity of the crop	EC dS m^{-1}	Na mmol l^{-1}	Cl mmol l^{-1}
1	susceptible	1	1.7	1.7
2	moderate susceptible	1.4	2.5	2.5
3	not susceptible	1.8	3.5	3.5
Class	**pH classes[1]**	**boundary**		
1	< 4.6	5.1		
2	4.6 – 5.4	5.9		
3	4.9 – 5.7	6.2		
4	5.2 – 6.0	6.5		
5	5.5 – 6.3	6.8		

[1] In mixtures with clay, the values should be increased with 0.5 unit.
Data after Straver et al. (1999).

The determination of the EC includes residual salts as well nutrients, thus the total osmotic potential. The concentrations of Na and Cl in the potting soil are defined also, since some crops are specific sensitive for Na or Cl (Sonneveld, 2000). A complication is that some crops show a distinct difference in response tot salinity dependent on the climatic conditions (Sonneveld, 2000). If this is the case, the crops are classified following the lowest acceptable level.

The third parameter concerns the pH requirements, in which five levels are distinguished. This is aimed at the pH conditions in the natural habitat of the plants and described more in detail in Section 14.5.

The combination of 11 nutrient, 3 salt sensitivity and 5 pH classes results to 165 categories in theory. However, many combinations are not required. The most important categories with the most representative species are listed in Appendix D.

The plant propagation is not categorised in this system. Usually the standard nutrient solution for the adult crop is used, with some incidental adjustments towards extra Ca and Fe. The management will be presented in section 14.10.

14.3 Potting Media

The growing media used for potted plants are mainly peat-based and consist of a mixture of different fractions white peat, black peat and a variety of constituents like perlite, clay, coir, fibrous materials and even artificial foams. The recipes for any mixture of these components have been developed by experiences of growers, substrate producers and research on the physical properties of the growing media (Klapwijk and Mostert, 1992; Kipp et al., 2002). For potted plants the use of different white peat fractions from sod turfs is essential, since this material together with other constituents offers opportunities for the production of substrates with specific characteristics due to the requirements of the crop and the growing system. For instance specific mixtures were developed for ebb and flood systems since they require both rapid water suction in the irrigation phase and rapid drainage in the drain-off phase and in the long run sufficient stability to maintain sufficient air space in the substrate, despite the intensive water movement. Last characteristic is brought by sufficient course fractions in the substrate, like from sod turf. For pressed pots, used for propagation of some soil grown crops and for bedding plants, special mixtures are required to give the blocks sufficient compaction, as merely black peat can be used for this purpose. Mixtures due to sowing, or rooting of cuttings, require a rather finer structure than mixtures used for plants which are transplanted with a well developed root system. The container size plays a role too, for lower pots often more course material is needed than for higher pot sizes. In practice many different recipes for substrates due to potted plants exists. The majority are developed in close cooperation between grower and substrate producer during many years and in some cases only based on experience, but widely used. For example clay addition in substrates is practised as it is claimed to prevent lush growth and improvement of plant quality for example for *Cyclamen* (Klapwijk and Mostert, 1992). The majority of potted plants are grown in mixtures with 40–60% peat moss, 40–60% coarse white peat fractions, like derived from sod turf. For bedding plants in trays, it is mainly 60–70% black peat and 40–30% peat moss. Orchids like *Phalaenopsis* require a very coarse mixture, for this crop only specific bark fractions are used, sometimes mixed with perlite or other coarse materials.

Nowadays, in some countries there is pressure on the use of peat and therefore, peat free substrates are developed. In many cases mixtures with compost or other renewable sources of organic material are used (Wever, 2002). In substrates mostly no more than 20% by volume of the peat could be replaced by compost without growth reduction or quality decline (Surrage and Carlile, 2008). The effects of the composition of the substrate mixtures on fertilization aspects are discussed in Chapter 11.

14.4 EC Control

The EC is an important parameter in view of fertilization and salinity aspects. This has been elaborated on in previous chapters. However, for potting and bedding plants the control of the EC and the effect of the EC show some typical characteristics. For short term crops, the EC is mainly determined by the base dressing of the substrate mixture and less by top dressings by the grower during the cropping period. For long term crops, accumulation of salts and nutrients can be a serious problem and is affected by both the nutrient management, the water quality as well as by the irrigation method. The small rooting volumes often practiced with potted plant strongly aggravate the accumulation of residual salts from the irrigation water as well from the nutrients, like shown in Fig. 8.1. Therefore a close control on the development of the EC in the root environment during the cultivation period is important.

14.4.1 EC Base Dressing

In Chapter 11, general aspects of the addition of the base dressing to peat mixtures are explained. For potted plants often compound fertilizers, like Pg-mix, are used in many European countries. Usually the base dressing is between 0.5 and 1.0 kg Pg-mix m^{-3} potting soil. Only when big plants are grown higher base dressings up to 1.5 kg m^{-3} are used. In case of cuttings directly planted into pots, only a limited base dressing of 0.25 kg m^{-3} is recommended, as otherwise formation of callus and initial root formation slowed down, because of too high salt concentrations. The low base dressing suffices, since young plant material requires only limited nutrients at the start. Another reason for a low base dressing is the often relative high proportion of NH_4 in the compound fertilizer, which can be toxic to plants and can rapidly decrease the pH value of the substrate, either by uptake or by nitrification. Generally, the EC in the potting mixture varies between 0.5 and 1.0 dS m^{-1} after the base dressing, which of course depends on the quantity of base dressing added and the constituents of the mixture, as discussed in Chapter 11.

14.4.2 EC Top Dressing

The majority of research of nutritional aspects in potted and bedding plants deals with the EC in the root environment during the cultivation period. The general aspects of EC on plant growth and quality have been explained and there is no reason to suppose that potted plants behave very differently. However, some effects of the EC are typical for potted plant species and a good example is the effect on plant quality. Potted plants are traded as a whole plant due to decoration and thus, any damage or deformation of any plant part is unacceptable. With some species leaf damage will occur if the EC in the substrate becomes too high. One of the effects often attributed to the EC is the phenomenon of necrotic margins or leaf tips, although this could not always be confirmed in experiments. This has been described for *Codiaeum* (Straver, 1991a) and *Dracaena* (Mulderij, 1999). Sometimes the climatic conditions also have an effect as was clearly demonstrated in trials with palm trees by Mulderij (1999a). The combined effect of the relative humidity and the EC of the nutrient solution were studied with *Chamaedorea* and *Chrysalidocarpus* (*Areca*). The effects on the plant weights were small and not significant, but a high EC clearly aggravated the development of necrotic leaf tips and also the necrotic spots were larger, especially in *Areca* (Table 14.2). Under lower humidity the EC effect was increased, mainly because of larger necrotic areas, shown by a higher necrosis index. The plants were followed throughout the post-harvest phase and the necrosis increased, however, in this case the increase was only affected by the EC treatments and not longer by the humidity treatment during the cultivation. The cause of the necrosis, however, could not be clarified as no indication was found in the analytical data of the tissue analysis. However, the leaf tips were not sampled separately which for example can specifically accumulate B or Mn as has been shown in Table 5.2.

Table 14.2 Plant weight and number of necrotic leaf tips per plant and the index for the disorder as affected by the EC of the nutrient solution supplied and the relative humidity with *Chamaedorea* and *Areca* as test crops. Index for the necrosis, 0-no and 10-serious symptoms

Factors studied		*Chamaedora*			*Areca*		
			Necrotic leaf tips			Necrotic leaf tips	
EC[1] dS m^{-1}	Humidity	Total weight g/plant	Number	Index	Total weight g/plant	Number	Index
1.2	High	29.5	0.45	0.01	94.7	6.4	2.7
2.5	High	26.8	0.81	0.04	83.9	7.8	6.4
1.2	Low	30.9	0.46	0.01	91.8	7.3	3.2
2.5	Low	25.8	2.17	0.19	84.5	8.6	7.7

[1] In the 1:1½ extract
After Mulderij (1999a).

Leaf colours are also important quality characteristics of many potted plants but are not well defined and rather different among the variety of plants grown and therefore comprise a complex phenomenon. In some cases the EC level affects the leaf colours, like in the experiment of Mulderij (2000) in which different EC levels in combination with the light level on the leaf colour of *Chamaedorea* was studied. An increased EC resulted in darker leaves. A complication in these trials was that with increasing EC the pH in the substrate was decreased too, which likely is caused by the increased NH_4 supply due to the extra nutrient addition with the higher EC treatments. So it is not yet clear whether the better leaf colour is either an EC or a pH effect. A lower pH induces a better condition for the uptake of most micro nutrients, which also can improve the leaf colour. Comparable results have been found with pelargonium (Van Leeuwen, 1992). Leaf and flower colour were improved with increasing EC level, but also in this experiment a decreasing pH was found in the substrate with increasing EC level, likely also induced by increasing NH_4 supply with the increased fertilizer addition.

Azalea is known to be sensitive for leaf scorch, which often is related to the salt sensitivity of this crop (Arnold Bik, 1965). This was confirmed in combined EC and NaCl trials by Van Leeuwen and Bulle (1992). *Azalea* was grown in an ebb and flood system with reuse of the drainage water. The addition of NaCl and the concentration of nutrients in the irrigation water were studied in six treatments as shown in the first column of Table 14.3. The weights of the aboveground plant parts were measured at the end of the vegetative and at the end of the generative period. In the vegetative period the plant growth was negatively affected merely by the high fertilization level and in the generative period mainly by the NaCl addition. This is understandable, because the NaCl in the root environment was highest in the second period. The plants grown at the low nutrient concentration showed a pale colour and leaf drop during the vegetative period, caused by nutrient deficiency. Plant quality and flowering at the end of the vegetative period was negatively affected by an increased EC value, caused as well by the higher nutrition as by the NaCl concentration in the

Table 14.3 Growth of *Azalea* as affected by fertilization levels and by NaCl concentrations in the irrigation water. The fertilization levels were complete nutrient solutions added to the irrigation water at concentration of 0.4 and 0.8 dS m^{-1}

Water supplied EC/NaCl-fertilizer	Vegetative period		Generative period	
	EC/NaCl[1]	Relative weight	EC/NaCl[1]	Relative weight
0.7/0.7-0.4	0.33/1.4	98	0.55/3.1	100
0.9/2.7-0.4	0.40/2.0	100	0.95/6.0	86
1.1/4.7-0.4	0.53/2.7	99	1.70/10.4	88
1.1/0.7-0.8	0.50/1.6	92	0.95/4.1	98
1.3/2.7-0.8	0.73/2.3	89	1.75/7.8	82
1.5/4.7-0.8	0.90/2.9	86	2.10/9.0	83

[1] In the 1:1½ extract of the substrate
Data from Van Leeuwen and Bulle (1992).

irrigation water. The results do not give the possibility to conclude about the character of the salinity effect, being an osmotic or specific NaCl effect. Result of Arnold Bik (1965) and of Brumm and Schenk (1991) give rise to the conclusion that specific effects play an important part, in view of the strong leaf scorch following the addition of NaCl. Many woody crop plants are specific sensitive to NaCl additions (Bernstein, 1976).

Another aspect is the use of the EC in the root environment to control growth. Bulle et al. (1996) compared the effect of different concentrations of a nutrient solution at EC values of 1.0, 2.2 and 3.4 with chrysanthemum. The total plant weight and the shoot weight were reduced, both at the lowest and the highest EC, while the flower weight was only slightly affected. The plant length and shoot length were reduced more strongly with the increasing EC. This resulted in an increasing compactness, expressed as the total fresh weight per cm length and this effect was stronger for shoots than for the total plant (Table 14.4).

Fertilization determines quality in many ways, which is demonstrated with an experiment of Boertje (1980). Bedding plants were grown in multiple pot trays in a well fertilized peat mixture. During cultivation the plants were supplied with 10 top

Table 14.4 Plant and shoot length in cm, plant and shoot weight and total flower weight in g/plant and the compactness expressed as g/cm of the shoots of pot chrysanthemum, as affected by different EC values of the nutrient solution

	Plant growth parameters					Compactness	
EC supplied dS m^{-1}	Plant height cm	Shoot length cm	Plant weight g	Shoot weight g	Flower weight g	Total plant g/cm	Shoot g/cm
1	14.2	7.9	91.6	35.3	39.0	6.5	4.5
2.2	15.2	8.3	113.8	53.9	41.4	7.5	6.5
3.4	14.0	7.3	108.2	52.0	37.3	7.7	7.2

Data after Van Leeuwen (1992a).

Table 14.5 Leaf scorch as affected by overhead fertilization of bedding plants with the mixed fertilizer 17-6-18. The leaf scorch was judged as following: 0-no, X-some, XX-substantial and XXX-severe

	Fertilizer concentration g l^{-1}					
Crop	0	0.5	1	1.5	3	4.5
Petunia	0	0	0	0	0	0
Impatiens	0	0	0	0	0	X
Dahlia	0	0	0	0	X	XX
Salvia	0	0	0	0	X	XX
Tagetes	0	0	X	X	XX	XXX
Ageratum	0	0	0	0	XX	XXX

Boertje (1980). *Reprinted by permission of the International Society Horticultural Science*

dressings of a fertilizer solution. After addition of the top dressing the plants were not rinsed with clean water. In the treatments where high fertilizer concentrations were used most plants showed serious leaf scorch (Table 14.5). However, there was a great difference among the sensitivity of the crops. *Tagetes* was most sensitive, while *Petunia* did not show any effect. In experiments with tomato leaf scorch especially was promoted by high NH_4 concentrations (Section 15.4). The N content in the fertilizer used for the top dressing in the experiment was 8% as NO_3 and 9% as NH_4. Leaf scorch started in the treatment of 1 g l^{-1} of the fertilizer which mean a NH_4 concentration of 6.4 mmol l^{-1}, and in the treatment with half of this concentration no any leaf scorch was found. This result is in good agreement with the data found for tomato as presented in Fig. 15.3.

Generally, the plant weight was linear increased with the concentration of the top dressing solution, like shown in Fig. 14.1. This especially was the case for the fast growing plant types, like *Petunia* and *Impatiens*. Such a relationship will be expected as long as the optimal nutrient levels not are crossed and osmotic stress occurs. Plant weight does not always reflect linearly plant quality. This is shown by the data in Fig. 14.2, where the relationship between plant weight and plant quality is shown. The relationships strongly differ for the plant type. With *Salvia* a sharp optimum for the quality is shown, while *Petunia* shows a broad range in optimum values. Negative optimum quality ratings in the low range often are connected with plant colour and in the high range a too wealthy vegetative development as well leaf damage by too high EC values in the root environment can play a part.

Sometimes the potential of growth regulation with EC gives conflicting results. Verberkt and De Jongh (1995) found with *Cyclamen* that an EC level of 1.7 resulted in larger plants of better quality than those grown at an EC of 1.1, but the flowering time was delayed. Moreover, the bulbs were bigger at the lowest EC.

As already mentioned, the EC is not only important for the results during the growing period in the greenhouse, but it also affects the consumer phase. This was demonstrated in experiments with bedding plants (Mulderij, 1998). Eight different EC treatments were compared with *Impatiens* and *Petunia* as test crops.

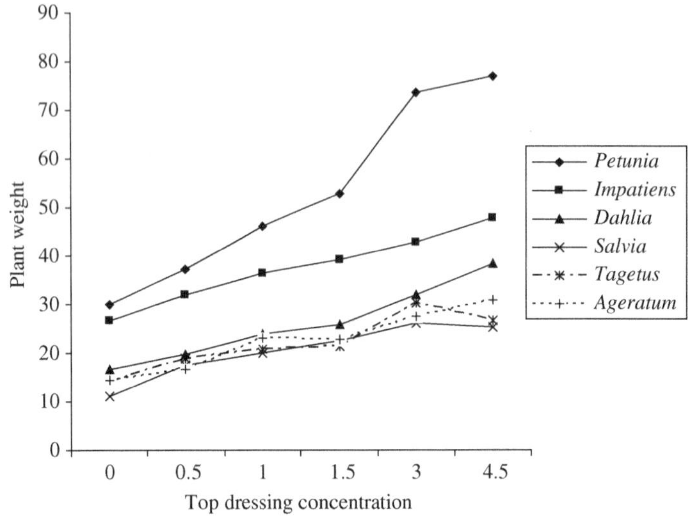

Fig. 14.1 Relationship between the concentration of the top dressing solution in g l^{-1} 17-6-18 and the plant weight of a series of bedding plants. Data after Boertje (1980). *Modified by permission of the International Society Horticultural Science*

In some treatments the EC was kept constant whereas in others the EC was changed between values from 0.3 to 2.2, like shown in Table 14.6 . *Petunia* showed improved growth when continuously supplied up till the highest EC value of 2.2. In all the treatments with a variable EC the growth was reduced in comparison with the treatment of the EC value of 2.2. It seems rather feasible that the plants react to an average EC. The *Impatiens* treatments showed less effect, and the highest plant weight

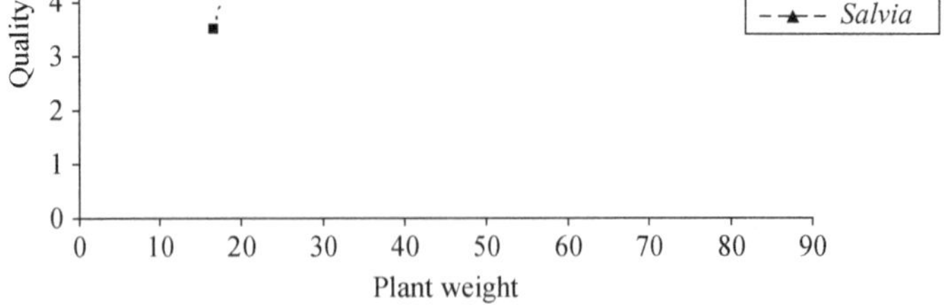

Fig. 14.2 The relationship between plant weight in g per plant and quality index of some bedding plants. The quality index rated from 0 – bad to 10 – excellent. Data after Boertje (1980)

Table 14.6 Plant fresh weights of *Petunia* and *Impatiens* at marketable stage and after 7 weeks of post harvest treatments in fertilized and non-fertilized balcony containers, as affected by 8 different EC regimes during the production phase

Treatments production phase[1]			*Petunia*			*Impatiens*		
			Marketable	After 7 weeks		Marketable	After 7 weeks	
1	2	3		Non fertilized	Fertilized		Non fertilized	Fertilized
0.6			2.6	5.0	24.1	3.6	6.7	31.8
1.1			3.8	8.0	26.4	4.4	11.1	31.1
2.2			4.6	13.1	36.4	4.1	16.8	39.3
0.6	2.2		3.6	13.1	22.8	3.8	12.7	33.5
2.2	0.6		3.8	8.3	32.6	4.0	12.9	31.3
0.3	2.2		3.2	7.7	25.4	3.9	10.8	31.3
2.2	0.3		3.5	7.1	24.9	4.2	10.3	35.5
0.6	2.2	0.6	3.4	7.1	29.9	3.6	9.9	36.2

[1] Phase 1, 2, and 3, weeks 1–2, 2–4 and 4–6 of the growing period, respectively.
After Mulderij (1998).

was obtained at an EC value of 1.1. The plants were monitored during a post-harvest period as well, when planted in balcony containers. Half of the plants were not fertilized in the container, while the other half received a base fertilizer in the containers. After 7 weeks the final weights of the plants were measured. It was remarkable that the effects of the production period were still apparent in the post-harvest period. This especially was the case in the non-fertilized treatment where large differences were shown in growth between the lowest and the highest EC treatment. These differences were also present in the fertilized containers but less distinct than in the non-fertilized containers. The *Impatiens* showed comparable results in the consumer phase, but less pronounced than with *Petunia*. Remarkably, the treatments with the highest EC in the production phase performed best in the consumer phase.

A specific problem with EC is the accumulation of salts at the top layer in the pot as has been shown in Fig. 8.1 (De Kreij and Straver, 1988). This problem arose with the introduction of the flooded bench systems, with water supply by ebb and flood instead of overhead irrigation. As long as the plants are flooded frequently, the plant development is less affected by a partially high EC value in the root environment, since plants escape from high salinity as discussed in Chapter 8. Effects of partial high EC values in the post-harvest phase will be discussed in Section 14.6.

14.5 pH Management

The pH of the substrate solution in the root environment is mainly determined by the substrate components in it. This is mainly peat of which the pH can be affected by liming as has been described in Section 11.4. The requirements for the target pH value in potted plants vary widely. For the majority of these plants no experimental data are available suitable for a classification into target pH ranges. The general effects of the pH value in the root environment as described in the Sections 11.3, 12.2 and 13.4 are also operative for potted plants. If no specific requirements are known for crops, the general recommendations for potted plants can be used as listed in Table 14.1. Nevertheless, some plant species have specific requirements for an acidic environment, like *Ericaceae* (Heathers) and *Rhododendrae*. The target value for the pH for these crops should be not higher than 4.6 and therefore often no liming at all is required (Arnold Bik, 1972). Even water without carbonate is recommended (Röber, 1987). Also *Hydrangea* spp. for blue flower production requires a low pH as will be discussed further in Section 14.7. Other species that positively react on low pH values are *Epipremnum aureum* and *Anthurium scherzerianum*. For these species the target pH values recommended varies between 4.6 and 5.4. Mulderij and Hüner (1997) found that with increasing pH in a range between 4.8 and 6.4 the number of misshapen leaves increased from 5 to almost 50%. Not only the number of affected leaves but also the extent of the damage increased dramatically. In other species of the *Araceae* this phenomenon is also known (Poole et al., 1984; Mulderij and De Jongh, 1995). So far the cause of the phenomenon is not yet clear, although some authors suggest that Mn is involved, as has been concluded from the decreased Mn content in the plant tissue (Mulderij, 1999a). However, this is not likely, since it is quite common that Mn contents decrease strongly with increasing pH value. A moderate low pH range, between 4.9 and 5.7 is required by a number of crops like palm trees *Chamaedorea* and *Dyctiosperma*, members of the *Orhidacea*, the *Gesnerceae* and the *Euphorbiaceae*. In this case the recommendation of the pH is mainly based on the conditions of the soils in the natural habitat of the plants rather than on experimental data. This is also true for a large group of species which require a high pH, with a target range between 5.5 and 6.3. All succulents like *Crassula, Euonymus, Echeveria, Sedum* and the *Cactaceae* belong to this group which also includes *Yucca, Pelargonium* and *Kalanchoë*. These plants all share a natural habitat in calcareous soils in arid and semi-arid zones, with naturally high pH conditions. Also *Hydrangea* for pink flower production requires a sufficient high pH level.

Trials with *Saintpaulia* illustrate that the pH can cause serious problems. Straver (1988) examined the effect of three lime application rates: 2.5, 4.5 and 6.5 kg limestone per m^3 of potting soil in combination with three NH_4:NO_3 ratios of 0:100, 25:75 and 50:50 in the nutrient solutions, with two different cultivars. With cultivar "Nr 83" severe incidence of malformed flowers appeared. The symptoms increased with decreasing lime applications and with increasing NH_4:NO_3 ratio, like shown in Table 14.7. This phenomenon is called "black-flower" and is characterized by small flowers, with glassiness margins of the petals, turning purple or black. With cultivar "Heidrun" no symptoms were found. The total fresh weight of the plants showed inconsistent effects. With "Nr 83" the growth was better with increasing lime application, however with "Heidrun" the best results were found at the lowest lime treatment. Both cultivars performed best at the lowest NH_4:NO_3 ratio. Tissue analysis showed that Mn and Fe contents were very high in some plant parts at the

Table 14.7 Total fresh weight and the % of black flowers with two cultivars of *Saintpaulia* as affected by liming levels in the base dressing (kg m^{-3}) and the NH_4/NO_3 ratio in the fertilizer application during top dressing

Lime addition	NH_4/N	Fresh weight		% "black flowers"	
		Nr 83	Heidrun	Nr 83	Heidrun
2.5	0	50.3	84.6	57.0	0
	0.25	49.5	80.5	78.8	0
	0.50	47.9	77.6	49.2	0
4.5	0	57.2	84.6	12.8	0
	0.25	52.0	80.5	39.3	0
	0.50	49.4	77.6	46.8	0
6.5	0	55.6	79.2	0.8	0
	0.25	55.3	74.0	1.6	0
	0.50	49.9	69.3	1.5	0

After Straver (1988).

lowest pH, in the damaged flowers however the contents were not extremely high. It shows that pH effects on plants not always could be related to nutrient uptake and also that there are extreme differences in susceptibility between genotypes.

With *Schefflera* often leaf yellowing and leaf senescence occur which often is associated with too low pH values in the substrate and with indications that Mn is involved. Straver (1997) examined the effects of lime application rates in combination with Mn levels and cultivars. The results showed that both a too low and too high liming reduced plant weight, with optimum results at a lime dosage of 4 kg m^{-3}, resulting in pH values on average of 5. However, leaf yellowing was not found, despite extreme Mn contents in the tissue. He demonstrated also that the Mn uptake can be rather extreme as a result of the Mn supply, more or less independent of the lime application. With increasing lime application, the Mn concentrations in the substrate analysis were reduced considerably, but the Mn uptake was scarcely reduced as shown in Fig. 14.3. The explanation is that the uptake merely is affected by the frequently supplied fresh nutrient solution which had standard pH conditions. Roots tend to concentrate at the bottom of the pots where with ebb and flood frequently fresh nutrient solution penetrates, while the samples mainly are gathered from the higher layers in the pots, whereas the Mn is oxidized especially at the high lime addition.

Basically, the pH in the potting soil as created by the lime addition can drop during crop cultivation if insufficient lime is added (Straver, 1997). This is also illustrated in Fig. 14.4, showing the pH developments with a *Schefflera* crop grown at three liming regimes. The pH development usually has a decreasing tendency, caused at one hand by processes due to the plant uptake, the nitrification of NH_4, decomposition of the organic matter and on the other hand by processes in the substrate by the dissolution and leaching of the lime buffer. It is complex to predict the pH development in a potting soil and this problem is enhanced when the particle size of lime varies (Fisher, 2006).

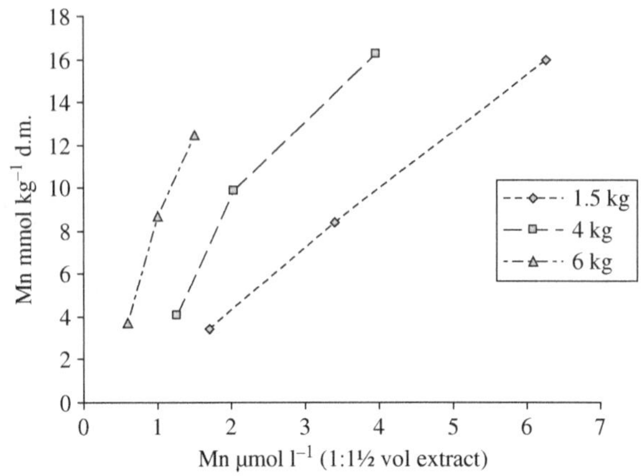

Fig. 14.3 Relationship between the Mn content in the potting soil (1:1½ extract) and the Mn content in *Schefflera* leaves, in an experiment with three lime application rates in the substrate of 1.5, 4 and 6.5 kg m^{-3} and three Mn concentrations in the nutrient solution supplied (After Straver, 1997)

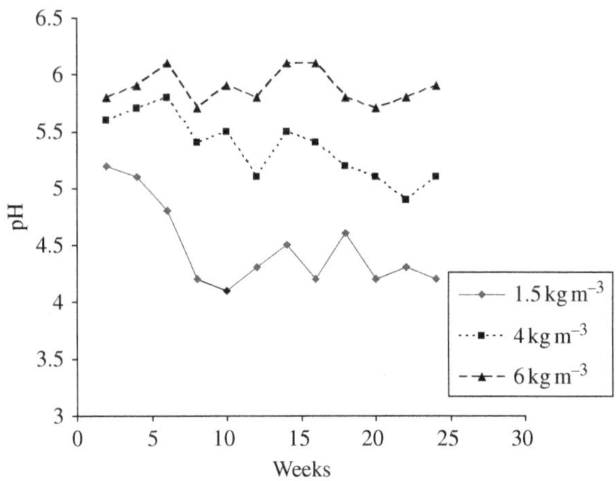

Fig. 14.4 pH values in the drainage water of *Saintpaulia* during the growing period at three levels of liming in the base dressing. After Straver (1997)

14.6 Nutrients

Although the uptake of nutrients can differ greatly between plant species, most plant types are not very sensitive for somewhat low or high concentrations of specific nutrients. This for example is shown with the results of an experiment of Straver (1991c). Two palm species *Howeia* and *Chrysalidocarpus* (*Areca*) were grown in substrate with a wide range of K:Ca ratios in the nutrient solution supplied. In the potting soil also equal K:Ca ratios were detected, while the plant weight and plant quality was unaffected, although the K:Ca in the plant changed seriously. The plant adjust the uptake more or less by increasing the K:Ca ratio in the plant above the external solution at low ratios and decrease this ratio in the plant below the K:Ca ratios in the external concentration at high K:Ca ratios as shown in Fig. 14.5. The same effect was found with *Spathyphyllum* (Straver, 1990). This merely will be true with an ample supply of nutrients and when not too extreme concentrations occur in the root environment. Obviously, with extreme low nutrient supply the growth as determined by shoot length, number of shoots and total fresh weight will be negatively affected, like is shown by Arnold Bik (1976). Thus, effects of marginal nutrient supply have perspectives for growth regulation. With the change over from overhead irrigation to ebb and flood irrigation of bedding plants it appeared that the traditional growth regulation by drought stress was less effective (Vogelezang et al., 1992). Baas et al. (1995) showed that with suboptimal P applications the growth in terms of plant fresh weight, plant height and leaf area could be reduced with advantages for the quality of a number of crops. However with *Poinsettia* and *Salvia* in some cases also the number of side shoots or the number of flowers were reduced, which downgraded the ornamental value. For *Impatiens* no negative side effects were determined. They also concluded that P stress could only be obtained and controlled in the top dressing if no P was supplied in the base dressing. This was confirmed in an extensive study by Warmenhoven and Van Noort (2005) who investigated 12 different bedding plants in a range of different P concentrations in the top dressing solution, including treatments with and without P in the base dressing. In these experiments the base dressing was carried out with either 0 P or 4% P in the compound fertilizer, in combination with P concentrations of 0.02, 0.1, 0.2 and 0.5 mmol l^{-1} in the nutrient solution supplied. P stress could hardly be obtained if the fertilizer used for base dressing of the substrate contained 4% P. In the treatments without P in the base dressing, the growth was strongly affected if the P supply was lower than 0.2 mmol l^{-1} in the nutrient solution supplied as is

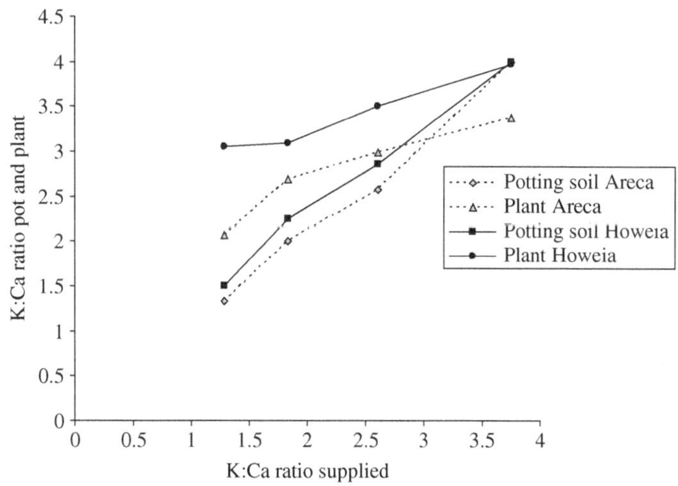

Fig. 14.5 K:Ca ratios in the 1:1½ extract of the substrate and in plant tissue as affected by the K:Ca ratio in the nutrient solution supplied in an ebb and flood system with *Howeia* and *Areca* (Straver, 1991c)

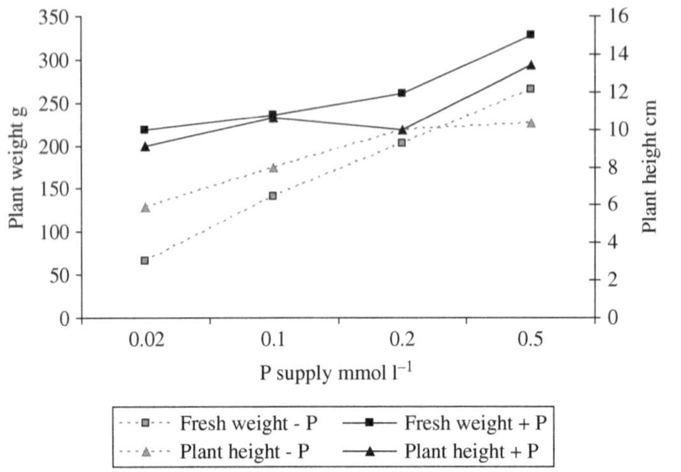

Fig. 14.6 Plant weight and plant height as affected by the P supply in the nutrient solution supplied with *Pelargonium* with and without P in the base dressing in the substrate. After Warmenhoven and van Noort (2005)

illustrated in Fig. 14.6 with *Pelargonium*. However, with some species even below 0.5 $mmol^{-1}$ growth reduction was found (*Begonia, Petunia*). The overall conclusion was that growth regulation with reduced P addition is possible but the bandwidth for the P supply to obtain at one hand a sufficiently reduced growth and an acceptable ornamental quality on the other hand is rather small. Quite easily thresholds are exceeded to either a too lush or a too retarded growth with a strongly visual P deficiency. However, from these results it also became clear that in case of bedding plants, the P-level in the base dressing often can be lowered, when P is added as base dressing. With potted chrysanthemum it appeared to be impossible to use P for growth regulation, indeed P reduction led to shorter plants and a reduced plant weight, however, the plant quality was also reduced (Van Leeuwen, 1992).

Specific research to the addition of micro nutrient elements in potted plant cultivation have been given less attention and the addition for top dressing is merely based on results gained with vegetables and cut flowers, as has been elaborated in Chapters 12 and 13. In the years a general recommendation for micro nutrient levels in potting soils has been developed. This is based on the general experience of micro nutrient requirements of crops and the effect of peat and the other constituents on fixation and complex formation of for instance Cu and Zn, as is presented in Chapter 11. The general recommendations for micro nutrients in potted and bedding plants grown in peat based substrates consist of fixed concentrations in the mixed fertilizers added with the base dressing and those in the nutrient solution and general target values for many crops and plant stages in the 1:1½ extract. In a series of experiments with *Spathiphyllum* (Verberkt et al., 1998), *Kalanchoë* (Verberkt et al., 1996) and *Epipremnum* (Mulderij, 1998) this generalization for micro nutrients was evaluated and it proved to work out quite well. Nevertheless, they drew attention to pH deviations, which can induce serious problems. Fe deficiency easily occurs in susceptible crops if the pH becomes too high and could hardly be cured by extra addition of Fe chelate (Mulderij and Hüner, 1997). Mulderij and De Jongh (1995) also concluded from trials with *Chamaedorea* and *Chrysalidocarpus* that the pH of the substrate much more affect the occurrence of Fe chlorosis than either the Fe concentration in the substrate or the type of Fe chelates used. With too low pH, Mn toxicity occurs in susceptible crops and at a high pH the uptake of Mn is strongly reduced. Mn supply to cure the symptoms is less effective at high pH values (Straver, 1997). Problems with Fe and Mn addition both are discussed more in detail in the Chapters 12 and 13.

Basically, the addition and recommendation for micro elements follow general rules for all categories of potted plants. One reason is that a major part of the micronutrients will be added by the base dressing of the substrate as has been discussed in Section 11.4 and the concentrations in the top dressing will have only limited effect. The other reason is the lack of information about specific micro nutrient requirements for the plant species grown. In Appendix D the general application and recommendation for micro element is listed.

14.7 Nutrient Management

The nutrient management of potted and bedding plants is complicated, since no direct information can be derived from the actual status of the EC, pH and nutrient concentrations in the root environment. In contradiction to rock wool cultivation with which the substrate solution easily can be sampled, it is difficult to gather such a solution from the substrate of potted and bedding plants. The circulating nutrient solution from ebb and flood systems is not suitable, since this solution is scarcely affected by the concentrations in the root environment. Suction of the substrate solution by means of porous cups has been tried but showed to be too laborious, as discussed in Section 4.8. For routine testing, sampling of the substrate of sufficient pots and analysis with the aid of 1:1½ extraction method as described in Section 4.5 is the best suitable and the commonly used method to get information about the nutrient status in the root environment during cultivation.

Apart from the lack of information on the nutrient status of the crop, the tools for nutrient management in growers practice are also limited. Compound fertilizers are mainly used for the fertigation of the crops as this is the most straightforward solution for growers and prevent the complexity of the calculation of nutrient solutions (De Jong, 1987). However, the fixed nutrient ratios in these fertilisers are a hindrance to meet the specific requirements of the great variation in the potted and the bedding plants. Nowadays, modern substrate producers can deliver substrates for specific requirement by using a mixture of different fertilizers and thus, it is also logical that growers prepare nutrient solutions closely focussed to the specific needs of crops. Base dressing of substrate is necessary, since the adsorption and fixation of some nutrients is substantial. Without a base dressing plants can suffer at the start by too low nutrient concentrations, despite a direct start of the top dressing. There will be interaction between the level of the base dressing and the top dressing, which can be illustrated by an experiment with *Saintpaulia*. In this experiment four levels of base dressing in the substrate viz. 0, 300, 600 and 1200 mg l^{-1} of a fertilizer mixture was added. In the 0 treatment only micro nutrients were supplied. The EC values in the 1:1½ extract of the substrate after fertilization were 0.3, 0.4, 0.6 and 0.8 respectively. The plants grown in 9 cm plastic pots filled with the different fertilized substrate were placed in benches with and ebb and flood irrigation in which two concentrations of the nutrient solution were compared with EC values of 1.1 and 1.7 dS m^{-1}. Two cultivars were included "Emi" and "Ramona" and two cropping periods were compared autumn and spring (Van Leeuwen, 1993). The total plant weights are listed in Table 14.8. Lowest weights often were found with combinations of low base dressing and low concentration in the nutrient solution used during cultivation as well in the combinations high base dressing and high concentration in the nutrient solution. With a high base dressing (0.6–0.8) mostly the EC value 1.1 is best and with a low base dressing (0.3–0.4) mostly the EC value of 1.7 is best. "Ramona" in autumn growing is an exception on this rule. A low base dressing requires a higher nutrient concentration in the irrigation water and upside down, being effects that could be expected. No consistent different effects of the fertilization regimes were noticed between the reaction of the cultivars and of the growing seasons.

The composition of the nutrient solution supplied is still important and will be tuned to the need of the crop, especially for crops with a long growing period to avoid accumulations or depletions during the growing period. However it is impossible to develop strict guide lines for the nutrient management for all different crops, so a general approach is chosen in the fertilization recommendation system as presented in Section 14.2. Especially for crops with a long growing period sampling and analysis of the substrate is recommended to trace unbalanced accumulations and depletions. Results can be compared with the guide values listed in Appendix D and the nutrient solution supplied can be adjusted following the algorithm presented in Chapter 12. As already mentioned, the fixed ratios of nutrients

Table 14.8 Plant weights (g per plant) of two cultivars of *Saintpaulia* as affected by different fertilization regimes in autumn and spring growing

Base dressing	Nutrient solution EC 1.1				Nutrient solution 1.7			
	Autumn growing		Spring growing		Autumn growing		Spring growing	
EC 1:1½	"Emi"	"Ramona"	"Emi"	"Ramona"	"Emi"	"Ramona"	"Emi"	"Ramona"
0.3	54.6	66.5	40.5	40.9	57.9	68.3	44.7	49.7
0.4	54.1	65.0	44.6	46.9	62.4	67.3	45.7	49.9
0.6	56.3	60.9	46.6	50.9	56.8	63.3	45.4	46.7
0.8	59.7	45.7	45.9	51.7	55.1	56.7	43.2	46.0

Data after Van Leeuwen (1993).

in compound fertilizers is a hindrance to a good tuning to the crop demand in relation to the analytical data of substrate analysis. Moreover, many of these fertilizers contain high levels of NH_4 or urea, which can cause an undesirable pH drop in the substrate with growth or quality problems as has been discussed.

In view of the aspects mentioned previously, the EC and the composition of the nutrient solution supplied at one hand and the uptake by the plant at the other hand, obviously will result to certain nutrient concentrations in the substrate. For any recommendation system based on substrate analysis it is important to know these interactions for an adequate crop specific nutrient management. It is unfeasible to achieve extensive studies for all potted plant species and these, generalisation is unavoidable. The suitability of the developed recommendation system as presented in Section 14.2 was studied in a series of experiments. (Straver, 1991; 1991a, b; Mulderij, 1993; 1994; Verberkt and De Jongh, 1995). The experiments were carried out with crops representative for different classes of the recommendation system, viz. *Saintpaulia, Codiaeum, Adiantum, Asplenium, Begonia, Nephrolepis, Poinsettia, and Cyclamen* in ebb and flood system. In the experiments nutrient solutions were compared in at least three concentrations, usually applied at EC values of 50%, 100% and 150% of the standard recommendations. Plant weight and plant quality were monitored as well as the development of the nutrients in the potting soil and in the plant. The results were evaluated for the different crops and compared with the established target values for the composition of the nutrient solution supplied as well for the composition of the 1:1½ extract of the substrate. The main conclusion was that in nearly all cases the established nutrient solution offered a satisfactory plant production, plant quality and mineral contents in the plant tissues. With *Asplenium* and *Nephrolepis* the EC level initially declared lead to strong accumulation of nutrients in the substrate and osmotic stress of the plants. Successive experiments with a reduced EC range gave better results for these crops. In all other crops no significant effect on plant weights were found within the range of EC values tested. A more or less linear relationship was found between the supplied concentration and the concentration in the substrate as determined in the 1:1½ volume extract, but the regression coefficient differed seriously among the crops. The effect of the supplied nutrient solution on the nutrient content in the plant was very different among the crops like shown in Fig. 14.7. The N-level in the plant tissue differed considerable, being with *Poinsettia* the highest and with *Asplenium* the lowest contents. With *Asplenium*, a significant increase in the N content in the plant was found with increasing N supply, but the content decreased at the highest additions. For *Poinsettia, Kalanchoë* and *Codiaeum* the increase was gradual, but still of significance and consistent. With Begonia, however, the uptake was hardly affected. The results obtained with K showed more differences than for N. Especially with *Asplenium* and *Codiaeum* the K contents are strongly increased in relation to the supply, while with the other crops the uptake was scarcely affected by increased concentrations in the root environment. Results of the experiments are used to adjust the target values for the concentrations in the root environment and also for the recommended adjustments in case of deviations from the target values for EC and nutrient concentration in the root environment. It should be realised that a low uptake of a mineral element

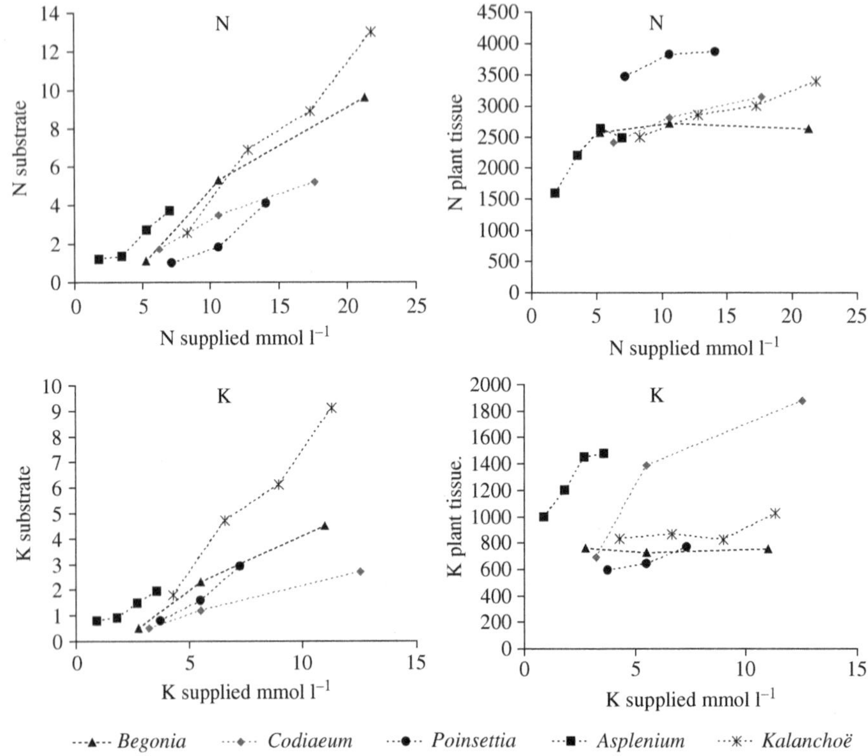

Fig. 14.7 N and K in the substrate (mmol l^{-1} in the1:1½ extract) and the N and K content in the plant (mmol kg^{-1} dry matter) as affected by the N and K concentration (mmol l^{-1}) in the supplied nutrient solution in an ebb and flood system with five different crops. Data derived from a series of experiments of Straver (1991, 1991a, b; Mulderij, 1993, 1994)

by a crop easily results in accumulation in the substrate, as found in closed rock wool systems with cut flower production (Sonneveld and Voogt, 2001). This effect will be aggravated by the small substrate volume used with potted plant production. This contradiction is unavoidable and should be considered in the judgement of the recommendations based on analytical data of substrate samples gathered during cultivation.

Obviously, the management of the nutrient supply is linked strongly to the irrigation. In ebb and flood systems this has some implications. The nutrient concentrations in the fertigation water are for various reasons often higher than the uptake concentrations of the nutrients (see Chapter 12). In contrast to closed systems with overhead irrigation, the excess of nutrients of potted plants in ebb and flood systems accumulate strongly specifically in the top layer of the substrate. Redistribution of these nutrients within the system is limited, especially Ca, K and P accumulate in the top layer (Otten, 1994). Therefore, the concentration of nutrients in the irrigation water should be closely attuned to the crop demand. However, despite a careful application, evaporation from the substrate surface easily causes accumulation of nutrients in the top layer, and reduces the availability to plants. The nutrient supply also should be adjusted to this phenomenon. Accumulation in the top layer is also affected by the irrigation practices. A regularly wet substrate surface on top of the containers aggravates the evaporation and thus, the accumulation of salts. This is demonstrated by Otten (1994) in an experiment with *Ficus Benjamini* grown in a flooded bench system with two irrigation frequencies and two nutrient levels. The P uptake was scarcely affected, while the quantity of P accumulated in the substrate is increased with frequent irrigation and the increased EC of the irrigation water, like shown in Fig. 14.8. Comparable results were gained also with all other nutrients. De Kreij and Straver (1988) also showed that increasing irrigation frequency with *Codiaeum* with flooded bench irrigation increased the total fresh weight of the plants but also the accumulation in the pot, especially in the top layer. This effect of irrigation frequency on nutrient accumulation was confirmed in an experiment with *Begonia,* (De Kreij, 1989). The pattern of salt accumulation is strongly affected by the irrigation method. This is demonstrated in an experiment with *Poinsettia,* Cox (2001) overhead irrigation and ebb and flood irrigation. At the lowest N concentration no accumulation and no difference in EC occured at all between the top and bottom sample of the substrate. However, the EC in the top layer of the pot increased tremendously with increasing N supply when irrigated by ebb and flood, as shown in Fig. 14.9, while growth and quality on the ebb and flood irrigation was not negatively affected. Consequently the high salt accumulation in the top layer does not affect plant growth, likely due to an escape from high salinity spots as discussed in Chapter 8. However, such a salt accumulation on the top of the pots can cause serious problems in the post-harvest phase at the consumer, when the plants are

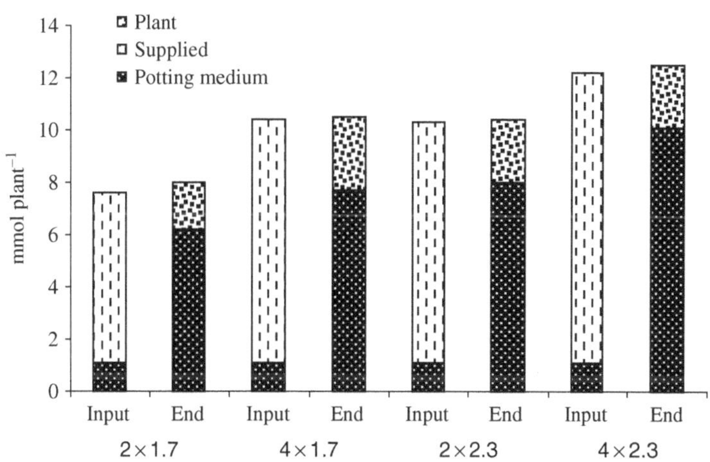

Fig. 14.8 The amount of the P accumulated in the substrate and in the plant and the cumulative amount added to the substrate as affected by the irrigation frequency (twice or four times a week) and the EC of the nutrient solution added (1.7 or 2.3 d S m^{-1}) after 11 weeks, with *Ficus benjamini* as test crop. Data after Otten (1994)

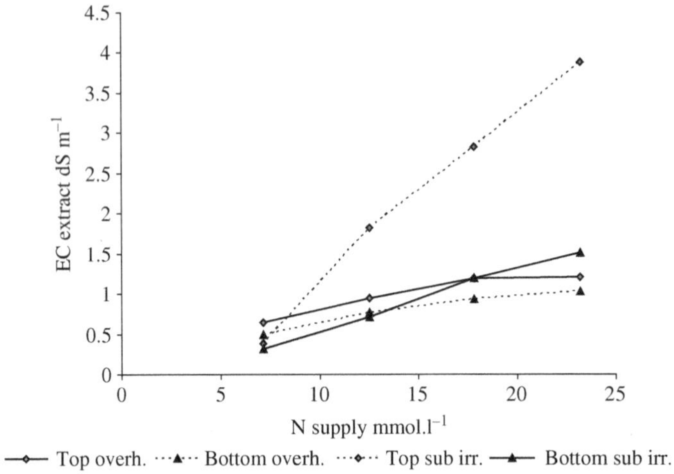

Fig. 14.9 Effects of N supply and irrigation method, viz. overhead sprinkler irrigation and sub irrigation on flooded benches, on the accumulation of nutrients in the substrate, (top = upper 2 cm, bottom = bottom 4 cm of the growth medium) expressed as EC in the 1:2 v/v extract of *Poinsettia* grown under increasing nutrient supply. After Cox (2001). *Modified by permission of Marcel Dekker*

irrigated as commonly from top and the salts will be moved down. In this case the extreme high salt concentrations from top reach the active roots developed at the low concentration at the bottom of the pot. These roots are not yet adjusted to such high concentrations and the plant can seriously suffer by osmotic shock. This has been demonstrated by Verberkt and van den Berg (1993). Impatiens was grown during 9 weeks in 12 cm plastic pots with a volume 0.64 l. The irrigation was carried out with nutrient solutions of different concentrations. The plant reaction on salt accumulation in the pot was studied in a post-harvest experiment, in which plants

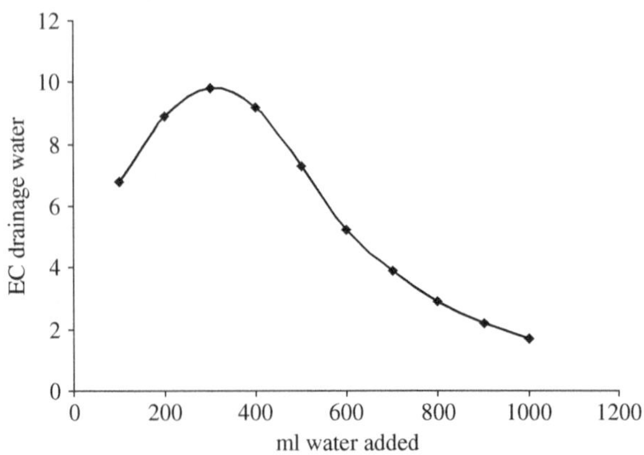

Fig. 14.10 Course of the EC dS m^{-1} of the drainage water during leaching of the substrate in the pots grown with *Impatiens* for the post harvest experiment. Treatment 6.45/2.83 as presented in Table 14.9. Data of Verberkt and Van den Berg (1993)

Table 14.9 Post harvest conditions of *Impatiens* plants as affected by the EC in the substrate and different post harvest treatments. Index: (–) no damage on the plants; (+) light symptoms of leaf burn and wilting; (++) moderate symptoms and (+++) severe symptoms

Post harvest treatments		EC substrate 1:1½ extract top/bottom			
Leaching	Water supply	0.79/0.42	1.95/0.57	4.84/2.08	6.45/2.83
Yes	Bottom	–	–	–	+
Yes	Top	–	–	–	+
No	Bottom	–	–	+	++
No	Top	–	+	+++	+++

Data Verberkt and Van der Berg (1993).

were compared grown at the different EC values during cultivation (Verberkt and Van den Berg, 1993). In the post-harvest experiment plants were compared after a leaching yes or no with 1 l of demineralised water per plant and with watering either from top or from bottom during the post harvest period. The course of the EC in the drainage water during the leaching treatment is shown in Fig. 14.10 and the results of the plant condition after 12 days are listed in Table 14.9. The maximum value of the EC in the drainage water occurred with the addition of 0.2 l water, being the moment that the substrate solution from the top layer arrived at the bottom. Watering from top induces more damage than watering from bottom, which is understandable because of the high EC from top that is washed to the lower part of the pot where the active roots are present. Leaching is effective, but cannot prevent the damage at all.

14.8 Environmental Aspects

The original growing system of potted plants characterized by individual plants grown in individual pots with restricted root volume inevitably leads to low water and nutrient use efficiencies, as discussed in Section 6.3 and induces environmental problems. The change over to flooded benches and concrete floors with the possible reuse of run-off and drainage water improved the efficiencies considerably. In an inventory on the water and fertilizer use of nurseries with different irrigation

Table 14.10 Average N and P use at nurseries with different irrigation systems, with three different crops

	N use kg ha^{-1} yr^{-1}			P use kg ha^{-1} yr^{-1}		
Crop	Flooded bench	Sprinkler	Drip irrgation	Flooded bench	Sprinkler	Drip irrgation
Kalanchoë	640	1105	814	220	266	208
Ficus	886	1282	861	255	342	267
Spathiphyllum	623	852	644	138	223	163

After Van Gemert and Ploeger (1993).

Table 14.11 N and P balance sheet calculated from the total N and P in the substrate at the start and at the delivery stage of the crop, the monitored fertilization and the N and P uptake by the crop. Five potted green plant species were grown, viz. *Dieffenbacchia, Dracaena, Hedera,* palms, and *Nephrolepis* in flooded benches or concrete floor (closed systems and almost closed systems) or with overhead sprinklers (open systems)

	N kg ha^{-1} yr^{-1}			P kg ha^{-1} yr^{-1}		
Factors	Closed	Almost closed	Open	Closed	Almost closed	Open
Intput						
Potting soil	461	473	789	35	26	58
Top dressing	531	471	988	146	72	297
Total	992	943	1777	181	98	355
Output						
Potting soil	566	554	958	55	38	85
Plant	251	167	305	53	29	83
Output	816	721	1262	108	67	168
Unexplained	176	223	515	73	31	187
Total	992	943	1777	181	98	355

After Van der Burg and De Kreij (2002, 2003).

systems it appeared that the differences in N and P use of nurseries with potted plants were considerable, as shown in Table 14.10 by Van Gemert and Ploeger (1993). With overhead sprinklers the N use was 25–80% higher than with ebb and flood. For P the differences were smaller, because the P supply was mainly added by base dressing. In a consecutive study Van Gemert (1995) concluded that even within a group of nurseries furnished with closed systems the differences are substantial. Nevertheless the fertilizer use decreased by 35–60% in case of reuse of run-off water. However, reuse of run-off and drainage water is not commonly practiced, because of the possibility of an out break of root diseases. Disinfestation of the run-off and drainage water is necessary, as discussed in Section 10.10. Big differences in use of fertilizer among nurseries were also reported by Van der Burg and De Kreij (2002; 2003) in an investigation of mineral balances of N and P with green foliage plants, the data of which are summarized in Table 14.11. They concluded that the differences were due to crop characteristics, irrigation regime and the growing system used. Remarkably, the quantity for N in the potting mixture at the start was already 40–60% of the total quantity made available to the crop, while for P this was only 20–30%. At the termination of the crop 70–80% of the total N output is present in the potting soil. For P this has increased to 50–60%. This indicates the strong accumulation of nutrients in the potting soil as discussed earlier. Mind that these data are based on total analysis, a major part of the N and P is part of the organic matter and not readily available to the plants. The gap between the total input and total output was large and for the open systems it amount for N to one third of the total input and for P even to 50%. The gap of 18% and 40% for N and P respectively for the closed system indicate that in this system also considerable losses occur. This gap could be partly due to denitrification for

N, as Agner and Schenck (2005) showed that denitrification losses in potted plant production can be considerable. However, part of the gap will be explained by leakages or unattended leaching. For the open system the gap between input and output of course will be merely explained by the drain to waste.

Apart from the direct leaching, it is relevant to note that due to the strong accumulation in the substrate, a great part of the nutrients escape literally from the plant reach and in this way escape from the nutrient management of the crop. Virtually, it causes a reduction of the nutrient use efficiency and it is questionable whether the non-used fertilizers should be seen as harmful to the environment or not. In fact the consumer could alleviate the problem as the plants will be irrigated from the top and some of the nutrients become available again.

14.9 Special Aspects

Some potted plants require specific treatments. The flower colour of the pink/blue varieties of *Hydrangea* is known to vary greatly from pink to red and from purple to blue. The blue colouring is promoted by the addition of Al to the substrate and is supplied as $Al_2(SO_4)_3$. However this is not always effective. Van Leeuwen et al. (1994) demonstrated that the pH also is of great importance. A reduced pH level in combination with the supply of Al better guarantees the blueing, which is possible either by a close control on the liming treatment or by the supply of extra NH_4 (Van Leeuwen, 1999). Beside the supply of the quantity of Al also the time of the addition is important, which was studied in an experiment with different supplies of Al during the growing period. The results clearly showed that an early supply gave the best results, like shown by the data listed in Table 14.12 (Van Leeuwen et al., 2003). From the research data, target values for Al in plant tissue were defined. At least 20 – 40 mmol kg^{-1} dry matter should be present in full grown leaves.

Bromeliads form an important group of potted plants and are present with many species, varying in botanical types. Among them are terrestrial plants as well as epiphytes. These plants all originate from very infertile soils and the growth rate is subsequently low. *Bromeliaceae* differ from other potted plants in many ways. Some of the types developed a so called tank, like *Guzmannia*, *Vriesea* and *Aechmea*, in which they collect water and nutrients, which are absorbed by the leaves (Endres and Mercier, 2001). However, when planted in a substrate, the plants develop a root

Table 14.12 Al content in leaf tissue and flower colour as affected by Al treatment during several phases in the growing period of blue *Hydrangea* species (after Van Leeuwen et al., 2003). Flower colour visually judged, ranging from 0 = bad, 4 = good

Application period in days after planting	5–15	5–25	5–35	15–25	15–35	15–45
Al supply g/plant	4.5	4.5	4.5	4.5	4.5	4.5
Al plant mmol kg^{-1} dry matter	18.8	15.6		12.2	6.1	
Flower colour	3.4	3.5	3.3	3.1	2.8	2.2

Table 14.13 Total fresh weight and flower weight in g per plant of *Aechmea, Guzmania* and *Vriesea* species as affected by EC in the nutrient solution applied to the substrate by ebb and flood and by foliar application

	Aechmea		*Guzmania*		*Vriesea*	
EC nutrient solution dS m^{-1}	Plant	Flower	Plant	Flower	Plant	Flower
By ebb and flood						
0.0	497	40.5	120	51	172	48
0.5	508	40.6	105	51	182	50
1.5	578	40.4	115	46	171	46
2.5	587	39.9	115	47	166	43
By foliar application						
0	561	40.4	106	47	161	46
1	518	39.8	116	48	173	44

After Mulderij (1994).

system, which has shown to be capable to absorb water and nutrients (Kämpf, 1982). Nevertheless, it is common practice to use foliar application by means of overhead sprinklers. It was questioned whether these crops can be grown in an ebb and flood systems with solely nutrient solution application to the substrate. An investigation was carried out with *Aechmea, Guzmania and Vriesea,* grown in a standard potting mixture with only micro nutrients as a base dressing in an ebb and flood system. As treatments nutrient solutions with EC values of 0, 0.5, 1.5 and 2.5 dS m^{-1} were compared in combination with two treatments of leaf applications, viz 0 and 1 dS m^{-1}. The results as presented in Table 14.13 show that bromeliads are able to absorb sufficient nutrients either by roots or by leaves or in a combination of both. The results between foliar and root applications were small. With *Guzmania* and *Vriesea* the growth was slightly better with leaf application; with *Aechmea* the opposite effect was found. With *Guzmania* and *Vriesea* increasing EC was slightly negative, with *Aechmea* higher plant fresh weights were found with increasing EC values. An advantage of root zone application is less risks on salt crystallization on the leafs, which often occur in *Bromeliaceae* (Mulderij, 1994).

14.10 Plant Propagation

Plant propagation occurs generally in very small substrate volumes. All types of substrates are used, dependent on the requirements of the cultivation method and the substrates used in the production phase. The combination of substrates with strong different pressure heads on the moisture in the substrates can induce problems during cultivation. When plants propagated in a substrate with a high pressure head on the moisture are placed in a substrate with a much lower pressure head on the moisture, the water is instantly sucked from the propagating cube or pot and the water storage in the propagating volume is strongly reduced, while the plant roots have penetrated not yet the substrate on which the plant cubes are placed. This requires specific attention of the water supply during the starting phase. Therefore, combination of propagation in rock wool cubes for peat mixture substrates is less desirable. Upside down, when plants are propagated in peaty substrate and are placed on rock wool slabs, the peaty material sucks much water from the slabs into the propagating volume. The propagating material becomes too wet during the growing season, which easily induces difficulties by stem rot. This occurs for example when plants raised in peaty substrate cubes are placed on rock wool slabs. A side effect of such a combination is the high evaporation from the continuously wet pot and tremendous salt accumulations in the material just around the stem of the plant, when a possible nozzle of the irrigation system is not placed on the propagating cube.

Mineral substrates are not fertilised beforehand, but saturated with a nutrient solution, before plants or seeds are brought in. The cubes are saturated with nutrient solution of a concentration agreeing with an EC value between 1.5 and 2.5 dS m^{-1}, dependent on the plant type. The composition of the nutrient solution varies less for plant types, because of the short duration of the propagating period. For the propagation of vegetable plants in rock wool cubes the composition presented in Table 14.14 (Sonneveld and Straver, 1994) is often applied and will be suitable for the propagation of many flower plants too. Adjustment of the NH_4 concentration is advisable dependent on the substrate characteristics, the crop requirements and the growing period to keep the pH within acceptable values. During cultivation the concentration can be adjusted dependent on the measurements of the EC in the solution in the cubes. For vegetable plants often higher EC values up to 4.0 dS m^{-1} are maintained, to prevent a too lush growth of the young plants. The relationship between the EC supplied and the weight of tomato plants in an experiment of Boertje (1981) is shown in Fig. 14.11 for a spring-summer propagation period. The plant weight was highest with the supply of a nutrient solution with an EC of 1.8, while the EC of the solution in the cubes was about 4.8 during the propagation period. In a comparable experiment with propagation in winter also highest plant weights were obtained at an EC of 1.8 dS m^{-1}, but the plants were too luxurious and pale of colour. Good quality tomato plants were obtained at an EC of 2.7 and 3.6 in the supplied solution under the poor light conditions during winter in The Netherlands (Boertje, 1980).

Table 14.14 Composition of the nutrient solution for the propagation of vegetable plants in rock wool cubes

Macro nutrients mmol l^{-1}		Micro nutrients μmol l^{-1}	
NH_4	1.25	Fe	25
K	6.75	Mn	10
Ca	4.5	Zn	5
Mg	3.0	B	35
NO_3	16.75	Cu	1
SO_4	2.5	Mo	0.5
H_2PO_4	1.25		

Data of Sonneveld and Straver (1994).

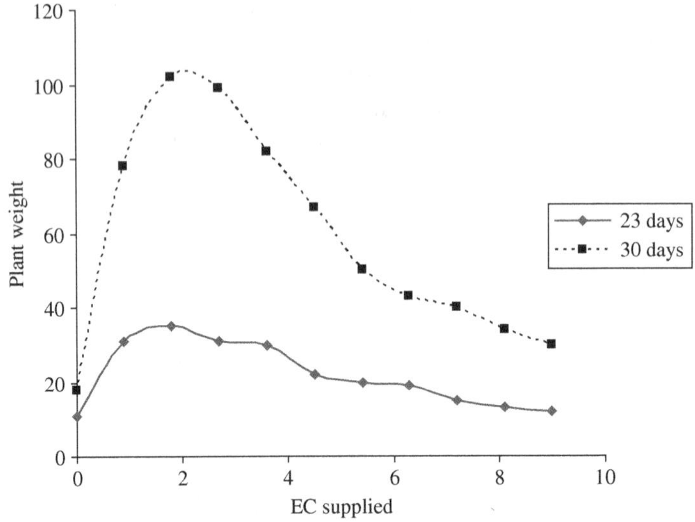

Fig. 14.11 Relationship between the EC of the nutrient solution supplied during propagation of tomato in rock wool cubes and the plant weight at 23 and 30 days after start. Data of Boertje (1981)

Comparable results are gained with the propagation of plants in substrates composed from natural organic materials (Boertje, 1975). Tomato plants grown in 1.2 l pots filled with organic potting compost under different fertilization regimes showed optimal growth with an application of 1–2 kg m^{-3} of a fertilizer mix N-P-K of 16-5-17. Beside this mix, the standard applications of lime, micro nutrients and P were added. Within the given range the fertilizer addition and the pot size can be adjusted to the expected plant size. Higher addition seriously decreased the growth. Under the given conditions tomato plants could be grown up to a plant weight of 50 g fresh weight, without top dressing. The use of small pots and low fertilizer applications quickly induce requirements for top dressings during propagation.

Peaty substrates are commonly fertilized with sufficient nutrients for the start and the first growing period of the propagation. Addition of fertilizers during the propagation can be necessary and depend on the ratio between the substrate volume used, the desired plant size and the plant type (Klapwijk and Mostert, 1992). For sowing and rooting of cuttings peaty substrates are lightly fertilized. For the raising of flower plants peaty substrates are moderately fertilized and for the raising of vegetable plants often heavily fertilized substrates are used. For definitions of the nutrient status see Table 11.12.

Some young plant types seem to be extremely sensitive for Mo deficiency, when propagated in peaty potting substrates. The addition of this element in peaty potting substrates surely is required for the raising of many young plants. Experience have been gained by tomato, lettuce and cauliflower (Van den Ende and Boertje, 1972) from which crops the last is best known as being most sensitive to Mo deficiency. See also the remarks about this in Section 11.4.3. The deficiency especially occurs if

Table 14.15 Index figures for Mo deficiency as affected by the pH of the growing medium and the addition of Mo as $(NH_4)_6Mo_7O_{24}.4H_2O$. Index: 0 – no symptoms and 10 – serious Mo deficiency

	Tomato μmol Mo l^{-1}		Lettuce μmol Mo l^{-1}	
pH	0	34	0	34
4.4 – 4.7	9	0	10	0
5.5 – 5.7	2	0	8	0
6.4 – 6.5	0	0	6	0

Data derived from Van den Ende and Boertje (1972). *Modified by permission of the International Society Horticultural Science*

the potting substrate is merely composed from high moor peat types (Roorda, 1965) and especially with low pH values (Boertje, 1979). Robinson (1987) even stated that healthy plants can be grown when the pH of the growing medium is sufficiently high. In Table 14.15 results are shown of an experiment with the propagation of young lettuce and tomato plants. Mo was yes or no added to the substrate at different pH levels. At a sufficient high value of the pH the symptoms of Mo deficiency completely disappeared for tomato, but not for lettuce. With the addition of 34 mmol m^{-3} Mo no symptoms were found at all pH levels. Lettuce seemed to be more sensitive than tomatoes. Thus, the quantities as mentioned in Table 11.15 suffice for nearly all crops.

References

Agner H and Schenk M K 2005. Peat properties and denitrification in cultures of potted ornamental plants. Europ. J. Hort. Sci. 70(3), 109–115.

Arnold Bik R 1965. De gevoeligheid van *Azalea indica* voor keukenzout in het gietwater. Meded. Dir. Tuinb. 28, 496–503.

Arnold Bik R 1972. The influence of water quality on the growth of azaleas (*Rhododendron simsii* Planch). Annual Report 1972 Experimental Station Floriculture Aalsmeer The Netherlands, 176 pp.

Arnold Bik R 1976. Quality in *Anthurium andreanum* and *Aechmea fasciata* grown in peat substrates as affected by nitrogen and potassium nutrition. Acta Hort. 64, 83–91.

Baas R Brandts A and Straver N 1995. Growth regulation of bedding plants and *Poinsettia* using low phosphorus fertilization and ebb-and flow irrigation. Acta Hort. 378, 129–137.

Bernstein L 1976. Physiological basis of salt tolerance in plants. Proc. Intern. Symp. Genetic Control Diversity in Plants Lahore Pakistan March 1976. Plenum Press New York, 283–290.

Boertje G A 1975. Potting composts. Annual Report 1971&1972 Glasshouse Crops Research and Experimental Station Naaldwijk The Netherlands, 35–38.

Boertje G A 1979. Molybdeengebrek in tomateplanten. Tuinderij 19(2), 24–25.

Boertje G A 1980. Results of liquid feeding in the production of bedding plants. Acta Hort. 99, 17–23.

Boertje G A 1981. Opkweek van tomaten in steenwolpotten in voorjaar en zomer. Tuinderij 21(9), 22–25.

Brumm I and Schenk M 1991. Chlor im Giesswasser von Azaleen. Deutscher Gartenbau 16/1991, 1030–1031.

Bulle A Verberkt H La Brijn L and De Jongh M A 1996. Invloed bemesting op groei en houdbaarheid potchrysant. Rapport 31 Proefstation voor Bloemisterij en Glasgroente Aalsmeer The Netherlands, 21, 68 pp.

Cox D A 2001. Growth, nutrient content, and growth medium electrical conductivity of *Poinsettia* irrigated by subirrigation or from overhead. J. Plant Nutr. 24(3), 523–533.

De Jong B 1987. Meer aandacht besteden aan bemesting en watergeven. Vakblad voor de Bloemisterij 42(11), 18–21.

De Kreij C and Straver N 1988. Flooded bench irrigation: effect of irrigation frequency and type of potting soil on growth of *Codiaeum* and on nutrient accumulation in the soil. Acta Hort. 221, 245–252.

De Kreij C 1989. Nutriëntenbalans, potgrond en gietfrequentie met eb – vloed watergeven bij Begonia. Rapport nr 84 Proefstation voor de Bloemisterij Aalsmeer The Netherlands, 8 pp.

Endres L and Mercier H 2001. Influence of nitrogen forms on the growth and nitrogen metabolism of bromeliads. J. Plant Nutr. 24(1), 29–42.

Fisher P R Huang J and Agro W R 2006. Modelling lime reaction in peat-based substrates. Acta Hort. 718, 461–468.

Kämpf A N 1982. Untersuchungen zu Düngung und Wachstum von zisternenbildenden Bromelien. Diss. Fac. Landw. und Gartenbau Technischen Univ. München, 42 pp.

Kipp J A Wever G and De Kreij C 2000. International Substrate manual. Elsevier International Business Information Doetinchem The Netherlands, 94 pp.

Klapwijk D and Mostert J 1992. Richtlijnen voor de productie van potgronden en substraten. Proefstation voor de Bloemisterij en Glasgroenten Naaldwijk The Netherlands Informatiereeks 73, 64 pp.

Mulderij G E 1993. EC trappen bij *Asplenium* and *Nephrolepis*. Proefstation voor Bloemisterij en Glasgroente Aalsmeer The Netherlands. Rapport 161, 31 pp.

Mulderij G E 1994. Bemesting bij *Bromeliaceae*. Proefstation voor Bloemisterij en Glasgroente Aalsmeer The Netherlands. Rapport 186, 28 pp.

Mulderij G E 1998. Eenjarige zomerbloeiers: EC en kwaliteit Proefstation voor Bloemisterij en Glasgroente Aalsmeer The Netherlands, Rapport 128, 37 pp.

Mulderij G E 1999. Uitval bij *Dracaena marginata,*: een inventarisatie: onderzoek naar het voorkomen van uitval en bladpunten bij *Dracaena marginata* in de winterperiode. Proefstation voor Bloemisterij en Glasgroente Aalsmeer The Netherlands. Rapport 91, 21 pp.

Mulderij G E 1999a. Bladpunten bij Palmen. Proefstation voor de Bloemisterij en Glasgroenten, Aalsmeer The Netherlands, Rapport 214, 44 pp.

Mulderij G E 2000. De invloed van licht en voeding op de badkleur van *Chamaedorea*. Proefstation voor Bloemisterij en Glasgroente Aalsmeer The Netherlands. Rapport 236. Rapport 91, 29 pp.

Mulderij G E and De Jongh M 1995. Fe bemesting bij palmen, de invloed van pH en Fe-chelaat op groei, gewassamenstelling en houdbaarheid van C*hamaedorea en Chrysalidocarpus*. Proefstation voor Bloemisterij, Aalsmeer The Netherlands. Rapport 208 Rapport 91, 15 pp.

Mulderij B and Hüner L 1997. Invloed van pH, Fe en EC op groei en ontwikkeling van *Epipremnum*. Proefstation voor Bloemisterij en Glasgroente Aalsmeer The Netherlands. Rapport 102, 36 pp.

Otten W 1994. Dynamics of water and nutrients for potted plants induced by flooded bench fertigation: experiments and simulation. Thesis, Wageningen University, 115 pp.

Poole R T Osborne L S and Chase A R 1984. Pothos. Foliage Digest 7: 2 1–3.

Röber R 1987. The influence of water quality on the growth of azaleas (*Rhododendron simsii* Planch). Gartenbauwissenschaft 52(2), 82–86.

Roorda van Eysinga J P N L 1965. Development of a pure peat mixture for raising plants with blocks. Publikatie no. 107 Proefstation voor de Groenten en Fruitteelt onder Glas Naaldwijk The Netherlands, 78 pp.

Robinson J B D 1987. Diagnosis of mineral disorders in plants volume 3, Glasshouse Crops. Ministry of Agriculture Fisheries and Food, London, 168 pp.

Sonneveld C 2000. Effects of salinity on substrate grown vegetables and ornamentals in greenhouse horticulture. Thesis Wageningen University, The Netherlands, 151 pp.

Sonneveld C and Straver N 1994. Nutrient solutions for Vegetables and Flowers grown in water or substrates. Research Station for Floriculture and Glasshouse Vegetables, Aalsmeer/Naaldwijk, The Netherlands, Series: Voedingsoplossingen Glastuinbouw, no. 8, 45 pp.

Sonneveld C and Voogt W 2001. Chemical analysis in substrate systems and hydroponics – use and interpretation. Acta Hort. 548, 247–259.

Straver N 1988. Groeiproblemen Saintpaulia houden verband met pH. Vakblad Bloemisterij 41, 40–41.

Straver N 1990. Kalium/Calcium- bemesting bij *Spathiphyllum*. Proefstation voor de Bloemisterij Aalsmeer The Netherlands. Rapport 122, 15 pp.

Straver N 1991. Bemestingsproeven bij *Euphorbia pulcherrima*. Proefstation voor Bloemisterij Aalsmeer The Netherlands. Rapport 107, 22 pp.

Straver N 1991a. Bemestingsproeven bij *Codiaeum*. Proefstation voor de Bloemisterij Aalsmeer The Netherlands. Rapport 108, 23 pp.

Straver N 1991b. Bemestingsproeven bij Begonia. Proefstation voor de Bloemisterij Aalsmeer The Netherlands. Rapport 110, 24 pp.

Straver N 1991c. Kalium/calcium bemesting bij palmen *(Chrysalidocarpus, Howeia)*. Proefstation voor Bloemisterij Aalsmeer The Netherlands. Rapport 120, 22 pp.

Straver N 1997. pH's in substraat en Mn bemesting bij *Schefflera*. Proefstation voor Bloemisterij en Glasgroente Aalsmeer The Netherlands. Rapport 91, 21 pp.

Straver N De Kreij C and Verberkt H 1999. Bemestingsadviesbasis potplanten. Proefstation voor Bloemisterij en Glasgroenten Naaldwijk The Netherlands, 55 pp.

Surrage V A, Carlile W R 2008. Variation in quality of composted green wastes of UK. Origin and their suitability for inclusion in growing media Acta Hort. 779, 631–636.

Van den Ende J and Boertje G A 1972. Molybdenum deficiency in young lettuce and tomato plants. Acta Hort. 26, 61–67.

Van der Burg A A M and De Kreij C 2002. Stikstof en fosfor bij groene en bonte planten. Praktijkonderzoek Plant en Omgeving Naaldwijk The Netherlands. Report no 563, 32 pp.

Van der Burg A A M and De Kreij C 2003. Stikstof en fosfor bij groene en bonte planten deel 2, onderzoek op bedrijven met *Calathea, Ficus en Codiaeum*. Praktijkonderzoek Plant en Omgeving Naaldwijk The Netherlands. Report no 16 3 96, 22 pp.

Van Gemert J and Ploeger C 1993. Milieu onderzoek; grote verschillen in middelengebruik. Vakblad Bloemisterij 48(4), 38–39.

Van Gemert J 1995. Niet recirculatie met druppelaars zelfde als recirculatie. Vakblad Bloemisterij 50(4), 30–31.

Van Leeuwen G J L 1992. Groeiregulatie bij *Pelargonium* door middel van EC. Proeftuin Noord Nederland Klazienaveen The Netherlands. Verslag no 30, 8 pp.

Van Leeuwen G J L 1992a. Groeiregulatie van potchrysant. Proeftuin Noord Nederland Klazinaveen The Netherlands Verslag nr 31, 13 pp.

Van Leeuwen G J L 1993. Voorraadbemesting en overbemesting bij *Saintpaulia*. Proeftuin Noord Nederland Klazinaveen The Netherlands. Verslag 32, 15 pp.

Van Leeuwen G J L 1999. Blauwen hortensia in trekfase doeltreffend : met de juiste strategie naar een goed resultaat. Vakblad Bloemisterij, 54(10), 66–67.

Van Leeuwen G J L and Bulle A A E 1992. Effecten EC and NaCl bij *Azalea*. Proeftuin Noord Nederland Klazinaveen The Netherlands. Verslag 33, 16 pp.

Van Leeuwen G J L Van Buren J Dijkstra T 1994. Met minder aluminium toch blauwe bloemen. Vakblad Bloemisterij 49(28), 32–35.

Van Leeuwen G J L Van Noort F R and Straver N A 2003. Teelt van *Hydrangea*. Rapport nr 585 Praktijkonderzoek Plant en Omgeving Aalsmeer The Netherlands, 109 pp.

Verberkt H and De Jongh M A 1995. Invloed bemesting op groei en kwaliteit *Cyclamen*. Proefstation voor Bloemisterij en Glasgroente Aalsmeer The Netherlands Rapport nr1, 34 pp.

Verberkt H De Jongh M A and Schaefer R 1996. Invloed voedingsconcentratie (EC) op groei en kwaliteit *Kalanchoe*. Rapport 61 Proefstation voor Bloemisterij en Glasgroenten Aalsmeer The Netherlands, 47 pp.

Verberkt H and Van den Berg Th 1993. Invloed EC op groei *Impatiens* New Guinea. Proefstation voor de Bloemisterij Aalsmeer The Netherlands. Proefverslag 1508–31, 19 pp.

Verberkt H Van den Berg D and Huner L 1998. Invloed voedingselementen op bladbeschadigingen bij *Spathiphyllum*. Rapport 124 Proefstation voor Bloemisterij en Glasgroenten Aalsmeer The Netherlands, 31 pp.

Vogelezang J Cuijpers L and De Graaf van der Zande M T 1992. Growth regulation of bedding plants by reversed day/night temperature only? Acta Hort. 305, 37–43.

Warmenhoven M and Van Noort F R 2005. Fosfaatbeperking in eenjarige zomerbloeiers. Praktijkonderzoek Plant en omgeving, Aalsmeer The Netherlands. Rapport 41717061, 44 pp.

Wever G 2002. Numerous options when considering a growing medium. Flower Tech. 5(7), 34–36.

Chapter 15
Fertigation in Soil Grown Crops

15.1 Introduction

The word fertigation is derived by a composition from the words fertilization and irrigation and the action expressed by it is exactly what the word suggests: fertilization and irrigation in one activity. Since long years fertigation is a common practise in greenhouse industry. The development of this method of fertilizer supply originated from the fifties of the 20th century. In these years a beginning was made with drip irrigation and the small spots wetted by this type of irrigation did not offer any possibilities for top dressings by hand. Therefore, with the introduction of drip irrigation also fertilizer diluters were introduced with which concentrated fertilizer solutions could be added to irrigation water streams. Different diluter systems have been developed, but the dilutions realised with these systems were not very precisely. This was not a strong handicap in the beginning, because the water irrigated was supplied by drip irrigation and did not touch the plant canopy. Thus, an accidental somewhat high concentration of fertilizers in the irrigation water did not affect the plant negatively. Afterwards, control and adjustments on the applied quantity of fertilizer always was possible and utmost, in that period the addition was still traditionally based on quantities of fertilizer per area. Later on, when overhead irrigation by sprinkler irrigation was developed, as a matter of course precise dilutions were required. This was essential for overhead sprinkling to prevent leaf damage by possibly high concentrations of fertilizers, as a result of an inaccurate function of the equipment. This precise addition of fertilizers was developed by on line measurement of the electrical conductivity (EC) of the irrigation water combined with injectors for the dosage of concentrated nutrient solution in the water stream. The increase of the EC in the irrigation water was used as a unit for the fertilizer concentration. The once attuned concentration is controlled by proportional adjustment of the injectors on basis of continuously measurements of the EC.

C. Sonneveld, W. Voogt, *Plant Nutrition of Greenhouse Crops*,

DOI 10.1007/978-90-481-2532-6_15,

15.2 Technical Equipment

The outline of an installation for fertilizer addition to irrigation water is shown in Fig. 15.1. Water and fertilizer solution is transported to the plant by suction and pressure of the pump installation. Mostly different reservoirs with concentrated fertilizer solutions are at disposal. A concentrated solution can be the solution of a single fertilizer as well a mixture of several fertilizers. Concentrated fertilizer solution is supplied by the injector from the reservoir from which the stopcock is opened. The injector is controlled by the command centre and realizes the required fertilizer concentration in the irrigation water by the results of continuous measurements of the EC. With the nowadays modern apparatus a more or less constant fertilizer concentration in the irrigation water is ensured. The concentration of fertilizers in the reservoirs is not critical, because the eventual concentration of fertilizers realised in the irrigation water for the plant is determined by the set point on the command centre. In common practice the fertilizer concentration of the stock solution in the basins mostly varies between 10 and 15% and preferably should not exceed the saturation point.

The control on the addition of the fertilizers to the irrigation water is based on the relationship between the EC of a solution and the concentration of specific mineral fertilizer or fertilizer mixture in it. This relationship reflects an increasing EC with an increasing concentration of fertilizer and can be linearly approximated within certain limits. The relationship depends on the type of the mineral salt(s) of which the fertilizer is composed and the temperature of the solution in which the EC measurement is carried out. The effect of the temperature on the EC is automatically compensated by the apparatus and the results are related to those at a temperature of 25°C. Thus, the EC reflected by the apparatus is directly connected with the concentration of each specific fertilizer. The missing link is the relationship between the

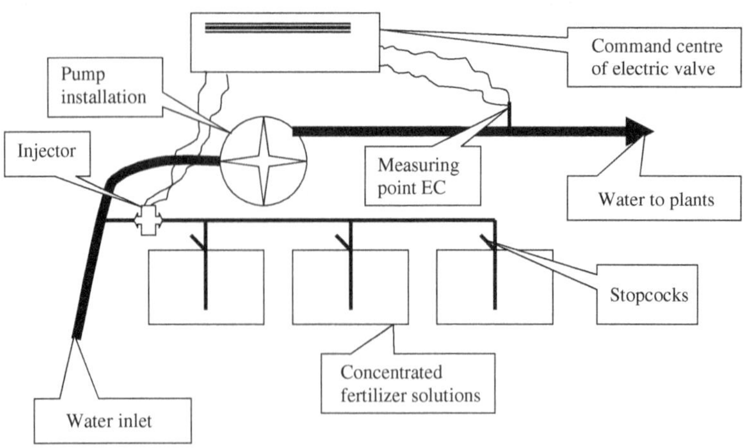

Fig. 15.1 Installation for addition of fertilizers to irrigation water

required concentration and the EC. This relationship, however, is separately published for different mineral salts and fertilizers (Sonneveld et al., 1966).

With the operation of the installation, the EC of the irrigation water without fertilizer addition is firstly measured. Secondly the grower should know what EC agrees with the fertilizer concentration that is required. The sum of both values is the EC set point on the command centre.

Sometimes, in the fertilizer solution mineral acids like HNO_3, H_2SO_4 or H_3PO_4 are incorporated in the nutrient stock solution to neutralize possible high carbonate concentrations in the primary irrigation water. For the chemical reaction and application, see Section 12.4. Such only can be recommended when an inline measurement of the pH is incorporated in the installation, to prevent accidentally too low pH values in the irrigation water. Such values are dangerous for the above ground part of the crop with overhead irrigation, but also for the roots with spot irrigation. It is experienced that the intensive water supply close to the roots in such cases seriously will damage the roots. The use of HNO_3 is most obvious in such cases, because S and P are often sufficiently available in greenhouses soils during cultivation. The neutralization of carbonate in water with acids makes this water aggressive, caused by the released CO_2 left behind in the water. Such water strongly promotes corrosion of metal parts of the equipment in the greenhouse. Therefore, the technical equipment of the full irrigation system, dosing unit and irrigation lines, preferable should be free of metal fittings.

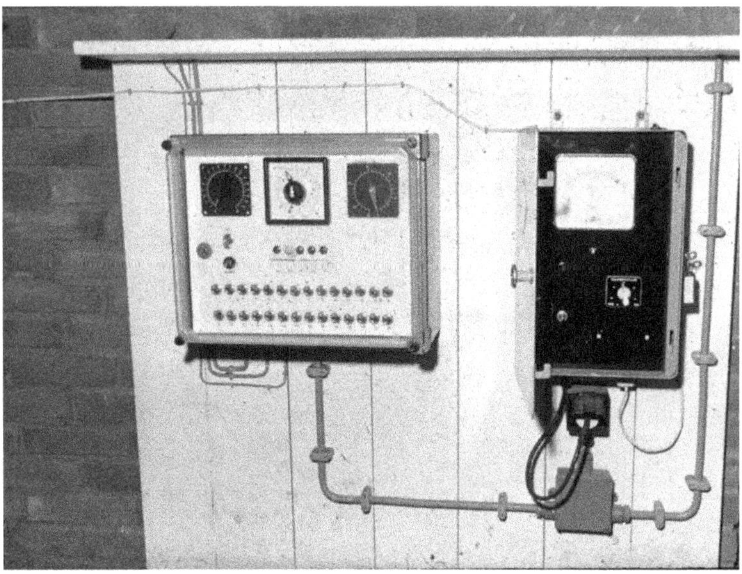

Picture 15.1 A simple instrument for measurement of the conductivity used with fertigation of soil grow crops

15.3 Fertilizers and Addition

Solutions of electrolytes conduct electricity. As mentioned before, this conductivity depends on the temperature of the solution, the concentration of ions and the character of the ions. Temperature differences are compensated automatically by the apparatus in function. It is widely agreed that the results are related to 25°C. With respect to the character of the ion the valence and the activity affect the electric conduction and when the concentration is expressed on mass/mass basis, also the atomic weights play an important part. In Fig. 15.2 the relationship between concentrations of KNO_3, K_2SO_4 and $MgSO_4.7H_2O$ (Epsom salt) solutions in mg l^{-1} on the one hand and the EC of these solutions on the other hand are shown. The relationships for KNO_3 and K_2SO_4 are more or less equal, but those for $MgSO_4.7H_2O$ differ much. This can be explained by the high quantity of crystalline water, connected to the $MgSO_4$ molecule and the low activity of both the Mg and the SO_4 ion, which activity strongly decreases with increasing concentration. At very low concentrations the relationship between concentration and EC is somewhat curve linear, which especially counts for salts containing ions with a low activity coefficient like Mg, Ca and SO_4. But for most fertilizers, the relationships for concentrations between 0.5 and 8 g l^{-1} can be linearly approximated. This offered the opportunity to introduce the concept of specific fertilizer EC values (EC_f), being the increase of the EC of a solution by addition 1 g of that specific fertilizer (Sonneveld, 1976 and 1982). Specific EC values of different fertilizers (EC_f) are listed in Table 15.1. Urea is not a mineral salt and thus, does not conduct electricity. However, it is a highly soluble compound and suitable for fertigation. Therefore, only urea mixed in a desirable ratio with a mineral fertilizer can be used for fertigation.

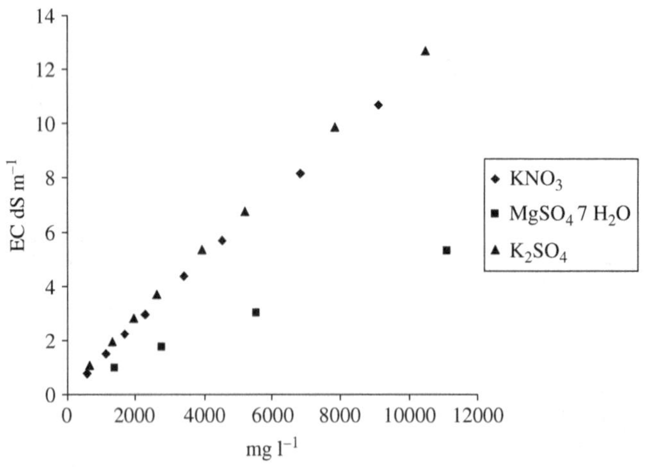

Fig. 15.2 Relationship between concentrations of KNO_3, K_2SO_4 and $MgSO_4.7H_2O$ (mg l^{-1}) and the EC (dS m^{-1}) of the solution

Table 15.1 Specific EC values of fertilizers, expressed as the increase of the EC followed by addition of every 1 gram fertilizer per litre water. The values are applicable to concentrations between 0.5 and 8 g l^{-1}

Fertilizer	Chemical composition	Specific EC value (EC_f)
Potassium nitrate	KNO_3	1.35
Calcium nitrate	$5(CaNO_3)_2.2H_2O.NH_4NO_3$	1.24
Ammonium nitrate	NH_4NO_3	1.64
Ammonium sulphate	$(NH_4)_2SO_4$	1.90
Urea	$CO(NH_2)_2$	0.00
Mono potassium phosphate	KH_2PO_4	0.68
Mono ammonium phosphate	$NH_4H_2PO_4$	0.86
Potassium sulphate	K_2SO_4	1.54
Magnesium sulphate	$MgSO_4.7H_2O$	0.94
Magnesium nitrate	$Mg(NO_3)_2.6H_2O$	0.84

After Sonneveld (1982).

Beside the traditional single fertilizers, there are numerous compound fertilizers suitable for fertigation. The EC_f values of these fertilizers are divers and depend on the mineral salts used for the composition. The EC_f values are determined by the producer and will be mentioned on the packing. Compound fertilizers also are obtained by mixing of traditional single fertilizers in the same basin. Not all fertilizers can be mixed together in the same stock solution. Ca containing salts must be separately kept from SO_4 or P containing types, because these components precipitate as $CaSO_4$, $CaHPO_4$ or $Ca_3(PO_4)_2$. In Table 15.2 some examples of the mixing of fertilizers are given. The data necessary for the calculations can be

Table 15.2 Examples of the preparation of compound fertilizers with determined ratios of nutrients by mixing of single fertilizers

		Contribution to nutrients in %			
Fertilizers used	Parts to the total	N	K	Mg	Contribution to EC_f of the mixture
Required composition N:K:Mg = 1:1:0.5					
KNO_3	0.275	3.6	10.5		0.371
NH_4NO_3	0.197	6.9			0.323
$MgSO_4.7H_2O$	0.528			5.2	0.496
Contribution to the total		10.5	10.5	5.2	1.190
Required composition N:K:Mg = 1:2:0.5					
KNO_3	0.469	6.1	17.8		0.633
NH_4NO_3	0.080	2.8			0.131
$MgSO_4.7H_2O$	0.451			4.5	0.424
Contribution to the total		8.9	17.8	4.5	1.188

found in Section 2.2, where the compositions of fertilizers are given and in Table 15.1, where the EC_f values of the single fertilizers are listed. The mixtures composed in this way are mostly cheaper than the compound fertilizers produced by factories.

Fertilizers suitable for fertigation must be easily soluble in water and may not contain significant insoluble residues, because of blocking if the irrigation system and pollution of the crop by overhead irrigation. Some fertilizers are available in various trade marks and qualities. For fertigation it is advisable to use high quality types, preferable without any insoluble component. Especially when the fertilizer is used with drip irrigation, the insoluble components easily block the narrow canals in the nozzles. This quickly causes an uneven water distribution by the drippers.

The addition of P to the irrigation water should be prevented as much as possible, especially if the pH of the irrigation water is above 6.5 and it contains some Ca, which often occurs in water used for irrigation of soil grown crops. P easily precipitates under these conditions and blocks the nozzles of drip irrigation systems. Moreover, P easily precipitates in the top soil layer and scarcely arrives at the roots. Therefore, the required P is preferably given as a base dressing. Only in specific cases P is supplied as a top dressing as will be discussed in Section 16.5.

15.4 Leaf Damage

With overhead irrigation leaves of crops can be damaged by necrosis followed by too high concentrations of mineral salts. This was already experienced with foliar sprays used to correct nutrient disorders during crop cultivation. Fertilizers containing NH_4 or urea were most aggressive when used for foliar application (Sonneveld, 1962). Overhead irrigation showed comparable results by the occurrence of leaf scorch, which was most evident with NH_4 containing fertilizers (Van der Post and Sonneveld, 1961). In advance it was concluded that fertilizers like $(NH_4)_2SO4$ could be applied with overhead sprinkling at a concentration of ½ atmosphere (50 kPa) osmotic pressure (Sonneveld and van den Ende, 1967). The NH_4 concentration of such a solution agrees with 5 mmol NH_4 l^{-1}. This is in good agreement with experiments carried out later on in which leaf scorch occurred at comparable NH_4 concentrations as shown in Fig. 15.3 (Sonneveld and Voogt, 1981). Up till a concentration of 4 mmol NH_4 l^{-1} there was hardly any leaf scorch. At higher concentrations the symptoms increased linearly with increasing concentrations. From this it will be concluded that with overhead sprinkling of NH_4 containing fertilizers a very precise control on the concentration is required to prevent leaf scorch. It is well known that the occurrence of leaf scorch is also affected by the type of crop, the crop condition and the climatic conditions. The effect of these factors on the occurrence of leaf scorch cannot be estimated quite well and therefore, the use of NH_4 containing fertilizers with overhead irrigation should be prevented as much as possible. Fertilizers containing N as NO_3 are much more safe with overhead fertigation than those containing N as NH_4.

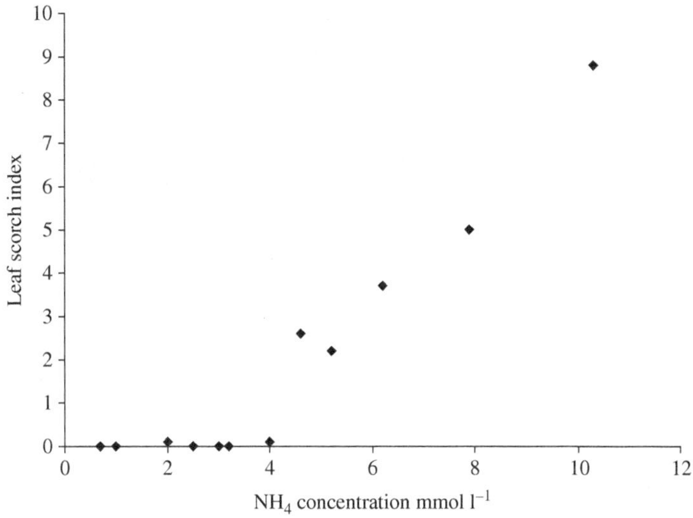

Fig. 15.3 Relationship between the NH_4 concentration in the sprinkling water with overhead irrigation and the leaf scorch of tomato. Index for leaf scorch: 0 – no and 10 – serious. Derived from Sonneveld and Voogt (1981). *Reprinted by permission of the Koninklijke Landbouwkundige Vereniging*

15.5 Irrigation Systems

The irrigation systems used for fertigation differ much in capacity and configuration (Heemskerk et al., 1997). Generally the systems can be divided in three groups, like following.

- Systems with which the whole soil surface in the greenhouse is irrigated. The overhead sprinkler irrigation system is the most representative system for this. Generally, the spray lines are highly placed in the greenhouse, some metres above the soil surface, to obtain as much as possible an equal distribution of water over the whole greenhouse area. This system is widely used for crops with a high planting density in combination with a short growing period, like many cut flowers and leafy vegetables. When crops grow up until the same level as the spray lines, the distribution of water in the second growth stage is hindered by the height of the crop. The spray lines then often are lowered to about 30 cm above the soil surface of the growing bed, between the plant rows. Generally, at that growth stage the lower leaves of the crop are removed, like for tomato, or sufficiently senescent as for sweet pepper and cucumber. In this way, sufficient space becomes available for the spray lines (Picture 15.2) of the irrigation system under the active growing part of the crop.
- Systems with which strips are irrigated by low levelled spray lines or lay-flat tubes. The width of the strip wetted by low levelled spray line irrigation mostly is larger than with lay-flat irrigation. In both situations the tube is situated between

Picture 15.2 Low level sprinkler irrigation used with strip irrigation of soil grown crops

the plant rows or in the growing beds. Sometimes the tubes are placed in shallow gullies to restrict the wetted area. These systems are used for fruit vegetables and for cut flowers with a long growing period.

- Systems with which spots are irrigated. Only small spots of the area are wetted by irrigation, like with drip irrigation systems. These systems often are used for crops with a low planting density like fruit vegetables.

With overhead sprinkling the whole area of the greenhouse is fertilized and high quantities of fertilizer are necessary to correct the nutrient status of the soil. With strip and spot irrigation only part of the greenhouse area is in use and adjustments of the nutrient status are easier to realise, because of the much smaller volume used by the plant roots. Thus, with respect to the management of the nutrient status the strip and sprinkler irrigation is preferable. Another advantage is the fact that during cultivation the crop is not wetted by the irrigation and paths remain dry. This is an advantage for the workers in the greenhouse, while the fact that the crop remains dry with the irrigation also is an advantage for the crop. In experiments with irrigation systems yields showed a tendency to be higher with strip and spot irrigation than with overhead irrigation (Van den Ende and De Graaf, 1974).

15.6 Nutrient Distribution and Irrigation Systems

In relation to the water distribution also the distribution of the nutrients in the soil shows great variability. This is clear from the data shown in Table 15.3, where the analytical data are shown of samples gathered from dry and wet spots with strip

Table 15.3 Contents of water soluble nutrients of a sandy soil over a depth of 0.25 m. Samples were gathered separately from dry and wet spots with strip irrigation of a tomato crop. Contents are expressed as mmol kg^{-1} dry soil

	N		K		Mg	
Weeks after planting	Wet	Dry	Wet	Dry	Wet	Dry
12	5.7	6.4	3.2	3.8	2.0	3.7
17	2.9	9.3	2.5	4.5	1.5	5.7
24	2.9	11.4	1.9	6.2	1.5	7.7

After Van den Ende and De Graaf (1974). *Modified by permission of the International Society Horticultural Science*

irrigation of tomato (Van den Ende and De Graaf, 1974). On the wet spots the nutrient concentrations in the soil are quickly adjusted to the concentrations in the nutrient solution added, while on the dry spots permanent accumulation occur. The distribution of NO_3 in a greenhouse soil with drip irrigation already is shown in Fig. 4.4 (Sonneveld et al., 1991). The wet spots are restricted to relatively small areas in which the concentration is low, while in the surrounding area strong accumulation occurs. The plant distance was 0.60 m, thus the NO_3 concentration in the 1:2 volume extract fluctuated between 0.7 and 9.8 in a volume of 0.3 m diameter and 0.4 m depth. Comparable results were found by Al-Harbi et al. (2005); Hoffman (1986); Oster et al. (1984) and Papadopoulos (1988).

Apparently, the nutrient concentrations on small distances with strip and spot irrigation differ that much, that it is not realistic to take a random sample for the determination of the nutrient status of the soil. It is likely that for sampling during the growing period two samples are necessary to get an impression of the nutrient status: one from the wet places and one from the dry places. These samples will represent the highest and the lowest nutrient concentrations available to plants. However, supposing that the concentration of the soil solution under the drippers is in equilibrium with the nutrient solution added, often one sample can be considered as enough. In that case, the samples will be gathered from the verges of the wet places, being the places where active roots are in contact with accumulated nutrients. Dependent on the situation other considerations will play a part too, as discussed in Section 4.14. Interpretation of the results presented, should be made on basis of the discussion about effects of unequal distribution of nutrients presented in Chapter 8. The water uptake is mainly determined by the low concentrated parts, while for the nutrient uptake the high concentrated parts are important. Thus, there should be equilibrium between the concentration supplied and the accumulation in the dry spots. On the one hand, an unlimited accumulation in the dry spots is senseless and induces high environmental pollution of nutrients when the soil is leached after the cropping period. On the other hand, too low concentrations in the drip solution to prevent accumulation in the dry parts can affect for example fruit quality of the produce. Details for interpretation will be presented in Chapter 16.

15.7 The Use of NH_4 and NH_2 Fertilizers

Fertilizers containing NH_4 are widely used in greenhouse culture, in case of fertigation, urea is also a suitable fertilizer. In greenhouse soils NH_4 supplied is quickly converted to NO_3, due to abundant microbial activity and high soil temperature. Noticeable concentrations of NH_4 only will be found during the first weeks after steam sterilisation of greenhouse soils (Sonneveld, 1979). With the nitrification process following conversion occurs.

$$NH_4 + 2\,O_2 + H_2O \rightarrow NO_3 + 2\,H_3O \tag{15.1}$$

Thus, with the conversion of NH_4 to NO_3 considerable concentrations of acid are released, which will lower the pH of the soil.

With the use of N in the form of urea – $CO(NH_2)_2$ – this compound is converted to NH_4 by hydrolyses as following.

$$CO(NH_2)_2 + 2H_2O \rightarrow NH_4 + CO_2 + H_2O \tag{15.2}$$

The CO_2 released with this reaction will be neutralized by the nitrification of the NH_4 that follows the on the hydrolysis step. Hydrolyses of NH_2 can be disturbed by absence of Ni (Marschner, 1997). Under such conditions high concentrations NH_2 can occur, which can be toxic to crops. Such low Ni is not logic under soil grown conditions. It never has been noticed in greenhouse crops.

It can be supposed that with frequent application of NH_4 containing irrigation water, the plant absorbs a substantial part of the N as NH_4. In this case the acidification of the soil will be equal to the acidification by nitrification of NH_4. Instead of NO_3 the plants absorbs NH_4, by which the cation absorption is increased and the plant releases H_3O instead of OH with the absorption of NO_3 (Van Beusichem, 1984). Thus, the absorption of one NH_4 is equivalent with two H_3O ions in comparison with the absorption of one NO_3 ion. Naturally, this only is true under the condition that the total N uptake of the crop is not affected.

Effects of different N forms on development of soil grown crops and on soil characteristics are shown by the results of an experiment with different flower and vegetable crops (Van den Bos, 1991). During a serie of years vegetables and flowers were grown with drip irrigation with a continuous supply of 8 mmol N per litre of water. In the different treatments the N was partly supplied as NO_3 and partly as NH_4 or NH_2, as shown in Table 15.4. In one treatment the N was completely supplied as NO_3, which treatment acted as control. The experiment was carried out in containers with calcareous loamy sand with about 3% $CaCO_3$ and a pH of 7.0. In a period of 7 years 4 different vegetable and 4 different flower crops were grown. The yields of most crops did not show significant differences. The gerbera crop, however, showed great differences in yield between the treatments in which the N was completely given as NO_3 and partly as NH_4 or NH_2. The total weight of the flowers harvested at the treatment with 50% NH_4 with the cultivar Bismut was about 40% higher than at the treatment with a full NO_3 fertilization. This may be explained

Table 15.4 Yield and leaf colour of two gerbera cultivars as affected by different N forms. The yield is expressed by the number of flowers per plant and the flower weight in g. The leaf colour is expressed by an index; 0 – completely yellow and 10 – completely green

N form	Bismut			Eoliet		
	Number	Weight	Colour index	Number	Weight	Colour index
100% NO_3	38	21	4.6	47	25	7.1
75% NO_3 and 25% NH_4	44	22	6.0	50	25	8.4
50% NO_3 and 50% NH_4	51	22	7.3	54	23	8.7
75% NO_3 and 25% NH_2	41	21	4.7	47	25	7.6
50% NO_3 and 50% NH_2	44	22	5.4	43	25	8.0

After Van den Bos (1991).

by the reduction of chlorosis in the leaves, caused by the NH_4 related pH decrease in the root environment. The effect of NH_2 addition is much lower than those of NH_4, which is understandable because of the restricted effect of this N form on the pH of the soil. The cultivar Eoliet is less affected by the N form, because this cultivar is not sensitive for chlorosis. With NH_2 supply no positive effect on the production is found with cultivar Eoliet.

The use of NH_4 accelerates the decomposition of carbonates in soils and suppresses the pH, while the concentrations soluble Ca and Mg increases, as shown in Table 15.5. With the use of NH_2 comparable effects have been found in the soil, but to a lesser extent. The uptake of Mn was surely improved by the NH_4 addition, while the Fe uptake was scarcely affected. However, the total Fe concentration in plants as determined in this experiment often is not a right measure for the activity of this element in the plant, see Section 5.4. The activity of Fe in plants can be affected by many factors and the supply of NH_4 as N form surely is one of them (Sonneveld and Voogt, 1994). Thus, the effect on the chlorosis can be explained by Mn uptake, but also the activity of Fe in the plant plays a part.

Table 15.5 The carbonate content in the soil, expressed as % $CaCO_3$ of the dry soil, the pH (water) during the growing period, the concentrations Ca and Mg (mmol l^{-1} in the 1:2 volume extract) and the concentration Mn and Fe in young gerbera leaves (mmol kg^{-1} dry matter)

N form	Soil				Young leaves			
					Bismut		Eoliet	
	$CaCO_3$	pH	Ca	Mg	Fe	Mn	Fe	Mn
100% NO_3	3.6	7.4	0.8	0.4	0.80	0.25	0.96	0.29
75% NO_3 and 25% NH_4	2.7	6.6	1.2	0.5	0.83	0.35	0.92	0.54
50% NO_3 and 50% NH_4	2.0	6.3	2.7	0.8	0.91	1.13	1.09	1.31
75% NO_3 and 25% NH_2	3.4	7.1	1.0	0.4	0.82	0.24	1.01	0.33
50% NO_3 and 50% NH_2	3.2	7.0	1.2	0.4	0.81	0.29	0.95	0.34

After Van den Bos (1991).

The data presented show that with crops sensitive to chlorosis grown in calcareous soils the yield and the leaf colour surely can be improved by the use of substantial quantities of NH_4. In soils rich on carbonate relatively high NH_4 concentrations have to be used to get sufficient effect. However, in such cases no overhead irrigation can be applied in view of the risk of leaf scorching, as discussed in Section 15.4. The risk of NH_4 toxicity in the root environment under these conditions is negligible because of the usually quick nitrification process in greenhouse soils.

15.8 Concentrations

The concentration of nutrients added to the irrigation water necessary for an optimal yield depend on factors like crop, irrigation system, base dressing and growing conditions. Required concentrations for different crops and growing conditions will be discussed in Chapter 16. In this section some general remarks are presented.

In a series of experiments with strip irrigation with fruit vegetable crops different fertilizer mixes, containing N, K and Mg, were added to the irrigation water (Sonneveld and Voogt, 1981). The nutrients were continuously supplied to the irrigation water and the EC was increased by the addition of different nutrients with values between 0.45 and 1.80 dS m^{-1}. The crop yield was highest in the range between 0.45 and 0.90 dS m^{-1}, which is comparable with concentrations of 0.3 and 0.5 g l^{-1} when no Mg was given and between 0.4 and 0.8 g l^{-1} when Mg was added. With higher concentrations up till 1.8 dS m^{-1} in the irrigation water often the yield was reduced varying from 0 until 18% compared to the lowest concentration. The EC in the wet strip was increased from 0.99 to 1.71 dS m^{-1} in the 1:2 volume extract at the 0.45 and 1.8 dS m^{-1} application in the irrigation water, respectively. Thus, relatively small increases of the EC in the soil by over fertilization seriously reduced the yield, caused by a decreased osmotic potential of the soil solution like occur with salinity.

The fertilizer application did not only affect the yield, but also the quality of the produce as has been found with salinity experiments. Higher application of fertilizer improved the quality of the fruits like a reduction of the uneven ripened tomato fruits and an improved colour index of cucumber fruits. On the other hand, the quality can be negatively affected by high concentrations of nutrients which will increase for example the incidence of blossom-end rot of tomatoes and sweet pepper, like shown in Table 15.6. It can be expected that beside the decreased osmotic potential with the increase of the fertilizer concentrations the increased addition of NH_4 play a part in the occurrence of blossom end rot. The ratio NH_4/N was not affected with the increased concentrations, but often NH_4 is preferentially absorbed by crops and can specifically affect in this way the Ca uptake. See also Chapter 9.

The addition of Mg to the nutrient solution is recommended, especially for crops sensitive to Mg deficiency, like tomato and eggplant. In the experiments of Sonneveld and Voogt (1981) the fertilizer additions with and without Mg did not always show significant yield differences. However, when there were significant

Table 15.6 Percentages blossom-end rot in tomatoes and sweet pepper as affected by the application of different fertilizer concentrations applied in the irrigation water with strip fertigation

Fertilizer concentration dS m^{-1}	Tomato 1979	Sweet pepper 1974	Sweet pepper 1976
0.45	0.10	2.2	1.3
0.90	0.16	2.7	1.4
1.35	0.56	3.9	5.6
1.80	1.38	6.2	9.3

After Sonneveld and Voogt (1981).

yield differences between the nutrient solutions, mostly the addition of Mg was favourable. Further, the fertilizer mixture with a low N:K ratio mol/mol 9/4 on the whole showed a higher yield than the ratio 10/2.5. The necessity of the addition of nutrients other than N, K and Mg will depend much on the chemical composition of the irrigation water and of the soil. Many types of irrigation water contain sufficient Ca and SO_4 to supply the crop, and in many greenhouse soils these elements including P are abundantly available. When P is required, because of too low a P status of the soil, a pre-planting application will cover mostly plant requirements of this element (Bar Yosef et al., 1995).

With respect to micro elements it was experienced in The Netherlands that the addition with fertigation is generally not required. Most of these elements are sufficient present in soil or irrigation water. The poor availability caused by high pH values in the soil is more likely the problem than the presence in the soil. This especially counts for Fe and Mn for crops sensitive to chlorosis, like rose and gerbera. The uptake of these elements depends strongly on the pH. Therefore, control of the pH by addition of NH_4 will be mostly sufficient to ensure the micro nutrient uptake by the crop. The addition of NH_4 to improve the uptake of these elements is very effective, not only with respect to the lowering of the pH of the bulk of the soil as discussed before, but especially with respect to the pH drop in the rhizosphere (Hinsinger et al., 2003; Junk, 1987). The supply of NH_4 as N form lowers the pH on the root surface much more than in the bulk soil and this strongly improves the micro nutrient uptake. The first thinkable micro element to be considered for addition will be B, if this element is not sufficiently available in the irrigation water and the natural background concentration in the soil is poor. B is easily washed out by an intensive water supply, which occurs in the wet spots with drip irrigation.

The elements and the concentrations of these to be added to the irrigation water connected to optimum yield depend not only on the factors already mentioned, but also on the definition of optimal yield. Especially in greenhouse production optimum yield is not a matter of maximum production, as the quality of the produce is at least important as the production in greenhouse industry. Such means that not always the concentrations connected with maximum yield are most favourable. Growers sometimes have to make a choice between yield and quality, which are not always in the same line. In the experiments with vegetable fruit crops presented by Sonneveld and Voogt (1981) a mixture of N, K, Mg and SO_4, mol ratios 9, 4, 1.6, and 1.6 at concentrations between 0.45 and 0.90 were mentioned to be optimal. This addition agreed with concentrations in mmol l^{-1} 3.7–7.4 N, 1.6–3.3 K, 0.7–1.3 Mg and 0.7–1.3 SO_4 for a continuous supply. The highest concentrations mentioned agreed rather well with the uptake concentration calculated for the crops grown in the experiments. The fact that with lower concentrations comparable results were obtained can be explained by utilization of the dressing by soil grown crops. Later findings showed much higher uptake concentrations, which can be explained by increasing yields over years (Sonneveld, 1997). However, with fertigation it is not necessary that the concentrations in the irrigation water always cover the uptake concentration. With soil grown crops plants always will utilize nutrients from the storage, which is substantial in the soil. Another factor that can be taken into account is the required over supply of water, which brings also extra nutrient in the soils. However, the ultimate decision on the concentration to be added, depends on the momentary crop condition and the equilibrium between yield and quality aimed at. The discussion so far is based on experiments with fruit vegetable crops. The uptake concentration of many ornamental crops is often lower than those of vegetables and it is logical that the addition of fertilizers for flower crops will be in agreement with this lower uptake.

The uptake of micro nutrients by greenhouse crops grown in soil mostly cannot be controlled by soil analysis, because the methods of soil analysis available insufficiently predict the uptake of these elements by crops. This is shown by the data of an investigation in which the results of soil analysis were compared with the micro nutrient concentrations in the crops (Sonneveld and Voogt, 2001). The results are shown in Table 15.7. The correlation coefficients are poor, only for Mn and B some significant values were found. Water and a solution of $CaCl_2$/DTPA were compared as extraction solutions. The correlation coefficients for the water extract were higher than those for the $CaCl_2$/DTPA extract. The relationship for B is shown in Fig. 15.4. The reason for the poor correlations will be the great variation of soil types, crops, cultivars and growing conditions in greenhouse industry.

Table 15.7 Correlation coefficient for the relationship between micro nutrient concentration in soil extracts; 1:2 volume extract and $CaCl_2$/DTPA extract, and in plant tissues

Elements	1:2 extract water	1:10 extract w/w $CaCl_2$/DTPA
Fe	–0.262	–0.152
Mn	0.518	0.206
Zn	–0.258	–0.189
B	0.714	0.465
Cu	–0.221	–0.045
Mo	0.152	–0.108

After Sonneveld and Voogt (2001).

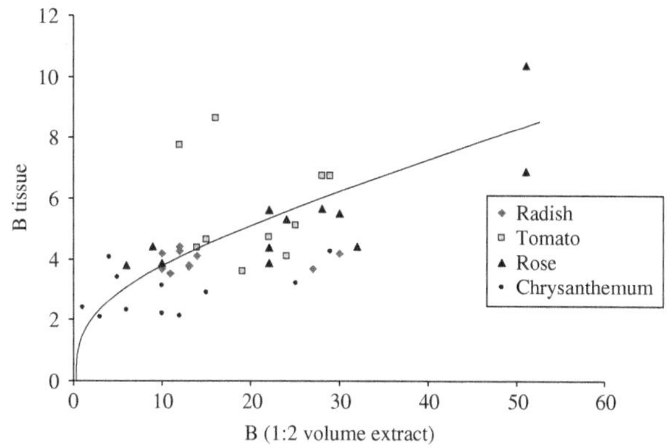

Fig. 15.4 Relationship between B concentrations in the 1:2 volume extract (μmol l^{-1} extract) and B concentrations of young fully grown leaves (mmol kg^{-1} dry matter) of different greenhouse crops. After Sonneveld and Voogt (2001)

15.9 Use of Tissue Tests

In view of the poor correlations commonly found between the results of soil testing methods on micro nutrients and the uptake of these elements by greenhouse crops tissue tests can be a helpful tool for the grower in the decision whether these elements should be included in the fertilization programme. In investigations of Sonneveld and Voogt (2001) mentioned in last section, micro nutrient concentrations of crops were determined to check the availability of these elements in soils used for greenhouse cultivation in The Netherlands. Four different crops were tested, for each crop on 10–12 different nurseries samples of soils and plant tissues were gathered. With the sampling the general instructions were followed. This means that for tomato, rose and chrysanthemum young fully developed leaves were sampled, while for radish the full top was used, except the oldest ring of leaves. The samples were rinsed with a detergent solution and afterwards washed with demineralised water. The average and extreme values found for the tissue samples of the different crops are listed in Table 15.8. The results for the crops concerned can be compared with the guide values as tabulated in Table 5.6 and Appendix B. The results show clear differences between crops and for different elements the average concentration is lowest for the rose crop. Striking differences are found between the lowest and highest values of the Mn concentrations for all crops. This can be explained by steam sterilisation, which at times occurs for all crops in the Dutch greenhouse industry. The first year after steam sterilisation the availability of Mn remains high and even can induce toxicity, as discussed in Section 10.4. The lowest Mn concentrations found for tomato and rose are rather low to ensure a healthy crop growth. This also is the case for Zn with tomato and B with radish, tomato and chrysanthemum. Among the Cu and Mo concentrations are also rather low values. However, for these elements restricted information is available about critical values for greenhouse crops. The Mo concentrations found with the rose crop are strikingly lower

Table 15.8 Average and extreme (in brackets) concentrations of micro nutrients of different greenhouse crops as has been found on Dutch nurseries. The concentrations are expressed as mmol kg^{-1} dry matter

	Crops			
Elements	Radish	Tomato	Rose	Chrysanthemum
Fe	2.20 (1.76–2.52)	1.86 (1.20–2.43)	0.92 (0.64–1.69)	1.83 (1.46–2.26)
Mn	1.82 (0.56–6.69)	2.58 (0.26–9.13)	1.65 (0.36–3.83)	4.05 (0.78–9.77)
Zn	1.24 (0.49–2.09)	0.39 (0.29–0.49)	0.42 (0.29–0.49)	1.11 (0.66–1.46)
B	3.96 (3.54–4.43)	5.55 (3.60–8.62)	5.34 (3.80–10.36)	2.92 (2.09–4.24)
Cu	0.106 (0.091–0.140)	0.138 (0.086–0.287)	0.077 (0.043–0.100)	0.208 (0.142–0.335)
Mo	0.022 (0.010–0.036)	0.019 (0.002–0.064)	0.004 (0.002–0.008)	0.034 (0.004–0.079)

After Sonneveld and Voogt (2001).

than those with the other crops. It is not yet clear whether for this crop Mo application is necessary. The commonly somewhat lower pH of the soils used for rose growing only can explain part of the difference in the uptake.

When the concentration of micro nutrients is too low, a higher replacement of NO_3 by NH_4 will be considered firstly, because the uptake of most of those elements depends on the pH of the soil.This, for example, is recommended for Mn preferably, as the uptake strongly reacts on the pH of the soil, while addition of Mn to the soil is very ineffective at a high pH. Secondly, the addition of micro nutrients to the irrigation water can be recommended. Fe will be added in chelated form. Mostly the form of EDDHA is preferred, because Fe deficiency in soil grown crops is commonly connected with a high pH of the soil.

When no deficiency symptoms are visible but plant tissue concentrations are low the concentrations recommended for fertigation for the different elements lay between 50 till 100% of the concentrations used for substrate cultivation as listed in Appendix C. With visible deficiency symptoms in the crop firstly 200% of these concentrations can be added for one or two weeks.

For macro nutrients the use of tissue tests is less obvious in the management of the fertilization of soil grown crops, because soil testing mostly inform sufficiently the availability of these elements to greenhouse crops.

References

Al-Harbi A R Al-Omram A M and Sheta A S 2005. Effects of growing media and water regimes on growth, yield and water use efficiency of squash (*Cucurbita pepo*). Acta Hort 697, 231–241.

Bar Yosef B Sagiv B Markovitch T and Levkovitch I 1995. Phosphorus placement effects on sweet corn. Growth, uptake and yield. *In*: Proc. of the Dahlia Greidinger Intern. Symposium on Fertigation. Technion – Israel Institute of Technology, Haifa, 25 March – 1 April 1995, 141–154.

Heemskerk M J Van Os E A Ruijs M N A Schotman R W 1997. Verbeteren van watergeefsystemen voor grondgebonden teelten. Proefstation voor Bloemisterij en Glasgroente Naaldwijk, Rapport 84, Deelrapport 1: Inventarisatie en verbetering regenleidingsystemen, 61pp.

Hinsinger P Plassard C Tang C and Jaillard B 2003. Origins of root-mediated pH changes in the rhizosphere and their responses to environmental constraints: A review. Plant Soil 248, 43–59.

Hoffman G J 1986. Salinity. *In*: Nukayama F S and Bucks D A (eds) Trickle Irrigation for Crop Production. Elsevier, Amsterdam, 345–362.

Junk A 1987. Soil-root interactions in the rhizosphere affecting plant availability of phosphorus. J. Plant Nutr. 10, 1197–1204.

Marschner H 1997. Mineral nutrition of higher plants. Academic Press, London.

Oster J D Hoffman G J and Robinson F E 1984. Dealing with salinity: Management alternatives: Crop water and soil. California Agric. 38(10), 29–32.

Papadopoulos I 1988. Nitrogen fertigation of trickle-irrigated potato. Fertilizer Research 16, 157–167.

Sonneveld 1962. Bladbespuiting met voedingselementen. Tuinderij 2, 504–505.

Sonneveld C Koornneef P and Van den Ende J 1966. De osmotische druk en het electrische geleidingsvermogen van enkele zoutoplossingen. Meded. Dir. Tuinbouw 29, 471–475.

Sonneveld C and Van den Ende J 1967. Bijmesten via de regenleiding met behulp van de concentratiemeter. Meded. Dir. Tuinb. 30, 54–60.

Sonneveld C 1976. Aanpassing concentratiemeters. Mededelingen Proefstation Groenten- en Fruitteelt onder Glas 30, 88–89.

Sonneveld C 1979. Changes in chemical properties of soil caused by steam sterilisation. *In*: Mulder D (ed) Soil Desinfestation, Elsevier, Amsterdam, 39–50.

Sonneveld C and Voogt S J 1981. Nitrogen, potash and magnesium nutrition of some vegetable fruit crops under glass. Neth. J. Agric. Sci. 29, 129–139.

Sonneveld C 1982. EC-waarden meststoffen. Mededelingen Proefstation Groenten- en Fruitteelt onder Glas 36, 65.

Sonneveld C Van den Bos A L Van der Burg A M M and Voogt W 1991. Fertigation in the greenhouse industry in The Netherlands. *In*: Fertigation/Chemigation. FAO, Rome 1991, 186–193.

Sonneveld C and Voogt W 1994. Effects of calcium and ammonium on the appearance of secondary shoot chlorosis in rockwool grown cucumbers. Proc. Sino Intern. Coll. Soilless Culture, Hangzou 1994, 88–94.

Sonneveld C 1997. Mineralenopname bij teelten onder glas. Proefstation voor Bloemisterij en Glasgroente Naaldwijk, The Netherlands, 35pp.

Sonneveld C and Voogt W 2001. Spoorelementen bepalingen in grond. Resultaten van een enquête onderzoek. Praktijkonderzoek Plant en Omgeving, Sector Glastuinbouw, 27pp.

Van Beusichem M L 1984. Non-ionic nitrogen nutrition of plants. Thesis Landbouwhogeschool Wageningen. 141pp.

Van der Post C J and Sonneveld C 1961. Beregening met mestoplossingen. Proefstation voor de Groente- en Fruitteelt onder Glas te Naaldwijk, The Netherlands. Jaarverslag 1961, 47–51.

Van den Bos A L 1991. Stikstofvormen bij intensive bemestingssystemen voor kasteelten. Proefstation voor Tuinbouw onder Glas te Naaldwijk, The Netherlands. Internal Report 1991 no. 21, 18pp.

Van den Ende J and De Graaf R 1974. Comparison of methods of water supply to hothouse tomatoes. Acta Hort. 35, 61–70.

Chapter 16
Nutrient Management of Soil Grown Crops

16.1 Introduction

The management of the fertilization of soil grown crops in greenhouses can be distinguished in the addition of fertilizers before cultivation, the base dressing and those added during the cultivations period of the crops, the top dressing. The growing period of the crops in greenhouse production varies strongly. Some vegetable crops like radish and lettuce have a growing period between 4 and 15 weeks mainly dependent on the growing season, whereby different crops can be grown successively in one year. Other vegetable crops like tomato, sweet pepper and cucumber can be grown year round in moderate climatic conditions, while in relatively hot climates, like the Mediterranean area, such crops can be grown for about 8 months, because of the high temperatures in the greenhouses in summer. The length of the growing period of most flower crops shows at least the same variation as those for vegetable in view of the great diversity of these crops grown in greenhouses. For some flower crops the growing period can cover several years, like roses. It will be clear that drastic changes in the soil, like the application of soil improvers, adjustment of the pH, base dressings, flooding and soil tillage only can be carried out between cropping periods in the greenhouse. Some handlings need to be carried out simultaneously, like the addition of less soluble fertilizers as base dressing and the soil tillage. Such fertilizers must be intensively mixed throughout the soil.

The management of the fertilization of greenhouse crops is based on relationships between optimal concentrations in the soil solution and maximum yield of a required quality. Just like discussed in Section 13.1, maximum yield and optimal quality can involve conflicting situations. Therefore, equal considerations are operative for the management of the fertilization of soil grown crops as for the nutrient management of substrate grown crops. The management of the fertilization of soil grown crops is based on the composition of the soil solution, just like with substrate growing. The composition of the soil solution will be estimated by sampling of the soil and analysis by the 1:2 extract as discussed in Section 4.2. A secondary advantage of this soil testing method is that beside the composition of the soil solution also a good estimation of the quantities of plant nutrients available in the root zone can be obtained. Upside down, it is possible to calculate the effects of fertilizers applied

C. Sonneveld, W. Voogt, *Plant Nutrition of Greenhouse Crops*,
DOI 10.1007/978-90-481-2532-6_16,

Table 16.1 Increase of the concentration of nutrient elements in the 1:2 extract (mmol l^{-1}) as expected by the addition of 1 kg per 100 m^2 greenhouse area of that element. The calculations are operative for a depth of the root zone of 0.25 m

Nutrient elements	Expected increase
N	1.79
S	0.78
K	0.64
Ca	0.62
Mg	1.03

on the increase of the nutrient concentrations in the soil extract and successively in the soil solution, as follows from the data presented in Table 4.2, but also can be calculated how much fertilizer will be added to cover together with the calculated storage in the soil the uptake of the crop grown. The effect of addition of 1 kg of a nutrient element per 100 m^{-2} on the analytical data of that element in the 1:2 extract is listed in Table 16.1. The calculations are carried out under conditions that no precipitation and adsorption occur. Deviations from this statement will occur with cations, K, Ca and Mg with respect to adsorption, which effects will be discussed further on in this chapter. P is not listed in the table, because of the low solubility of P compounds in soils.

The model of the relationship between the concentration of a nutrient element in greenhouse soils and the growth, which mostly can be defined as yield has a merely asymptotic character. Which means that yield increases strongly in the low concentration area and does not increase further on above the optimum value. Nutrient concentrations that exceed the optimum value substantially even can reduce the crop development, for example caused by too low osmotic potentials (high EC), toxicity or a reduced uptake of a different element by ionic competition. Therefore, a full asymptotic model satisfies only in an interval without yield reduction in the high range. For field crops an exponential response curve for the relationship between fertilizer application rate and yield has been developed (Neeteson and Wadman, 1987; Neeteson and Zwetsloot, 1989). Comparable response curves have been found between soil solution concentrations in substrate growing in rock wool as well in sand. In this comparison no differences were found between the response curves for rock wool grown and sand grown crops (Sonneveld et al., 2004), indicating that the growth response of crops on nutrient supply between soil and substrate does not differ essentially.

16.2 pH

For the determination of the pH in soil different methods are available. Best known is the determination in a soil:water suspension with a ratio 1:2 v/v. However, this method shows considerable fluctuations over the season, mainly caused by the changes in the salt status. Therefore, for a good estimation of the pH status of soils often a KCl solution of 1 mol l^{-1} is used, with which a more stable value is obtained

(De Vries and Dechering, 1960). Generally the values of these so called pH_{KCl} are lower than those pH determined in a water suspension. The difference for field soils amounted on average 0.7 pH unit. For different series of Dutch greenhouse soils (Roorda van Eysinga, 1966c, 1971a, b), the pH water values were on average 0.4 units higher than those in the KCl solution and in another investigation with 75 greenhouse samples an average difference of 0.2 units was found (Sonneveld and Voogt, 1986). The smaller differences detected for greenhouse soils can be explained by the higher salt status of these soils in comparison with field soils.

The recommended value of the pH of soils depends on the soil type, the requirements of the crop and possible growing conditions. The minimum guide pH_{KCl} values recommended for greenhouse soils vary from 6.0 till 6.7 for mineral soils, while for peaty soils a minimum value of 5.5 is recommended (Van den Bos, 1993). For soils with high Mn content like many marine and river clayey soils, a value of at least 6.5 is recommended, to prevent too high uptake of this element. This especially is true for steam sterilised soils, because of the strong release of Mn by this treatment, as discussed in Section 10.4. Soils with a high $CaCO_3$ content often have a pH value above 6.5. High pH values can induce chlorosis, mostly caused by an insufficient availability of Fe and Mn.

For the adjustment of the pH to control too low values generally different types of limestone are used. The quantity necessary depends on the characteristics of the soil, the depth of the root zone that will be adjusted, the bulk density of the soil and the pH difference that will be bridged. With the aid of these factors a method has been developed with which it is possible to calculate the quantities of limestone fertilizer to adjust too low pH values in the soil to values desired by the crop that will be grown (De Vries and Dechering, 1960). Results of such calculations for greenhouse crops are presented by Van den Bos et al. (1999). In Fig. 16.1 results of these calculations are presented for different soil types, dependent of the content of loss on ignition and clay. The quantities of base material that will be added are related to a pH increase of 0.1 units and presented as mol base (lime or hydroxide) per m^2. The adsorption capacity of organic matter is much higher than those of clay. Therefore, the quantities necessary by the loss on ignition is four times higher than those by clay. For example on a sandy soil with 5% organic matter lime fertilizer equivalent with 1.08 mol base m^{-2} is required to increase the pH with 0.1 unit. An example of the calculation related to addition of $CaCO_3$ is presented in Formula (16.1). The effective base content of fertilizers generally is expressed as % CaO. 1 mol base is equivalent with 28 g CaO or 20 g Ca.

$$Rq_{CaCO_3} = \frac{1}{2} \times 100.1 \times 1.08 = 54.1 \tag{16.1}$$

In which:

Rq_{CaCO_3} = required $CaCO_3$ in g m^{-2} $CaCO_3$ for an increase of the pH with 0.1 unit over a depth of 0.25 m

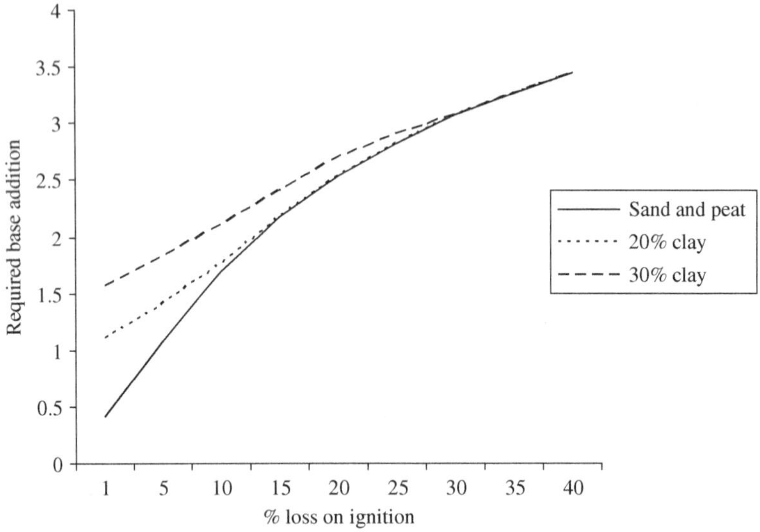

Fig. 16.1 The quantities of base material (mol m^{-2}) necessary to increase the pH value with 0.1 unit over a depth of 0.25 m, dependent on the organic matter (loss on ignition) and clay content of the soil (Van den Bos et al., 1999)

Addition of about 1.5 mol base material per m^2 per year is estimated as a maintenance dressing on soils poor in $CaCO_3$. Maximum applications added at once vary between 5 and 15 mol m^{-2} for sandy soils and peaty soils, respectively. When the pH is very low and applications are calculated higher than those mentioned, part of the required lime stone fertilizer will be given in the second year.

16.3 Flooding

The salt concentration in the root zone after crop cultivation can be at such a high level that thorough flooding is necessary before a next crop is planted in the greenhouse. This also can be necessary for a smooth out of a horizontal unequal salt distribution in the root zone. This for example especially needs attention when a row crop with walking paths is followed by a crop with a planting pattern that scatters the whole surface, as discussed in Section 6.4. In such cases the soil surface of the walking paths will be broken before flooding, to improve the vertical water transport. Comparable situation occur when a crop grown with drip irrigation is followed by a crop that covers the whole surface. Drip irrigation always bring about an unequal salt distribution, being worst at the end of the growing season, like shown for NO_3 in Fig. 4.4. The quantities of water necessary will be calculated following formula (6.5).

The salinity level at which flooding is necessary depends on the crop and the growing conditions aimed at. With crop rotations as described in the foregoing

Table 16.2 Limits for total salt (EC, dS m^{-1}), Na and Cl (mmol l^{-1}) concentrations in the 1:2 volume extract. With values higher than the limits presented, flooding is recommended

	Limits for flooding	
Crops	EC	Na and Cl
Cucumber	1.8–2.5	3–4
Tomato	2–2.5	4–5
Sweet pepper, gerbera	2	3.5
Carnation, rose	1.5–2	3–3.5
Lettuce, endive	1.5–2	3–4
Spinach	1.8–2.5	3–4
Bulbs	1	2
Other crops	1.5	3

paragraph, more or less always flooding is recommendable, to prevent an unequal start of the crop. Sometimes under poor light condition an increased salinity level is desired to prevent a too lush growth of the crop at start. The necessity of flooding is mainly based on the total salt, Na and Cl concentrations of the soil. From history rough maximum limits were given for total salt and Cl in the 1:2 volume extract (Sonneveld and Van den Ende, 1971), being an EC value of 2.1 dS m^{-1} for the total salt and 3.3 mmol l^{-1} for Cl. For Na no limit was given, but this will be estimated to be equal to the one for Cl. Later on more detailed limits were supplied by Van den Bos et al. (1999), as summarized in Table 16.2. For some crops a range is given, whereby a decision about flooding can be made within this range dependent on the growing conditions. When the crop will be grown under conditions favourable for salt damage, the low value of the range is maintained.

With leaching also nutrients are washed out, like shown in Table 16.3 (Sonneveld and Van Beusekom, 1974). The ultimate results also depend on the quality of the primary water used for flooding. In the presented situation the water used for flooding had following characteristics: EC 1.2 dS m^{-1}, Cl 5 and Mg 0.3 mmol l^{-1}. The concentration of any ion in the soil solution never can become lower than those in the

Table 16.3 The concentration of water soluble total salt (g kg^{-1} dry soil) and of water soluble ions (mmol kg^{-1} dry soil), as affected by flooding with various quantities of water. The walking paths and the cultivation strips were separately sampled

	Before flooding		After 150 mm water		After 450 mm water	
Determination	Path	Strip	Path	Strip	Path	Strip
Total salt	4.2	2.3	1.1	0.8	0.6	0.6
Cl	26.9	12.4	4.02	1.97	2.05	1.71
N (NO_3)	7.11	2.14	1.11	0.46	1.36	0.50
P	1.06	1.63	1.16	1.30	1.12	1.17
K	2.48	2.47	1.15	0.99	0.96	1.01
Mg	2.57	0.76	0.60	0.40	0.40	0.41

The data are derived from a loamy greenhouse soil over a depth of 0.3 m (Sonneveld and van Beusekom, 1974)

primary water used for the flooding. With this statement result of a flooding can be checked by sampling and analysis of the Cl concentration in the 1:2 volume extract following the equation presented in Table 4.1. The concentration in the soil solution will be estimated as being the concentration of the primary water used for leaching. Eventually the determination can be carried out in the dry soil and for the calculation Equation (16.2) will be used.

$$Cl_{ds} = \frac{Cl_{rw} wv_f}{\rho} \tag{16.2}$$

In which

Cl_{ds} = Cl content in mmol kg^{-1} dry soil
Cl_{rw} = the Cl concentration of the primary water
wv_f = volume fraction of water of the field moist soil
ρ = the bulk density in kg l^{-1}

In the example of Table 16.3, wv_f and ρ were 0.32 and 1.2, respectively. Thus, maximum result is obtained in the presented example with Cl_{ds} of 1.3 mmol kg^{-1} dry soil. This value is approximated after a flooding of 450 mm. The behaviour of NO_3 can be compared with that of Cl. The concentration water soluble P will not be affected significantly by leaching, because of the solubility equilibrium of P in the extract and possible release from the storage. The cation concentrations after flooding will be affected by the cations adsorbed on clay and organic matter.

16.4 Base Dressings with Nutrients

16.4.1 Nitrogen

Mineral N mainly occurs in greenhouse soils as NO_3. NH_4 supplied with the fertilizer dressings before crop cultivation is quickly converted to NO_3, because of the favourable conditions for this process in greenhouse soils. An exception on this rule happens on sterilised soils, where the bacterial activity is disturbed. This for example has been found after steam sterilisation. With this treatment not only NH_4 is released by an intensified decomposition of organic matter, like discussed in Section 10.5, but also the conversion of added NH_4 is delayed (Sonneveld, 1969a).

Research to the optimal level of mineral N in greenhouse soils mostly offered values between 3.5 and 7 mmol N kg^{-1} dry soil, which can be converted to values between 2.6 and 5.4 mmol l^{-1} 1:2 volume extract (Sonneveld and Van den Ende, 1971). Such a conclusion was based on extended studies of which results were published by Roorda van Eysinga (1972), who noticed for lettuce 6.5, for tomato 5.5 and for cucumber 7 mmol N kg^{-1} dry soil as optimum values. In different publications Spithost (1965) and Roorda van Eysinga (1971a) noticed an optimum of 3.5 and

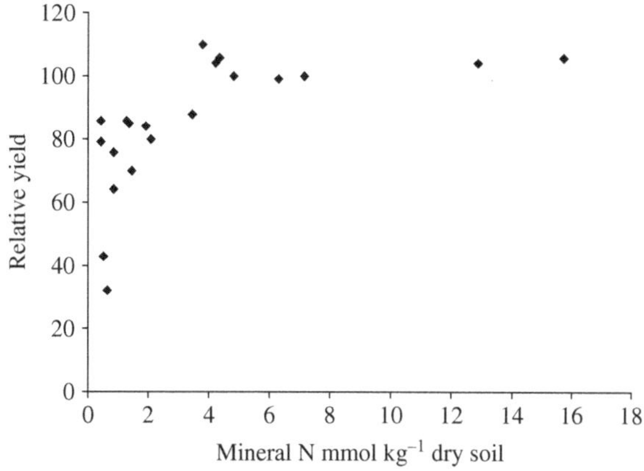

Fig. 16.2 Relationship between the mineral N content (mmol kg^{-1} dry soil) of different soils and the yield of tomato relative to the optimum yield. Data after Roorda van Eysinga (1971a)

5.5 mmol N kg^{-1} dry soil for tomato, respectively. Kohlrabi produced optimal yields at 7 mmol N kg^{-1} dry soil (Roorda van Eysinga and Mostert, 1972). In Fig. 16.2 an example is shown of the relationship between the mineral N concentration in the soil and the relative yield of tomato, as has been found in a series of experiments by Roorda van Eysinga (1971a). In these experiments the yield as has been gained in the unfertilized plots was related to the yield at optimal N fertilization in every experiment. The soil types in this research were merely of mineral origin, only one peaty soil was included. The figure shows sure enough, that optimum yields were gained in the experiments where the original N concentration was about 5 mmol kg^{-1} dry soil, which agrees with 3.5 mmol l^{-1} in the 1:2 volume extract. A specific effect of N supply is that an ample addition surely reduced the attack by *Botrytis cinerea* of tomato (Roorda van Eysinga, 1966a; Verhoef and Weber, 1965).

16.4.2 Phosphorus

Most greenhouse soils contain much P because of an abundant application of this element and the fact that it is scarcely washed out by an overdose of water. The solubility of P is low in relation to the total amount available. Therefore, the quantity of P determined with a water extraction depends much on the water to soil ration maintained during extraction. The quantities of P extracted by the customary soil testing method of the 1:2 volume extraction in the greenhouse cultivation is only a fraction of the total P in greenhouse soils. In the 1:2 extraction of 75 greenhouse samples on average 0.23 mmol P l^{-1} extract was determined (Sonneveld et al., 1990), while in the same samples with a 1:100 w/w extract 2.73 mmol l^{-1} soil was determined (Sonneveld and Voogt, 1986). On basis of the data presented in Section 4.2 can be calculated that with the 1:2 volume extraction on average 92 mmol m^{-2} and with the 1:100 w/w method 682 mmol m^{-2} greenhouse soil was extracted. The correlation coefficient between the results of both determinations ($r = 0.67$) was rather low,

Table 16.4 Coefficients of correlation for the relationships between P-water contents (1:5 w/w, 18°C) of greenhouse soils and the results of different other P determinations, following Roorda van Eysinga (1971). Pw-values are expressed on the volume and the results of the other determinations on weight of the dry soil

Determination	Extraction ratio	Mineral soils	Peaty soils
P-value water 50°C	1:10 w/w	0.96	0.84
Pw-value water 20°C	1:60 v/v	0.97	0.93
P-NH_4 lactate pH 3.75	1:20 w/w	0.89	0.71
P-citric acid 1%	1:10 w/w	0.90	0.51
P-total Fleischmann acid	1:4 w/w	0.67	0.29

which will be explained by the great variation of compounds in which P can occur in soils. Nevertheless, in many cases the results of different P determinations are reasonably correlated, like found by the extended research of Roorda van Eysinga (1971) with Dutch greenhouse soils, especially when mineral and peaty soil types were separated, like shown in Table 16.4. The correlation coefficients for the peaty soils are substantially lower than those for the mineral soils. Comparable results have been found recently with soils from elsewhere, with a different series of determinations (Indiati and Sing, 2001).

The quantities of total P in greenhouse soils are much higher than those available in the different water extracts. In Fig. 16.3 the relationship is shown between the results of the water soluble P contents determined in the 1:5 w/w water extracts and those of the total P determination (Roorda van Eysinga, 1971). With the relationship shown for the sandy soil can be calculated that the quantities of total P easily exceed the water soluble quantities with a factor of about 30. The quantities of total P for different soil types vary strongly and are mostly higher than those found for sand. Average values of 21, 31, 38 and 118 mmol kg^{-1}dry soil for sand, sandy loam, clayey loam and clayey peat, respectively were calculated for data of greenhouse soils (Roorda van Eysinga, 1971).

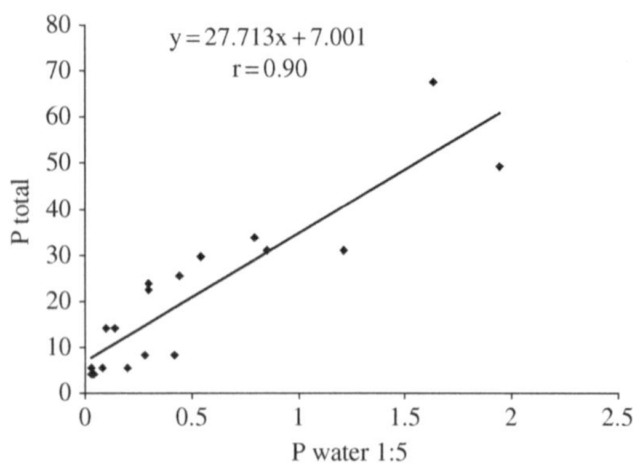

Fig. 16.3 Relationship between the results of the determination of water soluble P and total P content of a series of sandy greenhouse soils. The contents are expressed as mmol kg^{-1} dry soil. After Roorda van Eysinga (1971)

The analytical data of the different P determinations as mentioned in Table 16.4 reasonable reflect the P uptake. For lettuce the relationship was estimated by an equation of the model presented in Equation (16.3) (Roorda van Eysinga, 1971).

$$y = a \log x + b \tag{16.3}$$

In which

x = result of the P determination of the soil
y = result of the P determination in the crop

The relationship between the P concentration of the soil and the yield was also curve linear and showed a good agreement with the model for the P uptake. Therefore, the relationship between the P concentration in the soil and the yield was approximated by Roorda van Eysinga (1971) by a function according Equation (16.3). However, such a logarithmic model is not logic, because it suggests an increasing yield for values above an optimum nutrient uptake. In Fig. 16.4 the relationship is shown between the water soluble P content of the soil and the relative yield of lettuce for 19 different sandy soils on which experiments with P addition were carried out. The relative yield at the different experimental sites was calculated as mentioned for the N experiments. Maximum yield on the mineral soils was obtained at P concentrations of the soil of about 0.5 mmol kg^{-1} dry soil, determined in the 1:5 water extract. A further increase of the water soluble concentration does not increase the yield. For peaty soils this value was about twice as high. The quantity of dry soil per volume for the peaty soil types was about half of that for the mineral soils. Thus, the quantity of P required per volume of soil is more or less equal for all soils. This suggests that the interpretation will be based on the quantity of P per volume of soil. Thus, the interpretation presented for mineral soils on weight basis is true for all soils when expressed on volume, for the bulk density of mineral soils is roughly around 1. Such an interpretation is supported by Roorda van

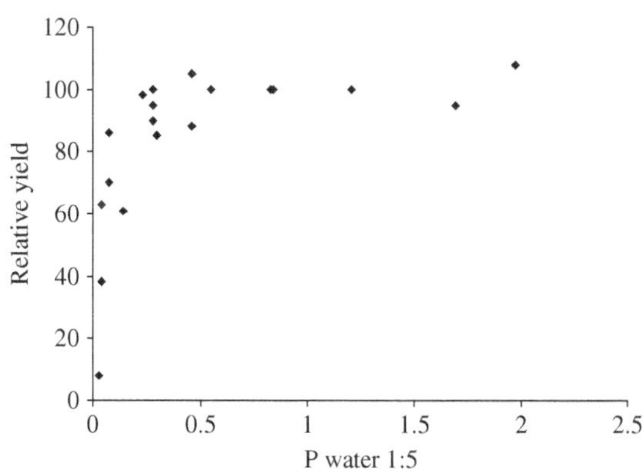

Fig. 16.4 Relationship between the P concentration of the soil (mmol kg^{-1} dry soil) soluble in 1:5 w/w water extract and the relative yield of lettuce on sandy soils. After Roorda van Eysinga (1971)

Eysinga (1971) with the remark in his report that interpretation of the results of the determination of water soluble P on basis of volume ratios as published by Van der Paauw (1969) and Sissingh (1969) showed less variation than those based on weight ratios. The at the time used data of the P 1:5 water method can be recalculated to data for the nowadays used specific 1:2 v/v method (Sonneveld and Van den Ende, 1971), which result for the 0.5 mmol kg^{-1} soil to a value of 0.14 mmol l^{-1} in the 1:2 extract. Roorda van Eysinga (1972) concluded that lettuce was most sensitive to P supply and the P supply can be omitted above a value of 70 mg P_2O_5 and for tomato above 50 mg P_2O_5 kg^{-1} dry soil, which can be recalculated to 0.24 and 0.14 mmol l^{-1} in the specific1:2 v/v volume extract.

When the soil is more or less saturated with P, the quantity of soluble P is relatively constant over long periods, like shown by the data of the loam soil in Fig. 16.5. However, when the soil contain less P heavy P applications are necessary to saturate the soil with sufficient P, like shown with the clay soil in Fig. 16.5. During the 3 years that the course of the P concentration in the soils was followed, a fertilizer dressing of 0.8 and 4.0 mol P per m^2 was applied on the loam and the clay soil, respectively, while the P applied by soil improvers was estimated on 0.6 and 3.6 mol per m^2, respectively. From these data will be concluded that in advance big quantities of P are necessary to bring field soils on the P level required for greenhouse cultivation. Therefore, Roorda van Eysinga (1966) has found huge effects of P application on newly reclaimed soils. He recommended an application of 1.3 mol P m^{-2} per year during the first years of greenhouse cultivation. Afterwards the application will be reduced later to 0.7 mol or lower dependent the on the development of the P status of the soil.

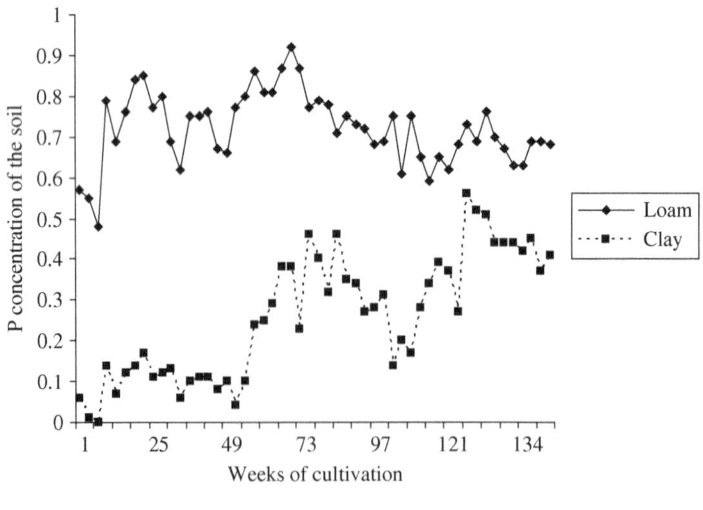

Fig. 16.5 Course of the P concentration in greenhouse soils (mmol kg^{-1} dry soil) as determined in a 1:5 water extract during 3 successive years. The loam soil was covered by a greenhouse for already many years, while on the clay soil a greenhouse was build from week 1. (Data Sonneveld, 1966, 1967, 1969)

Soils used for greenhouse cultivation during many years often contain much P, by addition of this element by fertilizers and by soil improvers that easily exceeds the uptake. Furthermore the leaching is small and in relation to other macro nutrient negligible as already discussed in Section 16.3. For this reason Roorda van Eysinga (1971) found for a specific group of Dutch greenhouse soils that the age of the greenhouse was a better indication for the determination of the optimum P application than the determination of P by different extraction methods. In older greenhouses P application is scarcely required and usually does not exceed 0.5 mol m^{-2} per year. This also was confirmed by recent experiments with lettuce, radish and chrysanthemum (Van den Bos, 2001, 2004, 2004a; Van Gurp, 1998) whereby comparisons were made with P dressing applied for every crop on average varying between 0 and 350 mmol m^{-2}. The experiments for lettuce and chrysanthemum were carried out on two different soils, a clayey soil and a sandy soil, while the experiment with radish crop was carried out on a sandy soil. The pH values of the soils varied between 6.0 and 7.5. In the experiments a great number of crops were successively grown to follow the crop development without P dressing in the long run. With lettuce on the clayey soil 13 crops were grown and on the sandy soil 23 crops, with the radish 25 crops were grown and in both chrysanthemum experiments 14 crops. In the treatments without P addition the P concentration in the 1:2 volume extract was 0.10 mmol l^{-1} in the lettuce experiment on the sandy soil and on the other soils it was about 0.03 mmol l^{-1}. The yields in the non fertilized treatments were incidental a shade lower than in the other treatments; on average the difference with the fertilized treatments was 1–4%. In the treatments with the lowest P application the P concentration in the 1:2 volume extract was 0.15 in the lettuce experiment on the sandy soil and varied in the other experiment from 0.03 till 0.07 mmol l^{-1}, which resulted to an optimal yield for all crops. Higher applications stimulated the P uptake, but did not increase the yield of the crops. In Table 16.5 some data of the experiments are presented. The lowest P applications in the experiments amounted to about 40 mmol m^{-2} per crop for radish and for chrysanthemum, and about 80 mmol m^{-2} per crop for lettuce. The average P uptake at optimum growth for lettuce, radish and chrysanthemum was 57, 22 and 64 mmol m^{-2} per crop, respectively. Thus, the additions easily exceeded the uptake.

Table 16.5 Average results of experiments with different P additions. 13 lettuce crops were successively grown on a clayey soil and 14 chrysanthemum crops on a sandy soil at. The P added is expressed as mmol m^{-2} per crop, the P concentration in the 1:2 volume extract ($P_{1:2}$) as mmol l^{-1} extract, the plant weight in g per plant and P concentration of the crop as mmol kg^{-1} dry matter

Lettuce				Chrysanthemum			
P added	$P_{1:2}$	Plant weight	P crop	P added	$P_{1:2}$	Plant weight	P crop
0	0.03	320	189	0	0.030	82.4	115
84	0.07	331	214	40	0.042	84.8	126
168	0.11	330	231	80	0.051	85.9	131
252	0.16	331	242	160	0.072	86.0	139
336	0.22	332	248				

Results of the former research by Roorda van Eysinga (1972) as mentioned before suggested optimum P concentrations in the 1:2 volume extract up to 0.24 mmol l^{-1} for lettuce, being the most sensitive crop for reaction on P fertilization. The more recent research presented in the former paragraph showed optimal yield at P concentrations between 0.04 and 0.07 mmol l^{-1} and in one case at a level of 0.15 mmol l^{-1}. Recommendations to growers by Van den Bos et al. (1999) showed guide values of 0.10 mmol l^{-1} in the 1:2 extract, for different crops and growing conditions and on this $P_{1:2}$ level addition of P is scarcely carried out only when the P storage is below the optimum. The concept of P storage is explained in following paragraph.

In greenhouse cultivation mostly correlation coefficients for the relationship between soil P and yield do not differ much for P when determined with water extracts or with mild acid extracts (Roorda van Eysinga, 1971). However, his research with a series of diluvial soils showed higher correlation coefficients for mild acid extraction than for water extraction (Roorda van Eysinga, 1961). Therefore, sometimes P is determined with the aid of NH_4-lactate buffered at pH 3.75 in a soil:solution ratio 1:20 w/w and used to determine the P storage (Egnér et al., 1960), the so called P-Al determination. The storage determined with this method is estimated low for concentrations below 10 mmol kg^{-1} dry soil and optimal for concentrations between 15 and 20 mmol kg^{-1} dry soil (Van den Bos, 1993). For greenhouse soils the P fertilization is based on both the storage and the solubility, determined by the P-Al and the P-1:2 extractions, respectively. The determination of P-Al is carried out once in two years, while those of P-1:2 with water is carried out more frequently.

16.4.3 Potassium, Calcium and Magnesium

The additions of K, Ca and Mg are mutually related, because these elements are on most soils for the greater part adsorbed on the clay and humus particles in the soils. In greenhouses for K the ratio water soluble/exchangeable mostly varies between 2 and 3 and is possibly higher for clayey soils than for sandy and peaty soils (Roorda van Eysinga, 1963). This means that the addition of one of the cations also affect the concentration of the other cations in the soil solution by exchange from the adsorption complex. Therefore, cations added with fertilizer applications will not always be found back in the soil solution in concentrations in agreement with the addition. A new equilibrium will occur, like shown by the data of Table 16.6 (Sonneveld, 1993). For the peaty soil the increase of the cations in the saturation extract is relatively in equilibrium with the supply, but for the clayey soil the increase of K in the saturation extract is relatively low, while those for the Ca and Mg is relatively high. This is in agreement with the characteristic of clayey soils, where K is strongly adsorbed and thereby is exchanged with Ca and Mg. For that reason often extra K is added in newly built greenhouses on formerly field soils poor on K when the soil contains a high content of clay. Mostly an extra base dressing of 2 till 4 mol K m^{-2} is added, dependent on clay content and the need of the crop, before

Table 16.6 Addition of nutrient cations (mol m^{-2}) with primary dressings of fertilizers and the increase of the cation concentrations in the saturation extract (mmol l^{-1})

	Clay soil		Peaty soil	
Elements	Addition	Increase	Addition	Increase
K	0.9	1.9	0.8	2.7
Ca	0.5	4.4	0.2	0.8
Mg	0.4	3.0	0.3	1.0

the start of the first cropping. For soils with a heavy K fixation comparable quantities up to 1600 kg K (4 mol m^{-2}) are reported in the literature as being favourable for tomato production (Doll and Lucas, 1973; Schäfer and Siebold, 1972). Insufficient availability of K in soils can strongly affect the fruit quality by an unequal colouring, like for example with tomato (Roorda van Eysinga, 1966a). Optimal yields are derived with water soluble K concentrations of 1.2 till 3.2 mmol kg^{-1} dry soil, for the mainly mineral soils studied in experiments (Roorda van Eysinga, 1966b). Those values can roughly be transformed to concentrations of 0.7 till 1.8 mmol l^{-1} in the 1:2 volume extract, following the information supplied by Sonneveld and Van den Ende (1971). For "high quality" produce higher concentrations are required up to 5.3 mmol kg^{-1} dry soil, which can be recalculated to 3.0 mmol l^{-1} in the 1:2 extract.

The concentrations of Ca and Mg are mostly adjusted to the K concentration maintained. Less specific research had been carried out for the supply of these elements for soil grown crops. There are significant differences between crops for the sensitivity to Mg deficiency, tomato and eggplant for example showed to be sensitive, while sweet pepper seldom shows Mg deficiency symptoms. In soils mostly Ca is sufficiently high as a result of the application with Ca containing fertilizers and soil improvers, the decomposition of $CaCO_3$ and the use of irrigation water that contains Ca. Therefore, the concentrations in the soil solution often exceed strongly those of the K (Sonneveld et al., 1990) and a known addition can be ignored. Mg also is applied to the soil by the sources mentioned for Ca. Nevertheless, less evidently than Ca and sometimes the addition of Mg by fertilization can be necessary. The Mg concentration in the soil mostly is maintained on a certain level in relation to the K concentration, which seems logical. In an investigation with tomato, samples were gathered from places where Mg deficiency was manifested in relation to places in the same greenhouses where the crop was more or less free from symptoms (Sonneveld, 1969b). The results of this investigation are summarized in Table 16.7 and it is indicated that for a sensitive crop like tomato a water soluble mol/mol ratio Mg/K of about 0.9 have to be maintained in the root zone. The general recommendation presented by Sonneveld and Van den Ende (1971) is somewhat lower, and amount to 0.65 for the mol/mol ration in the 1:2 volume extract. The higher ratio found for tomato is in agreement with the sensitivity for Mg deficiency of this crop. The Ca concentration is also important for the Mg uptake (Sonneveld and Voogt, 1991), but the Ca concentration was not included in the investigation discussed.

Table 16.7 Mg deficiency in tomato as affected by the Mg concentration in the leaves (mmol kg^{-1} dry matter), the Mg water concentration determined in a 1:5 water extract (mmol kg^{-1} dry soil) and the Mg/K mol/mol ratio determined with the 1:5 water extract

% chlorotic leaf area	Mg in leaves	Water soluble Mg	mol ratio water soluble Mg/K
≤10%	236	2.65	0.87
<10 and ≤ 45	208	2.45	0.69
> 45	129	2.10	0.64

Data after Sonneveld (1969b).

16.4.4 Addition of the Base Dressing

On basis of the experiments carried out with different crops and under different growing conditions guide values for analytical data in the 1: 2 volume extract are set up as recommendations to growers (Van den Bos et al., 1999). In Table 16.8 some examples are listed. Tomato always and radish just in winter are started with nutrient concentrations higher than required for an optimum nutrient uptake to prevent a too lush growth under poor light conditions in winter. With tomatoes at start such easily occur also under bright light conditions. The extra nutrients decrease the osmotic potential in the root environment, whereby the fruit setting and the bulb formation is promoted. The period indicated at the bottom of the table for radish is tuned on the weather conditions in North-West Europe. Under bright light conditions the nutrient status for tomato can be lowered dependent on the crop development and fruit quality. In Appendix E guide values are listed for different crops not presented in Table 16.8.

The quantities of nutrients necessary to bring the soil on the required nutrient level will be calculated with the aid of the analytical data of a soil sample and the data presented in Table 16.1. The calculations can be carried out with the aid of Equation (16.4).

Table 16.8 Guide values (mmol l^{-1} 1:2 volume extract) for the start of different crops grown in soil

Crops	K	Ca	Mg	N	SO_4	P	Cl
Chrysanthemum	1.0	1.5	0.8	2.0	1.5	0.10	
Rose	1.5	2.0	1.2	4.0	1.5	0.10	
Tomato	3.5	3.5	2.7	7.5	3.5	0.10	
Radish[1]	2.0	1.5	0.75	2.0	2.25	0.10	
Radish[2]	3.0	3.0	1.0	3.0	3.5	0.10	2.0

[1]March 15th –August 15th;
[2]August 15th – March 15th.
After Van den Bos et al. (1999).

$$F_M = \frac{A_r - A_a}{Q_e} \tag{16.4}$$

In which:

F_M = required fertilization in kg per 100 m^{-2} for different major elements, expressed as K, Ca, Mg, N or S
A_r = required analytical data in the 1:2 volume extract, see the recommendations in Table 16.8
A_a = the current analytical data as determined in the 1:2 volume extract, under the condition that $A_r > A_a$
Q_e = factor for nutrient addition as given for different elements in Table 16.1

The system is operative for the elements K, Ca, Mg, N and S. For the required analytical data indications are listed in Table 16.8 and Appendix E. For P a separate system is operative, based on the results of the P determination in the 1:2 volume extract, and possible on the P-Al determination. In Table 16.9 the P addition is listed as given by Van den Bos et al. (1999). The quantities of elements can be recalculated to additions of fertilizers listed in Chapter 2. Unequal requirements between anions and cations can be corrected by variations within the addition of NO_3 and NH_4. However, with these additions the soil pH will be taken into account. On calcareous soils addition of NH_4 is recommended, while an ample use of NH_4 on soils poor in CO_3 easily drops the pH value to unwanted levels. Under these conditions the use of N as NH_2 (urea) can be considered, because N applied in this form lowers the pH less than applied as NH_4. When the differences between anion and cation addition cannot be eliminated in the N form, mostly the addition of SO_4 is adjusted. This anion is often used as a residual factor.

When a soil improver is used, the minerals added with this material will be diminished on the nutrient addition calculated. This is mostly not the case for N, because the N in soil improvers is generally not directly available. The information about the nutrients available in soil improvers in Section 2.3 only gives a global impression.

Table 16.9 Addition of P (kg P per 100 m^2) based on the analytical data of the P-Al and P water (1:2 volume extract)

P 1:2 extract mmol l^{-1}	P-Al mmol kg^{-1} dry soil				
	0–2.8	2.9–5.6	5.7–11.3	11.4–16.9	>17.0
<0.05	4.0	3.0	2.0	1.0	0.0
0.06–0.10	3.0	2.0	1.0	0.5	0.0
0.11–0.15	*	1.0	0.5	0.0	0.0
0.16–0.20	*	0.5	0.0	0.0	0.0
>0.20	*	0.0	0.0	0.0	0.0

*very unlikely combination
After Van den Bos et al. (1999).

A chemical analysis of the material is very useful to get a right impression of the nutrients applied with the soil improver used.

16.5 Top Dressings

For many crops with a short growing period and limited quantities of nutrient uptake the nutrients available in the soil mostly are sufficient to supply the crop with nutrients until the end of the growing period. Examples of such a situation is the production of lettuce, radish and some flower crops. The uptake of N and K for radish as calculated from a great series of crops (Sonneveld, 1997) was on average 685 and 348 mmol m^{-2}, respectively. For a summer grown crop, the level recommended for N as well for K in the 1:2 extract is 2 mmol l^{-1}, like listed in Table 16.8. The quantities of N and K available can be calculated by the information presented in Section 4.2 and is at this level 800 mmol m^{-2} for both elements. The uptake of lettuce was on average 785 and 476 mmol m^{-2} for N and K, respectively and this crop is grown at the same nutrient level in the soil. Thus, for this crop the N available at the recommended level is scarcely sufficient. For high yielding year round grown vegetables the uptake easily exceed 7000 mmol N and 4000 mmol K per m^2 (Sonneveld, 1997). For such crops at the start of the growing period, only part of the nutrients is available in the root environment and a substantial part will be added by top dressings. Indeed, the uptake of nutrients of year round grown flowers is lower than those of year round vegetables, but higher than can be added at the start of the crop. Utmost, some flower crops survive for several successive years at the same place without the possibility for an in between base dressing.

Picture 16.1 Cucumber growing in soil with drip irrigation

The aim of the top dressing is the maintaining of a certain nutrient level in the soil for a longer period during crop cultivation. The levels in the soil recommended during crop cultivation are sometimes equal to those at the start, but differ in other cases. These differences can vary dependent on the growing conditions. This for example is the case with tomato grown under the climatic condition in North-West Europe. For this crop the most striking differences occur in the recommendations, as listed in Table 16.10. Values recommended for different crops during crop cultivation are listed in Appendix E. To maintain the nutrient status of the soil the concentrations of nutrients added to the irrigation water is standardized for different crops and is focussed on the uptake and the leaching during crop cultivation. In this way the withdrawal from the soil is more or less automatically compensated with the additions in the irrigation water. Nutrient solutions used for different crops are listed in Appendix E. Dependent on the requirements of crops and soils the NH_4 concentration can be adjusted. On very calcareous soils some crops suffer seriously by chlorosis and therefore, increased NH_4 concentrations are desirable. Furthermore, the concentrations added will be corrected on the nutrients already available in the primary water. The calculations necessary can be carried out by the system given for substrate growing in Chapter 12. In advance a concentration is added to the irrigation water in agreement with the standardized concentrations mentioned in Appendix E. Concentrations of the nutrients and ratios between them are adjusted during the growing period dependent on the results of soil samples gathered. Mostly every 3–6 weeks a sample is analysed following the 1:2 volume extraction presented in Section 4.2.

Adjustments on basis of the analytical data of soil samples are interpreted following the pathway shown in Fig. 16.6. In the area A-B the standard concentration of a nutrient is added to the irrigation water as mentioned in the standard nutrient solutions in Appendix E. The values A and B are –25% and +25% of the standard value for the 1:2 extract denote as S, respectively. Thus, in this area no adjustment on the standard concentration is applied. This is because of the sampling error estimated, as explained in Section 4.12. When adjustments are carried out in this area, they will be based earlier on the sampling error than on real changes of the concentrations in the root environment. In the area A-0 the concentration is linearly increased up to a maximum of 200%, while in the area B-D the concentration is linearly decreased to zero. The value D is commonly 200% of the standard guide value in the 1:2 extract. The concentration calculated can become very low and difficult to measure precisely by the fertigation system. In such cases the concentration is negligible as indicated

Table 16.10 Nutrient status as recommended for tomato growing in greenhouse cultivation dependent on the climatic conditions. The data are expressed as mmol l^{-1} of the 1:2 extract

Period	K	Ca	Mg	N	S	EC
Start	3.5	3.5	2.7	7.5	3.5	
Feb 15th – Dec 1st	2.2	2.5	1.7	5.0	2.5	1.4
Dec 1st – Feb 15th	3.3	3.8	2.6	7.5	3.8	2.1

After Van den Bos et al. (1999).

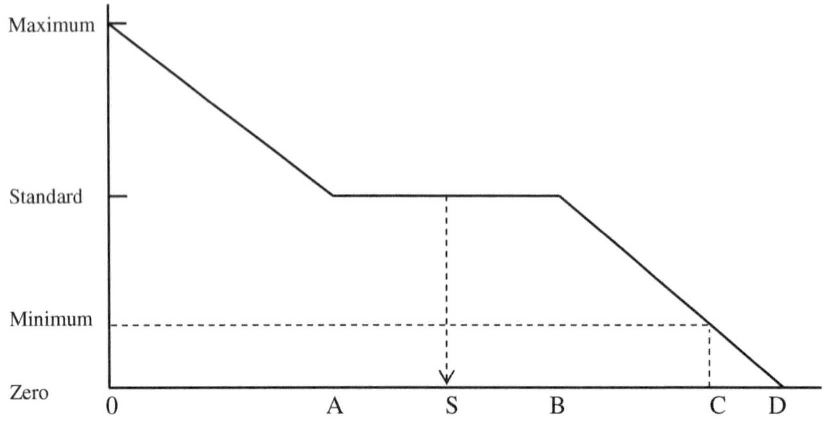

Fig. 16.6 Relationship between the concentrations added to the irrigation water and the analytical results of the soil sample during the growing period. The value S is the standard concentration in the 1:2 extract. The concentration added to the irrigation water is standard in the area A–B, increases in the area B–0, decreases in the area B–C and is negligible in the area C–D

by the area C-D. The concentration to be added to the crop will be calculated by Equations (16.5) and (16.6).

$$\text{Concentration below A: } X_{ad} = \left(1 + \frac{X_A - X_{an}}{X_A}\right) X_{st} \tag{16.5}$$

$$\text{Concentration above B: } X_{ad} = \left(1 - \frac{X_{an} - X_B}{X_D - X_B}\right) X_{st} \tag{16.6}$$

In which

X_{ad} = the concentration of nutrient X that will be added to the irrigation water
X_A, X_B, X_D = the concentrations of nutrient X in the 1:2 extract as referred in Fig. 16.6, for A, B, and D, respectively
X_{an} = the data in the 1:2 extract of the current soil analysis
X_{st} = the concentration of nutrient X in the standard nutrient solution

In the scheme of Table 16.11 an example of the calculations is presented. The difference between the sum of anions and those of cations of the nutrients calculated in the end can be compensated in this case by partly replacement of NO_3 by NH_4, which surely is an advantage for cucumber growing in soil, as discussed in Section 15.7.

The recommended nutrient solutions as given in Appendix E, are due to crops grown under conditions of irrigation with sprinkler systems. When grown with

Table 16.11 Calculation of the nutrient solution supplied with fertigation of a cucumber crop

Derived from Appendix E and calculated from text

	NH_4	K	Ca	Mg	NO_3	SO_4	EC
Guide values 1:2 extract		1.8	2.2	1.2	4.0	1.5	1.0
Nutrient solution	0.7	2.6	1.5	0.8	6.3	0.8	0.79
Current values 1:2 extract (example)	0.0	2.0	2.3	0.6	2.5	2.6	0.78
A-B area[1]		1.4–2.2	1.6–2.6	0.9–1.5	3.0–5.0	1.1–1.9	
D-value[1]		3.6	4.4	2.4	8.0	3.0	
Calculation[2] of the X_{ad} values	K, Ca, within the limits A-B no adjustment of the concentration Mg 0.3 unit below A, thus increase of the addition NO_3 0.5 unit below A, thus increase of the addition SO_4 0.7 unit above B, thus decrease of the addition						
Current nutrient solution[2]	0.8	2.0	1.5	1.0	7.3	0.1	
Composition primary water	0.0	0.2	1.0	0.5	0.5	1.0	
Nutrients to be added[3]	0.8	1.8	0.5	0.5	6.8	0.0	

[1] Calculated see text;
[2] Calculated following formulae (16.5) and (16.6);
[3] The current nutrients diminished with those in the primary water.

such irrigation systems, crops will utilize substantial quantities of the nutrients supplied with the base dressing. When drip irrigation is used to supply the irrigation water the utilization of the base dressing will be less, because the plant roots utilize restricted soil volumes and by this restricted quantities of the base dressing. Roots merely develop in the wet spots (Mmolawa and Or, 2000) and thus, have no possibilities for nutrient extraction from the dry spots. Therefore, the concentration recommended for drip irrigation systems are 25% higher than those for sprinkler irrigation.

Phosphorus is seldom applied with top dressings. Only when P concentrations <0.10 mmol l^{-1} in the 1:2 extract are determined 0.5 mmol l^{-1} is added and when a level <0.05 is found, 1.0 mmol l^{-1} is added to the irrigation water. This especially is necessary when the P storage (P-Al) in the soil is low, as indicated in Table 16.9. When the storage is high the benefit of such top dressings with P is doubtful.

16.6 Environmental Control

A precise and purposeful application of nutrients for soil grown crops is necessary to prevent environmental pollution. The nutrients added as an overdose are not absorbed by the crop and will be leached from the root environment with the irrigation surplus to the groundwater or the surrounding surface water. From the environmental concern special attention is focussed on the release of N and P from agricultural activities (Voogt, 2003). The control on the release of nutrients from soil grown crops commonly cannot be carried out by reuse of the drainage water, like is practiced with cultivation in substrate systems, because in soil grown crops the drainage water mostly cannot be gathered. In specific situations when greenhouses are situated in areas with a ground water table on a depth less than 1 m below the surface often the soil is equipped with a drainage system, from which the drainage water can be gathered and possibly reused. However, also in such situations a completely closed system cannot be realised, because of complication in the hydrology. Following situations can occur.

- When the ground water level by the drain system is lowered below the current ground water level, ground water from surroundings seep into the drainage system and mixes with the drainage water. For this situation see Fig. 16.7A
- As long as this ground water has a low salt content the mixing with the drainage water is no problem. However, when the ground water contains salts the reuse of drainage water is strongly restricted and mostly impossible.
- The influx of ground water can be more than the demand for irrigation and the residual must be discharged to the surface water.
- The current ground water level can be periodically below the drainage system, like shown in Fig. 16.7B In this situation the surplus water from irrigation penetrate the ground water and the drainage system does not function, like always happen in areas with a deep ground water level.
- When the irrigation water used contains substantial concentration residual salts reuse of drainage water is not effective, because of too high concentrations residual salts in the drainage water. See Section 7.8.

Thus, the reuse of drainage water for soil grown crops is not promising. In The Netherlands, where the groundwater table in greenhouse area is often close to the surface the application is restricted. In a survey (Voogt et al., 2008) only one third

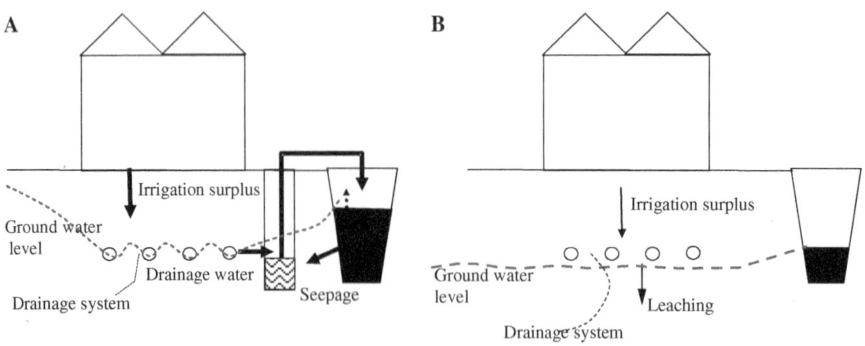

Fig. 16.7 Simplified schemes of the hydraulic situations in typical Dutch greenhouses with drainage systems. The ground water level is artificially lowered and the drainage discharge is affected by seepage from nearby water in ditches and canals (**A**). The ground water level lowered under the drainage tube level, with which the irrigation surplus penetrates the groundwater (**B**)

Picture 16.2 Year round chrysanthemum production in soil. In front plants and harvesting at the back-ground

of the greenhouses equipped with a drainage system could reuse substantial parts of the drainage water.

Therefore, to prevent environmental pollution as much as possible a well controlled supply of nutrients and water is very important with the cultivation of soil grown crops in the greenhouse industry. N is washed out quite easily from soils, because it mainly occur as NO_3 in soils and thus, completely present in the soil solution. It is transported from the root zone in concentrations equal to those in the soil solution, with which it partly will decompose by denitrification and for the other part is transported to the drain system or to the deeper soil layers. P is transported more difficult than N, because it precipitates with the Ca, Fe and Al in the root zone. However, when the root zone becomes oversaturated with P, it can be transported from the root zone just like N. However, the concentrations in the drainage water are much lower. When the soil once is oversaturated with P it offers problems for long periods, because huge quantities can be accumulated as a storage. For that reason the addition of P to soils is regulated by law in The Netherlands, even the quantities applied with the soil improvers. The quantities of soil improvers that can be added often are restricted by the quantity of P allowed to add to the soil by legislation. In soils with a big P storage, as often occur in greenhouse industry, the quantity of P added with soil improvers preferably will not exceed the uptake by the crop grown, to prevent in this way a further increase of the storage. In a study on five greenhouse holdings where organically grown vegetables were produced the P addition by means of organic fertilizers was on average 111 kg ha^{-1}, while the uptake on average was 60 kg ha^{-1} (Voogt, 1999). The great differences among the holdings

Table 16.12 Addition and uptake of P on five greenhouse holdings producing organically grown vegetables

Holding	Organic matter added[1] t ha^{-1}	P applied kg ha^{-1}	P uptake kg ha^{-1}
1	30	105	60
2	6	20	100
3	61	171	60
4	31	214	50
5	7	46	30

[1] expressed as dry organic material

After Voogt (1999). *Reprinted by permission of the International Society Horticultural Science.*

were notable, like shown in Table 16.12. In the organic greenhouse cultivation not only the overdosing of P is problematic, but also N and K often are supplied in too high quantities (Cuijpers et al., 2005). Therefore, organic growing is not a promising technique for an effective use of nutrients beforehand. With this growing method a careful supply of nutrients in relation to the uptake of the crop and a precise irrigation management are required as will be discussed for the regular methods of greenhouse crop production in soil.

The well controlled supply of water and nutrients is seriously hindered by an unequal distribution of water supply and water uptake like discussed in Section 6.3. From this discussion has to be concluded that an optimal water supply for all plants always results to leaching. With respect to the supply of nutrients the same conclusion will be drawn, because even when plants are grown on a sub optimum nutrient supply level, leaching of nutrients has been found, as demonstrated in experiments (Sonneveld and Voogt, 2001). An inquiry investigation in Dutch chrysanthemum greenhouse nurseries confirms this statement (Voogt et al., 2002). Thus, when under this condition nutrients escape from uptake, it surely will occur under conditions of an optimum supply.

However, a precise application of fertilizers sufficient for an optimum production can restrict the leaching of nutrients substantially. To this purpose the development of a model for irrigation and fertilization can be very helpful. The result of the development of such a model for irrigation is shown in Fig. 16.8, in which the relationship is shown between the evaporation measured and calculated following formula (6.1), during the cultivation of five successive chrysanthemum crops (Voogt et al., 2000). Also the fertilization can be approximated quite well by a model, like shown for the N and K uptake of chrysanthemum crops (Voogt, 2001). Best results of a modelled fertigation can be expected with crops grown with a high planting density and a high density of irrigation points. The variation in irrigation and water uptake by the crop is as discussed in Section 6.3 will be best met in this way by mutual equalizing. A fertigation model developed for chrysanthemum growing was evaluated on three greenhouse nurseries, growing year-round chrysanthemum (Voogt et al., 2006). On the nurseries modelled fertigation was compared with the standard fertigation in accordance with common practice. The fertigation following the model substantially increased the efficiency of the use of water and fertilizers. The water use efficiency

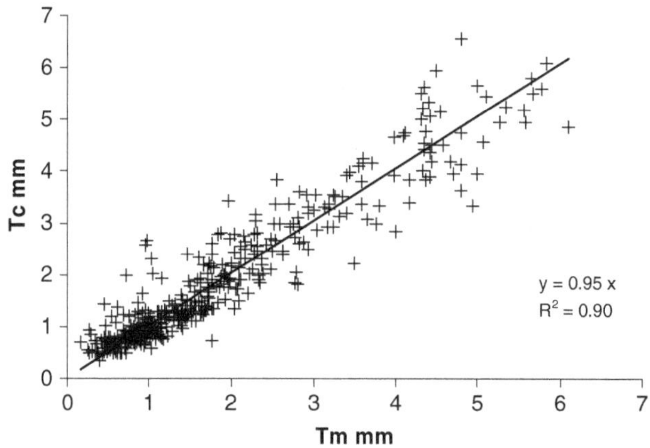

Fig. 16.8 Relationship between measured (Tm) and modelled (Tc) evapotranspiration, following formula (6.1). The intercept is constrained through the origin and the factor for maximum transpiration (m), plant height was 0.4m. Data from five chrysanthemum crops (Voogt et al., 2000). *Reprinted by permission of the International Society Horticultural Science*

Table 16.13 Nutrient balance sheet of N supply of an eustoma crop grown at different nutrient levels in a lysimeter experiment. The quantities are expressed as mmol m^{-2}

	N level in the irrigation water supplied mmol l^{-1}				
Factors	4.8	8.3	12.2	16.9	21.2
Supply[1]	+562	+971	+1427	+1977	+2480
Uptake crop	−654	−754	−762	−745	−823
Drainage water	−241	−372	−536	−729	−981
From soil storage	+726	+630			
Accumulated in soil			−108	−787	−486
Ratio +/−	0.69	0.70	0.99	1.14	0.92

[1]Supplied with irrigation water.
Data Van den Bos (1999).

increased on average from 0.78 to 0.87 and the efficiency of N addition increased on average from 0.74 to 1.00. However, the variation among the nurseries was substantial. The water use efficiency varied between 0.79–0.97 and 0.64–0.94 for the modelled irrigation and the standard fertigation, respectively. The N fertilisation surplus showed a variation of –0.26 to +0.24 and –0.15 to +0.86 for the modelled and standard fertigation, respectively. This resulted for the modelled fertigation to a surplus up to 210 kg N ha^{-1} and for the standard fertigation up to 740 kg N ha^{-1}. The surplus in this experiment was calculated from the difference between the N supplied with fertigation and the N uptake measured. Thus, by modelling the efficiency of water and nutrients can be improved substantially, but not completely prevented.

The N surplus as calculated in the investigation discussed, not always is discharged to the ground or surface water. As mentioned, part of the N surplus can get lost by denitrification, which especially occurs in the deeper soil layers (Heinen, 2006; Postma, 1996). Denitrification follows from the results of an experiment in which the optimization of the fertilization of eustoma was studied (Van den Bos, 1999). The crop was grown at different nutrient levels in a lysimeter system, with a full control on addition, uptake and leaching of water and nutrients. The lysimeters were filled with sandy loam soil over a depth of 0.45 m. The water in which the nutrients at different concentrations were added was supplied with drip irrigation, only a small part of the water was supplied with an overhead sprinkler system at start of the crop. The drip system was provided with about 15 nozzles per m^2. The results of the nutrient balance sheets of the different treatments are listed in Table 16.13. The total water supply in all treatments was $117\ l\ m^{-2}$ and the drainage was $29.5\ l\ m^{-2}$. Despite a high density of the drippers a leaching fraction of 0.25 occurred and therewith a discharge of N between 241 and 981 mmol m^{-2} was measured, agreeing with 34 and 137 kg ha^{-1}, respectively. The plant weight was not significantly affected, as shown in Fig. 16.9, while with the lowest supply the N concentration in the plant was lower than in the other treatments, indicating that the N supply was on the critical level in this treatment. The average ratio between the supply factors (+) and the discharge factors (−) in Table 16.13 is 0.89. This gap in the balance sheet will be explained by denitrification. The ratio varies strongly as shown in the last line of the table. Calculations made plausible that this can be explained by deviations with the estimating of the storages in the soil. Relatively small sampling errors reflect great difference in the estimations, because of the big soil volume in operation. Thus, in this experiment it was also proved, that the level of nutrient supply strongly affect the environmental pollution. Thus, a careful adjustment of the fertilization to the requirements of the crop can strongly reduce the environmental pollution by nutri-

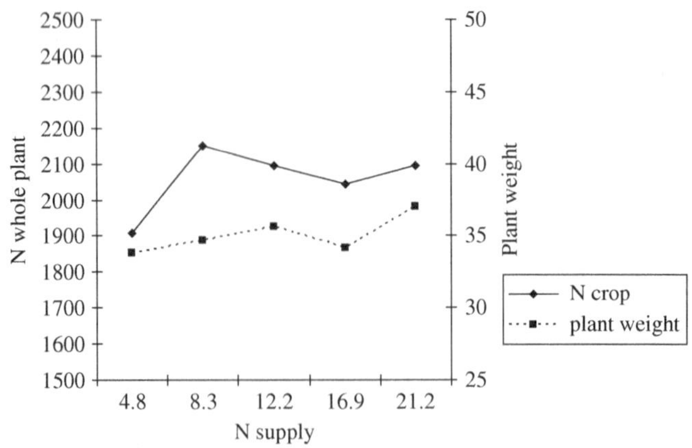

Fig. 16.9 N concentration of whole plants (mmol kg^{-1} dry matter) and the plant weight (g) of eustoma as affected by the N supply (mmol l^{-1}) in the irrigation water. Data Van den Bos (1999)

Table 16.14 Balance sheet of the water management as has been found in a typical Dutch greenhouse on a loamy soil grown with radish and cut flowers, during 4 successive years. The quantities of water are expressed as m^3 ha^{-1} $year^{-1}$

	Water supplied					Water withdrawn			
Factors	1996	1997	1998	1999	Factors	1996	1997	1998	1999
Irrigation	8654	10106	8168	8141	Transpiration[1]	7065	7114	6869	6996
Capillary rise	1328	324	388	1800	Via drainage system	5350	6319	4798	5826
Seepage	3273	3310	3250	3614	To ground water	841	307	139	734
Total	13255	13740	11806	13556	Total	13255	13740	11806	13556

[1]Including evaporation from soil surface
Data after Voogt et al. (2000a).

ents. However, it cannot completely prevent for soil grown crops, when the drainage water cannot be reused. This is shown with the balance sheet of the water supply of a greenhouse presented in Table 16.14. The situation in this greenhouse concerns a typical Dutch situation with the groundwater level close to the soil surface, which means that part of the year seepage occur, because of the a high water level in the canals and ditches. Under these conditions the water in the greenhouse soil is artificial lowered below the level of the surrounding area by a small pumping-engine placed on a drain tube system.

16.6.1 Heavy Metals

Besides the restrictions on the addition of soil improvers imposed by the P concentration, other restrictions are set by the heavy metals possibly present in such materials. Part of this problem is already discussed in Section 2.3, where acceptable total concentrations in soil improvers are discussed. In The Netherlands the contamination of soils with heavy metals is regulated on basis of the type of soil improver and the concentrations of heavy metals in soils itself. To this purpose the Dutch government developed reference values for total concentrations of heavy metals for soils related to the adsorption capacity of the soil (LAC werkgroep, 1991). They are listed in Table 16.15 and will be estimated as rough guide values, because the uptake by plants is only poorly reflected by the total concentrations.

The availability of total concentrations of heavy metals to plants, indeed, depends on the adsorption capacity of soils and thus, the availability to plant on sandy soils is much higher than those on loamy soils, like shown with the relationship in Fig. 16.10 (Smilde et al., 1992). However, the determinations of heavy metals in weak extraction solutions often reflect much better the availability to plants than the total concentrations, like shown by Smilde et al., (1992) with a solution of 0.1 mmol l^{-1} $CaCl_2$. The uptake of heavy metals depend on much more factors

Table 16.15 Calculation of the reference values for total concentrations of heavy metals and As, following the regulations of the Dutch government. The reference values are calculated on basis of the % clay (particles < 2μm) and % organic matter (OM) of the soils and are expressed as mg kg^{-1} dry soil

Elements	Reference value
Cd	0.4 + 0.007(Clay + 3OM)
Cr	50 + 2Clay
Cu	15 + 0.6(Clay + OM)
Hg	0.2 + 0.0017(2Clay + OM)
Ni	10 + Clay
Pb	50 + Clay + OM
Zn	50 + 1.5(2Clay + OM)
As	15 + 0.4(Clay + OM)

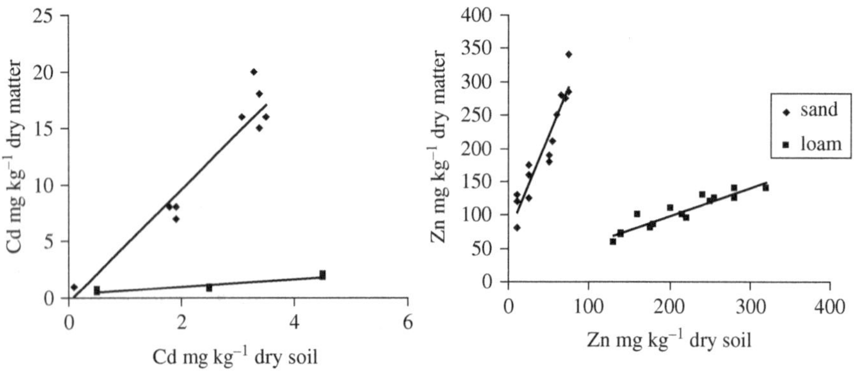

Fig. 16.10 Relationships between total Cd and Zn concentrations in soil on the one hand and those in plant tissues of maize on the other hand, dependent on soil type. Data after Smilde et al. (1992). *Modified by permission of Springer*

than just the clay and organic matter contents. The pH value of the soil for example is an important factor. The uptake of heavy metals by plants is often strongly aggravated by decreasing pH levels. The values calculated by the data of Table 16.15 are focussed on pH values estimated as being normal for a good agricultural practice (GAP). Furthermore, the crop involved plays an important part as well as the distribution within the plant (Smilde, 1976; Sonneveld and De Bes, 1984).

The type of soil improvers used and the quantities allowed to be added is controlled by regulations varying for different countries, but mostly based on the concentration of heavy metals in soils, comparable with data as listed in Table 16.15, and those in the soil improver, as discussed in Section 2.3.

References

Cuijpers W J M Voogt W and Koopmans C J 2005. Balancing crop requirements and fertilization strategy in organic greenhouse cultivation of sweet pepper. *In*: Research Sustainable production Systems. Proc. First Scientic Conference ISOFAR 2005, Adelaide Australia, 513–516.

De Vries O and Dechering F J A 1960. Grondonderzoek. Bedrijfslaboratorium voor Grond- en Gewasonderzoek Mariendaal-Oosterbeek. Ceres, Meppel.

Doll E C and Lucas R E 1973. Testing soils for potassium, calcium, and magnesium p 133–151. *In*: Walsh L M and Beaton J D (eds) Soil Testing and Plant Analysis. Soil Sci. Soc Amer. Madison, USA.

Egnér H Riehm H and Domingo W R 1960. Untersuchungen über die chemische Bodenanalysen als grundlage für die beurteilung des Nährstoffzustandes der Böden. II Chemische Extractionsmethoden zur Phosphor- und Kalibestimmung. K. Landbr. Högsk. Annlr. 26, 199–215.

Heinen M 2006. Simplified denitrification models: overview and properties. Geoderma 133(3–4) 444–463.

Indiati R and Singh B R 2001. Phosphorus extractability of heavily fertilized soils from Norway. Comm. Soil Sci. Plant Anal. 32, 1629–1641.

LAC werkgroep 1991. LAC – signaalwaarden. Ministerie Landbouw, Natuurbeheer en visserij's-Gravenhage, 42pp.

Mmolawa K and Or D 2000. Root zone solute dynamics under drip irrigation: A review. Plant Soil 222, 163–190.

Neeteson J J and Wadman W P 1987. Assessmant of economically optimum application rates of fertilizer N on the basis of response curves. Fertilizer Res. 12, 37–52.

Neeteson J J and Zwetsloot H J C 1989. An analysis of the response of sugar beet and potatoes to fertilizer ÿitrogen and soil mineral nitrogen. Neth. J. Agric. Sci. 37, 129–141.

Postma, 1996. Stikstofverlizen door denitrificatie op praktijkbedrijven met jaarrondchrysant. NMI, rapport c 96.11 juli 1996, 44pp.

Roorda van Eysinga J P N L 1961. Beoordeling van de fosfaattoestand van diluviale zandgrond voor de teelt van kropsla. Verslagen van Landbouwkundige Onderzoekingen, Centrum Landbouwpublicaties en Landbouwdocumentaties, Wageningen, 67.6, 26pp.

Roorda van Eysinga J P N L 1963. Vergelijking van K-water en K-HCl in bijmestmonsters van kassen. Instituut voor Bodemvruchtbaarheid Groningen, Rapport 20, 6pp.

Roorda van Eysinga J P N L 1966. Cultuurtechniek en bemesting, tuinbouwkundig bezien. Cultuurtechnisch Tijdschrift 5, 185–188.

Roorda van Eysinga J P N L 1966a. Mineral fertilization, yield and quality of vegetables. Proceedings of the 8th International Potash Congress, Brussels. International Potash Institute, Bern. 217–233.

Roorda van Eysinga J P N L 1966b. Bemesting van tomaten met kali. Verslagen van Landbouwkundige Onderzoekingen 677. Centrum Landbouwpublicaties en Landbouwdocumentaties, Wageningen, 37pp.

Roorda van Eysinga J P N L 1966c. Bemesting van Kropsla onder glas met bloedmeel en kalkammonsalpeter. Verslagen van Landbouwkundige Onderzoekingen 681. Centrum Landbouwpublicaties en Landbouwdocumentaties, Wageningen, 18pp.

Roorda van Eysinga J P N L 1971. Determination of the phosphate status of soils in the Naaldwijk area for growing lettuce in glasshouses. Agricultural Research Reports 753. Centre Agric. Publish. Doc. Wageningen, 25pp.

Roorda van Eysinga J P N L 1971a. Fertilization of tomatoes with nitrogen. Centre Agric. Publishing Doc., Wageningen, 17pp.

Roorda van Eysinga J P N L 1971b. Fertilisation of tomatoes with phosphate. Agricultural Research Reports 755. Centre Agric. Publish. Doc. Wageningen, 10pp.

Roorda van Eysinga J P N L 1972. Fertilization of glasshouse food crops with nitrogen, phosphorus and potassium. Colloquium Proceedings No 2. Potassium Institute Ltd Bern, 1–9.

Roorda van Eysinga J P N L and Mostert M 1972. De bemesting, in het bijzonder die met stikstof, van koolrabi onder glas. Bedrijfsontwikkeling 3, 391–393.

Schäfer P and Siebold M 1972. Influence of increasing potash application rates on yield and quality of the spring wheat "Kolibri". Results from a potash fixing location. Bayer. Landw. Jahrb. 49, 29–39.

Sissingh H A 1969. Die Lösung der Bodenphosphorsäureng bei waszriger Extraktion in Verbindung mit der Entwicklung einer neuen P-Wasser-Methode. Langw. Forsch. 23 Sonderheft 2, 110–120.

Smilde K W 1976. Toxische gehalyen aan zware metalen (Zn, Cu, Cr, Ni, en Cd) in grond en gewas. Instituut voor Bodemvruchtbaarheid, Nota 25, 36pp.

Smilde K W Van Luit B and Van Driel W 1992. The extraction by soil and absorption by plants of applied zinc and cadmium. Plant Soil 143, 233–238.

Sonneveld, 1966. Onderzoek naar het verloop van grondanalysecijfers (1964–1965). Proefstation voor de Groenten- en Fruitteelt onder Glas te Naaldwijk. Intern verslag 1966, 8pp.

Sonneveld, 1967. Onderzoek naar het verloop van grondanalysecijfers (1965–1966). Proefstation voor de Groenten- en Fruitteelt onder Glas te Naaldwijk. Intern verslag 1967, 7pp.

Sonneveld, 1969. Onderzoek naar het verloop van grondanalysecijfers (1966–1967). Proefstation voor de Groenten- en Fruitteelt onder Glas te Naaldwijk. Intern verslag 1969, 10pp.

Sonneveld C 1969a. De invloed van het stomen op de stikstofhuishouding van de grond. Tuinbouwmededelingen 32, 197–203.

Sonneveld C 1969b. Magnesiumgebrek bij tomaat. Praktijkonderzoek 1962. Proefstation voor de Groenten- en Fruitteelt onder Glas te Naaldwijk. Intern verslag 1966, 10pp.

Sonneveld C and Van den Ende J 1971. Soil analysis by means of a 1:2 volume extract. Plant Soil 35, 505–516.

Sonneveld C and Van Beusekom J 1974. Het doorspoelen van een lichte zavelgrond. Proefstation voor Groenten- en Fruitteelt onder Glas Naaldwijk. Intern verslag 1974, 677, 10pp.

Sonneveld C and De Bes S S 1984. Micro nutrient uptake of glasshouse cucumbers grown on rockwool. Comm. Soil Sci. Plant Anal. 15, 519–535.

Sonneveld C and Voogt W 1986. Grondonderzoek op basis van waterige extractie. Deel 1: methoden en resultaten. Proefstation voor Tuinbouw onder Glas Naaldwijk. Intern verslag 1986, 13, 36pp.

Sonneveld C Van den Ende J and De Bes S 1990. Estimating the chemical composition of soil solutions by obtaining saturation extracts or specific 1:2 by volume extracts. Plant Soil 122, 169–175.

Sonneveld C and Voogt W 1991. Effects of Ca-stress on blossom-end rot and Mg-deficiency in rockwool grown tomato. Acta Hort. 294, 81–88.

Sonneveld C 1993. Verschillen bij de bemesting van grond en substraat. *In*: Plantevoeding in de Glastuinbouw. Proefstation voor Tuinbouw onder Glas te Naaldwijk, Informatiereeks no 87, 34–39.

Sonneveld C 1997. Mineralenopname bij teelten onder glas. Proefstation voor Bloemisterij en Glasgroente, Naaldwijk. Intern verslag 81, 35pp.

Sonneveld C and Voogt W 2001. Chemical analysis in substrate systems and hydroponics – use and interpretation. Acta Hort. 548, 247–259.

Sonneveld C Van den Bos A L and Voogt W 2004, Modelling osmotic salinity effects on yield characteristics of substrate grown greenhouse crops. J. Plant Nutr. 27, 1931–1951.

Spithost L S 1965. Stikstofanalyse en stikstofbemesting van een zandgrond voor de teelt van tomaten. Verslagen van Landbouwkundige Onderzoekingen. Pudoc Wageningen 671, 23pp.

Van den Bos A 1993. Betekenis analysecijfers in kasgrond. *In*: Plantevoeding in de Glastuinbouw. Proefstation voor Tuinbouw onder Glas te Naaldwijk, Informatiereeks no 87, 89–95.

Van den Bos A L 1999. EC-niveau bij de teelt van Eustoma in kasgrond. Proefstation voor de Bloemisterij en Glasgroente Naaldwijk, The Netherlands, 41pp.

Van den Bos A L De Kreij C and Voogt W 1999. Bemestingsadviesbasis grond. Proefstation voor Bloemisterij en Glasgroente, Naaldwijk, Brochure PPO.171, 54pp.

Van den Bos A 2001. Minimalisatie Fosfaatbemesting – Onderzoek met botersla op Proeftuin Westmaas. Proefstation voor Bloemisterij en Glasgroente, Naaldwijk. Intern Rapport 235, 19pp.

Van den Bos A L 2004. Minimalisatie van fosfaatbemesting – Onderzoek met radijs op een praktijkbedrijf. Praktijkonderzoek Plant en Omgeving Wageningen UR. Rapport 32pp.

Van den Bos A L 2004a. Minimalisatie van fosfaatbemesting – Onderzoek met chrysanten op twee praktijkbedrijven. Praktijkonderzoek Plant en Omgeving Wageningen UR. Rapport 36pp.

Van der Paauw F 1969. Entwicklung und Verwertung einer nuen Wasserextraktionsmethode für die Bestimmung der pflanzenaufnehmbaren Phosphorsäure. Landw. Forsch. 23 Sonderheft 2, 102–109.

Van Gurp, H A J M 1998. Minimalisatie van fosfaatbemesting bij botersla 1992–1998. Proeftuin Zuid-Nederland, Rapport I-24, 32pp.

Verhoef K and Weber L 1965. *Botrytis* in tomaten. Proefstation voor de Groente en Fruitteelt onder Glas te Naaldwijk, The Netherlands. Jaarverslag 1965, 143–144.

Voogt W 1999. Water and mineral balances of organically grown vegetables under glass. Acte Hort. 506, 51–57.

Voogt W Kipp J A De Graaf R and Spaans L 2000. A fertigation model for glasshouse crops grown in soil. Acta Hort. 537, 495–502.

Voogt W De Graaf R Van der Burg A A M De Graaf R Van Oosten R 2000a. Water-, stikstof- en natriumbalans bij teelten in kasgrond. Intern Rapport Proefstation voor Bloemisterij en Glasgroente Naaldwijk The Netherlands, 40pp.

Voogt W 2001. Improving the N efficiency in soil grown glasshouse crops, by using the fertigation model. Proc. 2001 Dahlia Greidinger Symposium, Intern. Fert. Soc. Lisbon, 59–62.

Voogt W Assinck F Balendonck J Blom-Zandstra G Heinen M and De Zwart F H 2002. Minimalisering van de uitspoeling bij teelten in kasgrond. Praktijkonderzoek Plant & omgeving, sector glastuinbouw, Naaldwijk, The Netherlands, Rapport 543, 34pp.

Voogt W 2003. Nutrient management in soil and soil-less culture in The Netherlands: towards environmental goals. Proc. 529 Dahlia Greidinger Symposium Izmir Turkey, 2003. The Intern. Fertiliser Soc., 1–27.

Voogt W Steinbuch F and Van Winkel A (2006). Evaluation of the "fertigation model", a decision support system for water and nutrient supply for soil grown greenhouse crops. Acta Hort.718, 531–538.

Voogt W Sterk F and Vermeulen T 2008. Teelt in kasgrond; Nederland anno 2007. Rapport Wageningen UR glastuinbouw, Bleiswijk, The Netherlands, 43pp.

Chapter 17
Plant Nutrition in Future Greenhouse Production

17.1 Introduction

In the introduction chapter it was claimed that greenhouse production has no longer arguments as a supply market. The products of the greenhouse industry became in free competition with those from field production from all over the world. In this competition the greenhouse industry has developed into a branch operative to a consumer market and through that self-condemned to bring better and cheaper products on the market than those from the open field. Many greenhouse products have a luxurious image, which even more is the case for flowers than for vegetables. That will be the reason that the critical view as exists on the production methods in agriculture in general, possible is even more critical on those used in greenhouses. Therefore, presentation of such products on a consumer market does not mean only high standards for quality, a great diversity and low prices, but also the production methods play a prominent part. Thus, greenhouse production will survive in future under conditions of sustainable production methods. This means low energy use, an effective use of raw materials and low environmental pollution.

Last decennia innovations by growers and agro industries supported by the work of research institutes, have brought about strong impulsions for sustainable production methods in the greenhouse industry in Western Europe. The high energy use in this area has been met by the introduction of improved greenhouse constructions, the use of co-generation and heat storage in buffering tanks. Newest developments even offer opportunities to switch the greenhouse industry from a strong energy requiring activity to an energy producing one. The excess of energy in summer is put into storage in aquifers and is used in winter (Van Oosten and Koehorst, 2007). The excess in summer is more than the need in winter and thus, can be utilized elsewhere. Another example of a development towards sustainable production methods is the reduction of agro chemicals. Biological control of pests, improved climatic control and the use of substrates strongly reduced the need of pesticide application, and fumigation. Further, the biological control of root diseases is in development. When such systems become operative, the application in substrate systems may be more effective than in soil growing. Due to the restricted root volume the addition can be controlled better as well the conditions favourably for the development of such organism.

C. Sonneveld, W. Voogt, *Plant Nutrition of Greenhouse Crops*,

DOI 10.1007/978-90-481-2532-6_17,

Within the scope of this book it is logical to focus on the developments on the water and the fertilizer use in the greenhouse industry. It became clear from the contents of this book that a fruitful discussion about fertilizer use is only meaningful when the water use is taken into account. Reasons for such a combined view are following.

- Environmental pollution of ground water and surface water by nutrients merely occur by movement of the drainage water, when there is an irrigation surplus.
- The quantity of drainage water discharged from greenhouse production depends strongly on the quality of the raw water used for irrigation. High concentrations of residual salts require high leaching fractions to keep the osmotic potential within the acceptable limits.
- The concentrations of residual salts in the irrigation water can affect the nutrient levels maintained in the root environment.
- Finally, irrigation water can contain minerals which the plant can utilize as nutrients and these minerals have to be included in the calculations of the crop requirements.

The third reason points to the strong relationship between salinity and fertilization as discussed in this book. Salinity in greenhouse horticulture is not just a handicap, but it is discussed that it is a tool for growers to control plant development. It appeared that for some crops under specific conditions crop development can be favourably affected by lowering of the osmotic potential through increasing salinity. This tool has to be guided in close relation with crop, cultivar, rooting volume and growing conditions in the greenhouse, like global radiation, humidity and temperature. With respect to the rooting volume can be stated that growing in substrate offers better possibilities for the management of the osmotic potential than soil growing, because of the much smaller rooting volume.

In the public opinion sometimes sustainable production methods are connected with ecologic or organic production. However, these production methods mostly mean lower productions compared to the current methods, while the restrictions to the exclusive use of organic fertilizers and manures often force to high N and P inputs and henceforth potentially to higher environmental pollution per unit produce. It has been shown in modern greenhouse industry that with high technological inputs, the production of high yielding vegetables and flowers of an optimum quality can be realised in combination with a minimum of environmental pollution.

17.2 Water Requirements

The water requirement of greenhouse crops depend on different factors as discussed in Chapter 6. For a year round production programme the annual water requirements for transpiration for many greenhouse crops varies between 450 and 1000 mm in moderate climates. Main factors affecting these quantities are the crop type, the management of the climatic conditions in the greenhouse and the mineral composition of the water used. Therefore, it is understandable that the water

Table 17.1 Quantity of water in l used for the production of 1 kg tomatoes and sweet pepper in various countries and with different growing systems (Stanghellini, 2003), adjusted with recent data

Country and growing system	Tomato	Sweet pepper
Field production, Israel and Almeria	60	300
Unheated plastic greenhouse, Almeria	40	
Unheated glass greenhouse, Israel	30	
Improved unheated plastic greenhouse, Almeria	27	74
Climate controlled glass greenhouse with CO_2 enrichment, The Netherlands	22	33
As above with reuse of drainwater	15	25

required for crop production between countries differ strongly. This is shown by the data presented in Table 17.1 for a tomato and a sweet pepper crop (Stanghellini, 2003).

With the management of climatic conditions true enough restrictions in the transpiration are possible in standard greenhouse cultivation, but the management will be focussed earlier on optimal growing conditions for the crop than for restriction of the transpiration. Serious restrictions in the water use of crops without restrictions on optimization of climatic conditions only seem to be possible in the closed greenhouse concept. The air in these greenhouses is cooled artificially whereby an important part of the water evaporated in the greenhouse environment condensate and can be reused in the greenhouse. In the first greenhouses built following this concept in The Netherlands no exact measurements are carried out, but estimations indicate that 40% of the water can be won back in such greenhouses (Raaphorst, 2005).

The mineral composition of the irrigation water strongly determines the need for the drain to waste. In Chapter 7 this topic is extensively discussed. Drain to waste is necessary when the concentrations of residual salts accumulate in the root environment beyond the salinity threshold value, which occur when the concentrations of such salts in the primary irrigation water exceed the uptake concentration. As this easily occur for Na and Cl, the volume of the drain to waste is mainly determined by the concentrations of these ions. Generally, earlier by Na than by Cl, since the uptake of Na by most crops is lower than that of Cl. The leaching fraction of the irrigation water required also depends strongly on the accepted accumulation of the residual salts in the root environment. The accepted accumulation depends on the crop grown and the growing system employed. In Chapter 8 plant responses to spatial salt accumulation in the root environment are discussed and it was shown that plants are able to escape from spots with high EC values in the root environment. The results gained in the experiments indicate possibilities, but it is not yet clear what part of the root environment can be accepted as "accumulation" zone in relation to factors like crop, accumulation level, irrigation system and total root environment volume.

When the concentration of any ion in the irrigation water used does not exceed the uptake concentration of the crop grown it is possible to grow crops without any drain to waste because of harmful accumulations of residual salts. However, such is possible under conditions that the drainage water can be gathered and is reused in the growing system. Therefore, this procedure is more or less exclusively suitable for substrate cultivations and hydroponics. The need for drainage water exists under these conditions because of the discrepancy between water supply and water use by the crop. This discrepancy not only results from an unequal water supply, but also by an unequal water use of the crop. Both problems as yet cannot be solved and therefore an over supply of irrigation water of at least 20% is necessary to wash away the differences caused by this discrepancy.

Water of the quality intended rarely is available under natural condition, except rain water. Therefore many growers in North West Europe collect the rainwater from the greenhouse roofs and use it as irrigation water. The discrepancy between precipitation and use is merely covered by storage in big basins. Another possibility is desalinisation of the raw water from other origins.

The application of a system in which the drainage water is reused is safe under conditions that the drainage water is disinfestated before reuse, to prevent a general outbreak of a possible local infection. Such disinfestations will be recommended for the raw water too when an infection with plant pathogens is possible. Such surely can be supposed for the rain water from greenhouse blocks and from condensation of greenhouse cooling systems as mentioned before.

Under conditions of controlled salt accumulation and disinfestations as described, reuse of drainage water is common practice in The Netherlands. In experiments it was proved that crop development and yield in systems with reuse of drainage water were not affected in comparison with drain to waste systems (De Kreij and Runia, 2001). However, in recent years, some growers got the experience that despite disinfestations growth disturbances occur after a long duration of reuse of drainage water in substrate systems. Examples of such accidents are the occurrence of “corcky root” in tomato and the “thick root syndrome” in cucumber (Dechering and Kipp, 1995; Stallen, 1997; Verkerke, 1997). As long as no reasons for such disturbances could be traced, it hinders complete reuse of drainage water. Nowadays, cultivation of roses in closed systems sometimes shows growth disturbances by still unknown reason. Growers got the experience that renewal of the nutrient solution more or less solved the problem. This brings, indeed a reduced, but still a noticeable drain to waste and with that an environmental problem. Efforts to meet such accidental discharges will be discussed in Section 17.4.

17.3 Requirements of Minerals

17.3.1 Mineral Requirements of Crops

The uptake of minerals per area is much larger in the greenhouse industry compared to field production. Obviously, this is due to the high productions in the greenhouse industry and not to the high nutrient levels maintained in the root environment, as it is proved that plants do not absorb substantial extra nutrients under possible over-fertilised growing conditions. Some crops achieve a very effective utilization of their minerals absorbed at high yields. This occurs for example with vegetable fruit crops, which have an ineffective utilization of minerals during the vegetative stadium in the beginning of the growing period. In this period just stems and leaves are produced and no fruits. This pseudo ineffective absorption of minerals is expressed in the intercept of the relationships between yield and mineral uptake. An example of such a relationship is shown in formula (17.1) for N with tomatoes (Sonneveld, 1997).

$$y = 18.7x - 63.4 \tag{17.1}$$

In which

x = yield kg m^{-2}
y = N uptake kg ha^{-1}

From this formula can be concluded that at a low yield regime of about 10 kg fruits m^{-2}, 2.50 g N is absorbed per kg produced fruit, while with an average production of 50 kg m^{-2} in the modern greenhouse industry only 2.00 g N per kg produced fruit is absorbed. A further increase of the yield does not seriously reduce the use of nutrients per unit produce. With a production of 100 kg m^{-2} an N use is calculated of 1.93 g kg^{-1} fruits. Comparable results can be calculated for other nutrient elements. For the nowadays common yield of 50 kg tomatoes per m^2 can be calculated by the formulae found (Sonneveld, 1997) that the uptake of N, P and K is estimated at 998, 240 and 1744 kg ha^{-1}, respectively. With a future 100 kg m^{-2} fruit production the uptake will be 1933, 460 and 3244 kg, respectively. If the same total production will be gained in a low production system with yields of 10 kg m^{-2}, then ten times more area is necessary, while the mineral uptake of N, P and K will be increased to 2504, 643 and 5445 kg ha^{-1}, respectively. Thus, for fruit vegetable crops it is wise to strive for high yields in view of an effective utilization of minerals.

17.3.2 Availability of Minerals

Essential nutrients required are shown in Table 1.2 and concern 6 major nutrients and 6 micro nutrients and when Ni also is taken into account 7 essential micro elements. The quantities of micro elements required are low and for soil grown greenhouse crops mostly sufficiently available in the soil itself. For substrate cultivation the addition of micro elements is common practice, but the quantities required are less and therefore generally not a problem in relation to the storages on world scale. This is different for the major elements. The production of N fertilizers is no problem. The fixation of atmospheric N in fertilizers is an artificially process accomplished in industrial plants and is only restricted by economic factors. Ca and Mg are earth alkali metals available in inexhaustible quantities of calcium-magnesium

Table 17.2 World reserves and yearly production of P (rock phosphate) and K expressed as 10^3 metric tons rock phosphate and K_2O

	Rock phosphate	Potassium (K_2O)
Reserves[1]	18,000	8,310
Reserve base[2]	50,000	18,000
Production per year	145	30.8

[1]Currently economic extractable; [2]demonstrated resource.

stones. S is a residue of many industries and thus, the availability is over sufficient. The elements liable to question are P and K, which have to be withdrawn from restricted world storages. These storages are substantial. As found by the data summarized in Table 17.2, where a survey is given about world storages and use based on Jasinski (2008), Ober (2008), USGS (2006), and USGS (2006a). From the data can be calculated that the current extractable storage covers about the need for the coming 125 and 270 years for P and K, respectively. When the demonstrated resources were included the reserves cover about 350 and 600 years for P and K, respectively. These calculations are based on the nowadays productions. This means that there are no direct restrictions with respect to the available minerals and the supply is ensured for the coming centuries. Nevertheless, it was concluded that followed from estimated increases of P demand the supply of this element already at the turn of the 21st century can become problematic (Shu et al., 2006; Steen, 1998). Moreover, the sites of the P and K reserves are not equally spread over the world and are specific concentrated in a restricted number of countries. Part of them are political instable, which make the availability subject to speculations. Therefore, recycling from waste streams will be considered and a start is made with the use of green compost as a constituent of substrates. Up till now the use of such waste streams in greenhouse horticulture is problematic, not only from the viewpoint of unwanted components, like residual salts and unwanted minerals, but it is likely that also the organic compounds available can cause growth reduction, as discussed in Section 10.8. Therefore, the use of waste streams is surrounded with problems, like discussed in Section 11.3 and the research for utilizing these just has been started. The main problems of the use of such materials are the great variability of the material (Maher et al., 2008) and the risk on phytotoxic effects, which can induce huge growth reductions (Jarecki et al., 2005). Much research is required before it can be safely used in substrate cultures for the professional greenhouse industry.

17.3.3 Developments

From the foregoing paragraph is made clear that with increasing yields the nutrient requirements increase linearly by that. This increase can merely be attributed to the increase in volume of the crop, because the nutrient concentrations of crops show relatively little variation when grown at optimum production conditions. Therefore

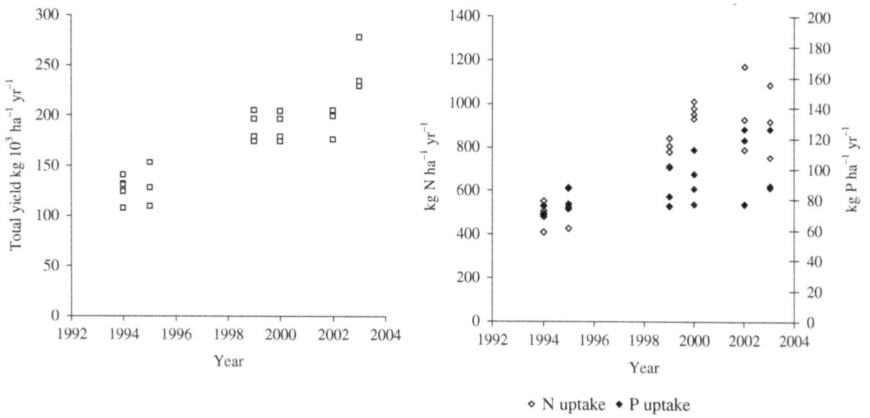

Fig. 17.1 Development of the yield of year-round chrysanthemum crops in commercial practice in the last decade and as consequence the total uptake of N and P by the crop. Data from Voogt (2004), Voogt (unpublished)

the increase of production results in an increase of nutrient use per unit area both for vegetable crops and for ornamentals. In greenhouse horticulture strong increases of yield in short periods occur, like for example shown for chrysanthemum over the period 1994–2003, shown in Fig. 17.1. Generally, improvement of the growing conditions is the main reason of such an increase of the crop volume, coming from 12.5 kg fresh weight to 22.5 kg m^{-2} over that period. This increase went together with an increase of N and P need of the crop from 500 to 950 and from 70 to 110 kg ha^{-1}, respectively. This means an increase of the production of 80% over a period of 10 years, with an increase of the nutrient uptake relatively of about the same magnitude. An example of vegetables is the yield development of tomato cropping in the greenhouse industry in The Netherlands in the last century. In the fifties of the 20th century a high level of the yield can be estimated at 10 kg m^{-2}. Nowadays, a high yield in the Dutch greenhouse industry for this crop will be estimated at least at 60 kg m^{-2} and under additional artificial lighting even a yield of 80 kg m^{-2} can be realised. With continuation of the innovations can be expected that yields will increase further on both for vegetables as well ornamentals, despite the nowadays realised high yields. Therefore, restriction of fertilizer use never should be based on quantities per area as was intended by the Dutch government some years ago, but have to be related to the production level.

In the developments towards higher yields all factors affecting crop growth have to be improved in a reasonable equilibrium. Factors that play an important part in this development are improvement of greenhouse constructions with a modelling of the climatic conditions in its track, breeding, the development of biological control of pests and diseases, improvement of the conditions in the root environment as realised with the switch from soil to substrates and last but not least improvement of the water supply with a revaluation of salinity considering the coherence between yield and quality. Within the objective of this book it is needless to say that such developments only can be realised with adjustment of the mineral nutrition of the crop to the need of the increased yields and modelling of the fertilizer supply in relation with the water uptake.

17.4 Future Plant Nutrition Research

It needs no argument that the future developments for sustainable productions in the greenhouse industry lay in cultivation in substrates. Setting aside the advantages for crop production, the statement holds considering the environmental consequences. The impossibility to have a close control on the water supply with soil grown crops as discussed in Section 6.3 and the thereby attendant environmental pollutions with minerals confirms this sufficiently. Therefore, research on the mineral plant nutrition for greenhouse crops will be focussed on substrate growing with reuse of drainage water.

The addition of major elements is merely modelled last decennia and many developments can be met with these models. However, there is a lack in the knowledge of the nutrient demands of some potted ornamentals. This can be related to the already existing great diversity, which evermore increases in this field. Other items are the interactions between major elements, which appeared to have specific opportunities in plant nutrition and such research need attention. Special attention should be paid to the development of new greenhouse types, like the closed greenhouse and other innovations connecting with energy management. The completely different growing conditions in such greenhouses can dramatically affect the relation between the nutrient supply and the uptake by the crop, specifically with respect to the supply and uptake of Ca.

In view of the supply of micro nutrients many studies in substrate are carried out in this field, but the addition of Cu in organic substrates is still unclear, while the recommended additions of Mo are abundantly in relation to the need of the crop. Just those both micro elements are the most suspicious with respect to environmental pollution. Another item subject to further research is the addition of an element like Ni, traced as an essential element, but so far scarcely included in plant nutrition studies, especially not in substrate where a possible shortage of such an element can be expected firstly. The risk that such essential elements become insufficiently available to crops increases, if substrates used are free of mineral traces and the fertilizers and the irrigation water added become more and more free from any background concentrations.

Substrate cultivation offers in this way excellent possibilities for a close control on unwanted minerals in the produce, like heavy metals. When the substrate and the materials added to the root environment are sufficiently free from such components the produce cannot be polluted by it. This put strong restrictions on the use of materials as substrate constituents, like composts from different origins. Such materials can be loaded with heavy metals and therefore, will be often unsuitable as constituent for substrates due to the production of food crops. However, also for the production of ornamentals studies to the possibilites and implications of the use of composts as constituents are not yet finished. The explanation for the growth reductions which often occur in substrate mixed with compost needs intensive studies before composts possibly will be introduced as a safe constituent in substrates, like discussed in the preceding section.

Plant nutrition also plays a role in the prevention of pathogen infection in crops. This offers possibilities for further development of biological control of greenhouse crops. We are aware that this item is insufficiently discussed in this book. Elements (compounds) as Si, Mn, Zn, NO_3 and Ca are known to be able to play a part in the plant resistance to pathogens. This subject need further attention in future research of plant nutrition in substrates, because the excellent possibility to control the uptake of mineral elements in such growing systems.

Last but not least, certain elements merely micro nutrients are too low in the diet of people (Cakmak, 2002) and several billion people globally suffer from deficiencies. Not only in developing, but also in well developed countries, like has been found for Se in Scandinavia (Hartikainen and Ekholm, 2001) and I already in different countries (Delange et al., 2002). It needs no argument that in greenhouse production such elements can be serviced to the crop and in this way included in the human diet. Especially for substrate grown crops an accurate supply and precise management of the uptake is within reach. However, thorough studies to uptake, distribution and acceptance of such element by plants are necessary in relation to human needs. In studies to plant adsorption of I, one of the elements that is too low in the diet, it appears that this element for example is not transported to generative plant parts, but surely to vegetative parts (Mackowiack et al., 1999; Dai et al., 2006). Such and many other considerations will be taken into account, before an effective policy in this direction can be developed.

In view of the experience gained from the current situation future greenhouse production will be focussed even more on sustainability and food safety, thus, future production systems will be characterized by systems with reuse of drainage water. Information technology will play a dominant role in the management of climate control as well as with fertigation management. When in future suitable sensors are developed for the determination of sufficient parameters *in situ*, like now possible with the determination of the EC and pH values, full computerized management of the nutrient supply can be realised.

Such a statement does not include that when the management of the fertilization can be taken over by a computer that nothing can be improved further on. Possibilities for example as mentioned in this section put questions to the public human health authorities to formulate the need of people in this direction, because they are insufficiently formulated. Greenhouse industry in substrate has to translate these formulations to crop concentrations, followed by the development of supply and uptake models for crops involved.

In future views it should be realised that water of the required quality and quantity will become increasingly scarce and definitely more expensive. For the near future, reduction of discharge of waste water is an issue in dense greenhouse areas like in The Netherlands. At present, this waste water is discharged to municipal waste water systems and into surface water. Relatively high concentrations of nutrients and residues of plant protection agents are observed in the waste water and in surface water in regions with intensive greenhouse cultivation (Baltus and Volkers-Verboom, 2005; Teunissen, 2005). Even in situations with closed growing systems sometimes drain to waste is inevitable as is stated in Section 17.2. Research to discover the reasons of the possible growth disturbances and a remedy to prevent such drainages to waste will be the best solution. However, as long as there is no cure for it, accidental drain to waste will remain. A solution for this in dense greenhouse areas will be found in closing of the water cycle on a regional scale. This can contribute to diminish or solve the water management problems described. Both irrigation water quality and quantity problems might be tackled if waste water streams are collected and purified to irrigation water (Van der Velde et al., 2008). This closed water cycle is also in line with the goal of the criteria of the European Water Framework Directive (WFD).

Other developments with respect to minerals will be definitely the combination with other agricultural branches to joint use of resources, area and energy and in some cases the mutual use of wastes. In particular aquaculture opens perspectives and is already surpassing the experimental phase (Savidov et al., 2007), but also the combination with cattle breeding and other agro-industries have been explored (Poot, 2004). In this way beside reuse of water reuse of minerals break through.

References

Baltus C A M and Volkers-Verboom L W 2005.Onderzoek naar emissies van N en P vanuit de glastuinbouw. RIZA Rapport 2005.007, 52 pp.

Cakmak I 2002. Plant nutrition research: Priorities to meet human needs for food in sustainable ways. Plant Soil 247, 3–24.

Dai J L Zhu Y G Huang Y Z Zhang M and Song J L 2006. Availability of iodide and iodate to spinach (*Spinacia oleracea L.*) in relation to iodine in soil solution. Plant Soil 289, 301–308.

Dechering A and Kipp J 1995. Meer wortelproblemen bij recirculatie. Groenten en Fruit/Glasgroenten 5(16), 16–17.

De Kreij K and Runia W 2001. Dezelfde productie bij recirculatie drainwater. Groenten en Fruit/Glasgroenten 11(3), 6–7.

Delange F Burgi H Chen-Zu P and Dunn J T 2002.World status of monitoring of iodine deficiency disorders control programs. Thyroid 12, 915–924.

Hartikainen H Ekholm P 2001. Selenium inclusion in fertilisers to correct an inadequate national dietary intake. Proc. Int. Fert. Soc. and Dahlia Greidinger Symp. Lisbon, 253–268.

Jarecki M K Chong C and Voroney P P 2005. Evaluation of compost leachates for plant growth in hydroponic culture. J. Plant Nutr. 28, 651–667.

Jasinski S M 2008. U.S. Geological Survey, Mineral Commodity Summaries, 124–125.

Mackowiack C L and Grossel P R 1999. Iodate and iodide effects on iodine and partitioning in rice (*Oryza sativa L.*) grown in solution culture. Plant Soil 212, 135–143.

Maher M Prasad M and Raviv M 2008. Organic soilless media components. *In*: Raviv M and Lieth J H (eds) Soilless Culture – Theory and Practice. Elsevier, Amstrerdam, 459–504.

Ober J A 2008. U.S. Geological Survey, Mineral Commodity Summaries, 128–129.

Poot E 2004. Innovatie voor intensivering. Praktijkonderzoek Plant en Omgeving, Naaldwijk, The Netherlands, Report 15 pp.

Raaphorst M 2005. Optimale teelt in de gesloten kas: teeltkundig verslag van de gesloten kas bij Themato in 2004. Praktijkonderzoek Plant & Omgeving, Business Unit Glastuinbouw, Naaldwijk The Netherlands, Rapport 58 pp.

Savidov N Hutchings E and Rakocy J E 2007. Fish and plant production in a recirculating aquaponic system: a new approach to sustainable agriculture in Canada. Acta Hort. 742, 209–222.

Shu L Schneider P Jegatheesan V and Johnson J 2006. An economic evaluation of phosphorus recovery as struvite from digester supernatant. Bioresource Technol. 97, 2211–2216.

Sonneveld C 1997. Mineralenopname bij teelten onder glas. Proefstation voor Bloemisterij en Glasgroente Naaldwijk, The Netherlands. Intern Verslag 81, 35 pp.

Stallen M 1997. Komkommer: stuurgroep intensiveert onderzoek naar dikke wortels. Groenten en Fruit/Glasgroenten 7(21), 17.

Stanghellini C 2003. Irrigation water: use efficiency and economics. *In*: Fernández-Fernández M Lorenzo-Mínguez P and Cuadrado-Gómez I M^a (eds) Improvement of Water use Efficiency in Protected Crops. Dirección General de Investigatión y Formación Agraria de la Junta de Andalucía.

Steen I 1998. Phosphorus availability in the 21st century. Management of a non-renewable resource. Phosphorus Potassium 217 (Sep-Oct 1998), 25–31.

Teunissen R J M 2005. Emissies van gewasbeschermingsmiddelen uit de glastuinbouw. RIZA Rapport 2005, 69 pp.

USGS 2006. 2006 Minerals Yearbook, Potash.

USGS 2006a. 2006 Minerals Yearbook, Phosphate Rock.

Van Oosten H and Koehorst H 2007. Toekomstsignalen. Stichting Innovatie Glastuinbouw en Innovatienetwerk, Bleiswijk en Utrecht, 77 pp.

Van der Velde R T Voogt W and Pieckhardt P 2008. Kasza: Design of a Closed Water System for the Greenhouse Horticulture, Water Science and Technology 58(3), 713–725.

Verkerke W 1997. Symptomen wortelverdikking komkommer. Proefstation voor Bloemisterij en Glasgroente, Naaldwijk The Netherlands. Intern Rapport, 16 pp.

Voogt W 2004. Nutrient management in soil and soilless culture in the Netherlands: towards environmental goals. Proc. Intern. Fertiliser Soc., York, 529, 1–27.

Voogt W (unpublished). Collection of data from water and mineral use monitoring projects with chrysanthemum 1995–2007.

Appendix A
Rounded Atomic Weights

The listed atomic weights (A_r) mentioned below in this appendix are used in this book for the chemical calculations. They are withdrawn from Aylward G H and Findlay T J V (1974). SI Chemical Data, Second edition. John Wiley and Sons.

Name	Abbreviation	A_r
Aluminium	Al	27.0
Arsenic	As	74.9
Boron	B	10.8
Bromine	Br	79.9
Calcium	Ca	40.1
Cadmium	Cd	112.4
Carbon	C	12.0
Chlorine	Cl	35.5
Chromium	Cr	52
Copper	Cu	63.6
Fluorine	F	19.0
Hydrogen	H	1.0
Iodine	I	126.9
Iron	Fe	55.9
Lead	Pb	207.2
Magnesium	Mg	24.3
Manganese	Mn	54.9
Mercury	Hg	200.6
Molybdenum	Mo	95.9
Nickel	Ni	58.7
Nitrogen	N	14.0
Oxygen	O	16.0
Phosphorus	P	31.0
Potassium	K	39.1
Selenium	Se	79
Silicon	Si	28.1
Sodium	Na	23.0
Sulphur	S	32.1
Zinc	Zn	65.4

C. Sonneveld, W. Voogt, *Plant Nutrition of Greenhouse Crops*,
DOI 10.1007/978-90-481-2532-6_BM2, © Springer Science+Business Media B.V. 2009

Appendix B
Guide Values for Tissue Analysis of Vegetables and Flower Crops

Optimum concentrations of mineral nutrients in greenhouse crops as given by De Kreij et al. (1992). The determinations are carried out by total destruction of dried material and expressed as mmol kg^{-1} dried material. For the references see Chapter 5.

	Vegetables			
Elements	Cucumber[1]	Sweet pepper[1]	Eggplant[1]	Radish[2]
K	800–1000	1400–1800	1200–1300	1500
Ca	600–800	500–600	600–800	900
Mg	150–300	200–300	100–200	240
N	3000–4000	2500–3500	3500	4500
P	200–300	125–240	150–300	135
S	100	150–250	100	230
Fe	1.5–2.0	2.0–4.0	1.5	2.6–3.8
Mn	1.0–3.0	1.0–3.0	1.0–3.0	0.6–0.8
Zn	0.75–2.20	3.00	0.70–1.00	0.70–1.90
B	5.0–7.0	5.0–7.0	2.0–5.0	5.0–6.0
Cu	0.16	0.14–0.20	–	–
Mo	0.01–0.10	–	–	–

[1] Young fully grown leaves;
[2] all leaves

Elements	Cut flowers					
	Carnation[1]	Anthurium[2]	Cymbidium[3]	Gerbera[4]	Bouvardia[4]	Hippeastrum[4]
K	800–1200	900–1000	600–750	1000–1280	500–850	1500
Ca	250–500	250–500	150–250	250–600	500–600	200–250
Mg	80–160	140–200	80–160	100–260	200–300	130
N	2000–3000	1400–1600	950–1450	1800–3500	3000–3500	2000–2200
P	60–200	50–100	50–100	80–200	200–300	90
S	–	70	–	–	100	–
Fe	1.0–2.0	0.5–2.0	0.5–1.0	1.0–2.0	2.3–4.0	1.0
Mn	0.6–5.5	0.7–2.0	0.55	0.7–2.7	0.6–1.7	1.0
Zn	0.30–1.50	0.70–2.00	0.30–0.80	0.5–0.8	0.60–1.00	0.60
B	2.0–9.0	5.0–7.0	2.0–6.5	2.8–3.7	3.0–5.0	3.0
Cu	0.10–0.20	0.10–0.20	0.08–0.16	0.06–0.20	–	0.10
Mo	–	–	–	–	–	–

[1]Fifth pair of leaves of young shoots;
[2]laeves from just harvested flowers;
[3]second full grown leaf from young shoots;
[4]young fully grown leaves.

	Potted flower plants[1]					
Elements	Azalea	Begonia	Pelargonium	Saintpaulia	Cyclamen	Hydrangea
K	200–500	500–750	640–1600	900–1500	1500	700–2000
Ca	300–500	250–500	300–600	300–500	200	300–500
Mg	70–135	150–250	80–210	250–350	125	100–250
N	1400–1600	2500–3500	2350–3400	1500–2500	1800	1600–4000
P	100–160	100–200	130–400	200–500	80	80–200
S	–	–	–	–	–	–
Fe	–	–	–	2.0–4.0	–	1.0–2.0
Mn	–	–	0.8–2.5	0.5–2.0	0.9	0.7–1.7
Zn	–	–	0.10–0.40	1.00–4.00	0.80	0.70–1.00
B	–	–	–	4.0–10.0	5.0	2.0–3.0
Cu	–	–	0.10–0.30	0.10–0.60	–	0.02–0.10
Mo	–	–	–	–	–	–

[1] Young fully grown leaves.

	Potted green plants[1]				
Elements	Dieffenbachia	Ficus	Nephrolepis	Yucca	Hedera
K	1000–1600	550–750	500–800	600–800	900–1050
Ca	400–600	300–800	100–200	400–500	320–400
Mg	200–300	100–160	200–300	200–300	200–270
N	2400–2800	1600–2500	1500–1800	1000–1300	–
P	200–300	80–100	100–200	80–110	190–290
S	–	–	–	–	–
Fe	1.0–2.0	1.0–2.0	0.5–1.5	0.5–1.5	2.5–6.5
Mn	1.5–2.5	0.5–1.5	0.5–2.0	0.3–0.9	2.0–3.0
Zn	1.20–3.00	0.40–0.80	0.50–1.00	0.50–0.60	0.60–1.10
B	3.0–5.0	2.0–4.0	2.0–4.0	1.0–2.0	2.8–3.1
Cu	0.05–0.10	0.08–0.016	0.10–0.50	0.04–0.20	0.05–0.12
Mo	–	–	–	–	0.02–0.04

[1]Young fully grown leaves

See also the data in Table 5.6.

Appendix C
Nutrient Solutions for Different Vegetable and Cut Flower Crops

Standard nutrient solutions (addition) as will be supplied for different crops and guide values for the chemical composition of the nutrient solutions in the root environment are published by De Kreij et al. (1999) and Sonneveld and Straver (1994). An extraction of such nutrient solutions is summarized in this appendix. For most crops different nutrient solutions will be given for free drainage systems and for closed systems. For free drainage a leaching fraction of about 0.3 is taken into account. For the references see Chapter 13.

	Tomato in rock wool			Cucumber in rock wool		
	Addition free drain	Addition closed	Root environment	Addition free drain	Addition closed	Root environment
EC dS m^{-1}	2.6	1.6	4.0	2.2	1.7	3.0
NH_4 mmol l^{-1}	1.2	1.0	< 0.5	1.25	1.0	< 0.5
K	9.5	6.5	8.0	8.0	6.5	8.0
Ca	5.4	2.75	10.0	4.0	2.75	6.5
Mg	2.4	1.0	4.5	1.375	1.0	3.0
NO_3	16.0	10.75	23.0	16.0	11.75	18.0
SO_4	4.4	1.5	6.8	1.375	1.0	3.5
H_2PO_4	1.5	1.25	1.0	1.25	1.25	0.9
Fe µmol l^{-1}	15	15	25	15	15	25
Mn	10	10	7	10	10	7
Zn	5	4	7	5	5	7
B	30	20	50	25	25	50
Cu	0.75	0.75	0.7	0.75	0.75	1.5
Mo	0.5	0.5	0.5	0.5	0.5	0.5

	Sweet pepper in rock wool			Eggplant in rock wool		
	Addition free drain	Addition closed	Root environment	Addition free drain	Addition closed	Root environment
EC dS m^{-1}	2.2	1.7	3.0	2.1	1.7	3.0
NH_4 mmol l^{-1}	0.75	0.75	< 0.5	1.5	1.0	< 0.5
K	6.5	5.75	5.0	6.75	6.5	6.2
Ca	5.0	3.5	8.5	3.25	2.25	6.2
Mg	1.5	1.125	3.0	2.5	1.5	4.5
NO_3	15.5	12.75	17.0	15.5	11.75	20.0
SO_4	1.75	1.0	3.0	1.5	1.125	3.0
H_2PO_4	1.25	1.0	1.2	1.25	1.0	0.9
Fe μmol l^{-1}	15	15	15	15	15	25
Mn	10	10	5	10	10	7
Zn	5	4	7	5	5	7
B	30	25	80	35	25	80
Cu	0.75	0.75	0.7	0.75	0.75	0.7
Mo	0.5	0.5	0.5	0.5	0.5	0.5

	Lettuce in circulating water		Chrysanthemum in circulating water	
	Addition	Root environment	Addition	Root environment
EC dS m^{-1}	2.6	2.5	1.8	1.7
NH_4 mmol l^{-1}	1.25	< 0.5	1.25	< 0.5
K	11.0	6.0	7.5	5.0
Ca	4.5	7.0	2.5	3.5
Mg	1.0	1.5	1.0	1.5
NO_3	19.0	19.0	12.75	10.0
SO_4	1.125	2.0	1.0	2.0
H_2PO_4	2.0	1.0	1.0	0.75
Fe μmol l^{-1}	40	40	60	80
Mn	5	1	20	10
Zn	4	5	3	5
B	30	50	20	20
Cu	0.75	1	0.5	1.0
Mo	0.5	0.5	0.5	0.5

	Rose in rock wool			Gerbera in rock wool		
	Addition free drain	Addition closed	Root environment	Addition free drain	Addition closed	Root environment
EC dS m^{-1}	1.6	0.7	2.2	1.7	1.1	2.2
NH_4 mmol l^{-1}	1.25	0.8	< 0.5	1.5	0.75	< 0.5
K	4.5	2.2	5.0	5.5	4.5	6.0
Ca	3.25	0.9	5.0	3.0	1.6	5.0
Mg	1.25	0.5	2.5	1.0	0.4	2.0
NO_3	11.0	4.3	12.5	11.25	7.25	13.0
SO_4	1.25	0.5	2.5	1.25	0.7	2.5
H_2PO_4	1.25	0.5	0.9	1.25	0.6	1.0
Fe µmol l^{-1}	25	15	25	35	25	40
Mn	5	5	3	5	5	3
Zn	3.5	3	3.5	4	3	5
B	20	15	20	30	20	40
Cu	0.75	0.5	1.0	0.75	0.5	1.0
Mo	0.5	0.5	0.5	0.5	0.5	0.5

	Cymbidium[1]		Anthurium[2]	
	Addition free drain	Root environment	Addition Free drain	Root environment
EC dS m^{-1}	0.8	0.8	0.8	1.0
NH_4 mmol l^{-1}	1.0(0.5)	< 0.5	0.5	< 0.5
K	2.8(3.0)	2.0	3.5	3.0
Ca	1.0(1.2)	1.5	1.25	2.0
Mg	0.75	1.0	0.9	1.3
NO_3	4.0(4.5)	3.0	6.0	5.0
SO_4	1.25(1.05)	1.5	0.8	1.25
H_2PO_4	0.8	0.6	0.7	0.6
Fe µmol l^{-1}	8	10	15	15
Mn	10	5	3	2
Zn	4	4	3	4
B	20	20	30	40
Cu	0.4	0.5	0.75	1.0
Mo	0.4	0.4	0.5	0.5

[1]In rock wool or urethane foam granules; in brackets when phenol foam is used;
[2]in rock wool or foam granules.

	Potted plants in expanded clay granules		Hippeastrum in pumice	
	Addition free drain	Root environment	Addition free drain	Root environment
EC dS m^{-1}	1.6	1.7	1.9	2.2
NH_4 mmol l^{-1}	1.1	< 0.5	1.25	< 0.5
K	5.5	4.5	7.0	6.5
Ca	3.0	4.0	3.0	5.0
Mg	0.75	1.0	1.0	2.0
NO_3	10.6	9.5	12.5	16.0
SO_4	1.0	2.0	1.25	2.0
H_2PO_4	1.5	1.0	1.25	1.0
Fe μmol l^{-1}	20	15	10	15
Mn	10	5	10	7
Zn	3	4	5	7
B	20	40	30	50
Cu	0.5	0.75	0.75	1.0
Mo	0.5	0.5	0.5	0.5

Appendix D
Nutrient Solutions and Guide Values for the 1:1½ Extract Recommended for Potted Plants

The crops are classified in different groups and some characteristic potted plants species, representative for the groups are added. The data are derived from Straver et al. (1999). See Chapter 14 for this reference.

Group 1			
	Nutrient solution	1:1½ extract	
	Veg. + Gen.[1]	Veg. + Gen.	Crops
EC ds m^{-1}	0.5	0.40	*Dionea*
NH_4 mmol l^{-1}	0.4	< 0.1	*Drosera*
K	1.8	1.0	*Asplenium*
Ca	1.0	0.8	*Cereus*
Mg	0.25	0.3	*Echinocactus*
NO_3	3.5	1.5	*Opuntia*
SO_4	0.35	0.4	(approx. 20
H_2PO_4	0.5	0.5	species)

[1] Veg.-vegetative growth phase; Gen.-generative growth phase.

Group 2					
	Nutrient solution		1:1½ extract		
	Veg.	Gen.	Veg.	Gen.	Crops
EC ds m^{-1}	1.0	1.0	0.50	0.50	*Chamaedorea*
NH_4 mmol l^{-1}	0.8	0.6	< 0.1	< 0.1	*Neoregelia*
K	3.7	4.4	1.2	1.3	*Osteospermum*
Ca	2.0	1.7	1.0	0.9	*Phalaenopsis*
Mg	0.5	0.5	0.3	0.3	*Saintpaulia*
NO_3	7.1	6.0	2.5	2.0	*Tagetes*
SO_4	0.7	1.2	0.6	1.0	*Verbena*
H_2PO_4	1.0	1.0	0.5	0.5	(approx. 280 species)

Group 3					
	Nutrient solution		1:1½ extract		
	Veg.	Gen.	Veg.	Gen.	Crops
EC dS m^{-1}	1.5	1.4	0.70	0.65	*Anthurium*
NH_4 mmol l^{-1}	1.1	1.0	< 0.1	< 0.1	*Areca*
K	5.5	5.5	1.6	1.6	*Calathea*
Ca	3.0	2.5	1.2	1.0	*Cordyline*
Mg	0.75	0.75	0.5	0.5	*Primula*
NO_3	10.9	8.5	4.0	3.0	*Dracaena*
SO_4	1.1	1.75	0.8	1.4	*Hedera*
H_2PO_4	1.0	1.0	0.5	0.5	*Fuchsia*
					Kalanchoe
					Zamioculcas
					(approx. 250 species)

Group 4					
	Nutrient solution		1:1½ extract		
	Veg.	Gen.	Veg.	Gen.	Crops
EC dS m^{-1}	2.0	1.5	0.90	0.75	*Bougainvillea*
NH_4 mmol l^{-1}	1.4	1.0	< 0.1	< 0.1	*Clerodendrum*
K	7.3	6.5	2.4	2.5	*Chrysanthemum*
Ca	4.0	2.5	1.4	1.0	*Hibiscus*
Mg	1.0	0.75	0.6	0.5	*Petunea*
NO_3	14.1	9.0	6.0	3.5	*Pelargonium*
SO_4	1.3	1.75	1.0	1.4	(approx. 20
H_2PO_4	2.0	1.5	0.5	0.5	species)

Group 5			
	Nutrient solution	1:1½ extract	
	Veg. + Gen.	Veg. + Gen.	Crops
EC dS m^{-1}	1.4	0.70	*Vriesea*
NH_4 mmol l^{-1}	1.0	< 0.1	*Aechmea*
K	6.5	2.4	
Ca	2.25	1.0	
Mg	0.75	0.5	
NO_3	9.5	3.5	
SO_4	1.25	1.0	
H_2PO_4	1.5	0.5	

Group 6

	Nutrient solution		1:1½ extract		
	Veg.	Gen.	Veg.	Gen.	Crops
EC dS m^{-1}	0.6	0.6	0.35	0.30	*Erica*
NH_4 mmol l^{-1}	1.5	1.5	< 0.1	< 0.1	*Rhododendron*
K	1.2	1.6	0.6	0.8	
Ca	1.2	1.0	1.0	0.5	
Mg	0.45	0.45	0.3	0.3	
NO_3	4.95	4.95	2.5	1.5	
SO_4	0.35	0.35	0.3	0.3	
H_2PO_4	0.35	0.35	0.2	0.2	

Group 7

	Nutrient solution		1:1½ extract		
	Veg.	Gen.	Veg.	Gen.	Crops
EC dS m^{-1}	1.5	1.5	0.70	0.70	*Begonia*
NH_4 mmol l^{-1}	1.25	1.25	< 0.1	< 0.1	
K	3.5	4.0	1.6	2.0	
Ca	4.0	3.75	1.2	1.0	
Mg	0.75	0.75	0.7	0.7	
NO_3	11.65	11.65	4.0	4.0	
SO_4	0.8	0.8	0.6	0.6	
H_2PO_4	1.0	1.0	0.5	0.5	

Group 8

	Nutrient solution	1:1½ extract	
	Veg. + Gen.	Veg. + Gen.	Crops
EC dS m^{-1}	1.6	0.70	*Euphorbia*
NH_4 mmol l^{-1}	2.25	< 0.1	*pulcherrima*
K	3.5	1.6	
Ca	3.75	1.4	
Mg	1.0	0.6	
NO_3	12.25	4.0	
SO_4	1.0	0.8	
H_2PO_4	1.0	0.5	

Group 9

	Nutrient solution			1:1½ extract			
	Veg. S.[1]	Veg. W.[1] Gen. S.	Gen. W.	Veg. S.	Veg. W. Gen. S.	Gen. W.	Crops
EC dS m^{-1}	2.1	1.5	1.4	1.10	0.90	0.65	*Ficus*
NH_4 mmol l^{-1}	1.4	1.1	1.0	< 0.1	< 0.1	< 0.1	
K	7.3	5.5	5.5	4.0	3.0	1.6	
Ca	4.0	3.0	2.5	2.0	1.7	1.0	
Mg	1.0	0.75	0.75	0.7	0.6	0.5	
NO_3	14.5	10.9	8.5	7.5	6.0	3.0	
SO_4	1.6	1.1	1.75	1.5	1.0	1.4	
H_2PO_4	1.0	1.0	1.0	0.5	0.5	0.5	

[1]S.-summer; W.-winter.

Group 10

	Nutrient solution	1:1½ extract	
	Veg. + Gen.	Veg. + Gen.	Crops
EC dS m^{-1}	1.0	0.70	*Chrysalidocarpus*
NH_4 mmol l^{-1}	0.8	< 0.1	
K	3.7	1.6	
Ca	2.0	1.2	
Mg	0.5	0.5	
NO_3	7.1	4.0	
SO_4	0.7	0.8	
H_2PO_4	1.0	0.5	

Group 11

	Nutrient solution		1:1½ extract		
	Veg.	Gen.	Veg.	Gen.	Crops
EC dS m^{-1}	2.0	1.5	0.90	0.75	*Spathiphyllum*
NH_4 mmol l^{-1}	1.4	1.0	< 0.1	< 0.1	
K	7.3	6.0	2.3	2.5	
Ca	4.0	2.5	1.4	1.2	
Mg	1.25	1.0	1.5	1.0	
NO_3	14.1	9.0	5.0	4.0	
SO_4	1.8	2.0	1.0	1.4	
H_2PO_4	1.5	1.0	0.8	0.7	

Micro element recommendation for potted plants, the standards for the nutrient solution supplied, and the guide values in the 1:1½ by volume extract with the range within no adjustments are recommended for the nutrient solution supplied. Concentrations expressed as $\mu mol\ l^{-1}$

	Nutrient solution supplied	1:1½ extract	
		Guide values	Range accepted
Fe	15	8	5–10
Mn	5	2	1–3
Zn	3	2	1.5–2.5
B	10	15	10–25
Cu	0.5	0.7	< 0.9
Mo	0.5	nd[1]	nd[1]

[1] nd = not defined

Appendix E
Guide Values for Soil Grown Crops

Guide values for the analytical data of the 1:2 volume extract for the base dressing. See also the data in Table 16.8

	K	Ca	Mg	N	SO_4	P	Cl
Crops	mmol l^{-1}						
Lily (Asian type)	1.0	1.5	0.8	2.0	1.5	0.10	
Lily (non Asian)	1.3	1.8	1.0	3.0	1.3	0.10	
Hippeastrum	1.3	1.5	1.0	2.5	1.5	0.10	
Carnation	1.5	2.5	1.2	4.0	1.5	0.10	
Cucumber	1.8	2.2	1.2	4.0	1.5	0.10	
Sweet pepper	2.0	2.5	1.2	4.5	2.0	0.10	
Eggplant	1.8	2.0	1.5	4.5	2.0	0.10	
Lettuce[1]	2.5	3.25	1.0	4.0	3.5	0.10	
Lettuce[2]	3.0	3.25	1.0	5.0	3.5	0.10	2.0

[1]March 15th–August 15th;
[2]August 15th–March 15th.

Guide values for the analytical data of the 1:2 volume extract maintained during crop cultivation. See also the data in Table 16.10

	K	Ca	Mg	N	S	EC
Crops	mmol l^{-1}					dS m^{-1}
Chrysanthemum	1.0	1.5	0.8	2.0	1.5	0.8
Rose	1.5	2.0	1.2	4.0	1.5	1.0
Hippeastrum	1.3	1.5	1.0	2.5	1.5	0.8
Carnation	1.5	2.5	1.2	4.0	1.5	1.2
Cucumber	1.8	2.2	1.2	4.0	1.5	1.0
Sweet pepper	2.0	2.5	1.2	4.5	2.0	1.1
Eggplant	1.8	2.0	1.5	4.5	2.0	1.2

Composition of nutrient solutions used with fertigation of soil grown crops irrigated with sprinkler irrigation systems. For drip irrigation the concentrations will be increased by 25%

	NH_4	K	Ca	Mg	NO_3	SO_4	EC_{irr}[1]
Crops	mmol l^{-1}						dS m^{-1}
Chrysanthemum	0.3	3.2	1.6	0.8	6.7	0.8	0.83
Rose	0.7	2.6	1.5	0.8	6.3	0.8	0.79
Hippeastrum	0.2	3.0	1.2	1.0	5.8	1.0	0.76
Carnation	0.3	3.2	1.6	0.8	6.7	0.8	0.83
Tomato	0.3	3.6	1.4	1.1	6.7	1.1	0.89
Cucumber	0.7	2.6	1.5	0.8	6.3	0.8	0.79
Sweet pepper	0.3	3.2	1.6	0.8	6.7	0.8	0.83
Eggplant	0.7	2.6	1.5	0.8	6.3	0.8	0.79

[1]Except the possible residual salts in the irrigation water.

Index

CPSIA information can be obtained
at www.ICGtesting.com
Printed in the USA
LVHW06*0731170618
580977LV00002B/11/P

9 789048 125319

$2\frac{3}{4}$

$2\ 6/8$

$1\frac{3}{4}$